GENSTAT 5 REFERENCE MANUAL

GENSTAT 5

Reference Manual

Planned and written by the

GENSTAT 5 COMMITTEE

*Statistics Department, Rothamsted Experimental Station,
Harpenden, Hertfordshire AL5 2JQ*

R.W. PAYNE (*Chairman*), P.W. LANE (*Secretary*)
A.E. AINSLEY, K.E. BICKNELL
P.G.N. DIGBY, S.A. HARDING, P.K. LEECH
H.R. SIMPSON, A.D. TODD
P.J. VERRIER, R.P. WHITE

Other contributing authors

J.C. GOWER
Rothamsted Experimental Station

G. TUNNICLIFFE WILSON
University of Lancaster

Technical writer

L.J. PATERSON
Heriot-Watt University

CLARENDON PRESS · OXFORD

Oxford University Press, Walton Street, Oxford OX2 6DP
Oxford New York Toronto
Delhi Bombay Calcutta Madras Karachi
Petaling Jaya Singapore Hong Kong Tokyo
Nairobi Dar es Salaam Cape Town
Melbourne Auckland

and associated companies in
Berlin Ibadan

Oxford is a trade mark of Oxford University Press

Published in the United States
by Oxford University Press, New York

© Lawes Agricultural Trust, 1987

First published 1987
Reprinted (with corrections) 1988
First published in paperback (with corrections) 1988

All rights reserved. No part of this publication may be reproduced,
stored in a retrieval system, or transmitted, in any form or by any means,
electronic, mechanical, photocopying, recording, or other wise, without
the prior permission of Oxford University Press

This book is sold subject to the condition that it shall not, by way
of trade or otherwise, be lent, re-sold, hired out, or otherwise circulated
without the publisher's prior consent in any form of binding or cover
other than that in which it is published and without a similar condition
including this condition being imposed on the subsequent purchaser

British Library Cataloguing in Publication Data

Genstat 5 reference manual.
1. Genstat (Computer system)
519.5'028'553 QA276.4
ISBN 0-19-852212-6
ISBN 0-19-852217-7 (pbk)

Library of Congress Cataloguing in Publication Data
Data available

[ISBN 0 19 852212 6]
ISBN 0-19-852217-7 (pbk)

Printed by St. Edmundsbury Press,
Bury St. Edmunds, Suffolk.

Preface

Genstat is a very general computer program for statistical analysis, with all the facilities of a general-purpose statistical package. All the usual analyses are readily available using the standard Genstat commands, or *directives*. However, Genstat is not just a collection of pre-programmed commands for selecting from fixed recipes of available analyses. It has a very flexible command language, which you can use to write your own programs to cover the occasions when the standard analyses do not give exactly what you want, or when you want to develop a new technique. Most users will need to do this only occasionally, since the standard facilities in Genstat are very comprehensive. However the ability to extend Genstat removes the temptation, that occurs with other packages, to use an inappropriate or approximate technique when an unusual set of data has to be analysed. Programs can be formed into procedures, to simplify their future use or to make them easily available to other users.

To use Genstat, then, you must first learn its command language. Chapter 1 describes the basic rules of the syntax. These are the same for every Genstat directive. Once you have learned these rules, you can learn the directives as you find that you need them.

Chapter 2 gives background information about the environment in which Genstat executes your program, and explains how you can modify it. It also describes how to obtain information about the syntax of directives from within Genstat. This is particularly useful when you are running Genstat interactively.

The data structures that are used to store information within Genstat are described in Chapter 3. There is a wide variety of structures, ranging from simple vectors of numbers or of text, to matrices and multi-way tables. There are also structures that point to other structures, allowing compound and hierarchical forms of data to be represented.

Chapter 4 describes input and output. Genstat can take input from several different files; it can have several output files too. There are also special files: for example, backing-store files allow you to store the contents of data structures, together with all their defining information, thus simplifying future analyses that examine the same set of data. You can also dump all the information about the current state of a Genstat run into a file, so that you can resume the run later on.

Genstat has many facilities for doing calculations and for manipulating data; these are covered in Chapter 5. You can do arithmetic calculations on any numerical data structure; you can also make logical tests on data values. Matrix operations include the calculation of inverses, eigenvalues, and singular value decompositions. You can form tables to summarize observations, for example, from surveys; these tables can be saved within Genstat for further calculations and manipulation.

Chapter 6 describes the facilities for programming in Genstat: for example, there are directives to loop over a sequence of commands, or to select between alternative sets of commands. It also explains how to write *procedures*; these are analogous to subroutines in other programming languages. A library of useful and reliable procedures is supplied with Genstat.

Chapter 7 covers graphics. For simple investigatory work, Genstat enables you to produce graphs, histograms, and contour plots on a terminal or a line-printer. These can also be plotted on graphics monitors or high-quality plotters, to give greater resolution or the higher quality required for publication. Pie charts can also be plotted.

These first seven chapters, then, describe all the non-statistical facilities in Genstat. They may be useful even to those who have no wish to do any statistics. For example you may want to use Genstat just as a graphics package, or to use it as a powerful language for calculations on vectors and matrices.

For statistical work, you will not need to know all the information in Chapters 1-7 straight away, although you will probably find most of it useful at some time or other. To get started, you need to read at least Section 1.1 in Chapter 1. You should probably also read the sections of Chapter 3 that describe the data structures required for the particular type of analysis that you want: for example, you need know only about variates for simple regression, or variates and factors for analysis of variance. Section 4.1 in Chapter 4 describes how to read data into Genstat. Later on you may want to learn how to print the contents of data structures, so that you can check their values; this is described in Section 4.2 in Chapter 4. The most useful part of Chapter 5 is Section 5.1, which describes how to do numerical calculations, for example to transform data. You will find loops useful (Subsection 6.2.1 in Chapter 6), once you start to analyse several sets of similar data in the same program. Graphics (Chapter 7) can give you useful insight into any patterns in the data, or allow you to present the results more clearly.

The statistical facilities in Genstat are divided into four areas: regression (Chapter 8), analysis of designed experiments (Chapter 9), multivariate and cluster analysis (Chapter 10), and time series (Chapter 11). Examples are given to illustrate how to use the techniques, but there is no attempt to teach statistics. References are given to suitable statistical text books.

The regression facilities in Genstat allow you to do simple and multiple linear regression, and to fit separate or related regression lines where your data are partitioned into groups. As well as the usual estimated regression coefficients, tables of residuals, fitted values, and so on, you can also obtain tables of predicted values; these fulfil the same purpose as tables of adjusted means in an analysis of variance. The output also includes regression diagnostics to warn about points with large residuals, or with high influence. The directives are designed to make it easy to explore a series of models, by adding and deleting explanatory variables. Genstat keeps a record of all the modifications that you make, and you can print a table summarizing the changes at any point in the series. The facilities are all available not just for ordinary linear regression, but also for the wider class of generalized linear models; these allow the fitting, for example, of probit and logit models for bioassay, or log-linear models for contingency tables. Non-linear models can be fitted: rational functions, exponential and logistic growth curves are available as standard options; you can also specify and fit more general curves.

Genstat can do analysis of variance for virtually all the standard experimental designs: for example, completely randomized orthogonal designs, randomized block designs, split plots, Latin and Graeco-Latin squares, repeated-measures designs, balanced incomplete blocks and other designs with balanced confounding, and lattices. The output includes the analysis-of-variance table, tables of means with standard errors, and estimates of polynomial and other contrasts. You can do analysis of covariance with any of these designs. Genstat can also cope with missing values.

Directives are available for all the standard multivariate techniques: principal component analysis, canonical variate analysis, factor rotation, principal coordinate analysis, and Procrustes rotation. Other techniques, for example correspondence analysis and canonical correlation analysis, are available as procedures that use Genstat's general facilities for operating on matrices and vectors. Hierarchical cluster analysis can be done using one of several methods: single linkage, average linkage, median, centroid, or furthest neighbour. Minimum spanning trees can be produced, and nearest neighbours can be listed. For non-hierarchical clustering Genstat uses an exchange algorithm, which aims to maximize one of four available criteria.

Time series can be analysed using the range of ARIMA and seasonal ARIMA models defined by Box and Jenkins. The relationship between series can be investigated by transfer-function models, relating one output series to several input series. Directives are provided to help with model selection and checking, as well as the estimation of parameters and forecasts from a model. There is a directive for calculating Fourier transforms, allowing you to do spectral analyses.

The final Chapter, 12, describes how you can extend Genstat by adding your own source code in Fortran, and how you can transfer results and data between Genstat and other programs. It also gives further details about the Genstat Procedure Library.

There are two Appendices: the first gives a concise summary of the syntax; the second lists the error messages that Genstat may give when you make a mistake.

The initials of the original authors are listed at the end of each Chapter; generally this also indicates responsibility for the corresponding source code of Genstat. The original text of the manual was rewritten by L.J. Paterson, to achieve a more uniform style; P.W. Lane and R.W. Payne acted as Editors.

This Manual is the definitive description of Genstat 5. However, for learning about Genstat, there are two other books (also published by Oxford University Press) which may be useful: "Genstat 5: an Introduction" (Lane, Galwey, and Alvey 1987) and "Genstat 5: a Second Course" (Digby, Galwey, and Lane 1988). A Newsletter is published twice a year by the Numerical Algorithms Group (address below); this describes new features of Genstat and novel ways of using it. There should also be local documentation, available at your Computer Centre, describing the particular details of how Genstat is implemented at your site: for example how to run Genstat on your computer, which graphics devices are available, information about the local procedure library, and so on.

Genstat 5 is currently developed (at Rothamsted) on a DEC VAX computer running the VMS operating system, but implementations are available for many other ranges of computer including mainframes, mini-computers, Unix-based Workstations and IBM-compatible PC's, from manufacturers including CDC, DEC, Honeywell, IBM, ICL, Prime, Sun and Unisys. For further information, contact.

Numerical Algorithms Group Ltd.
Mayfield House
256 Banbury Road
Oxford
United Kingdom OX2 7DE
Telephone: (0865) 511245
Telex: 83354 NAG UK G

or (for users in North America)

Numerical Algorithms Group Inc.
1101 31st Street, Suite 100
Downers Grove
IL 60515-1263
USA
Telephone: (312) 971 2337
Telex: 704743 NUMALGGRP UD

or (for users in Australasia)

Siromath Pty. Ltd.
Level 5
156 Pacific Highway
St. Leonards
NSW 2065
Australia
Telephone: (02) 436 0500
Telex: AA 26282

Although this version, 5, of Genstat has a completely redesigned syntax from the earlier versions, it nevertheless includes many of their concepts. These earlier versions were developed under the leadership of J.A. Nelder, who was also involved during the initial planning of Genstat 5, as was T.J. Dixon. Much of the code of Genstat 5 was adapted from code in the earlier versions, and we would like to acknowledge those who contributed to them but who are no longer involved with Genstat: N.G. Alvey, C.F. Banfield, R.I. Baxter, W.J. Krzanowski, J.A. Nelder, G.J.S. Ross, the late R.W.M. Wedderburn, and G.N. Wilkinson. We would also like to thank all those who have commented on drafts of parts of this manual: G.M. Arnold, C.J. Brien, A.W.A. Murray, J.A. Nelder, D.A. Preece, M.G. Richardson, K. Ryder, and K.I. Trinder. Thanks are also due to V. Payne and A.C. Piears for their patience in preparing and modifying the manuscript, and finally to our families for their tolerance and support.

R.W.P.

Rothamsted
January 1987

Contents

1: Introduction, terminology, and syntax 1

1.1 Genstat programs 1
    1.1.1 First example 1
    1.1.2 Second example 2
    1.1.3 Declarations 3
    1.1.4 Assigning values 3
    1.1.5 Calculations 3
    1.1.6 Printing 3
    1.1.7 Statements 4
    1.1.8 Punctuation 5
    1.1.9 Comments 5
    1.1.10 Running Genstat interactively 5
    1.1.11 Batch Genstat 6

1.2 Characters 8
    1.2.1 Letters 8
    1.2.2 Digits 8
    1.2.3 Simple operators 8
    1.2.4 Brackets 8
    1.2.5 Punctuation symbols 8
    1.2.6 Special symbols 9

1.3 Items 9
    1.3.1 Numbers 10
    1.3.2 Strings 10
    1.3.3 Identifiers 11
    1.3.4 System words 12
    1.3.5 Missing values 12
    1.3.6 Operators 12

1.4 Lists 13
    1.4.1 Number lists 14
    1.4.2 String lists 15
    1.4.3 Identifier lists 15
    1.4.4 How to compact lists 16

1.5 Expressions and formulae 18
    1.5.1 Functions 18
    1.5.2 Expressions 18
    1.5.3 Formulae 20

1.6 Statements 21
    1.6.1 Syntax of options and parameters 22
    1.6.2 Roles of options and parameters 24
    1.6.3 Types of option and parameter settings 25
    1.6.4 Repetition of a statement and its options 26

1.7 How to compact programs 27
    1.7.1 Procedures 27
    1.7.2 Macros 27

1.8 Conventions for examples in later chapters 29

2: The environment of a Genstat program 31

2.1 Information about the system 31
    2.1.1 The HELP directive 31
    2.1.2 Diagnostics 35
    2.1.3 The DISPLAY directive 38

2.2 How to set details of the environment 38
    2.2.1 The SET directive 38

2.2.2 The COPY directive 42
2.2.3 The UNITS directive 43

2.3 Accessing details of the environment of a program 45
2.3.1 The GET directive 45

2.4 Saving Space 48
2.4.1 The RESTRICT directive 49
2.4.2 The DELETE directive 53

2.5 Accessing details of data structures 54
2.5.1 The DUMP directive 54
2.5.2 The GETATTRIBUTE directive 56

3: Data Structures 59

3.1 Declarations 59
3.1.1 The VALUES option and parameter 60
3.1.2 The DECIMALS parameter 60
3.1.3 The EXTRA parameter 60
3.1.4 The MODIFY option 61

3.2 Single-valued data structures 61
3.2.1 Scalars 61
3.2.2 Dummies 62
3.2.3 Expression data structures 63
3.2.4 Formula data structures 64

3.3 Vectors and pointers 64
3.3.1 Variates 65
3.3.2 Texts 66
3.3.3 Factors 67
3.3.4 Pointers 69

3.4 Matrices 71
3.4.1 Rectangular matrices 72
3.4.2 Diagonal matrices 73
3.4.3 Symmetric matrices 74

3.5 Tables 76

3.6 Compound Structures 79
3.6.1 The LRV structure 80
3.6.2 The SSPM structure 82
3.6.3 The TSM structure 84

4: Input and Output 87

4.1 Reading data 87
4.1.1 Main features of the READ directive 89
4.1.2 Reading fixed-format data 93
4.1.3 Skipping unwanted data 95
4.1.4 Reading non-numerical data 97
4.1.5 Reading from a text structure 98
4.1.6 Reading data in batches, accumulation, scaling, and re-ordering 99
4.1.7 Errors while reading 101

4.2 Printing data 102
4.2.1 The PRINT directive 102
4.2.2 Interaction among options and parameters of PRINT 111

4.3 Getting access to external files 113
4.3.1 The OPEN directive 114
4.3.2 The SKIP directive 115
4.3.3 The PAGE directive 116
4.3.4 The CLOSE directive 117

4.4 Transferring input and output control 117

- 4.4.1 The INPUT directive 118
- 4.4.2 The RETURN directive 119
- 4.4.3 The OUTPUT directive 120

4.5 Storing and retrieving structures 122
- 4.5.1 Simple storing 122
- 4.5.2 Subfiles, userfiles, and workfiles 122
- 4.5.3 The STORE directive 124
- 4.5.4 The RETRIEVE directive 127
- 4.5.5 The CATALOGUE directive 129
- 4.5.6 The MERGE directive 133

4.6 Storing and retrieving data and programs in unformatted files 135
- 4.6.1 The RECORD directive 135
- 4.6.2 The RESUME directive 136
- 4.6.3 Storing and reading data with unformatted files 137
- 4.6.4 Communicating with other programs 138

5: Data Handling 139

5.1 Numerical calculations 139
- 5.1.1 The CALCULATE directive 140
- 5.1.2 Expressions with scalars and vectors 146
- 5.1.3 Expressions with matrices 149
- 5.1.4 Expressions with tables 153
- 5.1.5 Rules for implicit declarations 155
- 5.1.6 Rules for qualified identifiers 157
- 5.1.7 The INTERPOLATE directive 161

5.2 Functions for use in expressions 164
- 5.2.1 General and mathematical functions 165
- 5.2.2 Scalar functions 167
- 5.2.3 Variate functions 168
- 5.2.4 Matrix functions 170
- 5.2.5 Table functions 173
- 5.2.6 Dummy functions 174
- 5.2.7 Elements of structures 174
- 5.2.8 Statistical functions 176

5.3 Transferring and manipulating values 179
- 5.3.1 The EQUATE directive 179
- 5.3.2 The SORT directive 183
- 5.3.3 The RANDOMIZE directive 188
- 5.3.4 The COMBINE directive 193

5.4 Operations on text 197
- 5.4.1 The CONCATENATE directive 198
- 5.4.2 The EDIT directive 200
- 5.4.3 Commands for the EDIT directive 201

5.5 Operations on factors 205
- 5.5.1 The GENERATE directive 205

5.6 Operations on pointers 207
- 5.6.1 The ASSIGN directive 207

5.7 Operations on matrices and compound structures 208
- 5.7.1 The SVD directive 208
- 5.7.2 The FLRV directive 213
- 5.7.3 The FSSPM directive 218

5.8 Operations on tables 222
- 5.8.1 The TABULATE directive 222

5.8.2 The MARGIN directive 226

6: Job control 231

6.1 Genstat programs 231
- 6.1.1 The JOB directive 231
- 6.1.2 The ENDJOB directive 233
- 6.1.3 The STOP directive 233

6.2 Program control in Genstat 233
- 6.2.1 FOR loops 234
- 6.2.2 Block-if structures 236
- 6.2.3 The multiple-selection control structure 238
- 6.2.4 Exit from control structures 241

6.3 Procedures 243
- 6.3.1 Forming a procedure 244
- 6.3.2 Using a procedure library 249
- 6.3.3 Forming a procedure library 249

6.4 Debugging Genstat programs 250
- 6.4.1 Breaking into the execution of a program 250
- 6.4.2 Putting automatic breaks into a program 252

7: Graphical display 255

7.1 Line-printer graphics 255
- 7.1.1 The HISTOGRAM directive 256
- 7.1.2 The GRAPH directive 261
- 7.1.3 The CONTOUR directive 265

7.2 The environment for high-quality graphics 273
- 7.2.1 The AXES directive 275
- 7.2.2 The PEN directive 278
- 7.2.3 The DEVICE directive 283
- 7.2.4 The FRAME directive 283

7.3 High-quality graphics 284
- 7.3.1 The DHISTOGRAM directive 287
- 7.3.2 The DGRAPH directive 290
- 7.3.3 The DCONTOUR directive 297
- 7.3.4 The DPIE directive 301

8: Regression analysis 303

8.1 Simple linear regression 304
- 8.1.1 The MODEL directive 306
- 8.1.2 The FIT directive 309
- 8.1.3 The RDISPLAY directive 313
- 8.1.4 The RKEEP directive 314

8.2 Multiple linear regression 316
- 8.2.1 Extensions to the FIT and RDISPLAY directives 318
- 8.2.2 The TERMS directive 319
- 8.2.3 The ADD, DROP, and SWITCH directives 321
- 8.2.4 Extensions to output and the RKEEP directive following TERMS 324
- 8.2.5 The TRY directive 326
- 8.2.6 The STEP directive 327

8.3 Linear regression with grouped or qualitative data 329
- 8.3.1 Formulae in parameters of regression directives 332
- 8.3.2 How Genstat parameterizes factors 332
- 8.3.3 Parameterization of interaction, and marginality 334

- 8.3.4 The PREDICT directive 337
- **8.4 Generalized linear regression** 347
  - 8.4.1 Introduction to generalized linear models 350
  - 8.4.2 The deviance 355
  - 8.4.3 The RCYCLE directive 358
  - 8.4.4 Modifications to output and the RKEEP and PREDICT directives 359
- **8.5 Standard nonlinear curves** 361
  - 8.5.1 The FITCURVE directive 364
  - 8.5.2 Distributions and constraints in curve fitting 368
  - 8.5.3 Modifications to regression output and the RKEEP directive 369
  - 8.5.4 Modifying models fitted by FITCURVE 371
  - 8.5.5 Controlling the start of the search with the RCYCLE directive 374
- **8.6 General nonlinear regression, and minimizing a function** 375
  - 8.6.1 Fitting nonlinear models 376
  - 8.6.2 Nonlinear regression for models with some linear parameters 379
  - 8.6.3 Nonlinear regression models with no linear parameters 382
  - 8.6.4 General nonlinear models 385

9: Analysis of designed experiments 389

- **9.1 Designs with a single error term** 391
  - 9.1.1 The TREATMENTSTRUCTURE directive 394
  - 9.1.2 The ANOVA directive 397
  - 9.1.3 The ADISPLAY directive 401
- **9.2 Designs with several error terms** 410
  - 9.2.1 The BLOCKSTRUCTURE directive 410
- **9.3 Analysis of covariance** 418
  - 9.3.1 The COVARIATE directive 421
- **9.4 Missing values** 424
- **9.5 Contrasts between treatments** 426
- **9.6 Saving information from an analysis of variance** 433
  - 9.6.1 The AKEEP directive 433
- **9.7 Non-orthogonality and balance** 436
  - 9.7.1 Efficiency factors 437
  - 9.7.2 Balance 440
  - 9.7.3 Pseudo-factors 441
  - 9.7.4 Non-orthogonality between treatment terms 445
  - 9.7.5 The method of analysis 446

10: Multivariate and cluster analysis 449

- **10.1 Introduction** 449
  - 10.1.1 Rules for printing and saving results of multivariate analyses 450
- **10.2 Analyses based on sums of squares and products** 452

- 10.2.1 The PCP directive 452
- 10.2.2 The CVA directive 460
- 10.2.3 The FACROTATE directive 466

10.3 Forming measures of association 470
- 10.3.1 The FSIMILARITY directive 471
- 10.3.2 The REDUCE directive 475
- 10.3.3 Forming associations using CALCULATE 477

10.4 Ordination from associations 479
- 10.4.1 The PCO directive 481
- 10.4.2 The ADDPOINTS directive 490
- 10.4.3 The RELATE directive 493

10.5 Hierarchical cluster analysis 495
- 10.5.1 The HCLUSTER directive 496

10.6 Directives associated with hierarchical clustering 499
- 10.6.1 The HLIST directive 499
- 10.6.2 The HDISPLAY directive 502
- 10.6.3 The HSUMMARIZE directive 507

10.7 Non-hierarchical classification 509
- 10.7.1 The CLUSTER directive 511

10.8 Procrustes rotation 521
- 10.8.1 The ROTATE directive 525

11: Analysis of time series 533

11.1 Correlation 534
- 11.1.1 Autocorrelation 535
- 11.1.2 Partial autocorrelation 537
- 11.1.3 Crosscorrelation 539

11.2 Fourier transformation 541
- 11.2.1 Cosine transformation of a real series 543
- 11.2.2 Fourier transformation of a real series 545
- 11.2.3 Fourier transformation of a complex series 546
- 11.2.4 Fourier transformation of a conjugate sequence 547

11.3 ARIMA modelling 548
- 11.3.1 ARIMA models for time series 551
- 11.3.2 The ESTIMATE directive 554
- 11.3.3 Technical introduction to how Genstat fits ARIMA models 556
- 11.3.4 The TDISPLAY directive 560
- 11.3.5 The TKEEP directive 561
- 11.3.6 The FORECAST directive 563

11.4 Regression with autocorrelated (ARIMA) errors 569
- 11.4.1 The TRANSFERFUNCTION directive 573
- 11.4.2 Extensions to the ESTIMATE directive for regression with ARIMA errors 574
- 11.4.3 Extensions to the FORECAST directive for regression with ARIMA errors 576

11.5 Multi-input transfer-function models 577
- 11.5.1 Declaring transfer-function models 580
- 11.5.2 Extensions to the TRANSFER function directive for multi-input models 581
- 11.5.3 Extensions to the ESTIMATE

directive for multi-input models   582

11.5.4 Extensions to the TKEEP directive for multi-input models   582

11.5.5 Extensions to the FORECAST directive for multi-input models   582

11.6 Filtering time series   583

11.6.1 The FILTER directive   584

11.7 Forming preliminary estimates and displaying models   587

11.7.1 Preliminary estimation of ARIMA model parameters   587

11.7.2 Preliminary estimation of transfer-function model parameters   589

11.7.3 The TSUMMARIZE directive   590

11.7.4 Deriving the generalized form of a time-series model   592

12: Extending Genstat   595

12.1 Writing programs in the Genstat language   596

12.1.1 The Genstat Procedure Library   597

12.2 Adding Fortran subprograms with the OWN directive   598

12.2.1 The OWN directive   598

12.2.2 Modifying the OWN subroutine   599

12.2.3 Values of data structures and workspace   603

12.2.4 Extra diagnostic messages, and other output   604

12.2.5 Using OWN with the FITNONLINEAR directive   605

12.2.6 Relinking Genstat   606

12.3 Running an external program with the PASS directive   607

12.3.1 The PASS directive   607

12.3.2 Modifying the GNPASS program   609

12.4 Defining new directives   610

12.4.1 The DEFINE directive   611

12.4.2 Modifying the EXTRAD subroutine   612

12.5 Communicating with other programs   615

Appendix 1: Reference summary   619

A1.1 The Genstat language   619

A1.1.1 Terminology   619

A1.1.2 Data structures   626

A1.1.3 Control structures   633

A1.2 Data handling   639

A1.2.1 Input and output   639

A1.2.2 Graphics   648

A1.2.3 Backing store   656

A1.2.4 Calculation and manipulation   659

A1.3 Statistical analyses   680

A1.3.1 Regression analysis   680

A1.3.2 Analysis of designed experiments   692

A1.3.3 Functions for use in treatment formulae   697

A1.3.4 Multivariate and cluster analysis   698

A1.3.5 Analysis of time series   707

Appendix 2: Diagnostics   715

References   729

Index   735

1 Introduction, terminology, and syntax

Section 1.1 is an introduction to the rules and terminology of Genstat 5. The aim is to help you to begin to use Genstat to carry out statistical analyses; so we describe the ideas by means of two simple examples. The later Sections 1.2 to 1.7 give rigorous definitions of the terminology and syntax.

In Appendix 1 there is a condensed description of the syntax, known as the Reference Summary. Sections of this appear in Chapters 2 to 12, with examples and more detailed descriptions of the methods of use. As you get used to Genstat 5, you will probably find yourself using the Reference Summary more frequently, and Chapters 1 to 12 rather less. But remember always that the Reference Summary is very condensed: it is useful as a reminder of things that you already know from Chapters 1 to 12.

1.1 Genstat programs

Subsections 1.1.1 to 1.1.9 are about Genstat *programs*: that is, the sequences of instructions that you give to the computer when you want to use Genstat, analogously to the programs you might write in, say, Basic or Fortran or Pascal. Subsections 1.1.10 and 1.1.11 are about how to run Genstat: that is, how to get the computer to carry out the instructions in a program.

1.1.1 First example
Here is a simple example that calculates and prints the cost of a meal.

```
SCALAR IDENTIFIER=First,Dessert,Wine,Service; VALUE=8.50,1.50,3.00,1.10
CALCULATE Gcost=First+Dessert+Wine
CALCULATE Ncost=Gcost*Service
PRINT STRUCTURE=Gcost
PRINT STRUCTURE=Ncost
STOP
```

This program is a series of *statements*. Each of them uses a Genstat *directive* – a form of instruction to Genstat to do something. For example, the third line consists of one statement involving the directive that is used to do calculations with data. Later (Chapter 6) you will see that statements can use *procedures* instead of directives.

The example can be divided into three simple parts. First, data definition: the data values are specified together with the identifiers that will be used to refer to them. Then some calculations are done with the data, using these identifiers. And last, output: the results of the calculations are printed.

At the end of the program is a STOP statement: it tells the computer that you have finished using Genstat.

All Genstat programs can be thought of as being divided in this way. The division need not be as rigid as here. For example, the first PRINT statement could come before the second CALCULATE statement. But for the time being we shall keep this basic form.

The first line tells Genstat that there will be four items of data, stored in scalar data structures with identifiers First, Dessert, Wine, and Service. There are many different types of data structure in Genstat for storing the different sorts of information, numerical and textual, that you may need. The *scalar* stores a single number. After the semicolon in that line, values are given to these four scalars. The list of numbers is taken in parallel with the list of identifiers: thus the scalar called First gets the value 8.50; Dessert gets the value 1.50; and so on.

Next the program does two calculations. It finds the total cost before service has been added, and calls it Gcost (a name we have chosen to remind us that it is the gross cost). This statement itself implicitly defines Gcost as a scalar. Gcost gets the sum of the values in First, Dessert, and Wine: that is, it gets the value 13.

The other calculation changes gross cost into net cost by adding service. The result is stored in Ncost: it is 13 times 1.10, which is 14.3.

Finally the program prints out the results. In the first PRINT statement the value that is printed is in the scalar Gcost; in the second, it is in Ncost.

1.1.2 Second example

Now here is another example. The main difference is that it uses a more efficient way of storing the data. Instead of having a separate data structure for each piece of data, a single structure stores it all. This structure is called a *variate*, and is here called X. It stores 10 numerical values, so we say that X is of length 10.

```
VARIATE [NVALUES = 10] X
READ X
24.3 25.6 57.3 43.8 45.3
46.5 47.9 97.0 77.5 64.3 :
CALCULATE Xbar = MEAN(X)
PRINT STRUCTURE = Xbar; DECIMALS = 2
PRINT STRUCTURE = X; DECIMALS = 1
STOP
```

The first line tells Genstat that X will hold ten values. The next three lines cause ten values to be assigned to X. The end of the data for X is marked by a colon.

The calculation finds the mean of the ten values: Xbar here is implicitly defined to be a scalar, and is given the result of the function MEAN. This function is something that Genstat itself provides: for any variate X of any length, MEAN(X) calculates the mean of the values of X.

The output is again in two stages. First the mean is printed with two decimal places. Then the data themselves are printed with one decimal place, in a column down the page.

These two simple programs have already raised many of the essential features of Genstat. A Genstat program consists of a series of statements. Before describing the general syntax of statements (1.1.7), we shall use the two programs to summarize some of the basic things that can be done with them.

1.1.3 Declarations

A statement specifying the type and identifier of a data structure is called a *declaration*. Declarations can be explicit or implicit. Examples of explicit declarations are the first statements of each of these programs. Examples of implicit declarations are in the CALCULATE statements: the particular calculations done here produce a single-valued result, and so implicitly define a scalar structure.

Other kinds of calculation produce other kinds of results, thus implicitly defining other kinds of structures. Implicit definitions are called *default definitions*: the rules for the defaults are described in the manual at the same place as the directives that can make them.

1.1.4 Assigning values

Merely declaring a data structure does not assign any numerical information to it. You can do that in several ways. You can include the information in the declaration itself, as in the first program. Or you can read the values, as in the second. Or you can derive values as the results of calculations, as in each of these programs.

Later you will see that statistical analyses too can derive values to be assigned to data structures. You will see also that data can be read from files outside the Genstat program (that is, instead of listing the data as in the second program).

1.1.5 Calculations

Calculations can be done with many kinds of data structure. Genstat contains flexible tools for analysing data, ranging from the simplest (taking means, for example) to the advanced (such as multiple regression). But the essential point of all of them is that they do calculations with data held in data structures and referred to by their identifiers.

1.1.6 Printing

Many of the statistical directives contained in Genstat produce their own output. For example, the ANOVA directive will produce an analysis-of-variance table, tables of means, and so on. But frequently you will want to produce output of your own. PRINT is merely one way of doing that. Genstat can also, for example, produce tables, histograms, and graphs.

1.1.7 Statements

The one thing all these features of Genstat have in common is that you get access to them by means of the statements that make up a program. All statements have the same rules of syntax: first you give the name of a directive (or possibly of a procedure, see 1.6), then perhaps some options, and then usually some parameters.

The best way to become familiar with options and parameters is simply to use the directives. When you decide you would like to modify the behaviour of a directive in some way, look up its options and parameters. The way in which directives are used is simple:

(a) The name is compulsory, and is usually the best way of finding information about a directive in the index of the manual. Most names are intended to be natural, or to refer naturally to common statistical techniques. But since there may be many natural words for the same operation, you should become familiar with the particular ones used by Genstat. For example, there is no directive called PLOT but there is one called GRAPH.

(b) Options are enclosed in square brackets, as in the declaration of the variate in the second program. Each option has a name (for example NVALUES), and you can give it a setting (for example 10 here): the general form for setting an option is

    name = setting

Another example is the INDENTATION option in the statement:

    PRINT [INDENTATION = 7] STRUCTURE = X

This indents the printed values seven spaces from the left-hand margin of the page. Some options have default settings, that are the settings assumed by Genstat if you do not specify any explicitly. For example, the default for indentation is zero (values printed from the left margin). VALUES, on the other hand, has no default setting.

(c) Parameters are set in a similar way to the options, coming after the close of the square brackets.

One parameter that is nearly always part of a statement is a list of identifiers or an expression on which the statement is to operate. Some directives allow no more than this – for example CALCULATE; in such cases, there is no name for the parameter.

Other directives have several parameters. For example, in the first statement of the first program there are two parameters, IDENTIFIER and VALUE. Sometimes the names of the parameters can be left out, as we have done with VARIATE in the second program; the rules for this are described in 1.6.1.

Different parameters are separated by a semicolon. The lists inside the parameters are taken by Genstat in parallel, as for the values of the scalars in the first example.

Another example of a parameter setting is the DECIMALS portion of the following PRINT statement:

    PRINT A,B; DECIMALS = 1,2

This prints the values of the structure A with one decimal place, and the values of B with two decimals.

In this manual we mostly give names of directives, options, and parameters in capitals and in full. But small letters and abbreviations can be used instead (1.6.1).

1.1.8 Punctuation

Items in lists are separated by commas: see for example the list of identifiers in the SCALAR declaration in the first program, or the list of values in the same declaration.

Option settings are separated by semicolons, as are parameter settings.

The usual way of ending a statement is with the carriage-return key (⟨RETURN⟩) on the computer terminal. But you can also end with a colon, and thus get several statements on one line. The continuation symbol \ allows you to continue the statement onto the next line (1.6).

1.1.9 Comments

You should put comments into your programs to help other people to understand them, or to help you yourself understand them if you need to return to them later on. The series of comments is then a running guide to what the program is doing. You tell Genstat that you are making a comment by using the double-quote character ("); notice that this is not the same as two single quotes (' '). For example, you could put anywhere in the first program the comment

"This program calculates the total cost of a meal:
first without service charge, and then with 10% service."

You can type anything you like between the double quotes; Genstat simply ignores it. In longer programs you might want to put comments at several places in the program, to describe what different sets of lines are doing.

1.1.10 Running Genstat interactively

On most computers Genstat can be run either *interactively* or in *batch*.

In an *interactive* run, statements are carried out by Genstat as soon as you have typed them in. Thus you know the outcome of any particular step in the program before you go on to the next one: you can interact with the computer. The usual drawback of interactive computing is that it can take more of your time than batch computing: you may have to wait a long time before the computer finishes each statement.

Exact details of how you can run Genstat interactively are available from the people who manage your local computer. You should look for, or ask for, local documentation about the details. The main point to find out is how to invoke Genstat. Supposing now that we have invoked Genstat, this shows how our first program would look when it was run interactively.

EXAMPLE 1.1.10

```
Genstat 5   Release 1.0     (Vax/VMS4)
Copyright 1987, Lawes Agricultural Trust (Rothamsted Experimental Station)

> SCALAR IDENTIFIER=First,Dessert,Wine,Service; VALUE=8.50,1.50,3.00,1.10
> CALCULATE Gcost = First+Dessert+Wine
> CALCULATE Ncost = Gcost*Service
> PRINT STRUCTURE=Gcost

        Gcost
        13.00

> PRINT STRUCTURE=Ncost

        Ncost
        14.30

> STOP

******** End of job.   Maximum of 916 data units used at line 2 (41822 left)
```

The first thing that you get after invoking Genstat is the name and release number of Genstat; then there is a line saying that Genstat is the copyright of the Lawes Agricultural Trust. Next Genstat gives a prompt, here assumed to be >, but on some computers the prompt might be something else. You type the first statement after the prompt. At the end of the line (that is, after 1.10) you press the carriage-return key (<RETURN>), which tells Genstat to carry out the statement. Since there is no output from this statement, Genstat lets you know that it has finished by giving another prompt. You then type in the second line, and so on.

After the STOP statement, which ends this run of Genstat, you get a message that the job has finished. There is also information about how much of the workspace, available within Genstat, the program has taken up: some of it will have been occupied by data, some by identifier names, some by statements. You will not usually need to pay any attention to this information.

Sometimes you may want to include in your interactive run some statements that you have prepared in advance, and have stored in a computer file. You can do this with the INPUT directive: see Chapter 4. Likewise, data can be read in from a computer file instead of being typed in during the interactive run.

1.1.11 Batch Genstat

When a program is run in *batch*, all the statements are put in a file and the computer is told to execute them one by one. The main advantage is that you can prepare the file in advance, using the editor of your computer. Another advantage is that you can get on with something else while the computer is running the program. You can get details of how to run Genstat in batch from your local documentation.

1.1 Genstat programs

An example of output from a batch run is:

EXAMPLE 1.1.11

```
Genstat 5   Release 1.0     (Vax/VMS4)
Copyright 1987, Lawes Agricultural Trust (Rothamsted Experimental Station)
   1   "This program calculates the cost of a meal:
  -2    first without service charge, then with 10% service."
   3   SCALAR IDENTIFIER=First,Dessert,Wine,Service
   4   OPEN NAME='MEAL.DAT'; CHANNEL=2; FILETYPE=input
   5   READ [CHANNEL=2] First,Dessert,Wine,Service

   6   CALCULATE Gcost = First+Dessert+Wine
   7   CALCULATE Ncost = Gcost*Service
   8   PRINT STRUCTURE=Gcost

       Gcost
       13.00

   9   PRINT STRUCTURE=Ncost

       Ncost
       14.30

  10   STOP

******** End of job.   Maximum of 1084 data units used at line 6 (41654 left)
```

The file of statements contained the contents of the lines numbered 1 to 10. The first two were a comment: comments are especially helpful when you are running Genstat in batch, since you might not get the results from the run until long after you typed the program into the file. Notice that when a comment spills over onto more than one line, the lines after the first are preceded by a minus sign. Thus comments can easily be picked out.

This program assigns data in a different way to the example in 1.1.10. (But it need not have done so.) The data are read from another computer file, called MEAL.DAT, in which you have already put the information

    8.50 1.50 3.00 1.10 :

The program first has to open the file, attach it to a channel number, and tell the computer that it is to be used for input: all this is done by the parameters of line 4. Then the data are read from channel 2 into the four scalars. Thus, by the end of line 5, the scalars contain the same values as they did after the first statement of the example in 1.1.10.

Reading from a file like this would let you change the data easily without having to interfere with the program. For example, you could have a new file (in place of MEAL.DAT) for each customer.

The rest of the batch program is the same as the example in 1.1.10. Thus the file of statements for running Genstat in batch essentially mimics what you would type in if you were running Genstat interactively.

1.2 Characters

Sections 1.2 to 1.7 contain a rigorous definition of the Genstat language, starting with the simple aspects and moving gradually to the more complicated. You will probably not want to start reading this immediately: you might do best to come back to it once you have looked at the later chapters to see what you want to do with Genstat. There is a lot of cross-referencing among Sections 1.2 to 1.7, and there are also references forward to the rest of the manual.

The characters in Genstat statements are a subset of the ASCII set available on most computers. Inside Genstat they are classified as in Subsections 1.2.1 to 1.2.6.

1.2.1 Letters

A *letter* is any of the alphabetic characters A, B, up to Z, a, b, up to z, the underline character (_), and the percent character (%).

1.2.2 Digits

A *digit* is one of the numerical characters 0, 1, 2, up to 9.

1.2.3 Simple operators

These occur in arithmetic *expressions* or in the *formulae* that define statistical models. The *simple operators* are:

    +   −   *   /   .   =   <   >

Equals (=), less than (<), and greater than (>) occur only in expressions (1.5.2). Dot (.) occurs only in formulae (1.5.3).

The meanings of the simple operators, and of the *compound operators* made up of more than one character, are given in 1.3.6.

1.2.4 Brackets

There are two kinds of *bracket*.

Round brackets (or) are used in lists (1.4) and expressions (1.5.2), and to enclose the arguments of functions (1.5.1).

Square brackets [or] enclose option settings, and are used for suffixed identifiers (1.3.3). Left curly bracket { is synonymous with left square bracket [, and right curly bracket } with right square bracket]; these provide alternatives if square brackets are unavailable on your computer.

1.2.5 Punctuation symbols

Punctuation is used to separate different components of statements.

The *space* character can be used freely. Statements use *free format*: that is, there may be any number of spaces between items; items are described in 1.3. Spaces can be left out altogether if the items are already separated by another punctuation symbol, by a bracket, by an operator, or by an ampersand (1.2.6). Some computer terminals have a tab key ((TAB)), which has the effect of inserting spaces before subsequent characters on the terminal screen. This key should not be used in Genstat programs, since it is not stored in the computer as a series of spaces but as a specially coded control character.

Comma (,) is used to separate items in lists; lists are described in 1.4.

Equals (=) separates an option name or a parameter name from the list of settings.

Semicolon (;) is used to separate one list from another.

Colon (:) marks the end of a statement.

Newline is obtained by pressing the carriage-return key (⟨RETURN⟩). It is another way of marking the end of a statement. (But the SET directive can be used to request that newlines be ignored: 2.2.1.)

Single quote (') marks the start and finish of a string (1.3.2). On many computer terminals, there are two kinds of quote (' and `); these are synonymous.

Double quote (") marks the beginning and end of a comment (1.1.9).

1.2.6 Special symbols

Some characters have special meanings; details of how to use them are given later in this chapter.

Ampersand (&) indicates that the directive name or procedure name from the previous statement is to be repeated, together with any option settings that are not are not explicitly changed (1.6).

Asterisk (*) is used to denote a missing value (1.3.5).

Backslash (\) indicates that a statement is continued on the next line (1.6).

Dollar ($) is used to define subsets of the data in some structure. The dollar is followed by a list enclosed in square brackets, which specifies the subsets (1.5.2).

Exclamation (!) introduces an *unnamed* data structure (1.4.3). The vertical bar (|), available on some computers, is synonymous with exclamation.

Hash (#) is followed by the identifier of a data structure whose values are to be inserted at the current point of the program (1.4.4). On some computer terminals and printers, # is replaced by £.

1.3 Items

The characters in a Genstat statement are combined into *items* to make up pieces of information. There are six kinds of item, illustrated in these statements, that calculate

and print the area of a circle:
>   CALCULATE Area = 3.142 * Radius**2
>   PRINT [IPRINT=*] 'The area is',Area

The option here stops the name of the identifier being printed.

The words CALCULATE, PRINT, and IPRINT are *system words*, whereas Radius and Area are *identifiers* and the quoted characters make up a *string*. There are two *numbers* (3.142 and 2) and three *operators* (=, *, and **). The asterisk inside the square brackets is a *missing value*. The statements contain some other characters, which separate the items: these are the square brackets, the equals sign in the option, and the comma.

1.3.1 Numbers

A *number* conveys numerical information, and in its simplest form consists of digits only. For example,

>   0   245609

A number can also have a *sign* (+ or −) and a decimal point (.).

>   −2   4.5   +33.   −.2

However, a number must not contain any commas. Thus you must write one thousand as

>   1000

not

>   1,000

To avoid lots of zeroes in large or small numbers, you can use an *exponent*. For example

>   1E−20

means 10 to the power −20. Another example is

>   2D−5

which means 0.00002. D and E mean the same, and can also be replaced by d or e: these four are all called *exponent codes*. In general we have

> xEy

which means that the number x is to be multiplied by 10 to the power y. The number x can have a sign, as can the exponent y. There must not be any spaces between x and the exponent code, nor between a sign and the exponent. But there can be spaces between the exponent code and the exponent: for example

>   2d −5

again means 0.00002.

1.3.2 Strings

String is the technical name for what you can think of as ordinary English words or phrases. For example, some strings are

    apple;
    five apples;
    5 apples.

The spaces and punctuation are part of the string. One use of strings is in the output from a program, to help you to understand it.

More formally, we define a string to be a series of characters conveying textual information. In most places *quoted strings* are required: there, the characters are placed between single quotes ('): for example

    'apple'

There is also the *unquoted string*, which must have its first character as a letter and all its characters as letters or digits.

Upper- and lower-case letters are distinct within strings; so the strings

    Apple

and

    apple

are not the same.

If you want to put a single quote itself into a quoted string, you must put it in twice; otherwise Genstat thinks the string is ending. For example

    'don'' t do that'

will be interpreted as

    don't do that

Similarly, a quoted string cannot contain a double-quote character on its own, because this is interpreted inside a string as the beginning of a comment (1.1.9): a comment inside a string is not interpreted as part of the string but is ignored. So if you want a double quote in the string, you must put in two double quotes.

A continuation character (\) on its own in a quoted string continues the string onto the next line. However, a pair of continuation characters is interpreted as a single appearance of the character. For example

    'A is \\\\ to B'

is interpreted as

    A is \\ to B

If a quoted string contains a newline (⟨RETURN⟩) that does not follow an unduplicated continuation character, then it becomes a *string list* (1.4.2), unless you have used the SET directive to specify that newlines are to be ignored (2.2.1).

1.3.3 Identifiers

An *identifier* is the name used to refer to a data structure. An *unsuffixed* identifier is made up of letters and digits, starting with a letter. For example,

     Cost   Yield1985   Yield1986

Any characters beyond the first eight are ignored; thus the second and third identifiers here refer to the same structure. Capital letters are distinct from small letters, unless the SET directive (2.2.1) has been used to make them equivalent.

You can give suffixes to identifiers:

     Yield[1985]

The suffix is enclosed in square brackets, and is a number, or a quoted string, or an identifier. You can put spaces on either side of either of the square brackets.

A *suffixed* identifier is a value, or a set of values, of a pointer structure (3.3.4). Thus Yield[1985] and Yield[1986] are two structures which are pointed to by the pointer structure Yield: use of a suffixed identifier automatically defines the pointer (3.3.4).

1.3.4 System words

A *system word* is the name of a directive, or an option, or a parameter, or a function (1.5.1). The first character is a letter; subsequent characters are letters or digits. For example,

     PRINT   print   Log   Log10

You can use capital and small letters interchangeably: thus the first two system words here are equivalent. System words can be abbreviated: rules for names of directives and functions are in 1.6 and 1.5.1; those for names of options and parameters are in 1.6.1.

1.3.5 Missing values

A *missing value* represents unknown information. It is a single asterisk (*).

1.3.6 Operators

An *operator* stands for an arithmetic or logical operation, or some relationship between other kinds of item. Some operators have different meanings according to whether they appear in expressions or in formulae (1.5). Here is a list of all the operators and their names: more details are given in 5.1 and in 9.1.1.

    Arithmetic operators
addition	+
subtraction	−
multiplication	*
division	/
exponentiation	**
matrix product	*+

Assignment operator
assignment   =

Relational operators
equality   .EQ. or ==
string equality   .EQS.
non-equality   .NE. or /= or <>
string non-equality   .NES.
less than   .LT. or <
less than or equals   .LE. or <=
greater than   .GT. or >
greater than or equals   .GE. or >=
identifier equivalence   .IS.
identifier non-equivalence   .ISNT.
inclusion   .IN.
non-inclusion   .NI.

Logical operators
negation   .NOT.
conjunction   .AND.
disjunction   .OR.
exclusive disjunction   .EOR.

Formula operators
summation   +
dot product   .
cross product   *
nested product   /
deletion   −
crossed deletion   −*
nested deletion   −/
linkage of pseudo terms   //

Upper- and lower-case letters can be used interchangeably for relational and logical operators. However the characters making up any of these operators must be contiguous; thus, for example, there must be no spaces between the dots and the letters of a relational operator.

1.4 Lists

A *list* is a set of items that are to be treated in the same way in a statement. The items are usually separated by commas (but not always: see 1.4.4).

Here are some examples of the three kinds of list:

```
VARIATE [VALUES = 8.50,1.50,3.00] IDENTIFIER = Price
TEXT [VALUES = 'First course','dessert','wine'] IDENTIFIER = Item
PRINT STRUCTURE = Item,Price
```

The first set of values constitutes a *number list*, the second set a *string list*, and the STRUCTURE list for PRINT is an *identifier list*.

Missing values can occur in any of these lists. How to denote them, and what they mean, are described in 1.4.1, 1.4.2 and 1.4.3.

1.4.1 Number lists

Number lists appear in statements when values are put into a numerical data structure. Each item in a number list must be a number (1.3.1), a missing value, or the identifier of a data structure storing only one number; if an identifier, it stands for the value currently stored there. Missing values are interpreted as unknown observations in all directives that deal with numbers.

When numbers are to be listed in a repetitive or patterned series, you can save space and effort by compacting the lists as described in 1.4.4. Moreover, a set of numbers that form an arithmetic progression within a list can be written compactly using an *ellipsis*: this is three contiguous dots (...). For example,

$$1,2...10$$

means

$$1,2,3,4,5,6,7,8,9,10$$

In general, if k, m, and n are numbers and $d\,(=m-k)$ is the difference between m and k, then $k,m...n$ stands for k, $k+d$, $k+2d$, $k+3d$, up to n. If n is not in the progression defined by k and m, then the progression ends at the value beyond which n would be passed; this can be k itself. For example,

$$-2, -1.5 ... 0.4$$

means

$$-2, -1.5, -1, -0.5, 0$$

and

$$-2, -1.5 ... -1.6$$

means

$$-2$$

When the step length d is plus or minus 1, you can compact the list even further. For example,

$$10...1$$

means the same as $10,9...1$. In general, the construction $(k...n)$ is the same as $k,m...n$, where m is $k+1$ or $k-1$ depending on whether n is greater or less than k. If k equals n, the construction gives the single number k. You can leave out the brackets so long as there is no number preceding k that is not itself preceded by an ellipsis. For example,

    1...3,5...8   means   1,2,3,5,6,7,8
    1,2,3,5...8   means   1,2,3,5,7
    1,2,3,(5...8)   means   1,2,3,5,6,7,8

1.4.2 String lists

String lists appear in two places. They occur when values are assigned to a structure that is to store text (as opposed to numbers). And they occur when the setting of an option or parameter is one or more words chosen from the restricted set that is valid for the directive. The second of these uses is described in 1.6.3.

For the first, each item in the string list must be a string (1.3.2), or a missing value. The latter is equivalent to the empty string ''. An example of a string list with six items is

    one to five,'one-four','1to3','one to two',one,*

You can compact repetitive strings similarly to numbers (1.4.4).

Unless you have used the SET directive (2.2.1) to request that newlines be ignored, the intermediate quotes and commas in a list of quoted strings can be replaced by newlines. For example

    'Jack and Jill\
      went up the hill
    To fetch a pail of water.'

is the same as

    'Jack and Jill went up the hill','To fetch a pail of water.'

If the backslash (\) were left out, you would get

    'Jack and Jill',' went up the hill','To fetch a pail of water.'

while if you had previously specified SET [NEWLINE = ignored], this would give a single string:

'Jack and Jill went up the hillTo fetch a pail of water.'

1.4.3 Identifier lists

Identifier lists are needed by many options and parameters of statements; they name the structures that are to be operated on. Each item in an identifier list must be an identifier, a missing value (representing an unset item), or an *unnamed structure*. Examples are A,B in

    VARIATE A,B

and in

    PRINT A,B

The following types of structure, which are described in Chapter 3, can be unnamed: scalar, variate, text, pointer, expression, and formula.

An *unnamed scalar* is simply a number. The other types have a common form: they start with an exclamation mark, then comes a *type code*, and then comes a list enclosed in round brackets. The type codes are:

(a) V for an *unnamed variate*. For example,

!V(1...10)

is an unnamed variate containing the numbers 1 to 10. If you do not specify any type code, V is assumed by default. So this example is the same as

!(1...10)

(b) T for an *unnamed text*. (Each value of a text is a string: see 3.3.2.) For example,

!T(apples,pears)

is an unnamed text containing the two strings:

apples   and   pears

For a text containing a single string, an alternative is to give just the string within quotes. For example:

'apples'

(c) P for an *unnamed pointer* (3.3.4), when the list is of identifiers. For example,

!P(N,M,Q) is a pointer containing the identifiers N, M, and Q.

(d) E for an *unnamed expression*.

(e) F for an *unnamed formula*.

These last two are explained in 1.5.

You can compact unnamed scalars in progression with the ellipsis; other repetitive or patterned identifier lists can be compacted by the methods described in 1.4.4.

Furthermore, you can compact a list of identifiers with suffixes by using a *suffix list*. For example,

suffix list. For example,

A[1,2] is the same as A[1], A[2]

You can combine identifier lists and suffix lists:

(A,B)[1,2] is the same as A[1], A[2], B[1], B[2]

The lists are matched in lexicographic order: the items in the second list are matched in turn with the first item of the first list, then they are matched with the second item of the first list, and so on.

The empty suffix list [ ] stands for all suffixes of the identifiers preceding it. For example, if P[1], P[2], and P[3] are the only current suffixed identifiers involving P, then P[ ] is the same as P[1,2,3]. Further examples are described in 3.3.4.

1.4.4 How to compact lists

All three types of lists can be compacted by any of the methods described below, as well as those methods described individually for each type of list earlier in this section.

The values of a structure can be substituted into a list using the substitution symbol (#). If I is an identifier, then #I is a list whose items are the values of the structure identified by I. If I is a pointer, then any item in #I that is itself a pointer is replaced by the values of that pointer (3.3.4).

For example, suppose I is a variate holding values 3,4. Then

    1,2,#I

means

    1,2,3,4

If J is another variate with values 1,2, then the same list could be written as

    #J,#I

Notice that this list is quite different from the list of two identifiers

    J,I

You can do the same with lists of strings. For example, if Letters is a text containing the strings 'b', 'c', 'd', then

    'a',#Letters

is the same as

    'a','b','c','d'

If you put a dummy (3.2.2) in a list, it is automatically replaced by the identifier that it is currently storing. So if the identifier of the dummy is preceded by #, then the values put into the list are those of the structure that the dummy is storing.

When a list is to contain a set of items repeated several times, you can use a *multiplier*. A multiplier is a number without a sign; it can also be #identifier, where the identifier is of a structure storing one non-negative number. A pre-multiplier repeats each item in the list in turn:

    2(A,B,C)   is the same as   A,A,B,B,C,C

A post-multiplier repeats the whole list:

    ('a','b')2   is the same as   'a','b','a','b'

These can be combined:

    2(1...3)3   is the same as   1,1,2,2,3,3,1,1,2,2,3,3,1,1,2,2,3,3

If the multiplier has the value 0, the construction contributes no items to the list. A multiplier with value 1 can be left out to give the form

    (list)

You might want to use such a matched pair of brackets to indicate some grouping of items to someone else reading the program.

1.5 Expressions and formulae

Expressions contain arithmetic and logical operations, and are allowed only in directives that allow explicit calculations. For example

    CALCULATE Cost = First+Dessert+Wine

Formulae define the structure of a model in directives for some kinds of statistical analysis. For example

    TREATMENTSTRUCTURE Subjects*Rates

Both these constructions may contain *functions*. Details of functions are given in 5.2 for expressions and in 9.5 for formulae.

1.5.1 Functions

Functions have the form

    function name (sequence of arguments)

A function name is a system word of one of the standard functions (5.2 and 9.5). For example:

    SQRT(X)   SUM(Y + 4*Z)

The function name can be abbreviated to four characters. If you give further characters up to the eighth they must match the full form; characters beyond the eighth are ignored. The *arguments* of a function are either lists or expressions; if there are several arguments they are separated by semicolons. For example:

    CHISQ(2.5; 6)

1.5.2 Expressions

An *expression* consists of identifier lists, operators (1.3.6) and functions. Identifier lists must not include any missing identifiers, and the operators

    . −* −/ //

cannot be used in expressions.

The simplest form of expression is an identifier list by itself, or a function by itself. You build up expressions from identifiers and functions by mixtures of three rules. Let E and F be expressions. Then these are also expressions:

    (E)
    monadic operator E
    E dyadic operator F

The first means that putting brackets round an expression makes another expression. For the second, there are only two monadic operators:

    .NOT.
    −

The former negates a logical expression. The latter changes the sign of a numerical

1.5 Expressions and formulae

expression. All the other operators are dyadic (including – to mean subtract), and can be used with the third rule.

Other examples of expressions, illustrating the rules, are

    5,6
    A,B = −(C)
    SUM(X) .EQ. 4
    A = (B = C + 1) + 1

The precedence rules of operators are very similar to those of computer languages like Fortran, but are not exactly the same. The following list shows the precedence of all operators in expressions when brackets are not used to make the order of evaluation explicit:

    (1) .NOT. Monadic −
    (2) .IS. .ISNT. .IN. .NI. *+
    (3) **
    (4) * /
    (5) + −
    (6) Relational operators
    (7) .AND. .OR. .EOR.
    (8) =

Within each class, operations are done from left to right within an expression. For example,

    A > B+C/D*E   is the same as   A > (B+((C/D)*E))

An identifier list in an expression can contain *qualified identifiers*, picking out a subset of values of the structures. For example

    V$[3,4]

means take the third and fourth items of data from the variate V. This is the same as

    V$[3],V$[4]

In other words, a qualified identifier is

    identifier $ [sequence of identifier lists]

If there are two lists, their elements are taken in parallel: for example,

    M$[1,2; 4,5]

refers to a matrix M, and selects two elements: column 4 of row 1 and column 5 of row 2. So this is the same as

    M$[1;4],M$[2;5]

You can compact a list of qualified identifiers in an expression with a *qualified identifier* list:

    (A,B)$[1,2]   is the same as   A$[1],A$[2],B$[1],B$[2]

The general form is

(identifier list) $ [sequence of identifier lists]

If any list in the sequence is shorter than the others, it is repeated as often as is necessary; so for example

    M$[1,2; 4,5,6]

is the same as

    M$[1;4],M$[2;5],M$[1;6]

1.5.3 Formulae

A *formula* defines a statistical model; it consists of identifier lists, operators, and functions. The identifier lists must not include any missing identifiers, and only the operators

    +   −   *   /   .   −/   −*   //

can be used. The simplest form of a formula is an identifier list; also a formula can be a function by itself.

You build up other formulae by mixtures of two rules: if M and N are formulae, then so are

    (M)
    M operator N

For example

    Sex * Diet
    (Group / Variety) * Fertilizer
    Drug * POL(Dose; 2)

The operators in a formula have the following precedence:

    (1)   .
    (2)   //
    (3)   /
    (4)   *
    (5)   +   −   −/   −*

Within each class, operations are done from left to right within a formula.

A formula is expanded into a series of *model terms*, linked by the summation operator (+). A model term contains one or more elements, separated from each other by the operator dot (.), each element being either an identifier or a function whose arguments are single identifiers. For example, the expanded form of the first formula above is Sex + Diet + Sex.Diet. The interpretation of the terms is described in 8.3.1 and 9.1.1.

Identifiers in a list within a formula are treated as if they were separated by the summation operator and enclosed within brackets. For example

    A,B*C is the same as (A + B)*C

The following table shows how operators combine terms, using L and M to represent two sums of terms.

Construction	Expansion
L.M	Sum of all pairwise combinations of terms in L with terms in M using the dot operator, with the terms ordered as explained below. For example (A+B).(C+D.E) is the same as A.C+B.C+A.D.E+B.D.E
L*M	L+M+L.M ordered as explained below. For example (A+B)*C is the same as A+B+C+A.C+B.C
L/M	L + *L*.M where *L* is a term formed by combining all terms in L with the dot operator, ordered as explained below. For example (A+B)/(C+D.E) is the same as A+B+A.B.C+A.B.D.E
L−M	L without any terms that appear in M. For example (A+B)−(A+C) is the same as B
L−/M	L without any terms that consist of a term appearing in M combined with any other identifiers. For example (A+B+B.C)−/B is the same as A + B
L−*M	L−M−/M For example (A+B+B.C)−*B is the same as A

After expansions for the dot, slash, and star operators, the terms are rearranged in order of increasing numbers of identifiers. Terms with the same number of identifiers are arranged in lexicographical order with respect to the order in which the identifiers first occurred in the formula itself.

1.6 Statements

A *statement* is an instruction to Genstat, and has the general form:

    statement-name [option-sequence] parameter-sequence terminator

For example,

    READ [CHANNEL=2] STRUCTURE=First,Dessert,Wine

If there are no options, the square brackets can be left out; but there must then be at least one space between the statement name and the first parameter setting: for example

    PRINT STRUCTURE=Cost; DECIMALS=2

Some directives have options but no parameters: for example,

    SET [CASE=ignored]

makes upper-case and lower-case letters equivalent in identifiers (2.2.1). Others have neither options nor parameters. For example

STOP

The statement name is one of three things: the system word of one of the standard Genstat directives; or the name of a *procedure* (1.7.1 and 6.3); or the repetition symbol (&) (1.6.4).

The system word can be abbreviated to four characters. If you give further characters up to the eighth, they must match the full form; characters beyond the eighth are ignored. This allows you to define procedures whose names differ from those of standard Genstat directives only within the fifth to eighth characters. For example, the CALCULATE and STOP directives do not prevent you defining procedures called CALCULUS and STOP_IT.

The terminator of a statement is colon (:). Thus the line

SCALAR Cost : READ Cost

contains two statements.

Alternatively, you can end a statement by pressing the carriage-return key (⟨RETURN⟩): in other words, newline is normally synonymous with colon. You can cancel this synonymity using the SET directive (2.2.1): by specifying

SET [NEWLINE=ignored]

you can request that newlines be ignored in the rest of the program.

Even if newlines are not ignored, there are still three situations when a newline will not end a statement.

(a) When newline occurs within a string, it terminates that string and begins another (1.4.2).

(b) A newline within a comment is ignored (along with the rest of the comment): for example

OPEN NAME='MEAL.DAT'; "
"CHANNEL=2; FILETYPE=input

is a single statement.

(c) You can indicate that a statement is to continue onto the next line by putting a continuation symbol (\) before pressing carriage-return: for example,

OPEN NAME='MEAL.DAT';\
CHANNEL=2; FILETYPE=input

is again a single statement. Any characters between the continuation symbol and the end of the line are ignored. Genstat does however have the limitation that a statement must not exceed 2048 characters, after deletion of extraneous spaces.

1.6.1 Syntax of options and parameters

The sequences of options and parameters specify the items upon which the statement is to operate: these items are called the *arguments* of the statement. A sequence consists of one or more settings, each separated from the next by a semicolon (;). You can see an example of a sequence of parameter settings in the OPEN statement above. Each setting, whether of an option or a parameter, has one of the general forms:

    name = list
    name = expression
    name = formula

The list, expression, or formula can be null (length zero). Rules by which the "name=" can be left out are defined below; the types of setting are discussed further in 1.6.3.

An *option name* is a system word, which can be abbreviated to the minimum number of letters needed to distinguish it from the options that precede it in the prescribed order for the directive or procedure concerned. Characters up to the eighth must match the appropriate part of the full form; those after the eighth are ignored. For example, here are the options of the TABULATE directive (5.8.1), with the minimum form of each name printed in bold:

    PRINT, **CL**ASSIFICATION, **CO**UNTS, **S**EQUENTIAL, **M**ARGIN, **I**PRINT

Notice that the minimum for COUNTS is CO, since C on its own would not distinguish it from CLASSIFICATION which precedes it in this prescribed order.

A *parameter name* is also a system word, and has the same abbreviation rule as the option names. For example, the parameters for TABULATE are (with minimum forms in bold):

    DATA, **T**OTALS, **N**OBSERVATIONS, **ME**ANS, **MI**NIMA, **MA**XIMA, **V**ARIANCES

You usually need type no more than one or two characters for any option or parameter name; there are no directives that require more than four characters for their option and parameter names.

You can leave out the name and the equals character altogether by paying attention to the prescribed order of options, or of parameters, within the directive or procedure. The rules for parameters are the same as those for options, and are as follows:

(a) If the first option setting in a statement is for the first option defined for that directive or procedure, then "name=" can be left out.

(b) The "name=" can also be left out for later option settings if the preceding setting is for the option immediately before that option in the prescribed order.

For example,

    TABULATE [PRINT=totals,means; COUNTS=Rep; SEQUENTIAL=Sval]\
      DATA=Spending; MEANS=Meansp

can be abbreviated to

    TABULATE [totals,means; COUNTS=Rep; Sval] Spending; MEANS=Meansp

You can leave out "PRINT=" here by rule (a): it is the first option in the order prescribed for TABULATE. Likewise, rule (a) lets you leave out "DATA =".

You can leave out "SEQUENTIAL =" by rule (b), because COUNTS which precedes it here is also the option that precedes SEQUENTIAL in the definition of TABULATE. But you cannot leave out "COUNTS =", because in the prescribed order another option comes between it and PRINT. The same is true for "MEANS =".

An option or parameter setting can be *null*: that is, it can have a list of length zero, or a null expression, or a null formula. Thus, by putting a null setting for the CLASSIFICATION option and the TOTALS and NOBSERVATIONS parameters, all the names can be left out:

    TABULATE [totals,means; ; Repl; Sval] Spending; ; ; Meansp

If a directive has a single parameter, no name is defined. For example, there is no name for the expression that is the only parameter for the CALCULATE directive (5.1.1).

1.6.2 Roles of options and parameters

Parameters specify parallel series of arguments that are operated on in turn when the statement is carried out. For example, in

    TABLE [CLASSIFICATION = Age,Sex] IDENTIFIER = Income,Cars,Spending;\
    DECIMALS = 2,0,2

there are two parameters: IDENTIFIER and DECIMALS. The statement declares three tables: when they are printed later in the program Income will have two decimal places, Cars will have none, and Spending will have two.

The main information in a directive or procedure is usually given by the first parameter, and so this is said to define the series of *primary arguments*. In the example, they are Income, Cars, and Spending. Usually the parameter setting is an identifier list, and so the series is a list of identifiers of data structures.

Alternatively, the setting of the first parameter can be an expression, in which case the first identifier list in the expression is the series of primary arguments. For example, if A, B, M, N, P, and Q are variates, then in

    CALCULATE A,B = M,N + P,Q

the primary arguments are A and B.

Another possibility is that the setting may be a formula, in which case the expanded list of model terms (1.5.3) is the series of primary arguments. For example, in the regression statement

    FIT A∗B

the primary arguments are the terms A, B, and A.B.

Later parameters, or other lists within an expression, specify *secondary arguments* which run in parallel with the primary arguments, and provide ancillary information. Examples of secondary arguments are in the TABLE statement above, and on the right-hand side of the equals symbol in the CALCULATE statement.

1.6 Statements

The series of primary arguments should always be the longest; if a series of secondary arguments is longer, you are given a warning and elements beyond the length of the primary series are ignored. Any series that is shorter is recycled: that is, the series is traversed again, as many times as is necessary to match the length of the primary series. Thus the TABLE declaration above means exactly the same if it is written

    TABLE [CLASSIFICATION = Age,Sex] IDENTIFIER = Income,Cars,Spending;\
       DECIMALS = 2,0

Options, on the other hand, specify overall information that applies to all the primary arguments (with their corresponding secondary arguments). Thus in the TABLE example above, all three tables are classified by Age and Sex.

Many options have *default* values, namely values that are assumed if the option is not set explicitly in a statement. But some options have to be set, for example the OLDSTRUCTURE and NEWSTRUCTURE options of the COMBINE directive (5.3.4). Some secondary parameters also have defaults: for example, the METHOD option of GRAPH (7.1.2) assumes point plots.

1.6.3 Types of option and parameter settings

An option or parameter setting may need a formula, or an expression, or a list (1.6.1).

When the setting is an expression, and the parameter or option has a defined name, the name and its accompanying equals character cannot be left out if the expression begins with "unsuffixed identifier = ". This is because there would then be confusion between the name of the option or parameter and the unsuffixed identifier. For example, you could not leave out the name CONDITION in the statement

    RESTRICT STRUCTURE = Income; CONDITION = Agecond = Age > 30

That is, if you wrote

    RESTRICT Income; Agecond = Age > 30

Genstat would try to interpret Agecond as a parameter name. A message would be printed alerting you to this syntax error. But you could put the expression in brackets:

    RESTRICT Income; (Agecond = Age > 30)

No such problem arises with directives like CALCULATE (5.1.1), CASE, IF, and ELSIF (6.2), because in these the expression is the only parameter, and thus has no defined name. For example, you can write

    CALCULATE Agecond = Age > 30

Many options and parameters need lists of identifiers. Any restrictions on the types of identifiers for particular lists are mentioned along with the descriptions of the syntax in later chapters and in Appendix 1. For example, the specification of the CLASSIFICATION option of the TABLE directive (3.5) states

>   CLASSIFICATION = factors   Factors classifying the tables; default *

No structure other than a factor can be used here.

Apart from the VALUES option of the TEXT directive (3.2), all options or parameters that require string lists use them to choose one or more textual values from a set defined by Genstat for that option or parameter. For example, the PRINT options of ADISPLAY and ANOVA have possible values:

>   **a**ovtable, **information**, **c**ovariates, **e**ffects, **r**esiduals, **con**trasts, **means**, **%cv**, **mi**ssingvalues

These let you choose which components of output are to be printed from an analysis of variance (9.1.3). The rules for such sets of values defined by Genstat are exactly the same as those for option and parameter names (1.6.1): they may be typed in capital or small letters (or mixtures), and each one can be abbreviated to the minimum number of characters necessary to distinguish it from earlier values in the list. If more than that number is given, the extra characters must match the full form up to the eighth.

The minimum forms of the values for the PRINT options, above, are marked in bold. Thus

>   PRINT = Aovtable,Effects,MissingValues

is the same as

>   PRINT = aovtable,effects,missingvalues

and both of these can be abbreviated to

>   PRINT = a,e,mi

To prevent any printing at all, with the PRINT option of any directive, you give the missing string:

>   PRINT = *

or

>   PRINT = ''

Number lists are needed by the VALUES options of the directives that define numerical data structures (3.1.1), but not by the options or parameters of any other directive.

1.6.4 Repetition of a statement and its options

You can repeat a directive name by typing the ampersand character (&). At the same time you can reset as many options or parameters as you want; those options that you do not mention remain as in the previous statement. For example, after

>   READ [PRINT = data; CHANNEL = 2] Costs

the statement

>   & Profits

is equivalent to

    READ [PRINT = data; CHANNEL = 2] Profits

while the statement

    & [CHANNEL = 3; REWIND = yes] Profits

is equivalent to

    READ [PRINT = data; CHANNEL = 3; REWIND = yes] Profits

You need not type a colon or newline before an ampersand, as it automatically terminates the previous statement.

1.7 How to compact programs

You can store Genstat statements in two ways: in a procedure, or in a macro.

1.7.1 Procedures

A *procedure* is a series of complete Genstat statements. It is like a subroutine in Fortran, or a procedure in Basic or Pascal. These statements are self-contained, in that all the data structures that they use are accessible only within the procedure, apart from those explicitly defined as options or parameters of the procedure. Rules for writing and defining procedures are described in 6.3. The rules of syntax for using a procedure are the same as those of the standard Genstat directives; indeed, since you can get access to procedures automatically from libraries, you do not have to know whether a particular statement uses a directive or a procedure.

1.7.2 Macros

A *macro* is a Genstat text into which you have placed a section of Genstat program. The text must have an unsuffixed identifier. You can substitute the contents of the macro into the program by a contiguous pair of substitution symbols ##; the substitution takes place immediately after Genstat reads the statement that contains the substitution symbols.

A simple kind of macro would be a part of a Genstat statement. For example,

    TEXT [VALUES = '[PRINT = data,summary; CHANNEL = 2]'] Optset

assigns to a text with identifier Optset the string between the single quotes. If you later type

    READ ##Optset Patient,Sex,Weight
    READ ##Optset Calories,Wtgain

then Optset is treated as a macro and its contents are inserted into each of the two statements; so the named structures are read using the options for PRINT and CHANNEL defined in the string that has been put in Optset. The advantage of defining Optset in this way is that it saves effort in typing the two READ statements.

More complicated macros contain complete statements. For example, suppose that the computer file ALG.DAT contains three lines, each a quoted string (1.3.2):

```
'CALCULATE Previous = Root'
'& Root = (X/Previous + Previous)/2'
'PRINT STRUCTURE = Root,Previous; DECIMALS = 4' :
```

We are now going to put these three statements into a text called Estsqrt to be used as a macro. For example, a simple program for calculating the square root of 48 (without using the standard function SQRT) is:

```
SET [INPRINT = statements,macros]
SCALAR IDENTIFIER = X,Root; VALUE = 48
TEXT [NVALUES = 3] Estsqrt
OPEN NAME = 'ALG.DAT'; CHANNEL = 2
READ [CHANNEL = 2] STRUCTURE = Estsqrt
##Estsqrt
##Estsqrt
##Estsqrt
PRINT [IPRINT = *] '3 iterations calculate the square root of 48 as',Root
STOP
```

Output from running this program in batch is shown below.

EXAMPLE 1.7.2

```
Genstat 5   Release 1.0     (Vax/VMS4)
Copyright 1987, Lawes Agricultural Trust (Rothamsted Experimental Station)

   1  SET [INPRINT=statements,macros]
   2  SCALAR IDENTIFIER=X,Root; VALUE=48
   3  TEXT [NVALUES=3] Estsqrt
   4  OPEN NAME='ALG.DAT'; CHANNEL=2
   5  READ [CHANNEL=2] STRUCTURE=Estsqrt

   6  ##Estsqrt
     1  CALCULATE Previous = Root
     2  & Root = (X/Previous + Previous)/2
     3  PRINT STRUCTURE=Root,Previous; DECIMALS=4

          Root     Previous
       24.5000      48.0000

   7  ##Estsqrt
     1  CALCULATE Previous = Root
     2  & Root = (X/Previous + Previous)/2
     3  PRINT STRUCTURE=Root,Previous; DECIMALS=4

          Root     Previous
       13.2296      24.5000

   8  ##Estsqrt
     1  CALCULATE Previous = Root
     2  & Root = (X/Previous + Previous)/2
     3  PRINT STRUCTURE=Root,Previous; DECIMALS=4

          Root     Previous
        8.4289      13.2296

   9  PRINT [IPRINT=*] '3 iterations calculate the square root of 48 as',Root
```

```
 3 iterations calculate the square root of 48 as       8.429
10    STOP
******* End of job.   Maximum of 1148 data units used at line 1 (41590 left)
```

The first statement arranges to print statements and contents of macros. Then X and Root are defined as scalars, and both are given the value 48. Estsqrt is defined as a text with three values (or lines), and read from ALG.DAT in lines 4 to 5. The macro is substituted into the program three times: because of SET, its contents are printed each time, with line numbers indented by two characters. The IPRINT option of PRINT in line 9 prevents printing of the identifier Root: all you get is the value stored in Root.

As you can see from the output, the value is still some way from convergence. Methods of testing for convergence in iterative algorithms like this are described in 6.2.4.

1.8 Conventions for examples in later chapters

You have now seen that you can lay out your programs in many ways. You can put in spaces to make them more readable, or you can leave spaces out to make them compact (1.2.5). You can type statements spread over several lines, or you can have more than one to a line (1.6). You can write system words in capital letters, or in small letters, or in a mixture, and you can use the full or the abbreviated forms (1.5.1 and 1.6). You can do the same with strings in options and parameters (1.6.3). You can write identifiers with capital letters or with small letters, or in a mixture, and you can control whether or not these are equivalent (1.3.3).

However in this manual we have imposed some conventions. The use of spaces is standardized. System words are given in full and in capitals; the only exception is that the name, and corresponding equals character, of the first defined parameter of a directive will usually be left out in later chapters. Option strings are given in full and in small letters. Identifiers will begin with a capital; any other letters are small. There is usually only one statement per line, unless this is very wasteful of space; continuation lines are indented.

We hope these conventions will help you to recognize what each item is, both in descriptions of syntax and in examples. But you should develop the style of programming that you find most convenient.

P.W.L.
R.W.P.

2 The environment of a Genstat program

The word "environment" refers to the conditions or rules that apply to the statements you write in a program. For example, the way a statement operates may depend on how many data structures have been set up already: if there is not enough space, the job may fail. Some of the rules of syntax can be modified, such as whether lower-case and upper-case letters are taken to be synonymous: this affects how you write identifiers in statements. The directives described in this chapter allow you to find out what the environment is at any time within a job, or to modify it where Genstat lets you do that.

2.1 Information about the system

Genstat can be used either in batch mode or in interactive mode. For batch work you can find useful guidance in this manual and in the Genstat Introductory Guides. For interactive work you will still find these documents useful. However, the HELP directive is available within Genstat; it gives information on how to use the language and on the environment of the current job.

If you type an incorrect statement in a job (either in batch or interactively) Genstat will give a diagnostic error-message. How to understand these messages is described in 2.1.2. You can prevent diagnostics being printed by using SET (2.2.1) or JOB (6.1.1). Even when you have done this, however, you may sometimes want to know what the latest diagnostic was. Likewise, in interactive mode, you may want to remind yourself of the latest diagnostic after the message has disappeared off the top of the screen. The DISPLAY directive (2.1.3) lets you print the most recent diagnostic.

2.1.1 The HELP directive

HELP

Prints details about the Genstat language and environment. When no parameters are given, HELP gives information on how to use it. When used interactively, at each stage HELP responds with a series of menus to allow choice of information. Responding with a star (∗) causes information on all the displayed words to be printed, carriage-return (⟨RETURN⟩) causes the previous menu to be displayed and colon (:) is used to exit from HELP.

No options

> **Parameter**
>
> strings   Directive names or keywords indexing the desired details.

You start by typing HELP, or HELP followed by a list of allowed strings (see below, in (a)). Genstat replies with some information, a new menu of allowed strings and the prompt

    HELP⟩

In response to the prompt you can do one of four things.

(a) Type one of the given strings to obtain further information on that subject. You need to include only enough characters to distinguish the string from those earlier in the list; but any further characters up to the eighth must match the given form. Genstat then gives you information about the subject, followed by another set of allowed strings and the prompt ready for you to give further input. You can always skip a level in the prompting hierarchy by forming the allowed strings into lists, using commas. For example,

    HELP⟩ text,options

skips the information and the menu that you would get if you asked for 'text' only. You would therefore immediately be given information on the options of the TEXT directive. You could, indeed, have done this at the very start, when you first typed HELP:

    HELP text,options

The case of the letters that you type does not matter.

(b) Type an asterisk (*) to see information on all the strings displayed in the menu.

(c) Type carriage-return (⟨RETURN⟩) to move back to the previous menu in the hierarchy.

(d) Type a colon (:) to exit from HELP.

The top level of HELP contains information on all the individual directives, and also on these subjects:

EXAMPLE 2.1.1a

Subject	String for HELP
Syntax and terminology	DEFINITION
Data structures	STRUCTURE
Control of execution of programs	PROGRAMCONTROL
Calculations and manipulation	DATAHANDLING
Functions for use in calculations	SYSTEMFUNCTIONS
Input, output and backing store	COMMUNICATION

Graphs, histograms and contour plots	PICTURES
Analysis of balanced experiments	AOV
Linear and nonlinear regression analysis	REGRESSION
Multivariate and cluster analysis	MULTIVARIATE
Time series analysis	TIMESERIES
The current environment	ENVIRONMENT
Libraries and subprograms	LIBRARY
Text-editor commands	TEDITCOMMANDS

Whenever there is more than a screenful of information, HELP pauses and gives a question mark (?) as a prompt. You should respond either by pressing carriage-return (⟨RETURN⟩) to continue with the current information, or by typing a list of allowed strings for further information, or by typing a colon (:) to exit from HELP. Allowed strings in this context are the strings that would be permitted if the current information had been completed: these are the strings in the menu at the next level down if one exists, or from the bottom level menu if you are already receiving information at that level.

Here is an example of a typical HELP session; note that HELP continually reminds you how to use it, and also gives many useful cross-references.

EXAMPLE 2.1.1b

```
> HELP INPUT

INPUT

    Specifies the input file from which to take further commands.

Options:    PRINT, REWIND
Parameters: Unnamed

Further information available on
Options, Parameter

    :  exits from HELP, *  gives information on all the above words
  and <RETURN> moves up a level

HELP>

    The top level of HELP contains information on the following subjects
    and on all individual directives:

    Syntax and terminology                          DEFINITION
    Data structures                                 STRUCTURE
    Control of execution of programs                PROGRAMCONTROL

    Calculations and manipulation                   DATAHANDLING
    Functions for use in calculations               SYSTEMFUNCTIONS
    Input, output and backing store                 COMMUNICATION
    Graphs, histograms and contour plots            PICTURES

    Analysis of balanced experiments                AOV
```

```
    Linear and nonlinear regression analysis      REGRESSION
    Multivariate and cluster analysis             MULTIVARIATE
    Time series analysis                          TIMESERIES

    ... Press <RETURN> to continue, : to exit from HELP
        or a list of allowed words for further information
? COMM

COMMUNICATION

    The following directives control input and output of data:

    File handling                         OPEN, CLOSE
    Switching between files               INPUT, OUTPUT, RETURN
    Reading and printing                  COPY, READ, SKIP, PAGE, PRINT
    Printing current/system information   DUMP, HELP, DISPLAY

    Storing/retrieving data on backing store  STORE, RETRIEVE
    Combining and listing stored data         MERGE, CATALOGUE
    Storing/retrieving a job                  RECORD, RESUME
Further information available on
OPEN, CLOSE, INPUT, OUTPUT, RETURN, COPY, READ, SKIP, PAGE, PRINT,
DUMP, HELP, DISPLAY, STORE, RETRIEVE, CATALOGUE, MERGE, RECORD,
RESUME

    : exits from HELP, * gives information on all the above words
 and <RETURN> moves up a level

HELP> RE,*



COMMUNICATION, RETURN

Option
     NTIMES          = scalar       Number of streams to ascend; default 1
Parameter
           None

Further information available on
Options, Parameter

    : exits from HELP, * gives information on all the above words
 and <RETURN> moves up a level

HELP>

Further information available on
OPEN, CLOSE, INPUT, OUTPUT, RETURN, COPY, READ, SKIP, PAGE, PRINT,
DUMP, HELP, DISPLAY, STORE, RETRIEVE, CATALOGUE, MERGE, RECORD,
RESUME

    : exits from HELP, * gives information on all the above words
 and <RETURN> moves up a level

HELP> READ

COMMUNICATION, READ

    Reads data from an input file, unformatted file or a text.


Options:      PRINT, CHANNEL, SERIAL, SETNVALUES, LAYOUT, END, SEQUENTIAL,
              MISSING, SKIP, BLANK, JUSTIFIED, ERRORS, FORMAT, QUIT,
              REWIND
Parameters:   STRUCTURE, FIELDWIDTH, DECIMALS, SKIP, FREPRESENTATION
```

```
Further information available on
Options, Parameters

    : exits from HELP, *  gives information on all the above words
  and <RETURN> moves up a level

HELP> O

COMMUNICATION, READ, Options

    PRINT           = strings      What to print (data, errors, summary);
                                   default e,s
    CHANNEL         = identifier   Channel number of file, or text structure
                                   from which to read data; default current file
    SERIAL          = string       Whether structures are in serial order, i.e.
                                   all values of the first structure, then all
                                   of the second, and so on (no, yes);
                                   default n, i.e. values in parallel
    SETNVALUES      = string       Whether to set number of values of structures
                                   from the number of values read (no, yes);
                                   default n
    LAYOUT          = string       How values are presented (separated,
                                   fixedfield); default s
    END             = text         What string terminates data (* means there is
                                   no terminator); default ':'
    SEQUENTIAL      = scalar       To store the number of units read (negative
                                   if terminator is met); default *

    ...  Press <RETURN> to continue, : to exit from HELP
         or a list of allowed words for further information
?
    Exit from HELP
```

In an interactive run, a HELP statement continues until you type a colon; newlines generate prompts from HELP. In batch this is not so and newline will be interpreted as ending the statement in the usual way, unless you have used the SET directive to ignore newlines (2.2.1). Otherwise you use HELP in the normal way, by typing HELP followed by a list of strings indicating what information is required. You can still use the asterisk to get information on all the strings at a specified level. For example,

> HELP Text
> HELP REGression,fit,options
> HELP AOV,∗

2.1.2 Diagnostics

Diagnostics are produced when Genstat finds an error in your program. Errors are of two types: *faults*, which immediately stop execution of the current statement, and *warnings* which produce diagnostics but allow execution to continue. Some directives also produce *messages*. These are either errors from which Genstat knows how to recover, as for example in HELP (2.1.1); or else they point out some unusual feature of the data, such as outliers detected by the FIT directive (8.1.2). Messages do not give diagnostics.

When a fault or a warning is found, Genstat produces a diagnostic giving the code number of the error and an indication of where it has occurred. This will be the number of the statement where an error has been found: counting from the start of the program, or from the start of a FOR loop (6.2.1) if the statement is within a loop, or from the start of a procedure if the statement is within a procedure (1.7.1, 6.3). You then get a listing of the relevant part of the statement. The next line gives a definition of the error code (for a complete list of which see Appendix 2), and it is followed by further information if relevant.

As an example, suppose you want to run in interactive mode a slightly modified version of the example given in 1.1.1, but that you are rather careless and make three mistakes:

EXAMPLE 2.1.2a

```
  1    SCALAR IDENTIFIER=First,Dessert,Wine,Service; VALUE=8.50,1.50,3.00,1.10
  2    CALCULATE Gcost = First+Dessert+Wine
  3    PRINT STRUCTURE=Gvost; DECIMALS=1
******* Warning (Code VA 4). Statement 1 on Line 3
Command: PRINT STRUCTURE=Gvost; DECIMALS=1

Values not set
Gvost has no values.

  4    PRINT STRUCTURE=Gcost; DECIMALS=!
******* Warning (Code SX 50). Statement 1 on Line 4
At...    ; DECIMALS\=\!:
Unexpected characters after end of command

  5    PRINT STRUCTURE=Gcost; DECIMALS=1

       Gcost
        13.0

  6    CALCULATE Ncost = Gcost*
******* Fault (Code SX 12). Statement 1 on Line 6
At...    t = Gcost*\:\
Incompatible adjacent elements (e.g. comma missing)

  7    CALCULATE Ncost = Gcost*Service
  8    PRINT STRUCTURE=Ncost; DECIMALS=2

       Ncost
       14.30
```

When you are using Genstat interactively nearly all statements are *executed* – that is, carried out by Genstat – immediately you finish typing them. The only exceptions are within FOR loops and during the definition of procedures.

2.1 Information about the system

If an error occurs while *interpreting* a statement – that is, when Genstat checks that it has correct syntax – the second line of the diagnostic shows by backslashes (\) where in the statement the error has been found. The SX 12 fault on line 6 above is of this type. An error found while executing a statement results merely in an echoing of the relevant statement of the program: for example, VA 4 on line 3 above.

If you ran the above example in batch mode you would get the output

---

EXAMPLE 2.1.2b

```
   1  SCALAR IDENTIFIER=First,Dessert,Wine,Service; VALUE=8.50,1.50,3.00,1.10
   2  CALCULATE Gcost = First+Dessert+Wine
   3  PRINT STRUCTURE=Gvost; DECIMALS=1

******** Warning (Code VA 4). Statement 1 on Line 3
Command: PRINT STRUCTURE=Gvost; DECIMALS=1

Values not set
Gvost has no values.

   4  CALCULATE Ncost = Gcost*

******** Fault (Code SX 12). Statement 1 on Line 4
At...    t = Gcost*\:\
Incompatible adjacent elements (e.g. comma missing)


A fatal fault has occurred - the rest of this job will be ignored
   5  PRINT STRUCTURE=Ncost; DECIMALS=2
   6  STOP
```

When a fault is found in batch mode, a diagnostic is printed out and execution stops immediately (for example, SX 12 on line 4 above). The remainder of the job is also printed out, but there will be only elementary checking for errors in statements after the one in which the error was found. The detection of a fault will not affect the execution of following jobs in the same program (6.1). If a warning occurs, a diagnostic is given but execution continues (for example, VA 4 on line 3 above).

In interactive mode, there are faults that prevent execution of any statements other than ENDJOB (6.1.2) and STOP (6.1.3). This happens when a system fault (one with the prefix SY, standing for system) occurs. Other faults allow you to continue.

Sometimes you may want to prevent diagnostics being printed – for example, warnings that you may already know about, or faults that you may want to intercept yourself within a procedure. You can use the DIAGNOSTICS option of SET (2.2.1) or JOB (6.1.1) to suppress either warnings, or both faults and warnings; the option also allows extra information to be produced after a fault, which may be useful to developers of Genstat.

Some errors in programming give multiple diagnostics. You can use the ERRORS option of SET or JOB to set the maximum number of errors that is permitted before

38 2 *The environment of a Genstat program*

Genstat abandons execution of a statement.

2.1.3 The DISPLAY directive

DISPLAY

Reprints the last diagnostic.

Option

PRINT = string What information to print (diagnostic); default d

No parameters

You should use the DISPLAY directive only once between faults, since using it deletes the stored information about the diagnostic.

Currently there is only one setting of the option PRINT: namely, 'diagnostic'. The option is present so that further information can be included in later releases of Genstat.

2.2 How to set details of the environment

The output from the Genstat programs in Chapter 1 was produced in the standard environment. For example, the Genstat statements were not echoed with line numbers when a program was run interactively, but they were when it was run in batch; newlines in the programs were taken as terminators of statements unless a continuation symbol was given; upper-case and lower-case letters were treated as distinct in identifiers. You can change these and other details of the environment of a job by the SET directive (2.2.1).

Two further directives set specific details of the environment. The COPY directive (2.2.2) lets you form and control a transcript of a job; and you use the UNITS directive (2.2.3) to name a structure that labels the units of data being analysed: these labels are output automatically by many of the directives for analysis.

2.2.1 The SET directive

SET

Sets details of the "environment" of a Genstat job.

Options

INPRINT	= strings	Printing of input as in PRINT option of INPUT (statements, macros, procedures, unchanged); default u
OUTPRINT	= strings	Additions to output as in PRINT option of OUTPUT (dots, page, unchanged); default u
DIAGNOSTIC	= strings	Output to be printed for a Genstat diagnostic (warnings, faults, extra); default w,f
ERRORS	= scalar	Number of errors that a job may contain before it is abandoned; default * i.e. no limit
PAUSE	= scalar	Number of lines to output before pausing (interactive use only); default * i.e. no pause
NEWLINE	= string	How to treat newline (significant, ignored); default s
CASE	= string	Whether lower- and upper-case (small and capital) letters are to be regarded as identical in identifiers (significant, ignored); default s
RUN	= string	Whether or not the run is interactive (interactive, batch); default * i.e. the current setting is left unchanged
UNITS	= identifier	To (re)set the current units structure; default * i.e. leave unchanged
BLOCKSTRUCTURE	= identifier	To (re)set the internal record of the most recent BLOCKSTRUCTURE statement; default * i.e. leave unchanged
TREATMENTSTRUCTURE	= identifier	To (re)set the internal record of the most recent TREATMENTSTRUCTURE statement; default * i.e. leave unchanged
COVARIATE	= identifier	To (re)set the internal record of the most recent COVARIATE statement; default * i.e. leave unchanged
ASAVE	= identifier	To (re)set the current ANOVA save structure; default * i.e. leave unchanged

RSAVE	= identifier	To (re)set the current regression save structure; default * i.e. leave unchanged
TSAVE	= identifier	To (re)set the current time-series save structure; default * i.e. leave unchanged

No parameters

The default of SET is to do nothing: that is, each option by default leaves the corresponding attribute of the environment unchanged. Of course you have to start somewhere: at the first use of SET in a program, the default of each option is defined by Genstat in the ways outlined in this section; these are called the *initial defaults*.

The INPRINT option controls what parts of a Genstat job supplied in the current input channel are recorded in the current output file: the input channel can be either an input file or a terminal. Three parts are distinguished: explicit statements; statements, or parts of statements, that you have supplied in macros using the ## notation (1.7.2); and statements that you have supplied in procedures. The initial default of SET is taken from the setting of the PRINT option in the latest INPUT statement (4.4.1). If there has been no INPUT statement, the initial default is the setting of the INPRINT option of the latest JOB statement (6.1.1). If no JOB statement has been given, the initial default is to record only the statements, unless output is to a terminal in which case the default is to record nothing.

The OUTPRINT option controls how the output from many Genstat directives starts: the output can be preceded by a page throw, or by a line of dots beginning with the line number of the statement producing the analysis, or by both. If output is directly to a terminal, no page throws are given. The initial default is the setting of the PRINT option in the latest OUTPUT statement (4.4.3). If there has been no OUTPUT statement, the initial default is the setting of the OUTPRINT option of the latest JOB statement. If no JOB statement has been given, the initial default is to give both dots and page throws, unless output is to a terminal, in which case the default is to give neither.

You get the line of dots from directives for regression analysis, analysis of designed experiments, multivariate analysis and analysis of time series; also from the FLRV, FSSPM and SVD directives (5.7). If you give an analysis statement within a FOR loop (6.2.1), the line number preceding the line of dots usually corresponds to the ENDFOR statement rather than to the analysis statement: the exception is for the first pass of the loop when you have set the option COMPILE=each in the FOR directive.

You get the page throw with any of the above, and with the GRAPH, HISTOGRAM and CONTOUR directives (7.1).

The DIAGNOSTIC option lets you switch off the reporting of faults. You might do this within procedures, to prevent faults being reported to a user who does not need

2.2 How to set details of the environment

to know in detail what is going on inside the procedure. By initial default, both warnings and faults (2.1.2) are printed, along with information to help you understand the reason for the message. You can switch off either or both of these types of message. The 'extra' setting gives you extra information, in the form of a dump of the current state of the job; but this is likely to be useful only for developers of Genstat (12.4).

The ERRORS option controls what Genstat does when many faults happen within a single job. By initial default, up to five errors are reported.

The PAUSE option lets you specify how many lines of output are produced at a time; you might, for example, want to read the output on a terminal screen before more output replaces it. Obviously this is relevant only in interactive mode. By initial default, all output is sent to the current output channel as soon as it is available. Some terminals can store it, and let you scroll forward and back to read it at leisure: others just provide keys to freeze the output while you are reading a section, and then to continue to the next segment of output. If you set PAUSE = n, then after every n lines of output Genstat gives a prompt:

Press RETURN to continue

If you have specified that Genstat should echo all the lines that you input, then the echoed lines are counted among the n. After you have read the section of output displayed, press the RETURN key to get the next n lines. Once all the output has been exhausted, Genstat prompts for further statements.

The NEWLINE option allows you to cancel the initial default whereby a newline (⟨RETURN⟩) is a terminator both for strings within a string list (1.4.2) and for a statement (1.6). Thus, for example, if you give

SET [NEWLINE = ignored]

you no longer have to use a backslash (\) to continue a statement onto a new line, since ⟨RETURN⟩ is no longer interpreted as the end of a statement.

The CASE option specifies whether upper- and lower-case letters are to be treated as the same in identifiers. The initial default is that upper and lower case are not the same; thus, an identifier X is distinct from an identifier x. If CASE is set to 'ignored', then in later statements, both x and X are treated as the same identifier, X. Thus the structure with identifier x cannot be referenced, unless CASE is later reset to 'significant'.

The RUN option controls whether Genstat interprets the program as being in batch or in interactive mode; note that this assumed mode is independent of whether the program really is being run in batch or interactively. Initially, a program is taken to be in interactive mode only if the first input channel and the first output channel are both connected to a terminal. The setting of the assumed mode has two effects – on recovery from faults (2.1.2), and on how HELP (2.1.1) and EDIT (5.4.2) operate.

The UNITS option provides a way of setting the *units structure*. (The UNITS directive described in 2.2.3 also does this.) The setting can be the identifier of a variate or text structure; this will become the default labelling structure of other variates, texts or factors with the same length, in those directives that use such labels. The setting can also be a scalar to specify the default number of units. For further details, see 2.2.3. The setting of the UNITS option is lost at the end of each job within a program.

The last six options of the SET directive specify special *save structures* for the analysis directives described in Chapters 8, 9 and 11. You can set the options only to an identifier that you have previously established by the SPECIAL option of the GET directive (2.3.1) or by the SAVE options in the various analysis directives themselves. For example, if two sets of regression analyses are in progress in one job, the SET directive can be used to switch between them:

```
MODEL [SAVE=S1] Y1
FIT X1
MODEL [SAVE=S2] Y2
FIT X1
SET [RSAVE=S1]
FIT X1,X2
```

This program fits the regression of Y1 on X1, using save structure S1, then the regression of Y2 on X1 with save structure S2. Finally, the regression of Y1 on X1 and X2 is fitted, because the current regression save structure is changed to S1 before the last FIT statement.

The settings of these last six options are lost at the end of a job.

2.2.2 The COPY directive

COPY

Forms a transcript of a job.

Option

| PRINT | = strings | What to transcribe (statements, output); default s |

Parameter

| | scalar | Channel number of output file |

If you set the PRINT option to 'statements' only, a record is kept in the specified output file of the statements in a job. You could later edit the resulting file to get a job that you could run again in batch: you may have to remove Fortran control characters and the line numbers added to the statements. If you set the PRINT option to 'output' only, a record of the output of the job is kept. The record is then the same

2.2 How to set details of the environment

as is produced on the primary output in interactive mode, or in batch mode if you give the statement SET [INPRINT = *] (2.2.1). The output does not include the contents of macros or even procedures unless you have set the INPRINT option of the SET directive to 'macros' or 'procedures' respectively. If you set the PRINT option to 'statements' as well as to 'output', the transcript file is a full record of the job; you can use the transcript to print a copy of a job that you have run interactively on a terminal.

If you start a new job, copying continues as in the previous job but with the initial default setting of the PRINT option of COPY: that is, only statements are copied. You can switch off copying at any time by giving a COPY statement with no parameter. If the channel number in the COPY statement is the same as the current output channel, output will be duplicated on that channel, whether it is a file or a terminal.

2.2.3 The UNITS directive

UNITS

Defines an auxiliary vector of labels and/or the length of any vector whose length is not defined when a statement needing it is executed.

Option

NVALUES	= scalar	Default length for vectors

Parameter

	variate or text	Vector of labels

The UNITS directive gives you an alternative mechanism to the UNITS option of the SET directive (2.2.1) for defining a set of labels, or the default length of *vectors* in a job, or both. A *vector* is a variate, a text, or a factor. Thus the statement

    UNITS [NVALUES = 10]

is equivalent to

    SET [UNITS = 10]

Either has the effect that the length of vectors subsequently used in the program will be taken to be 10 if you do not otherwise define them. This happens in several directives. It happens if you use the READ directive (4.1) to read a previously undeclared structure, provided the SETNVALUES option of READ has the default value, 'no'. It happens in the directives VARIATE, TEXT, and FACTOR (3.3) when you have not otherwise defined the length. It happens if you use the RESTRICT directive (2.4.1) to restrict a previously undeclared structure. And it happens when you use the CALCULATE directive with the URAND or EXPAND functions, if you do not also specify the secondary argument (5.2).

Thus, for example, this program sets the structure Data to be a variate with 10 values, reading them from a file called FILE.DAT:

```
OPEN 'FILE.DAT'; CHANNEL=2
UNITS [NVALUES=10]
READ [CHANNEL=2] Data
```

Similarly, this defines Group as a factor with 20 values:

```
UNITS [NVALUES=20]
FACTOR [LEVELS=2] Group
```

If you specify a variate or a text structure in the parameter of the UNITS directive, you must already have defined its length. This length will be taken to define the default length for other vectors, as above. If you set the NVALUES option here as well, its setting must match the length of the vector in the parameter; but if you have not already defined the length of the vector the option setting will do that for you. For example, if you have not yet defined the length of Plots, then the statement

```
UNITS [NVALUES=30] Plots
```

sets Plots to have 30 values. But if you had previously specified the length of Plots to be, say, 20, then this statement would fail.

The values of the units structure are later used as labels for output from regression or time-series directives, provided the vectors concerned have the same length as the units structure and provided also that these vectors do not have labels associated with them already. For example, if you ask for the fitted values to be printed from a linear regression, the values of the units structure are used as labels; they are listed in the left-most column:

EXAMPLE 2.2.3

```
  1  TEXT [VALUES=a,b,c,d,e] Code
  2  UNITS Code
  3  VARIATE X,Y; VALUES=!(1...5),!(2,5,6,8,9)
  4  MODEL Y
  5  FIT [PRINT=fittedvalues] X

  5.........................................................

  ***** Regression Analysis *****

  *** Fitted values and residuals ***

                              Standardized
         Code     Response  Fitted value  residual  Leverage
            a         2.00          2.60     -1.57      0.60
            b         5.00          4.30      1.38      0.30
```

```
              c           6.00         6.00         0.00         0.20
              d           8.00         7.70         0.59         0.30
              e           9.00         9.40        -1.04         0.60

Mean                      6.00         6.00        -0.13         0.40
```

However, if Y already had labels associated with it, then these would be output instead of the values of the units structure Code.

You can cancel the effect of a UNITS statement by

    UNITS [NVALUES = *]

This means that statements that require a units structure will fail, which is the situation at the start of each job in a program. READ, for example, fails if you give in its first parameter a vector of undefined length.

The statement

    UNITS *

cancels any reference to a units structure, but retains the default length if you have already defined it.

2.3 Accessing details of the environment of a program

We now describe two further useful things that you can do with the environment. You can check what the current settings are; and you can save the current settings in order to re-impose them later on. Saving is particularly useful in procedures (1.7.1 and 6.3). Sometimes the environment inside a procedure has to be different from that in the rest of the program. So you would save the environment settings at the beginning of the procedure and then re-impose them just before leaving it.

You do both these things with the GET directive. (Another way of getting information about the current environment is with the HELP 'environment' facility: 2.1.1.)

2.3.1 The GET directive

GET

Accesses details of the "environment" of a Genstat job.

Options

ENVIRONMENT = pointer   Pointer given unit labels inprint, outprint, diagnostic, errors, pause, newline, case and run, used to save

		the current settings of those options of SET; default *
SPECIAL	= pointer	Pointer given unit labels units, blockstructure, treatmentstructure, covariate, asave, rsave and tsave, used to save the current settings of those options of SET; default *
LAST	= text	To save the last input statement; default *
FAULT	= scalar	To save the last fault code; default *
EPS	= scalar	To obtain the value of the smallest x (on this computer) such that $1+x > 1$; default *

No parameters

The ENVIRONMENT and SPECIAL options of GET are used to access and save the current settings of the options of the SET directive (2.2.1). The options of SET are divided into two groups. Those that apply to the general environment can be saved using the ENVIRONMENT option: these are INPRINT, OUTPRINT, DIAGNOSTIC, ERRORS, PAUSE, NEWLINE, CASE, and RUN. Those that apply only to the save structures associated with particular directives are saved by using the SPECIAL option: these options are UNITS, BLOCKSTRUCTURE, TREATMENTSTRUCTURE, COVARIATE, ASAVE, RSAVE, and TSAVE.

When you use the ENVIRONMENT option, Genstat sets up a pointer (3.3.4) with units identified by the labels of the corresponding options of SET: these labels are 'inprint', 'outprint', and so on. Each unit of this pointer contains one or more strings, or a scalar, to show what the current setting is. Thus, if the values of the options of SET had not been changed from their default values, typing the statement

    GET [ENVIRONMENT=Env]

would set up a pointer called Env with elements Env['inprint'], Env['outprint'], and so on. Note that you must use the correct case, here lower case, to refer to labels of a pointer. Each element can also be referred to by its position in the pointer; for example, Env['inprint'] could be referred to as Env[1].

    PRINT Env[ ]

would then give the following output:

2.3 Accessing details of the environment of a program

EXAMPLE 2.3.1

```
  1  GET [ENVIRONMENT=Env]
  2  PRINT Env[]

Env['inprint']  Env['outprint']  Env['diagnostics']  Env['errors']
    statements      dots,page       warnings, faults          5


Env['pause']  Env['newline']  Env['case']   Env['run']
      0       significant  significant  interactive
```

Thus you do not have to remember how you have set the environment; you can use GET to find out about it, and SET to change it. For example, suppose that the OUTPRINT option of SET was set to 'dots' and 'page'. Then you could use this program to alter it to 'dots' alone, and later to return to 'dots' and 'page':

    GET [ENVIRONMENT=Env]
    SET [OUTPRINT=dots]
    (more statements)
    SET [OUTPRINT=#Env['outprint']]

The SPECIAL option likewise sets up a pointer to save the relevant information. The unit labels of this pointer are 'units', 'blockstructure', 'treatmentstructure', 'covariate', 'asave', 'rsave' and 'tsave'. The first element of the pointer is the units structure, or, failing that, the number of units if you have defined it for the current job. Printing the contents of the other elements is not usually informative, as the information is stored in coded form. You use the last six elements of the pointer to access the special save structures of the analysis directives in Chapters 8, 9, and 11. They are most useful for recovering information about an analysis when you have forgotten to save it. (Otherwise you would have to do the analysis all over again.) For example, in the example at the end of 2.2.1, if you forgot to save S1 while modelling Y1, you could put instead

    MODEL Y1
    FIT X1
    GET [SPECIAL=S1]
    MODEL [SAVE=S2] Y2
    FIT X1
    SET [RSAVE=S1['rsave']]
    FIT X1, X2

Furthermore, GET [SPECIAL] provides the only way of accessing the save structures associated with the analysis-of-variance directives BLOCKSTRUCTURE (9.2.1), COVARIATE (9.3.1), and TREATMENTSTRUCTURE (9.1.1). In other words, there is no SAVE alternative here.

The LAST option is used to save the latest statement that you input. You can then give the statement again later in the job without having to retype it. The option has the same effect as setting up a macro (1.7.2) containing a single statement, and is accessed in the same way. For example, the statements

```
PRINT [SERIAL=yes; IPRINT=*; SQUASH=yes] !T('New Data'), Y
GET [LAST=Prdat]
(statements)
READ Y
(data)
##Prdat
```

would print the data, Y, under the title New Data and save the PRINT statement in a text called Prdat. After the next data set is read, the 'New Data' heading and the new data set are printed out in the same format as the previous data set. (The options of PRINT are described in 4.2.)

The FAULT option is used to save the last fault code as an integer value. A complete list of fault code definitions is in Appendix 2. This option is particularly useful in procedures, in combination with the ERRORS option of SET, to control the printing of diagnostics.

The EPS option is used to obtain the smallest number, ε, such that $1.0 + \varepsilon$ is recognized by your computer to be greater than 1.0.

2.4 Saving Space

Genstat baulks if you use too much space. Although you may be able to increase the amount available depending on the implementation on your computer, you should first try to be more economical. In this section we describe two directives that help you to write programs that do not waste space.

When you are writing a Genstat program you will sometimes find that you want to work on only a subset of the units of a data structure. For example you may wish to tabulate only those results that are larger than a particular value, or to do a regression only with the results recorded on a particular day. If you copied the relevant units into a smaller data structure, you would take up extra space on the computer. So the RESTRICT directive (2.4.1) enables you to operate on subsets of vectors. The way in which the RESTRICT directive affects the operation of other directives is described in detail in the chapters that are devoted to these directives.

Another example of when you can save space is in a large program, in which you finish using some data structures part way through the job. The values of the data structures remain clogging up the available space. Moreover, Genstat actually sets up unnamed structures in addition to those that you declare explicitly or implicitly: it uses them as workspace, but they quickly become redundant. The DELETE directive (2.4.2) lets you delete values of named or unnamed structures.

2.4 Saving space

Yet another example is if you want a data structure to have different attributes in different parts of a job: you might, for example, want to change the length or the type. Again you use DELETE: it lets you delete the attributes of named and unnamed data structures so that the space can be re-used. Doing this saves space compared with setting up a whole new structure.

In all uses of DELETE, the amount of space recovered depends both on the size of the data structures deleted and on the number and complication of operations that have been performed with them.

2.4.1 The RESTRICT directive

RESTRICT

Defines a restricted set of units of vectors for subsequent statements.

No options

Parameters

VECTOR	= vectors	Vectors to be restricted
CONDITION	= expression	Logical expression defining the restriction for each vector; a zero (false) value indicates that the unit concerned is not in the set
SAVESET	= variates	List of the units in each restricted set

Not all directives take account of RESTRICT. For those that do, usually only one vector in the list of parameters has to be restricted for the directive to treat them all as being restricted in the same way. A fault is reported if any vectors in such a list are restricted in different ways. A summary of which directives do obey restrictions and which do not is given below. You might find helpful the general guideline that RESTRICT is obeyed by all directives that operate on vectors except in those statements where explicit identification of elements is possible: for example, qualified identifiers and ELEMENTed identifiers in CALCULATE (5.1.6), EQUATE (5.3.1), and READ (4.1). But this guideline does not always operate, so you should check each directive in this manual to confirm precisely what happens. Restrictions on texts are obeyed only by PRINT (4.2), CALCULATE (5.1.2), CONCATENATE (5.4.1), and EDIT (5.4.2).

The effect of RESTRICT on directives that have vectors as parameters:

RESTRICT obeyed   **RESTRICT not obeyed**

Data structures
  SSPM   [TERMS= ; GROUPS=]   FACTOR
                                                                       TEXT
                                                                           VARIATE

Control structures
  CASE   expression
  ELSIF   expression
  EXIT   expression
  IF   expression

Input and output
  PRINT   STRUCTURE=   DUMP
                                                                           READ

Graphics
  CONTOUR   GRID=   AXES
  DCONTOUR   GRID=   PEN
  DGRAPH   Y= ; X=   SKIP
  GRAPH   Y= ; X=
  DHISTOGRAM   DATA=
  HISTOGRAM   DATA=

Backing store
  RETRIEVE
  STORE

Calculation and manipulation
  CALCULATE   expression   ASSIGN
                    (except using qualified identifiers   COMBINE
                    or ELEMENTed identifiers)   DELETE
  CONCATENATE   EQUATE
  EDIT   GENERATE
  FSSP   [WEIGHTS=]   UNITS
  INTERPOLATE   OLDVALUES= ; OLDINTERVALS= ;
                    NEWVALUES= ; NEWINTERVALS=
  RANDOMIZE   parameter
  RESTRICT   VECTOR= ; CONDITION=
  SORT   OLDVECTOR=
  TABULATE   [CLASSIFICATION=] DATA=

Regression analysis
  ADD   parameter
  DROP   parameter

2.4 Saving space

RESTRICT obeyed   **RESTRICT not obeyed**

FIT	parameter
FITCURVE	parameter
FITNONLINEAR	parameter
MODEL	[WEIGHTS= ; OFFSET= ; GROUPS=]
	Y= ; NBINOMIAL=
PREDICT	CLASSIFY=
RKEEP	RESIDUALS=; FITTEDVALUES=; LEVERAGE=
STEP	parameter
SWITCH	parameter
TERMS	parameter
TRY	parameter

Analysis of variance

ANOVA	Y=	BLOCKSTRUCTURE
		COVARIATE
		TREATMENTSTRUCTURE

Multivariate and cluster analysis

FSIMILARITY	DATA=	
HLIST	DATA=	CLUSTER
HSUMMARIZE	DATA=	PCO
PCP	DATA=	REDUCE
RELATE	DATA=	

Time-series analysis

CORRELATE	SERIES= ; LAGGEDSERIES=	FORECAST
ESTIMATE	SERIES=	FTSM
FILTER	OLDSERIES=	
FOURIER	SERIES= ; ISERIES= ;	
	TRANSFORM= ; ITRANSFORM=	
TRANSFER	SERIES=	

The VECTOR parameter specifies the vector or vectors that are to be restricted. These can be variates, factors, or texts, but all the vectors listed must be of the same length.

The CONDITION parameter defines the nature of the restriction: that is, it specifies which units of the vectors are to be used in later operations with these vectors. There are four main ways in which CONDITION can do restrictions.

In the most widely useful way, the parameter is a logical expression, like this:

    RESTRICT A; CONDITION=A.EQ.2

which restricts the vector A to units with the value 2. In all such uses of the parameter, the logical expression causes Genstat to construct internally a variate of zeroes and ones, of the same length as the vectors being restricted. A zero means that the corresponding unit is to be excluded. You can use these logical expressions to deal with the values of variates and of texts and with the levels of factors.

Now a more elaborate example: to restrict the variate B and the factor F to the units for which F has levels 1 or 2 or 4, you could use the statement

    RESTRICT B,F; CONDITION=(F.LE.2).OR.(F.EQ.4)

With text to define the restrictions, you cannot use the logical operators like .EQ. and .NE.; you should use operators .IN. .NI. .EQS. .NES. instead. Thus

    TEXT [VALUES='This is a line','So is this','And this too'] T
    RESTRICT T; CONDITION=T.IN.'So is this'

restricts the text T to the second line of text only.

Logical expressions and operators are described in detail in 5.1.

In the second way of restricting, the condition is a variate of the same length as the vectors to be restricted. Again a zero indicates that the corresponding unit in the vector to be restricted is excluded; any non-zero entry causes inclusion. If T was a text with six values the statement

    RESTRICT T; CONDITION=!(1,0,1,0,0,0)

would restrict T to the first and third lines of text. The same effect could be achieved by using the EXPAND function (5.2.7):

    RESTRICT T; CONDITION=EXPAND(!(1,3))

The third way of using the CONDITION parameter is to carry across the restrictions from one vector to others. For example,

    RESTRICT A,B; CONDITION=RESTRICTION(C)

restricts the vectors A and B to the same units as those to which C is restricted.

The fourth use of the CONDITION parameter is in fact to end all restrictions: if you do not set the parameter at all, then the restrictions on the vectors are removed. Thus

    RESTRICT A

removes all restrictions that have been set on A.

Note that if the vectors used in the CONDITION parameter are themselves restricted these restrictions will remain in force for the current calculation of the condition. A danger here, therefore, is that you accidentally end up restricting out all the elements of a vector, by using RESTRICT repeatedly. The safest way to avoid that is to remove the restrictions on any vectors to be used in the CONDITION parameter before using them to restrict vectors in a different way.

The SAVESET parameter is used to save the list of unit numbers that you have specified not to be excluded by the restriction. Genstat sets up a variate with one value for each unit retained by the restriction: the values are the numbers of the units. For example, in the illustration with text T above, we could have written

    RESTRICT T; CONDITION=!(1,0,1,0,0,0); SAVESET=S

S is created here to be a variate of length 2, with values 1 and 3.

2.4.2 The DELETE directive

DELETE

Deletes the attributes and values of structures.

Options

REDEFINE	= string	Whether or not to delete the attributes of the structures so that the type etc can be redefined (no, yes); default n
LIST	= string	How to interpret the list of structures (inclusive, exclusive, all); default i

Parameter

identifiers	Structures whose values (and attributes, if requested) are to be deleted

---

If you set the REDEFINE option to 'yes', all the attributes of the structures specified by the parameter are deleted. The only things that remain in the computer are the identifier and the internal reference number of the structure. The default setting of this option is 'no', in which case only the values of the structure are deleted. For example, suppose we have defined a scalar Dose by

SCALAR IDENTIFIER = Dose; VALUE = 6

This gives Dose the value 6. If we then put

DELETE Dose

only the value of Dose is deleted; so we could now assign it a new value:

READ Dose

10:

Dose remains a scalar, but it now has the value 10.

You do not necessarily have to delete the values of a data structure before giving it new values, but it is sometimes safer to do so, as, for example, if you use READ (4.1) in a FOR loop (6.2.1) when the same structure is given different values each time around the loop.

On the other hand, if we used the other setting of REDEFINE in the above example:

DELETE [REDEFINE = yes] Dose

then we could completely redefine Dose as (for example) a variate with 16 values:

VARIATE [NVALUES=16] Dose

Dose would now be ready to receive any 16 values that we wanted to give to it.

Once you have defined the type of a structure in a job (as variate, factor, or whatever), you cannot redefine it without using DELETE with the REDEFINE option set to 'yes'.

The LIST option defines how the parameter list is to be interpreted. If LIST=inclusive, the default setting, attributes or values are deleted only from structures in the list and from unnamed structures set up by Genstat. If there was no parameter list, then only unnamed structures are deleted. LIST=exclusive means that the parameter list is the complement of the set of structures that are deleted. That is, all named or unnamed structures that are not in the list are deleted. LIST=all causes the attributes or values of all structures to be deleted. If LIST=all, any parameter list is ignored; and LIST=exclusive with no parameter is equivalent to LIST=all.

2.5 Accessing details of data structures

You can easily forget details about data structures. For example, in a long interactive session you might forget the identifiers of certain structures, or the attributes that you have given them. The DUMP directive (2.5.1) helps you by displaying lists of structures or lists of their attributes or lists of their values. (Developers of Genstat might also find DUMP useful in batch mode.) Despite its name, DUMP does not actually lose the information about the structures: it merely prints it for you.

If you are writing a procedure, you may need to store and not just display the attributes of structures. The GETATTRIBUTE directive (2.5.2) lets you do this.

2.5.1 The DUMP directive

DUMP

Prints information about data structures, and internal system information.

Options

PRINT	= strings	What information to print about structures (attributes, values, identifiers); default a
CHANNEL	= scalar	Channel number of output file; default current output file
INFORMATION	= string	What information to print for each structure (brief, full); default b
TYPE	= strings	Which types of structure to include in addition to those in the parameter list (all, diagonalmatrix, dummy,

2.5 Accessing details of data structures 55

COMMON	= strings	expression, factor, formula, LRV, matrix, pointer, scalar, SSPM, symmetricmatrix, table, text, TSM, variate); default * i.e. none Which internal Fortran commons to display (all, banks, compl, diagpk, direct, fncon, inout, input, mainac, output, periph, print, root, syscon, wsp); default *
SYSTEM	= string	Whether to display Genstat system structures (no, yes); default n
Parameter		
	identifiers	Structures whose information is to be printed

DUMP displays information about the structures specified by the parameter; what it tells you is controlled by the options. You can get extra information by setting the options TYPE and COMMON; if you have not put any identifiers in the parameter list, you get any TYPE and COMMON information that you have requested on its own.

The PRINT option lets you specify what is to be printed for the dumped structures: you can ask for just the identifiers, or values and identifiers, or attributes (the identifier is an attribute), or for all three. For example, to get all three for the structures A and B you would put:

    DUMP [PRINT=attributes,values] A,B

EXAMPLE 2.5.1a

```
  5  DUMP [PRINT=attributes,values] A,B

***** DUMP *****

Identifier      Type    Length    Values Missing   Ref.No.
         A   Variate         9   Present       1     -1030
    1.0000             2.0000          3.0000        4.0000      5.0000
    6.0000             7.0000          8.0000             *

         B    Factor         9   Absent        *     -1033
No values
```

The CHANNEL option specifies the output channel to which the information is sent; you will usually want this to be the first channel, giving output on a terminal screen in an interactive run.

By setting INFORMATION = full, all the attributes of each structure can be displayed rather than the selection displayed by default. This selection, however, usually contains all the most important ones. In the 'full' form, the output is much more terse, and is unlikely to be useful except to developers of Genstat.

The TYPE option lets you display lists of all structures of a particular type, or of several types. For example, if you have forgotten the identifier of a factor, give the statement

    DUMP [TYPE = factor; PRINT = identifiers]

EXAMPLE 2.5.1b

```
  6  DUMP [TYPE=factor; PRINT=identifiers]

***** DUMP *****
 List of structure names
3          F1         F2
```

This lists all the current factors. When PRINT = attributes or values (or both), the setting TYPE = all provides a list of all named and unnamed structures, except system structures. "PRINT = identifiers; TYPE = all" lists only named structures.

The COMMON option is provided to let developers of Genstat display important internal information. Similarly, the SYSTEM option allows all the system structures to be dumped: there are many of these, so it is not a good idea to set this option frivolously.

2.5.2 The GETATTRIBUTE directive

GETATTRIBUTE

Accesses attributes of structures.

Option

ATTRIBUTE	= strings	Which attributes to access (nvalues, nlevels, nrows, ncolumns, type, levels, labels, nmv); default * i.e. none

Parameters

STRUCTURE	= identifiers	Structures whose attributes are to be

		accessed
SAVE	= pointers	Pointer to store copies of the attributes of each structure, these are labelled by the ATTRIBUTE strings

The GETATTRIBUTE directive stores in unnamed scalar structures information about each of the structures that are listed with its STRUCTURE parameter. It refers to the lists of scalars by pointers, which are set up by the SAVE parameter. You must always set the option and both parameters. For example,

    VARIATE [VALUES = 1,2,*,4] X
    GETATTRIBUTE [ATTRIBUTE = nvalues,nmv] X; SAVE = P

sets up P to be a pointer referring to the two scalars mentioned in the option.

The first is P['nvalues'], alternatively referred to as P[1], storing the value 4; and the second is P['nmv'], or P[2], storing the value 1:

EXAMPLE 2.5.2a

```
  9  PRINT P[ ]

P['nvalues']     P['nmv']
      4.000        1.000
```

If you include in the option an attribute that is not relevant to a structure, it is omitted from the pointer. Thus the 'nlevels', 'levels' and 'labels' settings are relevant only for factors, and 'nrows' and 'ncolumns' only for matrices. The references to those attributes that you do specify are always stored in the order shown in the definition of the ATTRIBUTE option at the beginning of this subsection. If an attribute of a structure has not been set then the corresponding scalar contains the missing value. For factors, the attributes 'levels' and 'labels' are not scalars, but are references to vectors of levels and labels – if indeed you have associated any with the factor. If the factor has not been declared with levels other than the default integers, or if it has no labels vector (3.3.3), then the corresponding entry in the pointer is set to the missing value. For example,

    VARIATE [VALUES = 4,8,12] Lev
    FACTOR [LEVELS = Lev] F
    GETATTRIBUTE [ATTRIBUTE = levels,labels] F; SAVE = P

sets up P as a pointer with two values, the first being Lev and the second missing:

EXAMPLE 2.5.2b

```
 13  DUMP [PRINT=attributes,values] P,Lev

***** DUMP *****
Identifier      Type  Length   Values Missing   Ref.No.
         P   Pointer       2  Present       0     -1046
     -1053         *

       Lev   Variate       3  Present       0     -1053
    4.0000            8.0000          12.0000
```

The setting 'type' produces a scalar value denoting the type of structure, according to the code:

1	scalar	5	matrix	11	expression	15	LRV
2	factor	6	diagonal matrix	12	formula	16	SSPM
3	text	7	symmetric matrix	13	dummy	17	TSM
4	variate	8	table	14	pointer		

A.E.A.
P.W.L.

3 Data structures

Data structures store the information on which a Genstat program operates. Examples are data for statistical analyses, coordinates for graphs, text for annotation, and so on. You can also store almost anything that can be printed in an analysis. This enables you to extend the range of facilities that Genstat offers, by taking information from one directive and using it as input for another (12.1). To let you do this, Genstat has a wide range of different structures, but you do not need to know about all of them for most analyses.

The simplest structures store a single piece of information or *value*, the most important being the scalar which stores a single number (3.2.1).

More useful are the structures that store several values. For example a variate contains a list of numbers (3.3.1); you might use it for the response and explanatory variables in a linear regression (Chapter 8), or for the variables in an analysis of variance (Chapter 9).

There are three different kinds of matrix: rectangular, symmetric and diagonal (3.4). You would use these mainly in multivariate analysis (Chapter 10).

There are also multi-way tables (3.5), which can be filled with various sorts of summaries; for example, means, totals, minima, and maxima (5.8).

Not all structures store numbers. Other *modes* for their values are strings (3.3.2), identifiers (3.2.2 and 3.3.4), expressions (3.2.3), and formulae (3.2.4). The combination of the shape of a data structure and the mode of its values is determined by its *type*. There are 15 types available. Your program can contain as many data structures of each type as you like, limited only by the total amount of workspace that they occupy within Genstat.

3.1 Declarations

Most data structures have a name; the exceptions are called *unnamed structures* and are described in 1.4.3. The name is called an *identifier* (1.3.3), and you use it to refer to the structure within your program. The only restrictions on how you form identifiers are the rules given in 1.3.3. You can define the identifier of a structure, together with its type, with a type of statement known as a *declaration*. There is a directive available to declare each type of structure. For example the declaration

    SCALAR Length

uses the SCALAR directive to define a scalar with identifier Length.

You can declare several structures in a single statement: for example

    SCALAR Length,Width,Height

declares Length, Width, and Height all to be scalars. The only things that you must define in a declaration are the identifier and the type. But you can also specify values for the data structures (3.1.1), as well as various attributes that carry ancillary information about the structures.

Some attributes must be specified before the structure can be given values: an example is the number of rows and columns of a matrix (3.4.1). Others need be set only if you choose to use them; for example, the number of decimal places (3.1.2).

Options and parameters that apply generally to several different directives are described in this section; the others are described with the directive concerned, later in this chapter.

3.1.1 The VALUES option and parameter

In any declaration, you can assign values either by an option or by a parameter. The same name is used for both purposes: it is VALUE if the structure is of a type that stores a single value, and VALUES if it stores several. The option defines a common value (or set of values) for all the structures in the declaration, while the parameter allows the structures each to be given different values.

With the option you must supply a list of values. With the parameter, however, you must give a list of identifiers of data structures of the appropriate mode; for this the unnamed structures described in 1.4.3 are particularly useful. Thus, to declare variates X and Xsq each with its own set of values, you can put:

VARIATE X,Xsq; VALUES=!(1,2,3,4),!(1,4,9,16)

X then contains the values 1 up to 4, and Xsq contains 1, 4, 9, and 16. If both the option and the parameter are specified, the parameter takes precedence. For example,

SCALAR [VALUE=12.5] Length,Width,Height; VALUE=*,*,200

gives Length and Width the value 12.5 and Height the value 200. (The asterisk in the identifier list for the VALUE parameter means an omitted entry: see 1.4.3.)

3.1.2 The DECIMALS parameter

You can use the DECIMALS parameter to define the number of decimal places that Genstat will use by default whenever the values of the structure are printed: this applies to output either by PRINT or from an analysis. For example,

SCALAR Length,Width,Height; VALUE=12.5,6.25,120; DECIMALS=1,2,0

specifies that Length, Width, and Height should in future be printed with one, two, and zero decimal places respectively, although you can of course override this within the PRINT directive itself (4.2). This parameter occurs only in the declarations of structures that contain numbers.

3.1.3 The EXTRA parameter

You can associate a text with each data structure by means of the parameter EXTRA.

3.2 Single-valued data structures

This text is then used by many Genstat directives to give a fuller annotation of output. For example:

    SCALAR Length,Weight; EXTRA = 'in centimetres','in grams'

3.1.4 The MODIFY option

Normally if you declare a data structure for a second time, you will lose all its existing attributes and values. But you can retain them by setting option MODIFY = yes. Thus, to redeclare the scalar Length, changing only its number of decimals to two, you would need to put

    SCALAR [MODIFY = yes] Length; DECIMALS = 2

The one attribute that you cannot readily redefine is the type. Before you can redeclare an identifier to refer to a structure of a different type, you must delete all its attributes. (See 2.4.2, in which there is an example redeclaring a scalar as a variate.)

3.2 Single-valued data structures

3.2.1 Scalars

A scalar data structure stores a single number (1.3.1). The SCALAR directive which declares scalars has only the general options and parameters already described in 3.1.

SCALAR

Declares one or more scalar data structures.

Options

VALUE	= scalar	Value for all the scalars; default is a missing value
MODIFY	= string	Whether to modify (instead of redefining) existing structures (no, yes); default n

Parameters

IDENTIFIER	= identifiers	Identifiers of the scalars
VALUE	= scalars	Value for each scalar
DECIMALS	= scalars	Number of decimal places for printing
EXTRA	= texts	Extra text associated with each identifier

SCALAR is the one type of declaration where values are defined by default: if you do not define a value explicitly for a scalar, Genstat gives it a missing value.

Examples are given in 3.1. Unnamed scalars are described in 1.4.3.

3.2.2 Dummies

A dummy is a data structure that itself stores the identifier of some other structure. You will find this useful in identifier lists, where in nearly all cases Genstat replaces a dummy by the identifier that it stores. The only two exceptions are the IDENTIFIER parameter of the DUMMY directive itself (see below), and the STRUCTURE parameter of ASSIGN (5.6.1).

Dummies are particularly useful when you want the same series of statements to be used with several different data structures. By referring to a dummy structure within the statements you can make them apply to whichever structure you require. The dummy structure then becomes a sort of hole into which you can plug the structure you need, and the important point is that you can then plug in another one without changing the statements themselves.

The most obviously useful instances of this are in loops and procedures, where dummies are declared automatically: for examples, see 6.2.1 and 6.3.1.

To declare a dummy explicitly, you use the DUMMY directive. This has only the general options and parameters already described in 3.1.

DUMMY

Declares one or more dummy data structures.

Options

VALUE	= identifier	Value for all the dummies; default *
MODIFY	= string	Whether to modify (instead of redefining) existing structures (no, yes); default n

Parameters

IDENTIFIER	= identifiers	Identifiers of the dummies
VALUE	= identifiers	Value for each dummy
EXTRA	= texts	Extra text associated with each identifier

For example:

    DUMMY Xdum,Ydum; VALUE=Growth,Time

3.2 Single-valued data structures

3.2.3 Expression data structures

The expression data structure stores a Genstat expression (1.5.2), for example

    Hours = Minutes/60

Usually you will find it easiest to type out an expression like this explicitly whenever you need it. The main use, then, for this rather specialized data structure is to supply an expression as the argument of a procedure.

Options and parameters of the EXPRESSION directive, which declares expressions, are already described in 3.1.

EXPRESSION

Declares one or more expression data structures.

Options

VALUE	= expression	Value for all the expressions; default *
MODIFY	= string	Whether to modify (instead of redefining) existing structures (no, yes); default n

Parameters

IDENTIFIER	= identifiers	Identifiers of the expressions
VALUE	= expressions	Value for each expression
EXTRA	= texts	Extra text associated with each identifier

Here are two examples using the VALUE option:

    EXPRESSION [VALUE = Length*Width*Height] Vcalc
    EXPRESSION [VALUE = Dose = LOG10(Dose)] Dtrans

These put the expression Length*Width*Height into the identifier Vcalc, and the expression Dose=LOG10(Dose) into Dtrans. Both expressions could be declared simultaneously, using the VALUE parameter, by putting

    EXPRESSION Vcalc,Dtrans; VALUE = !E(Length*Width*Height),
    !E(Dose = LOG10(Dose))

Rules for omitting "VALUE=" when the expression contains an assignment are described in 1.6.3. Unnamed expressions like !E(Length*Width*Height) are described in 1.4.3.

3.2.4 Formula data structures

The formula data structure stores a Genstat formula (1.5.3). As with the expression data structure (3.2.3), its main use is to give a formula as the argument of a procedure (6.3). The FORMULA directive which declares formulae has only the general options and parameters described in 3.1.

FORMULA

Declares one or more formula data structures.

Options

VALUE	= formula	Value for all the formulae; default *
MODIFY	= string	Whether to modify (instead of redefining) existing structures (no, yes); default n

Parameters

IDENTIFIER	= identifiers	Identifiers of the formulae
VALUE	= formulae	Value for each formula
EXTRA	= texts	Extra text associated with each identifier

For example:

```
FORMULA [VALUE=Drug*Logdose] Model
FORMULA BModel,Tmodel; VALUE=!F(Litter/Rat),!F(Vitamin*Protein)
```

The construction !F(Litter/Rat) is an example of an unnamed formula, as described in 1.4.3.

3.3 Vectors and pointers

Most Genstat directives operate on structures that store several values. The most important of these contain a series of values, which you can imagine as being arranged as a *vector* in a column. Genstat has three different types of vector: variates (3.3.1), texts (3.3.2), and factors (3.3.3). Also, the pointer structure (3.3.4), which stores a series of identifiers, is treated like a vector in some directives.

The directives that declare vectors and pointers all have an option called NVALUES,

3.3 Vectors and pointers

with which you can specify a scalar to define the number of values to be stored in the vector or pointer. Alternatively, you can use NVALUES to specify another vector which some Genstat directives will then use to label the elements of the vector or pointer (2.2.3). If NVALUES is omitted in the declaration of a vector, Genstat takes the value or vector specified by the preceding UNITS statement if you have given one (2.2.3). In Genstat we call the elements of a vector its *units*.

For vectors, the labelling structure can be either a text or a variate, but pointers can be labelled only by a text (3.3.2). If you define values in the declaration and omit the NVALUES option, Genstat will deduce the appropriate setting from the number of values specified. However, a good safety device is to define both, since this makes Genstat check that you have specified as many values as you intended. Thus, for example, if you were to type

VARIATE [NVALUES = 5; VALUES = 1,2,3.4,5] X

Genstat would be able to tell you that X has been given only four values instead of the five that it should have (or that you had set five values instead of four). Further examples are given with the different types of structure below.

3.3.1 Variates

The variate is probably the structure that you will use most often in Genstat. You can think of this as being just a series of numbers – a vector, in mathematical language. Variates occur for example as the response and explanatory variables in regression (Chapter 8), as covariates and y-variables in analysis of variance (Chapter 9), and can be used to form the matrices of correlations, similarities, or sums of squares and products required for multivariate analyses (Chapter 10). Unnamed variates, for example !(1,2,3,4,5), are described in 1.4.2. To declare a variate you use the VARIATE directive.

VARIATE

Declares one or more variate data structures.

Options

NVALUES	= scalar or vector	Number of units, or vector of labels; default * takes the setting from the preceding UNITS statement, if any
VALUES	= numbers	Values for all the variates; default *
MODIFY	= string	Whether to modify (instead of redefining) existing structures (no, yes); default n

Parameters

IDENTIFIER	= identifiers	Identifiers of the variates
VALUES	= identifiers	Values for each variate
DECIMALS	= scalars	Number of decimal places for output
EXTRA	= texts	Extra text associated with each identifier

For example:

        VARIATE Weight; EXTRA='in grams'
        VARIATE Volume,Price; VALUES=!(60,125,200),!(0.6,1.2,1.75); DECIMALS=0,2

3.3.2 Texts

Each unit of a Genstat text structure is a string (1.3.2) which you can regard as a line of textual description. Texts can be used to label vectors and pointers (2.2.3 and 3.3), for captions or pieces of explanation within output (4.2.1), to store Genstat statements (1.7.2), and to store output (4.2.1). All the things that you can do with texts are described in 5.4. You declare texts with the TEXT directive.

TEXT

Declares one or more text data structures.

Options

NVALUES	= scalar or vector	Number of strings, or vector of labels; default * takes the setting from the preceding UNITS statement, if any
VALUES	= strings	Values for all the texts; default *
MODIFY	= string	Whether to modify (instead of redefining) existing structures (no, yes); default n

Parameters

IDENTIFIER	= identifiers	Identifiers of the texts
VALUES	= texts	Values for each text
CHARACTERS	= scalars	Numbers of characters of the lines of each text to be printed by default
EXTRA	= texts	Extra text associated with each identifier

3.3 Vectors and pointers

For example:

TEXT [NVALUES=5] Name; VALUES=!T(Triumph,Lotus,'Aston Martin',Ferrari,MG)

Unnamed texts, like that in the VALUES parameter in this example, are described in 1.4.3. Notice that the third value has to be enclosed in single quotes as it contains a space. The rules governing when strings need to be quoted and when the quotes can be omitted are described in 1.3.2.

You may be unable to define all the values of a long text in its declaration, because of the restriction on the total length of a statement (1.6). One possibility is to read the values (4.1). Alternatively, you could define several texts each containing several lines of the full text and then use EQUATE (5.3.1) to join them together. Or you could form the values from within the editor (5.4.2).

3.3.3 Factors

Factors are used to indicate groupings of units. The commonest occurrence is in designed experiments (Chapter 9). For example, suppose you had 12 observations in an experiment, the first four on one treatment, the next four on a second treatment, and the last four on a third. Then you could record which treatment went with which observation by declaring a factor with the values

1,1,1,1,2,2,2,2,3,3,3,3

Thus a factor is a vector that has only a limited set of possible values, one for each group; this limitation distinguishes factors from variates and texts. In Genstat, the groups are referred to by numbers known as *levels*. Unless otherwise specified these are the integers one up to the number of groups, as in our example; but you can specify any other numbers by the LEVELS option of the FACTOR directive (see below). You can also give textual labels to the groups, using the LABELS option of FACTOR: these might, for example, be mnemonics for the biochemical names of treatments in an experiment. The full syntax of FACTOR is:

FACTOR

Declares one or more factor data structures.

Options

NVALUES	= scalar or vector	Number of units, or vector of labels; default * takes the setting from the preceding UNITS statement, if any
LEVELS	= scalar or variate	Number of levels, or series of numbers which will be used to refer to levels in the program; default *

VALUES	= numbers	Values for all the factors, given as levels; default *
LABELS	= text	Labels for levels, for input and output; default *
MODIFY	= string	Whether to modify (instead of redefining) existing structures (no, yes); default n

Parameters

IDENTIFIER	= identifiers	Identifiers of the factors
VALUES	= identifiers	Values for each factor, specified as levels or labels
DECIMALS	= scalars	Number of decimals for printing levels
CHARACTERS	= scalars	Number of characters for printing labels
EXTRA	= texts	Extra text associated with each identifier

Use of the VALUES parameter to assign values has the advantage that you can refer either to labels or to levels; the VALUES option lets you refer only to levels. So, to summarize, the LEVELS and LABELS options list the groups that can occur, while the VALUES option or parameter specifies which groups actually do occur, and in what pattern over the units.

Our simple explanatory example would therefore be:

     FACTOR [LEVELS=3; VALUES=4(1...3)] Treatment

Other examples are:

     FACTOR [LEVELS=!(2,4,8,16); VALUES=8,4,2,16,4,2,16,8,2] Dose
     FACTOR [LABELS=!T(male,female)] Sex; VALUES=!T(4(male,female))
     FACTOR [LEVELS=!(0,2.5,5); LABELS=!T(none,standard,double)] Rate; \
       VALUES=!(0,5,2.5,5,0,2.5)

Notice that if we had assigned the values using the VALUES option in the second of these, we would have needed to use the (numerical) levels:

     FACTOR [LABELS=!T(male,female); VALUES=4(1,2)] Sex

Conversely, in the VALUES parameter in the declaration of Rate, we can use either the labels or the levels; so the following statement gives Rate exactly the same values:

     FACTOR [LEVELS=!(0,2.5,5); LABELS=!T(none,standard,double)] Rate; \
       VALUES=!T(none,double,standard,double,none,standard)

Apart from their use in designed experiments, you might also use factors to define groups for tabulation (5.8.1), to specify groups for parallel regression lines (8.3), and to store groupings derived from cluster analysis (10.5 and 10.7).

3.3.4 Pointers

A pointer is a data structure that points to other structures: that is, each of its elements is the identifier of some other Genstat data structure. You use pointers in Genstat wherever you have to specify a collection of structures; for example in EQUATE (5.3.1), COMBINE (5.3.4), in some functions (5.2.3), and for a data matrix specified via the variates forming its columns (10.2). You can use them as a convenient means of compacting lists (1.4.4), and they are also involved in the use of suffixed identifiers (see below). You can declare pointers using the POINTER directive.

POINTER
Declares one or more pointer structures.

Options

NVALUES	= scalar or text	Number of values, or labels for values; default *
VALUES	= identifiers	Values for all the pointers; default *
SUFFIXES	= variate	Defines an integer number for each of the suffixes; default * indicates that the numbers 1,2,... are to be used
MODIFY	= string	Whether to modify (instead of redefining) existing structures (no, yes); default n

Parameters

IDENTIFIER	= identifiers	Identifiers of the pointers
VALUES	= pointers	Values for each pointer
EXTRA	= texts	Extra text associated with each identifier

Thus, for example,

    POINTER [VALUES=Yield,Costs,Profit] Info

sets up a pointer Info with values Yield, Costs, and Profit. These three are themselves data structures, which can be assigned values, operated on, and so forth. You can refer to individual elements of pointers by suffixes, enclosed in square brackets (1.3.3): so Info[3] is Profit, and Info[1,2] is the list of structures Yield, Costs. Thus if Yield held the values 5.6 and 6.1, then

    PRINT Info[1]

would print the values of Yield, as shown below:

EXAMPLE 3.3.4

```
  1  VARIATE [NVALUES=2] Yield,Costs,Profit
  2  READ Yield,Costs,Profit

    Identifier    Minimum       Mean    Maximum     Values    Missing
         Yield      5.600      5.850      6.100          2          0
         Costs       1200       1365       1530          2          0
        Profit      455.0      537.5      620.0          2          0

  4  POINTER [VALUES=Yield,Costs,Profit] Info
  5  PRINT Info[1]

      Yield
      5.600
      6.100
```

In fact, when Genstat meets a suffixed identifier, it sets up a pointer automatically if necessary. For example if your program contains a suffixed identifier Data[4], Genstat first checks whether or not a pointer called Data already exists and, if not, creates it; then if there is no element for suffix 4 it creates one. If the pointer Data already exists but does not have a fourth element, then an appearance of Data[4] automatically extends Data. So you can add elements to pointers without redeclaring them.

The suffixes need not run from one, nor be a complete list, although they must be integers; if you give a decimal number it will be rounded to the nearest integer (for example, −27.2 becomes −27). You specify the list of suffixes that you require by the SUFFIXES option; if you omit this, they are assumed to run from one up to the number of values. You can also label the elements of pointers by supplying a text in the NVALUES option (3.3): for example

POINTER [NVALUES = !T(saloons,hatchbacks,estates)] Sales

allows you to refer to Sales['hatchbacks'], Sales['saloons','estates'], and even to Sales[1,2,'estates']. The suffix list within the square brackets is a list of identifiers, so the strings must be quoted: they are then treated as unnamed texts each with a single value (1.4.3).

The identifiers in a suffix list can be of scalars, variates, or texts; this of course includes numbers and strings as unnamed scalars and texts respectively. If one of these structures contains several values, it defines a sub-pointer: for example Info[!(3,2)] is a pointer with two elements, Profit and Costs. You can also give a null list to mean all the elements of the pointer: for example Info[ ] is Yield,Costs,Profit. You must be careful not to confuse a sub-pointer with a list of some of the elements of a pointer: for example Info[!(3,2)] is a single pointer with two elements, whereas Info[3,2] is a list of the two structures Profit and Costs.

Elements of pointers can themselves be pointers, allowing you to construct trees of structures. For example

    VARIATE A,B,C,D,E
    POINTER R; VALUES=!P(D,E)
    & S; VALUES=!P(B,C)
    & Q; VALUES=!P(A,S)
    & P; VALUES=!P(Q,R)

defines the tree

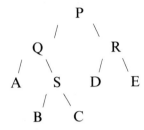

You can refer to elements within the tree by giving several levels of suffixes: for example P[2][1] is R[1] which is D; P[2,1][1,2] is (R,Q)[1,2] or D,E,A,S. The special symbol # (1.2.6 and 1.4.4) allows you to list all the structures at the ends of the branches of the tree: #P replaces P by the identifiers of the structures to which it points (Q and R); then, if any of these is a pointer, it replaces it by its own values, and so on. Thus #P is the list A,B,C,D,E.

3.4 Matrices

A matrix stores a set of numbers as a two-dimensional array indexed by rows and columns. For example, the array

    1   2   3   4
    5   6   7   8
    9 10 11 12

is called a three-by-four matrix.

You specify the size of the matrix by saying how many rows and columns it is to have; the total number of values is obtained by multiplying the number of rows by the number of columns. In the example there are 12 values. If the numbers of rows and columns are equal the matrix is said to be square.

Any matrix can be stored as an ordinary rectangular matrix. Genstat also has special ways of storing symmetric matrices (3.4.3) and diagonal matrices (3.4.2): these are needed in many statistical contexts.

You can assign values to matrices when they are declared, just as with vectors. But you must also set the size of the matrix, and it must correspond exactly to the number of values that you assign. Genstat stores the values of the matrix in row order: that is, all of the first row, followed by all of the second row, and so on. So you must assign the values in this order.

You are most likely to use matrices in multivariate analysis (Chapter 10), where you may need them either to input data, or to save the results of an analysis. Genstat also provides many facilities for matrix calculations: for example, you can add and multiply matrices, find their inverses, and decompose them to diagonal form (5.2.4 and 5.7).

3.4.1 Rectangular matrices

The Genstat matrix structure is a rectangular array. You declare it using the MATRIX directive.

MATRIX

Declares one or more matrix data structures.

Options

ROWS	= scalar, vector, or pointer	Number of rows, or labels for rows; default *
COLUMNS	= scalar, vector, or pointer	Number of columns, or labels for columns; default *
VALUES	= numbers	Values for all the matrices; default *
MODIFY	= string	Whether to modify (instead of redefining) existing structures (no, yes); default n

Parameters

IDENTIFIER	= identifiers	Identifiers of the matrices
VALUES	= identifiers	Values for each matrix
DECIMALS	= scalars	Number of decimal places for printing
EXTRA	= texts	Extra text associated with each identifier

You use the ROWS and COLUMNS options to specify the size of the matrix. The simplest way of doing this is to use scalars, giving explicitly the numbers of rows and columns. But you can also use vectors or pointers. In that case, the length of the vector or pointer implicitly defines the number of rows (or columns), and its values are used as labels when the matrix is printed. Here is an example:

3.4 Matrices

EXAMPLE 3.4.1

```
  1  TEXT [VALUES=Beer,Lager,Orange] Drink
  2  VARIATE [VALUES=0.5,1.0] Quantity
  3  MATRIX [ROWS=Drink; COLUMNS=Quantity; \
  4     VALUES=1.1,0.6,1.2,0.65,0.8,0.45] Cost
  5  PRINT Cost; DECIMALS=2
```

```
                    Cost
    Quantity        0.50       1.00
       Drink
        Beer        1.10       0.60
       Lager        1.20       0.65
      Orange        0.80       0.45
```

Genstat will in some contexts interpret a variate as being equivalent to a matrix with a single column; this is described with each directive, such as CALCULATE (5.1.3).

3.4.2 Diagonal Matrices

A square matrix that has zero entries except on its leading diagonal is called a diagonal matrix: for example,

$$\begin{array}{ccc} 2 & 0 & 0 \\ 0 & 1 & 0 \\ 0 & 0 & 3 \end{array}$$

Another example is the identity matrix, which has a diagonal of ones. To save space, Genstat has a special structure for diagonal matrices. You will probably use them most often to store latent roots in multivariate analysis (5.7.1 and 10.1). To declare diagonal matrices you use the DIAGONALMATRIX directive.

DIAGONALMATRIX

Declares one or more diagonal matrix data structures.

Options

ROWS	= scalar, vector, or pointer	Number of rows, or labels for rows (and columns); default *
VALUES	= numbers	Values for all the diagonal matrices; default *

MODIFY	= string	Whether to modify (instead of redefining) existing structures (no, yes); default n

Parameters

IDENTIFIER	= identifiers	Identifiers of the diagonal matrices
VALUES	= identifiers	Values for each diagonal matrix
DECIMALS	= scalars	Number of decimal places for printing
EXTRA	= texts	Extra text associated with each identifier

Because a diagonal matrix is square, you specify only the number of rows. You can use either a scalar or a labels vector or a pointer with the ROWS option, as in the MATRIX directive (3.4.1).

When you give the values of a diagonal matrix, either in a declaration or when its values are read, you should specify only the diagonal elements. (Genstat does not store the off-diagonal elements, but assumes them to be zero.) Similarly, when a diagonal matrix is printed it appears as a column of numbers; Genstat omits the off-diagonal zeros. For example:

EXAMPLE 3.4.2

```
  1  DIAGONALMATRIX [ROWS=3; VALUES=2,1,3] D
  2  PRINT D

             D
      1   2.000
      2   1.000
      3   3.000
```

3.4.3 Symmetric Matrices

A symmetric square matrix is symmetric about its leading diagonal: that is the value in column i of row j is the same as that in column j of row i. For example:

```
    1  2  3
    2  1  4
    3  4  1
```

You often come across symmetric matrices in statistics. Suppose, for example, that we have n random variables $X_1 \ldots X_n$. Then the covariance of X_i with X_j is the same as the covariance of X_j with X_i. The covariance matrix of the random variables is therefore symmetric: the off-diagonal elements of the matrix are the covariances (and the diagonal elements are the variances).

Because of this symmetry, Genstat stores only the diagonal elements and those below it; this is called the *lower triangle*. So you must specify only these values, whether in the declaration or in a READ statement (4.1). (As always, you give them in row order: so if there are *n* rows, then for the first you supply one value, for the second two, and so on.) Likewise, Genstat prints only the lower triangle in output, for example with PRINT (4.2).

The syntax for the declaration of symmetric matrices is as follows:

SYMMETRICMATRIX

Declares one or more symmetric matrix data structures.

Options

ROWS	= scalar, vector, or pointer	Number of rows, or labels for rows and columns); default *
VALUES	= numbers	Values for all the symmetric matrices; default *
MODIFY	= string	Whether to modify (instead of redefining) existing structures (no, yes); default n

Parameters

IDENTIFIER	= identifiers	Identifiers of the symmetric matrices
VALUES	= identifiers	Values for each symmetric matrix
DECIMALS	= scalars	Number of decimal places for printing
EXTRA	= texts	Extra text associated with each identifier

The ROWS option defines both the number of rows and the number of columns. You can use a vector or pointer to specify row and column labels, as with MATRIX (3.4.1). For example:

EXAMPLE 3.4.3

```
  1  VARIATE Weight,Height,Reach
  2  POINTER [VALUES=Weight,Height,Reach] Vars
  3  SYMMETRICMATRIX [ROWS=Vars; VALUES=1.0,0.68,1.0,0.43,0.72,1.0] Correl
  4  PRINT Correl
```

```
            Correl
Weight      1.0000
Height      0.6800    1.0000
 Reach      0.4300    0.7200    1.0000

            Weight    Height    Reach
```

3.5 Tables

You use a table to store numerical summaries of data that are classified into groups. With Genstat, the classification into groups is specified by a set of factors (3.3.3). The table contains an element, called a *cell*, for each combination of the levels of the factors that classify it.

You can specify the values of a table when you declare it. More often, you may wish to calculate the values within Genstat. The TABULATE directive (5.8.1) allows you to summarize observations, for example from surveys. The observed values are supplied in a variate, and the levels of the factors classifying the table indicate the group to which each observed unit belongs. The table can contain, in each of its cells, either the total of the observations with the corresponding levels of the classifying factors, or you can have the mean, or the minimum value, or the maximum value, or the variance. In an analysis of variance, you can save tables of means and tables of replications by the AKEEP directive (9.6.1). Calculations with tables are described in 5.1.4. The full list of facilities available for tables is given in 5.8. To declare tables you use the TABLE directive.

TABLE

Declares one or more table data structures.

Options

CLASSIFICATION	= factors	Factors classifying the tables; default *
MARGINS	= string	Whether to add margins (no, yes); default n
VALUES	= numbers	Values for all the tables; default *
MODIFY	= string	Whether to modify (instead of redefining) existing structures (no, yes); default n

Parameters

IDENTIFIER	= identifiers	Identifiers of the tables

VALUES	= identifiers	Values for each table
DECIMALS	= scalars	Number of decimal places for printing
EXTRA	= texts	Extra text associated with each identifier
UNKNOWN	= identifiers	Identifier for scalar to hold summary of unclassified data associated with each table

The example below shows a table called Classnum which stores numbers of children of each sex in the classes of a school. Here there are two factors defined in lines 1 and 2: Class with levels 1 to 5 and Sex with levels labelled 'boy' and 'girl'. The CLASSIFICATION option of the TABLE declaration (line 3) defines them to be the factors classifying the table, and the VALUES option defines a value for each of the 10 cells (two sexes × five classes) of the table. As you can see from the printed form of the table, the cells are arranged with both levels of Sex for Class 1, then both levels of Sex for Class 2, and so on. If there were three classifying factors, the table would have cells for all the levels of the third factor at level 1 of the first and second factors, then cells for all the levels of the third factor at level 1 of the first factor and level 2 of the second factor, and so on. In other words, the right-most factor in the classification rotates fastest, followed by the second from the right, and so on. This is illustrated by the second table, Schoolnm, which has a further factor School before Class and Sex in the list of classifying factors. Tables can be classified by up to nine factors.

EXAMPLE 3.5a

```
  1   FACTOR [LABELS=!T(boy,girl)] Sex
  2   FACTOR [LEVELS=5] Class
  3   TABLE [CLASSIFICATION=Class,Sex; VALUES=15,17,29,31,34,30,33,35,28,27] \
  4       Classnum
  5   PRINT Classnum; DECIMALS=0

              Classnum
      Sex          boy            girl
    Class
      1             15              17
      2             29              31
      3             34              30
      4             33              35
      5             28              27

  6   FACTOR [LEVELS=2] School
  7   TABLE [CLASSIFICATION=School,Class,Sex; \
  8       VALUES=15,17,29,31,34,30,33,35,28,27,18,16,33,31,35,36,34,33,31,32] \
  9       Schoolnm
 10   PRINT Schoolnm; DECIMALS=0
```

```
                    Schoolnm
            Sex       boy         girl
   School  Class
     1       1         15          17
             2         29          31
             3         34          30
             4         33          35
             5         28          27
     2       1         18          16
             2         33          31
             3         35          36
             4         34          33
             5         31          32
```

A table can also have *margins*. There is then a margin for each classifying factor; this contains some sort of summary over the levels of that factor. For example, if you have a table in which the cells contain totals of the observations, you would want the marginal cells to contain totals across the levels of the factor: see the next section of the example. You can define a table to have margins when you declare it, using the MARGINS option of the TABLE directive. Or you can add margins later by the MARGIN directive (5.8.2), as shown below.

EXAMPLE 3.5b

```
  11  MARGIN Classnum,Schoolnm
  12  PRINT Classnum; DECIMALS=0

            Classnum
    Sex       boy         girl       Margin
   Class
     1         15          17          32
     2         29          31          60
     3         34          30          64
     4         33          35          68
     5         28          27          55

  Margin      139         140         279
```

The margin row of Classnum contains the total numbers of boys and girls in the school (totalled over classes), and the margin column contains the total numbers (boys plus girls) in each class. The cell where this column and row coincide contains the total number in the school. With Schoolnm, there are marginal summaries over each classifying factor individually, over each pair of factors, and over all three factors. Thus the margin over a single factor is itself a two-dimensional array, classified by the other two factors: this is shown at the end of the example.

EXAMPLE 3.5c

```
 13  PRINT Schoolnm; DECIMALS=0
                       Schoolnm
              Sex         boy       girl     Margin
    School    Class
    1         1            15         17         32
              2            29         31         60
              3            34         30         64
              4            33         35         68
              5            28         27         55

              Margin      139        140        279

    2         1            18         16         34
              2            33         31         64
              3            35         36         71
              4            34         33         67
              5            31         32         63

              Margin      151        148        299

    Margin    1            33         33         66
              2            62         62        124
              3            69         66        135
              4            67         68        135
              5            59         59        118

              Margin      290        288        578
```

Tables also have an associated scalar which collects a summary of all the observations for which any of the classifying factors has a missing value; these observations cannot be assigned to any cell of the table itself. This scalar can be given an identifier, so that you can refer to it, using the UNKNOWN parameter of the TABLE directive.

3.6 Compound structures

You can use the pointer structure (3.3.4) to group together related data structures, so that you can refer to them as a single structure. Some Genstat directives expect standard combinations of data structures for their input or output; in these cases you use special pointers called *compound structures*. These differ from ordinary pointers in that they have a fixed number of elements which must be of the correct types, and must form a consistent set (in terms of their sizes and so on).

You can refer to elements of these structures in exactly the same way as the elements of pointers: for example if L is an LRV (3.6.1) then L refers to the set of structures L[1], L[2], L[3]. The suffixes run from 1 upwards, and you cannot change that. Neither can you change the labels that Genstat gives to the structures; details of these labels come later in this section. You can give the individual elements of a compound structure identifiers in their own right, just as with pointers. Indeed, you can use all the features of pointer syntax: for example, you can use the null list, or the substitution symbol #, to list all the elements of the structure (3.3.4).

When you declare a compound structure, you conveniently declare, simultaneously and automatically, a whole collection of structures. At the same time you ensure that they match the requirements of whatever form of analysis you want to use.

3.6.1 The LRV structure

You use the LRV structure to store latent roots and vectors resulting from the decomposition of a matrix (5.7.2), or produced in multivariate analysis (Chapter 10). You need not store all the latent roots; usually Genstat will select the largest ones. The LRV structure points to three structures (identified by their suffixes):

[1] or ['Vectors'] is a matrix whose columns are the latent vectors: the word "vector" is used here in its mathematical sense rather than in the more specific Genstat sense; in fact, latent vectors are most conveniently stored in matrices rather than in Genstat vectors;

[2] or ['Roots'] is a diagonal matrix whose elements are the latent roots;

[3] or ['Trace'] is a scalar holding the trace of the matrix, which is the sum of all its latent roots.

To declare an LRV you use the LRV directive.

LRV

Declares one or more LRV data structures.

Options

ROWS	= scalar, vector, or pointer	Number of rows, or row labels, for the matrix; default *
COLUMNS	= scalar, vector, or pointer	Number of columns, or column labels, for matrix and diagonal matrix: default *

Parameters

IDENTIFIER	= identifiers	Identifiers of the LRVs
VECTORS	= matrices	Matrix to contain the latent vectors for each LRV
ROOTS	= diagonal matrices	Diagonal matrix to contain the latent roots for each LRV
TRACE	= scalars	Trace of the matrix

3.6 Compound Structures

You specify the length of each latent vector by the ROWS option; this then defines the number of rows in the 'Vectors' matrix. You set the number of latent roots to be stored using the COLUMNS option; this is also the number of latent vectors, and so defines the number of columns in the 'Vectors' matrix and the number of elements in the 'Roots' matrix. If you do not specify the number of columns Genstat will set it to be the same as the number of rows. The value of COLUMNS can be less than the value of ROWS, but if it is greater than that of ROWS, Genstat gives an error diagnostic. Row and column labels can be defined, as in the declaration of matrices (3.4).

You can specify identifiers for the three individual elements of the LRV by using the VECTORS, ROOTS, and TRACE parameters. If you have declared them already they must be of the correct type. (You can also have given them values.) If you have given these identifiers row or column settings, then these will be used for the LRV declaration and must match any of the corresponding options of LRV that you choose to set.

This example declares an LRV, and then forms its values (see 5.7.2).

EXAMPLE 3.6.1

```
  1   POINTER [VALUES=stem,leaf,root,petal,pollen] Vars
  2   SYMMETRICMATRIX [ROWS=Vars] Symm
  3   READ Symm

      Identifier    Minimum        Mean     Maximum      Values    Missing
           Symm     -0.9820      0.1974      1.0000          15          0

  9   PRINT Symm

                       Symm
        stem         1.0000
        leaf        -0.6550      1.0000
        root        -0.9450      0.8660      1.0000
        petal       -0.7560      0.0000      0.5000      1.0000
        pollen       0.5000     -0.9820     -0.7560      0.1890      1.0000
                       stem        leaf        root       petal      pollen

 10   LRV [ROWS=Vars;COLUMNS=2] Latent; VECTORS=Lvecs
 11   FLRV Symm; Latent
 12   PRINT Latent['Vectors','Roots']

                      Lvecs
                          1           2
        Vars
        stem        -0.4875      0.3372
        leaf         0.4875      0.3372
        root         0.5335     -0.0770
        petal        0.2227     -0.7383
        pollen      -0.4366     -0.4707

                Latent['Roots']
                          1           2
                      3.482       1.518
```

3.6.2 The SSPM structure

The SSPM structure stores a matrix of corrected sums of squares and products, and associated information, as used for regression (Chapter 8) and some multivariate analyses (Chapter 10). You can form values for SSPM structures by the FSSPM directive (5.7.3). However most multivariate and regression analyses can be done without declaring and forming an SSPM explicitly.

An SSPM comprises four structures (identified by their suffixes).

[1] or ['Sums'] is a symmetric matrix containing the sums of squares and products. The number of rows and columns of this matrix will equal the number of elements of the expanded terms list: that is, the number of variates plus the number of dummy variates generated by the model formula. (See the TERMS directive: 8.2.2.)

[2] or ['Means'] is a variate containing the mean for each term.

[3] or ['Nunits'] is a scalar holding the total number of units used in constructing the sums of squares and products matrix. If the SSPM is weighted, this scalar will hold the sum of the weights.

A within-group SSPM has one additional element:

[4] or ['Wmeans'] is a pointer, pointing to variates holding within-group means. There is one variate for each term of the SSPM plus one extra. They are all of the same length, namely the number of levels of the GROUPS factor. The extra variate holds counts of the number of units in each group.

The syntax for the declaration of SSPM structures is as follows:

SSPM

Declares one or more SSPM data structures.

Options

TERMS	= model formula	Terms for which sums of squares and products are to be calculated; default *
FACTORIAL	= scalar	Maximum number of vectors in a term; default 3
FULL	= string	Full factor parameterization (no, yes); default n
GROUPS	= factor	Groups for within-group SSPMs; default *
DF	= scalar	Number of degrees of freedom for sums of squares; default *

Parameters

IDENTIFIER	= identifiers	Identifiers of the SSPMs

3.6 Compound Structures

SSP	= symmetric matrices	Symmetric matrix to contain the sums of squares and products for each SSPM
MEANS	= variates	Variate to contain the means for each SSPM
NUNITS	= scalars	Number of units or sum of weights
WMEANS	= pointers	Pointers to variates of group means

The TERMS option defines the model for whose components the sums of squares and products are to be calculated. In the simplest case the model is just a list of variates. But you can use more complex model formulae, involving variates and factors: this is done in conjunction with the FACTORIAL and FULL options. See the TERMS directive (8.2.2) for a full description of how model formulae are interpreted.

You form a within-group matrix of sums of squares and products by supplying a factor in the GROUPS option.

Sometimes you may already have calculated values for the matrix of sums of squares and products. You can then assign them to the component structures of the SSPM for example by READ (4.1). You would still, however, need to set the number of degrees of freedom associated with the matrix, and for that you use the DF option.

The parameter lists let you specify identifiers for the four components of an SSPM. You can have declared them previously (and you can have given them values); if so they must be of the correct type.

This example shows the declaration and formation (5.7.3) of an SSPM.

EXAMPLE 3.6.2

```
  1   READ [SETNVALUES=yes] V[1...5]

      Identifier    Minimum       Mean    Maximum    Values    Missing
           V[1]       1.000       2.667     4.000        3          0
           V[2]       0.000       2.000     4.000        3          0
           V[3]       1.000       3.000     7.000        3          0
           V[4]       0.0000      0.6667    1.0000       3          0
           V[5]       0.000       1.333     3.000        3          0

  5   SSPM [TERMS=V[1...5]] Ssp
  6   FSSPM [PRINT=sspm] Ssp
```

*** Degrees of.freedom ***

Sums of squares: 2
Sums of products: 1

*** Sums of squares and products ***

V[1]   1   4.6667

```
V[2]         2        -4.0000       8.0000
V[3]         3       -10.0000      12.0000     24.0000
V[4]         4        -1.3333       0.0000      2.0000     0.6667
V[5]         5         2.3333      -6.0000     -8.0000     0.3333     4.6667
                                       1            2           3          4          5
```

*** Means ***

```
V[1]                2.667
V[2]                2.000
V[3]                3.000
V[4]                0.6667
V[5]                1.333
```

*** Number of units used ***

```
           3
```

3.6.3 The TSM structure

The TSM structure stores a time-series model which you can use in Box-Jenkins modelling of time series (see Chapter 11). The information that you give to specify the model is stored in two variates, called the *orders* and the *parameters*; an optional third variate contains *lags*. A complete description of how these structures are defined and assigned values is given in Chapter 11.

The elements of a TSM are:

[1] or ['Orders'];
[2] or ['Parameters'];
[3] or ['Lags'].

To declare a TSM you use the TSM directive.

TSM

Declares one or more TSM data structures.

Option

MODEL	= string	Type of model (arima, transfer); default a

Parameters

IDENTIFIER	= identifiers	Identifiers of the TSMs
ORDERS	= variates	Orders of the autoregressive, integrated and moving average parts of each TSM
PARAMETERS	= variates	Parameters of each TSM
LAGS	= variates	Lags, if not default

3.6 Compound Structures

The TSM directive sets up a compound structure pointing to the variates that will later be used to define the model. You set the type of model by the MODEL option. You can use the parameters of TSM to supply previously declared identifiers as the elements of the TSM, just as with the LRV and SSPM. In this way you can specify a variate of lags, to give the TSM three elements rather than the default of two.

Here are some examples:

    TSM [MODEL=arima] T1
    TSM [MODEL=transfer] T2; ORDERS=!(1,0,1)
    TSM T3; ORDERS=O; PARAMETERS=P; LAGS=L

S.A.H.
R.W.P.

4 Input and output

This chapter describes how to transfer data and programs between structures in a Genstat program and files on a computer. The commonest needs are to take data values stored in a file and put them in a data structure, and to put the values of a structure into a file. When working interactively (1.1.10), the "file" may be the terminal, so that you can type the values that you want to be stored in a structure, and display the values of a structure on the screen. Computers allow many types of file. Usually, data and programs are stored in *character files*; these can easily be communicated to a person by printing them on a line printer or by displaying them on the screen of a terminal. But when data are to be used many times, it is more efficient to store them in a *binary file*, in which the values are represented by a coded form that is faster to communicate to a computer program.

Section 4.1 describes how to *read* data; that is, how to transfer values stored in a character file into a data structure. Then Section 4.2 describes the reverse process: *printing*. Section 4.3 deals with the details of referencing files on a computer, though the simplest methods are illustrated in Section 4.1; and Section 4.4 shows how to transfer control of input for a Genstat program from one file to another. Section 4.5 describes how to transfer data into binary files and to retrieve them later. Finally, Section 4.6 deals with an alternative and quicker method of communicating with a binary file, when you want to store the whole of the current job, or when there is no need to keep any record of the individual data structures apart from their values.

4.1 Reading data

The simplest way to put data into data structures is to assign them directly in the declaration of the structures, as described in Chapter 3. This method was used in the first example of Chapter 1 (1.1.1):

    SCALAR First,Dessert,Wine,Service; VALUE=8.50,1.50,3.00,1.10

In the second example of Chapter 1 (1.1.2) we showed that you could read data that were recorded immediately after a READ statement:

    VARIATE [NVALUES=10] X
    READ X
    24.3 25.6 57.3 43.8 45.3
    46.5 47.9 97.0 77.5 64.3 :

Before giving a READ statement, you usually need to declare the size and type of the data structures that will hold the values. However, the READ directive will assume that an undeclared structure is to be a variate, and, unless otherwise specified, that the length of a variate, factor, or text structure is the length of the units structure (2.2.3). The SETNVALUES option of READ can be set to specify that the length of a structure is to be determined by the number of data values found.

You indicate the end of the data set by a *terminator*, which by default is a colon (:). Genstat checks that the number of values read is the same as the size of the structure. If too few values are found, then you get a warning, and the remaining values in the structure are set to the missing value. If too many values are found you get a fault. Note that any characters after the terminator on the same line are ignored.

When reading values for more than one structure, you can present values either *serially* or *in parallel*. In serial, all the values for one structure come after all the values for another; for example:

$$x_1 \quad x_2 \quad x_3:$$
$$y_1 \quad y_2:$$
$$z_1 \quad z_2 \quad z_3 \quad z_4 \quad z_5 \quad z_6:$$

Here all the values of X come first, followed by all the values for Y, and then all the values for Z. The set of values for each structure is ended by a terminator.

To be read in parallel, the structures must have the same size. For example:

$$a_1 \quad b_1 \quad c_1$$
$$a_2 \quad b_2 \quad c_2$$
$$a_3 \quad b_3 \quad c_3$$
$$a_4 \quad b_4 \quad c_4:$$

Here A, B, and C are in parallel, each with four values.

In free format you separate the values by a separator or by the end-of-line character (usually <RETURN>). The default separator is one or more spaces; you can change that to comma, semicolon, or colon. If you use colon, you cannot also use it to mark the end of the data: you must then change the terminator by setting the END option of READ.

Remember that it may be much easier to assign values in declarations than by reading, particularly when the values are repetitive or in an arithmetic progression. For example, it is much easier to write

    VARIATE [VALUES = 3(1...7)4] Day

than to write

    VARIATE [NVALUES = 84] Day
    READ Day

followed by 84 separate values.

The file containing both program and data in this example is treated by Genstat as input file number 1. Remember that a terminal may be treated as a "file", for example when you are running Genstat interactively. You may wish to put data in another file, separated from the program. This is particularly recommended when running Genstat interactively: it is usually better to create a data file first and then get Genstat

4.1 Reading data

to read it, because you then have a permanent copy of the data for use in a second program, or for a re-run of your first program. You tell Genstat about the second file by opening it, either explicitly with an OPEN statement or by attaching the file when you give the initial operating-system command to run Genstat. We do not describe the latter method here, because the details differ from one computer to another; read the local documentation about the implementation of Genstat on your computer.

Suppose the data are stored in a file called DATA. In other words, DATA contains the numbers and the data terminator:

```
24.3   25.6   57.3   43.8   45.3
46.5   47.9   97.0   77.5   64.3:
```

First you must open this file, and attach it to the program using what is called a *channel number*; this is just a convenient reference with which Genstat records which file is which:

```
OPEN 'DATA'; CHANNEL=2; FILETYPE=input
```

You cannot use channel number 1 here because the program itself will be in a file attached using channel number 1.

Next, you read the data as before, but specifying the channel number in the READ statement:

```
VARIATE [NVALUES=10] X
READ [CHANNEL=2] X
```

There is a limit on the number of input channels that you can have open simultaneously; you can give the statement

```
HELP environment, channel
```

to find out what it is.

4.1.1 Main features of the READ directive

READ

Reads data from an input file, an unformatted file, or a text.

Options

PRINT	= strings	What to print (data, errors, summary); default e,s
CHANNEL	= identifier	Channel number of file, or text structure from which to read data; default current file
SERIAL	= string	Whether structures are in serial order, i.e. all values of the first

		structure, then all of the second, and so on (no, yes); default n, i.e. values in parallel
SETNVALUES	= string	Whether to set number of values of structures from the number of values read (no, yes); default n
LAYOUT	= string	How values are presented (separated, fixedfield); default s
END	= text	What string terminates data (* means there is no terminator); default ':'
SEQUENTIAL	= scalar	To store the number of units read (negative if terminator is met); default *
ADD	= string	Whether to add values to existing values (no, yes); default n (Only available in serial read)
MISSING	= text	What character represents missing values; default '*'
SKIP	= scalar	Number of characters (LAYOUT=f) or values (LAYOUT=s) to be skipped between units (* means skip to next record); default 0 (Only available in parallel read)
BLANK	= string	Interpretation of blank fields with LAYOUT=f (missing, zero, error); default m
JUSTIFIED	= string	How values are to be assumed justified with LAYOUT=f (right, left, both, neither); default r
ERRORS	= scalar	How many errors to allow in the data before reporting a fault rather than a warning, a negative setting, $-N$, causes reading of data to stop after the N'th error; default 0
FORMAT	= variate	Allows a format to be specified for situations where the layout varies for different units, option SKIP & parameters FIELDWIDTH & SKIP are then ignored (in the variate: 0 switches to fixed format; 0.1, 0.2, 0.3, or 0.4 to free format with space, comma, colon or semi-colon

			respectively as separators; * skips to the beginning of the next line; in fixed format, a positive integer N indicates an item in a field width of N, −N skips N characters; in free format, N indicates N items, −N skips N items); default *
QUIT	=	scalar	Channel number of file to return to after a fatal error; default * i.e. current input file
UNFORMATTED	=	string	Whether file is unformatted (no, yes); default n
REWIND	=	string	Whether to rewind the file before reading (no,yes); default n

Parameters

STRUCTURE	=	identifiers	Structures into which to read the data
FIELDWIDTH	=	scalars	Field width from which to read values of each structure (LAYOUT = f only)
DECIMALS	=	scalars	Number of decimal places for numerical data containing no decimal points
SKIP	=	scalars	Number of values (LAYOUT = s) or characters (LAYOUT = f) to skip before reading a value
FREPRESENTATION	=	string	How factor values are represented (labels, levels, ordinals); if omitted, levels are assumed

As with all Genstat directives, the best way to understand READ is to start simply and build up an understanding of the options and parameters from there. Since there are many options and parameters of READ, we have divided the information into subsections dealing with various aspects. As usual, the more complex and less commonly required aspects come last.

Here is a simple example of reading the values of three variates.

```
VARIATE [NVALUES = 7] Yield[1...3]
READ [CHANNEL = 2] Yield[]
```

This first declares three variates, called Yield[1], Yield[2], and Yield[3], each with seven values. The READ statement then puts values into these variates from a file already attached to channel 2, as described above. These values must be in parallel; for example, they can be presented as three columns of seven figures. The READ statement gives you some summaries:

EXAMPLE 4.1.1

```
    Identifier    Minimum      Mean   Maximum    Values    Missing
      Yield[1]      22.00     45.43     75.00         7          0
      Yield[2]      12.00     50.20     95.00         7          2
      Yield[3]       2.00     58.71     98.00         7          0
```

You get this summary information by default, but it is produced only for variates, matrices, and tables. The summary for each structure includes the message "Skew" if its values have a markedly skew distribution; to be precise, the message is given if the difference between mean and minimum is more than three times, or less than a third of, the difference between maximum and mean.

You also get warnings on errors in the data: these are errors such as not having the right number of values. Genstat obviously cannot detect, in this example, a 46 misrecorded as a 64. If Genstat does find errors, READ will keep going to the end, so that you get a report on all of them. At the end you will get a fatal error unless you have used the ERRORS option: that is, the whole operation of READ will be aborted. You can use the PRINT option to prevent this output appearing; but you should do that only if you know there are no errors.

PRINT can produce a copy of the input data as they are read; this is a good way of searching for errors. If you set PRINT=*, you get no output from READ at all.

Normally you will already have set the number of values in the structure being read, perhaps by a UNITS statement. Genstat will look for this number of items for this structure in the file. But you can change that by the SETNVALUES option. The length will then be set to be whatever number of values is actually read: any previous setting of the number of values will be ignored. You should be wary of using this option, since it loses the automatic check on the number of values. But you may have to use it when you are writing a general program to analyse different sets of data, or when you want to read large data sets whose size you do not know (as in many surveys).

You can mark the end of data by any string of up to eight characters; a mnemonic such as EOD is useful. If you do not set the option, Genstat takes the end to be a colon (:).

By default, a missing value is indicated by an asterisk (*); but you can use the MISSING option to change this to any other single character. Be careful, however: any data item that begins with this character is treated as a missing value. Thus if the default is in force, then the three "values"

4.1 Reading data

    *   ***   *789

are all treated as missing.

An example that sets both these options is:

    READ [CHANNEL=3; SETNVALUES=yes; END='END'; MISSING='−']
    \Crop,Yield

This treats END (without single quotes) as marking the end of the data, and sets the length of Crop and Yield to be the number of items of data that happen to be in the input file. Notice that all negative values will be set to the missing value.

You are not allowed to set the END string to be the same as the missing value character. Moreover, neither must be the same as the separator in free format: see 4.2.3. Case is significant for both END and MISSING: for example, end is different from End. Neither the missing-value character nor the separator may be part of the END string.

A file can contain several sets of data: for example, it might contain 50 measurements on heights of plants, followed by 50 values of weights. You could read the first 50 by one statement, and the next 50 by another. After reading the first 50, Genstat keeps a note of where it has got to in the file, and so returns to the correct place for the second READ. But occasionally you may want to go right back to the beginning of the file. To do this you set the REWIND option to 'yes'.

4.1.2 Reading fixed-format data

The easiest way to read data is in free format. The advantage of this is that typing data into the data file is easy: usually you simply enter one after another, separated by a space.

You can use the LAYOUT option to change to fixed format; and you must then actually specify the format. There are two ways of setting the format. If the values for each structure always occupy the same number of character positions, then you should use the FIELDWIDTH parameter. For example,

    READ [CHANNEL=2; LAYOUT=fixed] Apple,Pear; FIELDWIDTH=3,5

takes data from channel 2 in fixed format. The data are in parallel: that is, reading across lines of the file, Apple and Pear values appear alternately. Each piece of Apple data takes up three spaces, and each piece of Pear data takes up five. Suppose there are 80 characters per line in the file; each pair of Apple and Pear values takes up 8, and so you get 10 pairs per line. So the first line looks like:

    $Apple_1 \ Pear_1 \ Apple_2 \ Pear_2 \ ... \ Apple_{10} \ Pear_{10}$

Suppose that the first two values for Apple were 1 and 200, and that the first two for Pear were 10 and 1200. Then, using | to represent the beginning of the line, the first four items on this line would be:

    | 1   10200   1200

Note that Genstat knows where the 200 starts because it knows that there are five characters in the previous field, the one containing "10". (In this example each item of data is flush with the right of the field: see the JUSTIFIED option.)

Blank characters at the end of a record are not ignored. Thus you must have 10 pairs per line in the example, since otherwise some data items would be set to the missing value (but see the BLANK and SKIP options below).

This simple form works only in special cases. For example, suppose there were 11 pairs per line and that you set FIELDWIDTH=3,4. Then you would either have to set the WIDTH option of OPEN to 77 (4.3.1), or use an appropriate setting of the FORMAT option (4.1.3); the latter causes FIELDWIDTH to be ignored.

In fact you use FORMAT more generally if the values for each structure do not always occupy the same number of spaces. The general idea is that the variate that you use in the FORMAT option specifies how many spaces are taken up by each value. To read the values of Apple and Pear given 11 to a line, you would need to put

> VARIATE [VALUES=(3,4)11,*] Line11
> READ [LAYOUT=fixed; FORMAT=Line11] Apple, Pear

The non-missing values of this format variate are matched successively with the values read, regardless of which structure is being read. Thus, after 22 values have been read from the first line, with field-widths alternately 3 and 4, the missing value is encountered which causes a skip to the next line. Then the whole format is recycled, and so on until all the values have been read.

There are more details about FORMAT at the end of 4.1.3, where it is used to skip characters.

When you are using fixed format, the data terminator must begin within the next field position after the last data value.

Normally Genstat treats a blank field in fixed-format data as a missing value; and you do not get a warning. You can get a warning by setting option BLANK=error. Alternatively, you can make blanks be interpreted as zeroes by setting BLANK=zero.

Normally data in fixed format are taken to be right-justified: that is, their right-hand ends are flush with the right-hand end of the field; you can either have blanks or leading zeroes in the redundant spaces at the left of the field. You can change the default by setting the option JUSTIFIED. For example, using | to represent the edges of a field, the value 123 can appear in a field of width 5 as

> | 123| Right-justified: there may be leading blanks
> |123 | Left-justified: there may be trailing blanks
> |00123| Left- and right-justified: there must be no blanks
> | 123 | No justification: there may be leading or trailing blanks

You can set the JUSTIFIED option to check the blanks in each field. The settings 'right', 'left', 'both' and 'neither' correspond to the four examples above: any blanks not allowed by a setting will be reported as errors. Note that |1 2 3| always gives an error.

4.1 Reading data

Now an example: suppose that we want to read the values of five scalars using a fixed format. The values are left-justified in their fields, and the data terminator is EOD. Then we could put:

SCALAR V,W,X,Y,Z
READ [LAYOUT = fixed; JUSTIFIED = left; END = 'EOD'] V,W,X,Y,Z; \
  FIELDWIDTH = 4,5,7
1.235.62 678.9 3.7810.31EOD

This will assign values 1.23, 5.62, 678.9, 3.78, and 10.31 to V, W, X, Y, and Z respectively. Note that the setting of the FIELDWIDTH parameter is recycled so that 4 corresponds to Y as well as to V, and 5 to Z as well as to W.

4.1.3 Skipping unwanted data

When data are punched in parallel, you may want to avoid reading the values for some of the structures. How to do this depends on whether or not you have set the FORMAT option.

Firstly, take the case when the FORMAT option is not set. You then have a further choice, depending on whether or not you are using free format.

For free format (LAYOUT = separated), you can use the SKIP option and parameter.

The SKIP parameter specifies how many parallel columns are to be ignored. For example, suppose we have a file with data as follows, again using | to mark the start of each line:

| 11 81 21 31 91
| 12 82 22 32 92
| 13 83 23 33 93
| 14 84 24 34 94
| 15 85 25 35 95
| 16 86 26 36 96
| 17 87 27 37 97:

If we wanted only the first, third, and fourth columns, we would put:

READ [SKIP = *] Xx,Yy,Zz; SKIP = 0,1,0

The SKIP parameter here tells Genstat to ignore one value between Xx and Yy. Note that the * for the SKIP option tells READ to skip to the next line after taking a value for each structure from the current one. If you did not set the option (so that it took its default 0), READ would continue on the same line: for example, the second value for Xx would be taken to be 91.

The SKIP option sets how many values are to be ignored between one unit and the next. Thus an alternative to the example above is:

READ [SKIP = 1] Xx,Yy,Zz; SKIP = 0,1,0

This tells READ to skip the value in the final column before assigning the next value to Xx.

If the format is fixed (LAYOUT=fixed), the values of SKIP specify numbers of characters to be ignored, instead of numbers of values. Take as an example the data above. You can get exactly the same effect as the statement above by

READ [LAYOUT=fixed; SKIP=*] Xx,Yy,Zz; FIELDWIDTH=3; SKIP=0,3,0

Note that you now have to skip three characters between Xx and Yy, since each item in the second column takes one field of three characters.

If you have set the FORMAT option, Genstat ignores the SKIP option and parameter. Then READ is controlled entirely by the values of the FORMAT option. These values are not in parallel with the list of structures: they apply to data values in turn, recycling from the beginning when necessary.

How Genstat deals with a value from the format variate depends on that value itself, say r, and also on the nearest integer to r, say n:

$r = *$: skip to the beginning of the next line;
$n < 0$: in free format the next $|n|$ values are skipped,
in fixed format the next $|n|$ characters are skipped;
$n > 0$: in free format the next n values are accepted,
in fixed format the next n characters are accepted as a single value;
$r = 0.0$: subsequent data values are in fixed format;
$0 < r < 0.5$: subsequent data values are in free format with separator space ($r=0.1$), comma ($r=0.2$), colon ($r=0.3$), semicolon ($r=0.4$);
$n = 0$: the next format element is picked up and the process repeated until a non-zero value of n is found.

Here are some examples. Suppose we have data:

```
|13 14 15
|16 17 18 19
| :
```

That is, we have seven two-digit numbers separated by a single space; a single space also precedes the values 13 and 16.

Let X be a variate of length four; then

READ [FORMAT=!(*,4)] X

causes X to take the values 16, 17, 18, and 19.

READ [FORMAT=!(−1,2)] X

causes one value to be ignored before every two values: thus X takes 14, 15, 17, and 18.

READ [FORMAT=!((−1,2)2,*); LAYOUT=fixed] X

causes X to take 13, 14, 16, and 17 because the "−1" refers now only to the single

4.1 Reading data 97

spaces.

If the data were instead

```
| 13 14 15
| 16;17;18;19
| :
```

then

```
READ [FORMAT=!(-1,2,0.4)] X
```

again causes X to take 14, 15, 17, and 18: the "2" takes two values in the first line; then "0.4" switches to the separator being a semicolon; and then the FORMAT list cycles back to "-1,2", causing two values to be taken from the second line.

4.1.4 Reading non-numerical data

The rules for the interpretation of strings while reading are different from those in string lists in a statement (1.4.2). Double quotes and backslashes are accepted as ordinary characters, and the strings cannot be continued over a line; the missing-value character, unquoted, is accepted as a missing value.

With free format, you must put strings in quotes unless they begin with a letter and contain only letters or digits. Any contiguous pair of single quotes is replaced by a single quote, and a quoted empty string is treated as a missing value. For example:

```
UNITS [NVALUES=3]
TEXT S,T
VARIATE V
READ S,V,T
'Molotov Cocktail'  23.4  John
'T.N.T.'  45.6  James
'Bomb "Disposal" Unit'  67.8  Jeremy :
```

Here, the quotes are all necessary: in the first item because of the blank; in the fourth because of the dots; and in the seventh item because of the blanks and the quotes. None of the three names read into T need quotes, because each contains only letters. Notice that if quotes had not been placed around Molotov Cocktail, Genstat would have tried to interpret it as two items, which would have produced an error since the variate V cannot accept textual data. Also note that the third item of S is

Bomb 'Disposal' Unit

In fixed format, by contrast, the contents of each field are taken exactly as they appear in the input file. The only way of representing the missing value is by a blank field. For example,

```
READ [LAYOUT=fixed; SKIP=*] T,V; FIELDWIDTH=5,2
'   A    C '  1 1    x x x
    2 3        2 2   y y y
^   '    '  ^  3 3   z z z
:
```

sets the values of T to

|'A C'|, | 23 |, |'''|

The values of factors are normally represented by their levels. However you can change this by using the FREPRESENTATION parameter. If you set it to 'labels', READ will accept as values the labels of the factor. The setting FREPRESENTATION=ordinals causes READ to expect an integer *n* in the range 1 up to the number of levels.

If you are using labels, the field−width has to be the same as the number of characters that you gave in the declaration of the factor: for example, |T| is not same as |T|. This means that all the labels of the factor must contain the same number of characters, unless you use the FORMAT option to change the field-width from unit to unit. Here is an example:

```
UNITS [NVALUES=6]
FACTOR [LABELS=!T(a,b,c,d,e); LEVELS=!(11,22,33,44,55)] F
FACTOR [LEVELS=3; LABELS=!T(Abc,Bca,Cab)] G
READ F,X,G; FREPRESENTATION=*,*,labels
33    −73.7   Bca
22     65.3   Abc
55    −33.2   ABC
11     77.2   Cab
22    −99.3   Bca
44     88.2   Cab :
```

The values of G as read must match the strings declared as labels exactly. Thus an error is reported during this read because the value ABC is not in the labels list.

You can read "values" (and labels) for pointers. You must set the length of the pointer in advance, and the "values" must be unsuffixed identifiers. For example:

```
POINTER [NVALUES=3] Pp
TEXT [NVALUES=3] Tp
READ Pp,Tp
xXx Apples  YyY Pears  XyZ Lemons :
POINTER [MODIFY=yes; NVALUES=Tp] Pp
```

The last statement establishes the text structure as the labels vector of the pointer. The pointer then has three levels, Pp[1] or Pp['Apples'] referring to the structure xXx, and so on.

4.1.5 Reading from a text structure

You can set the CHANNEL option of READ to point to a text structure. READ treats each string of the text structure as if it were a record from an input file. These strings need not all be of the same length. Thus you will find that fixed-format read is

4.1 Reading data 99

difficult to get right, and is perhaps better avoided here.
For example:

```
TEXT [VALUES = '
22.3 "John Smith" 56.7
45.6 "Jane Doe" " 93.1:
22 33 44 55
66 55 77 88 :'] Vtv
TEXT T
READ [CHANNEL = Vtv; SETNVALUES = yes] V,T,W
& X
```

This gives V, T, and W each two values, and X eight.

Be careful about strings that appear in the text, because sets of the single-quote, double-quote, or backslash characters will be halved in length when they are assigned to the text structure (1.3.2). In the example above, the first line that is stored in Vtv and read from it is actually

22.3 'John Smith' 56.7

4.1.6 Reading data in batches, accumulation, scaling, and re-ordering

Sometimes you may have more data to read than can be stored in the space available to Genstat. You may be able to increase the storage space – see your local documentation for this – but you can also use the SEQUENTIAL option of READ to process the data in batches.

First declare the structures to be of some convenient size; then set the SEQUENTIAL option of READ to the identifier of some scalar:

```
SCALAR N
VARIATE [NVALUES = 1000] X
READ [CHANNEL = 2; SEQUENTIAL = N] X
```

After reading 1000 units the value of N is set to 1000; or if the data terminator has been met, N will be set to minus the number of values found. This device is really only useful in conjunction with a loop. Construct a loop (6.2.1) with the NTIMES option of FOR large enough to cover all the data values. The loop should contain the READ statement and any other statements you need for processing the data. For example:

```
SCALAR N
VARIATE [NVALUES = 1000] X[1...7]
FOR [NTIMES = 9999]
  READ [CHANNEL = 2; PRINT = *; SEQUENTIAL = N] X[ ]
  (Other statements processing the current set of values)
  EXIT N .LE. 0
ENDFOR
```

You can use this option to read data from more than one channel simultaneously, but you cannot read data sequentially from the same channel as that containing the READ statement. If you want to read several structures sequentially from the same file, you must read them in parallel.

Both TABULATE (5.8.1) and FSSPM (5.7.3) also have a SEQUENTIAL option to allow tables and SSPM structures to be formed sequentially from large numbers of units. You can carry out linear regression based on the SSPM structures (8.2.2); they are also used in multivariate analysis (10.2).

The ADD option allows you to add values to those already stored in a structure. This allows you to form cumulative totals without having to store all the individual values that contribute to the totals. You must set SERIAL=yes if you use ADD=yes; and it works only for variates. For example:

```
VARIATE [NVALUES=6] A
READ [ADD=yes; SERIAL=yes] 3(A)
5  12  9  *  *   9 :
8   1  3  *  2  10 :
3   4  0  *  11  * :
```

This starts by assigning the values 5, 12, 9, *, *, and 9 to A. Then A is read again, and its values become 13, 13, 12, *, 2, 19: with ADD=yes, and only with ADD=yes, missing values are interpreted as zeroes when being added to non-missing values. Finally A contains the values 16, 17, 12, *, 13, 19.

You can scale values with the DECIMALS parameter. For example, suppose you put

```
READ [SETNVALUES=yes] A; DECIMALS=3
2523  2.1  376  0.78  :
```

Then the values of A would be 2.523, 2.1, 0.376, 0.78. In general, then, the parameter specifies the power of 10 by which a value that does not contain a decimal point is scaled down. The power must not be negative.

You can use the units structure to re-order data that are read in parallel. For example:

```
VARIATE [VALUES=1,2,3] Plot
UNITS Plot
```

we get

```
READ W1,Plot,W2
22.2 2 2.22
33.3 3 3.33
11.1 1 1.11:
```

This does not change the values of Plot, but its values are used to re-order the values of the other structures. Thus if we now put

```
PRINT Plot,W1,W2
```

we get

Plot	W1	W2
1	11.1	1.11
2	22.2	2.22
3	33.3	3.33

4.1.7 Errors while reading

The READ directive makes a distinction between serious and minor errors.

There are two kinds of serious error. Firstly, there are those that inhibit any attempt to execute the statement. For example, you cannot read structures of different lengths in parallel; neither can you exceed the available space for holding the data. Secondly, there are those that cast doubt on the validity of the data after they have been read. For example, suppose that values for five structures of length 10 are being read in parallel. If Genstat finds only 47 values, it does not know what has happened to the other three. This is not a fail-safe check, however. If 45 values are found, then the last element of each of the five structures is set to be missing, and Genstat gives a warning only; obviously, then, you should pay attention to such warnings.

Genstat treats serious errors as faults; what happens next is not affected by the setting of the ERRORS option.

Minor errors are concerned with individual data elements. You get comments about them under a heading of the form:

Unit	S.N.	Input
i	j	*string* Code ....

Here, i is the number of the dubious unit (counting from 1), j the position of the structure in the STRUCTURES parameter (again counting from 1), and *string* is the relevant part of the input record. READ will continue reading, commenting on all such errors; but when it reaches the data terminator, a fault is caused, unless you have set the ERRORS option to override the default action.

You can prevent the error reports from appearing by setting the PRINT option appropriately.

If you set the ERRORS option to an integer n, then if n is positive, up to n minor errors are allowed before a fault happens; if n is negative, after $-n$ minor errors no further reading of the data occurs. You should use a negative n in conjunction with the QUIT option.

With a large set of data, an error for example in a format would generate many minor errors. So a sensible thing to do is to set the ERRORS option to be negative, and to specify a channel number to which control will be switched when the fatal fault is reported. (For a full discussion of switching control between channels see 4.4.)

4.2 Printing Data

Most of this section is a description of the PRINT directive (4.2.1). In 4.2.2 we describe how the options and parameters interact with each other.

4.2.1 The PRINT directive

PRINT

Prints data in tabular format in an output file, unformatted file, or text.

Options

CHANNEL	= identifier	Channel number of file, or identifier of a text to store output; default current output file
SERIAL	= string	Whether structures are to be printed in serial order, i.e. all values of the first structure, then all of the second, and so on (no, yes); default n, i.e. values in parallel
IPRINT	= string	What identifier (if any) to print for the structure (identifier, associatedidentifier), for a table associatedidentifier prints the identifier of the variate from which the table was formed, IPRINT = * suppresses the identifier altogether; default i
RLPRINT	= strings	What row labels to print (labels, integers); default l
CLPRINT	= strings	What column labels to print (labels, integers); default l
RLWIDTH	= scalar	Field width for row labels; default 13
INDENTATION	= scalar	Number of spaces to leave before the first character in the line; default 0
WIDTH	= scalar	Last allowed position for characters in the line; default width of current output file
SQUASH	= string	Whether to omit blank lines in the layout of values (no, yes); default n
MISSING	= text	What to print for missing value; default '*'

ORIENTATION	= string	How to print vectors or pointers (down, across); default d, i.e. down the page
PERMUTE	= vector	Permutation of table classifiers; default *
INTERLEAVE	= scalar	Level of classification at which structures are to be parallel; default highest level, e.g. values for tables are interleaved within all the classifying factors
NDOWN	= scalar	Number of table classifiers to be printed down the page; default all but one unless there is only one, when it is printed down
PUNKNOWN	= string	When to print unknown cells of tables (present, always, zero, missing, never); default p
UNFORMATTED	= string	Whether file is unformatted (no, yes); default n
REWIND	= string	Whether to rewind unformatted file before printing (no,yes); default n

Parameters

STRUCTURE	= identifiers	Structures to be printed
FIELDWIDTH	= scalars	Field width in which to print the values of each structure (a negative value $-N$ prints numbers in E-format in width N); if omitted, a default is determined (for numbers, this is usually 12; for text, the width is one more character than the longest line)
DECIMALS	= scalars	Number of decimal places for numbers; if omitted, a default is determined which prints the mean absolute value to four significant figures
SKIP	= scalars	Number of spaces to leave before each value of a structure (* means newline before structure)

JUSTIFICATION	= strings	How to position values within the field (right, left); if omitted, right is assumed
MNAME	= strings	Name to print for table margins (Margin, Total, Nobserved, Mean, Minimum, Maximum, Variance, Count, Median); if omitted, "Margin" is printed

Very often you will only need to put

PRINT identifiers

in order to display the contents of the identifiers listed. The structures are fully annotated with their identifiers, and with row and column labels or numbers, where appropriate. Factors are represented by their labels if available, and otherwise by their levels. Genstat also does a lot of convenient things implicitly for you. The layout of the values is determined by the size and shape of the structures being printed, and by the space needed to print individual values. The output is arranged in columns; the structures are split if the page is not wide enough, so that one set of columns is completed before the next is printed.

In this section we use a single running example. It relates to a family Christmas party for eight people. The basic PRINT can be seen if we print the names, ages and sexes of these people:

EXAMPLE 4.2.1a

```
  1    TEXT [VALUES=David,Sarah,Christopher,Audrey,Simon,Andrew, \
  2        Susan,John] Name
  3    FACTOR [LABELS=!T(Male,Female); VALUES=1,2,1,2,1,1,2,1] Sex
  4    VARIATE [VALUES=11,32,45,60,27,16,24,58] Age
  5    PRINT Name,Sex,Age

            Name           Sex         Age
           David          Male       11.00
           Sarah        Female       32.00
     Christopher          Male       45.00
          Audrey        Female       60.00
           Simon          Male       27.00
          Andrew          Male       16.00
           Susan        Female       24.00
            John          Male       58.00
```

This output is headed by the identifiers. There is no further labelling here; but with other types of structure you get more. For example, rows and columns of tables are labelled by the labels or levels of the classifying factors.

4.2 Printing data

No matter what the type of structure, Genstat works out a field-width that will be adequate to allow the values of the structure to be aligned in columns. For numerical structures, moreover, Genstat chooses a format that will print the mean absolute value to at least four significant figures.

Now we describe the many modifications that you can make to these default actions of PRINT.

If you want to impose a format, you use either or both of the FIELDWIDTH and DECIMALS parameters. For example, if you know that the values of some numerical structure are always small whole numbers, you should set DECIMALS = 0: this will give a much more readable output than the default of four significant figures, which would have several zeroes after the decimal point. The default field-width (12 for numerical structures on most computers) is often wasteful: you may prefer, for example, FIELDWIDTH = 8. Both of these parameters are used in the second example:

EXAMPLE 4.2.1b

```
  6    PRINT Name,Sex,Age; FIELDWIDTH=14,8,6; DECIMALS=0

           Name     Sex   Age
          David    Male    11
          Sarah  Female    32
    Christopher    Male    45
         Audrey  Female    60
          Simon    Male    27
         Andrew    Male    16
          Susan  Female    24
           John    Male    58
```

Notice that DECIMALS is here actually associated with each of the three structures, but it is ignored for the texts. Thus you would have got exactly the same effect if you had set DECIMALS = 1,2,0 for example.

Another convenient way of altering the layout is by the SKIP parameter. For example, SKIP = 5 puts five spaces before the structure values in addition to those caused by the field-width. By default, no extra spaces are inserted unless a value fills the field completely, when a single space will be inserted. In the default PRINT you get a blank line before the first printed line, but if you set SKIP then you lose that. You can reinstate it by putting, for example, SKIP = !(*,5) instead of SKIP = 5. The * means "put a newline before this structure". There are further details later about how to use SKIP.

Since the human eye is used to reading typescript, you may want to left-justify text. You do this by setting JUSTIFICATION = left. For example, suppose our mythical family have voted for three games each; the result of the poll is:

EXAMPLE 4.2.1c

```
  7    VARIATE [VALUES=2,5,6,7,4] Votes
  8    TEXT [VALUES=Chess,Whist,Draughts,Tiddleywinks,Monopoly] Game
```

```
  9  PRINT Votes,Game; FIELDWIDTH=8; DECIMALS=0; SKIP=!(*,1),3; \
 10     JUSTIFICATION=right,left
```

```
   Votes  Game
       2  Chess
       5  Whist
       6  Draughts
       7  Tiddleywinks
       4  Monopoly
```

You can use left-justification for numerical structures too, but you are likely to find this less useful.

When printing tables, the margin will be labelled "Margin" by default. You can change the label by setting the MNAME parameter. When printing tables in parallel, the margins of each will have the same name, the first you have set. But in serial printing, you can use a different margin name for each table. (For serial printing, see the SERIAL option below.) When printing a mixture of tables and other structures, you must insert dummy names or missing values so that the margin names keep step with the table identifiers.

For an example of margin names, suppose our family has been divided into two teams, Red and Blue, by age, the four youngest playing against the four oldest. You can record this information by:

EXAMPLE 4.2.1d

```
 11  FACTOR [LABELS=!T(Red,Blue); VALUES=1,1,2,2,1,2,1,2] Team
 12  TABULATE [CLASSIFICATION=Team; MARGIN=yes] Age; MEAN=Meanage
 13  PRINT Meanage; MNAME=mean
```

```
              Meanage
    Team
     Red       23.50
     Blue      44.75

     Mean      34.12
```

To illustrate the remaining aspects of PRINT, we now imagine that each team was split into two pairs who played each pair in the opposing team once. The teams then changed partners (keeping to the same teams) and the new pairs played the opposing team once. Although they played as pairs they kept their individual scores, which you will see printed in later examples.

You can switch output to a channel other than the current one either by using the OUTPUT directive (4.4.3) or by using the CHANNEL option of PRINT; and you can do this, if you want, for each individual PRINT statement. You must previously have opened the channel for output (4.3.1); for example:

```
OPEN 'RESULTS'; CHANNEL = 4; FILETYPE = output
```

Genstat will use a standard length for this channel, to limit the number of characters output in any line. But you can set the WIDTH parameter of the OPEN directive to specify a smaller limit.

Instead of a channel number, you can send the output to a text structure. In this case you do not have an OPEN statement, and instead of setting the CHANNEL option to an output channel number you set it to the identifier of a text structure. There is no need to declare this text structure in advance, as PRINT will declare it as a text for you. If, however, you have declared it, it must be a text, and if it already has values they will be replaced. Each line of output becomes one value of the text. You are most likely to want to do this in order to manipulate the text further. Remember, however, that if you later print the text, its strings will be right-justified by default, and you will need to set JUSTIFICATION = left in the later PRINT statement to achieve the normal appearance of your output.

The maximum (and default) line length of this text is the length of what is called the *output buffer*. This is likely to be 200 on most computers. If you intend to print it to an output file, set the WIDTH option as appropriate.

If you try to print several structures together, PRINT, by default, will try to print them in parallel (that is, SERIAL = no). All structures except symmetric matrices can be considered as rectangular, and so if they are all the same size and shape then they can be printed in parallel columns. If you want to print just a single vector or pointer you will get it down the page (ORIENTATION = down).

However, if not all the structures are compatible they will be printed one-by-one (even if some of them do match). In this case, moreover, vectors and pointers will be printed across the page as if you had set option ORIENTATION = across (see below). If you want to avoid getting vectors or pointers across the page, set option SERIAL = yes. You can then print several incompatible structures in the same statement, as if they had been printed individually, and ORIENTATION will take its usual default value.

The IPRINT, RLPRINT, and CLPRINT options all control the labelling of structures. You can omit the identifier of the structure by setting IPRINT = *. For tables you might want to print the identifier of the variate from which the table was filled, instead of the identifier of the table itself: you do this with IPRINT = associatedidentifier. (You can fill a table with the values of a variate by TABULATE, 5.8.1, or by AKEEP, 9.6.1.)

RLPRINT and CLPRINT control the printing of row and column labels. For symmetric matrices you can give rows and columns serial integer labels as well as, or instead of, the labels provided with the structure: set, say, RLPRINT = integers or CLPRINT = integers,labels. If the structure identifiers form labels for columns, as they do by default, then they will be omitted along with any other column labels if CLPRINT is set to *. Some settings of the NDOWN and INTERLEAVE options cause the identifiers to be treated as row labels: they are then omitted if RLPRINT = *. In both cases, the setting of IPRINT is overridden.

Suppose, for example, that the table Mage holds the mean age of the members of each Team × Sex group, and that you have formed it from the variate Age using TABULATE (5.8.1). The next example shows two things: how to put the output from PRINT into a text, and how to set IPRINT = associatedidentifier. The table Mage is put into the text Tt, which is then printed without its identifier (since IPRINT = *):

EXAMPLE 4.2.1e

```
 14   VARIATE [VALUES=16,13,13,12,14,12,15,14] Score
 15   TABULATE [CLASSIFICATION=Sex,Team; MARGINS=yes] Age,Score; \
 16     MEANS=Mage,Mscore; VARIANCES=Vage,Vscore; \
 17     NOBSERVATIONS=Nage,Nscore
 18   PRINT [CHANNEL=Tt; IPRINT=associatedidentifier] Mage
 19   PRINT [IPRINT=*] Tt; JUSTIFICATION=left
```

	Age		
Team	Red	Blue	Margin
Sex			
Male	19.00	39.67	31.40
Female	28.00	60.00	38.67
Margin	23.50	44.75	34.13

You control the space allowed for row labels by RLWIDTH. By default this is set to 13, but you might want something else if the labels are very small. If the width provided (by you, or implicitly) is inadequate, PRINT automatically resets it to accommodate the longest row label.

You set left-hand and right-hand page margins by the INDENTATION and WIDTH options. For example, INDENTATION = 5 gives five blank characters before each line is printed; then WIDTH = 75 causes printing to be in the next 70 character positions. For an 80-character file this would leave at least five blanks at each side of the page.

By default, PRINT puts a blank line before and after the material being printed, and also between structures printed with SERIAL = yes. It may also insert one between parts of a structure, particularly a multi-way table or a table with margins. If you want to use the structure later as data for the same or another program, these extra blank lines may be inconvenient. You can get rid of them by setting option SQUASH = yes; but this does not omit blank lines that are part of a text structure.

In general, Genstat prints missing entries in a structure as * for numerical structures, and as blank for text or for factor labels. You can use the MISSING option to specify a character string instead. For example, you could set MISSING = 'Not Known' or MISSING = ' '.

You use the ORIENTATION option to specify whether vectors and pointers are printed across or down the page. It is ignored when other structures are being printed. The default for vectors and pointers is to be printed in columns (ORIENTATION = down). You will find ORIENTATION particularly useful for printing a single

4.2 Printing data

structure with many values.

If you set ORIENTATION = across, PRINT puts a blank line between each block of values to make the output more readable. So for a single very large structure you may also want to set SQUASH = yes.

The PERMUTE, INTERLEAVE, NDOWN and PUNKNOWN options are mainly for tables. A multi-way table need not have its printed layout dictated by the way in which it was declared: for example, you can interchange rows and columns. The next example shows what happens by default when you are printing two-way tables in parallel: it gives details of the number of people in each Team × Sex group, along with their average score after the tiddleywinks game is finished.

EXAMPLE 4.2.1f

```
 20    PRINT Nscore,Mscore; FIELDWIDTH=8; DECIMALS=0,1
          Team         Red             Blue            Margin
                    Nscore  Mscore  Nscore  Mscore  Nscore  Mscore
          Sex
          Male         2     15.0      3     13.0      5     13.8
          Female       2     14.0      1     12.0      3     13.3

          Margin       4     14.5      4     12.8      8     13.6
```

Suppose we declare

   TABLE [CLASSIFICATION = A,B] T

Then if A has only three levels and B has 25, say, the layout on the page will be very fragmented and difficult to read if we print T as it stands. However,

   PRINT [PERMUTE = B,A] T

will give a compact rectangular layout with three columns and 25 rows of values. The order in which the values of T are stored is not affected. In the next example Team and Sex both have two levels, so there is no improvement in the layout; but it shows what happens in a simple case.

EXAMPLE 4.2.1g

```
 21    PRINT [PERMUTE=Team,Sex] Mscore,Vscore; FIELDWIDTH=8; DECIMALS=2
          Sex         Male           Female          Margin
                   Mscore  Vscore  Mscore  Vscore  Mscore  Vscore
          Team
          Red       15.00    2.00   14.00    2.00   14.50    1.67
          Blue      13.00    1.00   12.00       *   12.75    0.92

          Margin    13.80    2.20   13.33    2.33   13.62    1.98
```

When you are printing structures in parallel, you can think of the list as a table having an extra classifying factor: the "list of structures". This factor would normally be represented by the columns across the page, but you can change its position in the extended classifying set by setting the INTERLEAVE option. By default it comes last, thus giving the standard parallel columns; but if you give INTERLEAVE a positive integer value the extra factor will be given that position in the extended set. For example, when printing several three-way tables in parallel, the default is INTERLEAVE = 4. INTERLEAVE has no effect on vectors or pointers. An example is:

EXAMPLE 4.2.1h

```
 22   PRINT [INTERLEAVE=1] Mscore,Vscore,Nscore; FIELDWIDTH=7,8,6; \
 23        DECIMALS=1,2,0

         Sex     Male                       Female
        Team      Red    Blue  Margin         Red    Blue  Margin

      Mscore     15.0    13.0    13.8        14.0    12.0    13.3
      Vscore     2.00    1.00    2.20        2.00       *    2.33
      Nscore        2       3       5           2       1       3

         Sex   Margin
        Team      Red    Blue  Margin

      Mscore     14.5    12.8    13.6
      Vscore     1.67    0.92    1.98
      Nscore        4       4       8
```

The decimal points do not line up. This is because the statement has specified the number of decimal places. The maximum field-width has been taken to make the columns line up.

By default, tables and matrices are printed so that the columns are labelled by the extra classifying factor "list of structures" and by the last genuine classifying factor; and all other factors label rows. These last are thought of as running "down" the page. You can alter the number of "down" factors by the NDOWN option. Thus NDOWN = 0 gives a single row of values with all associated labelling above them. At the other extreme, suppose we have two tables with three classifying factors; then NDOWN = 4 will give a single column of values to the right of all their labels. In the above example there are only two classifying factors, and so NDOWN = 2 is sufficient to obtain a single column:

EXAMPLE 4.2.1i

```
 24   PRINT [NDOWN=2] Vage
```

```
                         Vage
Sex          Team
Male          Red        128.0
             Blue        462.3

            Margin       391.3

Female        Red         32.0
             Blue           *

            Margin       357.3

Margin        Red         80.3
             Blue        411.6

            Margin       339.8
```

Setting NDOWN = 0 would give the single values strung out across the page, with labels above them.

PUNKNOWN stands for "print unknown", and controls the printing of the "unknown" cell of tables (see 3.5). The default action is to print this cell, labelled with the table identifier, but only if it contains a value other than missing value or zero. You can select one of five settings:

present (default)	print value if not missing or zero
always	print the unknown cell regardless of value
zero	print unless the value is zero
missing	print unless the value is missing
never	do not print the "unknown" cell whatever its value

You use the last two options of PRINT, UNFORMATTED and REWIND, only if you want to use unformatted output files (4.6.3). The REWIND option has no effect on character files.

4.2.2 Interaction among options and parameters of PRINT

Obviously there is a lot of scope for interaction among the options, which may complicate the simple actions of the individual parameters and options. Equally obviously, describing all possible interactions would take up an enormous amount of space. In this section we give a summary of the most important ones.

If SERIAL = no, which is the default, you can use the SKIP parameter to put blank lines before the print and to control the spaces before elements of each structure. For example, in

   PRINT A,B,C ; SKIP = !(3(*),5)

the "3(*)" puts three blank lines at the start of the output, and has no effect on B and C, but the three variates are still separated by the five spaces specified. SKIP = 5,3,2 will ensure five, three, and two blanks respectively before values of A, B, and C, but no blank lines above or below the output.

If you have set SERIAL = yes, you can use SKIP also to control the number of blank lines between structures. If SERIAL = yes and you use the SKIP parameter, then the layout is entirely in your hands (except that the whole output will start on a new line). You must provide a separate element in SKIP for each structure; the element can be * or an integer number, or it can be an unnamed variate of the form !(n(*),m). SKIP = * starts each structure on a new line, and SKIP = 2, for example, precedes each structure value by two spaces. SKIP = !(*,2) starts each structure on a new line and precedes each value by two spaces.

Genstat sets SKIP to 0 by default, though one space will be inserted before a value that completely fills its field. But if you provide any skip list at all, you must set what space there will be between fields, even if you are really only interested in controlling newlines: thus SKIP = * and no more would put no spaces at all between fields.

An example of the default is:

EXAMPLE 4.2.2a

```
 25   PRINT [SERIAL=yes; ORIENTATION=across] Votes,Game; \
 26       FIELDWIDTH=4,10; DECIMALS=0

       Votes     2    5    6    7    4
       Game     Chess     Whist  Draughts Tiddleywinks  Monopoly
```

If the list of structures contains only vectors, pointers, or scalars, and if ORIENTATION = across, then you will get no newlines between structures except those provided in the SKIP parameter; note that a scalar is treated here like a vector of length one:

EXAMPLE 4.2.2b

```
 27   PRINT [SERIAL=yes; ORIENTATION=across] Votes,Game; \
 28       SKIP=!(2(*),1),0; DECIMALS=0; FIELDWIDTH=4,10


       Votes     2    5    6    7    4
       Game     Chess     Whist  DraughtsTiddleywinks  Monopoly
```

You can use SERIAL = yes with IPRINT = * to mix named and unnamed structures; this is particularly convenient when you want to form headings:

EXAMPLE 4.2.2c

```
 29   PRINT [SERIAL=yes; IPRINT=*] 'These are the votes(',Votes,')', \
 30       'for these',5,'(',Game,') games'; DECIMALS=0; \
 31       FIELDWIDTH=1,5,1,1,2,3(1); SKIP=!(2(*),0),0,1,!(*,1),0,1,1,0

       These are the votes(    2    5    6    7    4 )
        for these 5 ( Chess Whist Draughts Tiddleywinks Monopoly) games
```

If the DECIMALS parameter is not set, PRINT uses any setting of the DECIMALS attribute that has been set in the declaration of a structure. PRINT also uses some other attributes of structures in the same way.

(a) PRINT respects any restriction that you have applied to its vectors. Thus if you are printing vectors in parallel, then those that are restricted must all have the same restriction. If they have different restrictions then Genstat automatically switches printing to serial.

(b) If you have set the EXTRA parameter of a structure in its declaration, then the extra text is normally used when Genstat prints identifiers; but that might not happen if the extra text would upset the layout. The extra text is ignored when the identifiers are used as column headings.

(c) If you have set the CHARACTERS parameter of a factor or text, then PRINT restricts accordingly the number of characters that are printed.

4.3 Getting access to external files

Every time you use Genstat, you must specify – implicitly or explicitly – the file from which Genstat is to start taking your statements. This file is automatically attached to the first input channel. The terminal can itself be treated as a file; if you specify that the terminal is on the first input channel, Genstat will prompt for a statement at the terminal. The method of specifying the file on the first input channel depends on your computer. Usually, if you give no filename when invoking Genstat, it will assume that the terminal is to be on the first channel; but read your local documentation to find out the details.

Similarly, you must specify the file to which Genstat is to start sending output; this is automatically attached to the first output channel. This also can be the terminal; if both the first input and first output channels are attached to the terminal, Genstat will start running interactively (1.1.10).

On some computers you can open other files when you invoke Genstat (and to find out if yours is one of these, see your local documentation). But whether or not you can do this, you always have the alternative of using the OPEN directive; this also lets you set some options concerning the shape of the file.

You can omit lines in a file by the SKIP directive, and you can arrange for a section of output to start on a new page with the PAGE directive. When you have finished with a file, you should tell Genstat to CLOSE the file (but remember that all files are automatically closed by the STOP directive).

4.3.1 The OPEN directive

OPEN

Opens files.

No options

Parameters

NAME	= texts	External names of the files
CHANNEL	= scalars	Channel number to be used to refer to each file in other statements; numbers for each type of file are independent
FILETYPE	= strings	Type of each file (input, output, unformatted, backingstore, procedurelibrary, graphics); if omitted, input is assumed
WIDTH	= scalars	Maximum width of a record in each file; if omitted, 80 is assumed for input files, the full line-printer width (usually 132) for output files
PAGE	= scalars	Number of lines per page (relevant only for output files)
ACCESS	= string	Allowed type of access for unformatted or backing-store files (readonly, writeonly, both); if unspecified, b is assumed

---

For example, suppose we want to analyse some data that are held in a file called Mydata, and that we want to store the results in a file called Results. We can open these by:

   OPEN 'Mydata','Results'; CHANNEL=2,3; FILETYPE=input,output

This attaches the file Mydata to input channel 2 and the file Results to output channel 3; these numbers are arbitrary, and the range you are allowed may be restricted by your local computer; you can find out what the range is by putting

   HELP environment,channel

What names you are allowed to give to files will also depend on the rules set by your local computer: see your local documentation.

You should be very careful not to open the same file on more than one channel at the same time. The consequence of doing so would be chaotic. You cannot have more than one file open on the same channel at the same time.

You must already have created an input file before you open it: otherwise there will be nothing there for your program to read. Output files, on the other hand, are automatically created when you open them.

You can also open files for backing storage (4.5) and for unformatted input and output (4.6). When you open one of these, you can both read from it and send output to it, unless you set the ACCESS option (see below). Graphics files are discussed in 7.3, and procedure library files in 6.3.

Genstat reads input and unformatted files sequentially from the beginning. You can start again at the beginning by setting the REWIND option of READ (4.1.1) or INPUT (4.4.1). Your operating system is not likely to allow you to overwrite output files; so if you want to re-use an output file you must delete it before you invoke Genstat.

Genstat sets default values for the WIDTH parameter that differ between input and output.

For output files, the default is the largest number of characters that can usually be displayed in a single line (the technical name for which is a *record*). This number is typically 80 for terminals, and 120 or 132 for actual files, which are likely to be displayed on a line printer. You can use the WIDTH parameter to restrict the number of output characters to a smaller number, or to a larger number up to 200; this parameter applies only to character files (input or output). The statement

    HELP environment,channel

gives information about line length as well as the number of channels.

For input files, the default is normally 80. You can change this to read either fewer characters from each line, or longer lines. For the latter, remember that if the file to be read has been output by a Fortran program, then each line will probably begin with a carriage-control character (usually space). It will therefore be one character longer than you might expect (133, rather than 132, for example). In particular, all output files produced by Genstat have carriage-control characters.

The PAGE parameter is relevant only for output files. You may not want graphs, for example, to overlap the perforations on the output paper. The 'page' setting of the OUTPRINT option of SET (2.2.1) and JOB (6.1.1) specifies whether some output is started on a new page: the size of the page is controlled by the PAGE parameter in OPEN.

The ACCESS parameter may not be fully implemented on your computer; for example, the 'readonly' setting is not always available. Check with your local documentation if you want to set this parameter.

4.3.2 The SKIP directive

SKIP

Skips lines in input or output files.

Options

CHANNEL	= scalar	Channel number of file; default current file
FILETYPE	= string	Type of the file concerned (input, output); default i

Parameter

	identifier	How many lines to skip; for input files, a text means skip until the contents of the text have been found, these contents are then the next characters to be read

The effect of this directive depends on whether it is being applied to input or output files.

For input, you can get Genstat to ignore n lines by setting the parameter to n. You can also skip everything up to a certain point by setting the parameter to be a text with a single element that marks the point. For example,

SKIP [CHANNEL = 2] 'EOD'

will skip the contents of the input file attached to the second input channel until the string EOD is found.

For output files,

SKIP [FILETYPE = output] 5

causes five blank lines to be sent to the current output channel.

4.3.3 The PAGE directive

PAGE

Moves to the top of the next page of an output file.

Option

CHANNEL	= scalars	Channel number of file; default * i.e. current output file

No parameters

4.4 Transferring input and output control

You use PAGE if you want to make a section of output from some statement start on a new page. Some directives will automatically start a new page, unless you have specified otherwise (2.2.1). If you have cancelled the automatic paging, or want to start a new page with output from other directives like PRINT, then use PAGE. For example:

    PAGE
    PRINT X,Y

PAGE has no effect if the output file is the terminal. Otherwise it puts into the output file a line consisting just of the Fortran control character "1", which is the code for a new page. The effect of PAGE is thus not modified by setting the page size of the output file with the OPEN directive (4.3.1).

4.3.4 The CLOSE directive

CLOSE

Closes files.

No option

Parameters

CHANNEL	= scalars	Numbers of the channels to which the files are attached
FILETYPE	= strings	Type of each file (input, output, unformatted, backingstore, procedurelibrary, graphics); if omitted, input is assumed

---

When you have finished with a file, you should tell Genstat to close the corresponding channel. This releases a computer resource, and may well save you money. You can then also re-use the channel to attach another file of the same type.

You cannot close a channel to which the terminal is attached, nor the current input or output channel – these are the first channels initially, but they can be changed by the INPUT and OUTPUT directives.

4.4 Transferring input and output control

Genstat always starts to take statements from the file in the first input channel (4.3). However, you can use the INPUT directive or macro substitution (1.7.2) to change this within a job – to take statements either from another file or from a text structure defined in the job. Subsequently, you can use the RETURN directive to go back to the original file of statements.

Similarly, you can change the file to which output is sent by using the OUTPUT directive. Many directives, like PRINT, have a CHANNEL option that lets you specify where the output from the directive is to go. If this is not set, or if the directive has no CHANNEL option, any output will go to the current output file as defined by OUTPUT.

4.4.1 The INPUT directive

INPUT

Specifies the input file from which to take further statements.

Options

PRINT	= strings	What output to generate from statements in the file (statements, macros, procedures, unchanged); default s
REWIND	= string	Whether to rewind the file (no, yes); default n

Parameter

scalar	Channel number of input file

You can get input from two kinds of *streams*: from files, which means either the terminal or existing external files; or from text structures. The computer limits the number of input files that you can use at the same time. You can give the statement

    HELP environment,channel

to find out what these are. There is no formal limit to the number of text structures, but the input buffer-space may become full.

You can readily switch control to a channel that you have opened by putting:

    INPUT channel

You can switch to a text structure T by using a macro (1.7.2):

    # #T

You cannot use a suffixed identifier here: for example, # #P[3] will not read from a text structure P[3], but will try to read from a text structure P, return to the current channel, and try to interpret [3] as following the contents of P.

The job maintains a stack of input streams. Suppose you specify channel k:

    INPUT k

This statement has one of three effects:

(a) If k is the current stream, the statement is ignored.

(b) If k is not in the stack, it is added to it.

(c) If k is already in the stack, there must be a chain of successive input streams in the stack:

$$k \rightarrow k_1 \rightarrow k_2 \rightarrow ... \rightarrow k_n$$

The intermediate streams k_1 ... k_n are suspended at their current positions and removed from the stack. Control is then passed to the beginning of stream k, if it has never been used before, or to the point at which it was last suspended. A subsequent

    INPUT k_i

will follow the same rules and continue stream k_i from the point at which it was suspended.

4.4.2 The RETURN directive

RETURN

Returns to a previous input stream (text vector or input channel).

Option

NTIMES   = scalar   Number of streams to ascend; default 1

No parameters

If you set the NTIMES option to n:

    RETURN [NTIMES = n]

then, if $n = 0$, the statement has no effect. If $n > 0$ control returns through n streams. If you do not set the option, n is assumed to be 1; so control is passed back to the stream from which the current stream was called. If you try to return from the primary input, your statement will be ignored.

    If Genstat meets the end of the current input stream, it will try to return control to the channel from which it was entered. This is called an *implicit return*. The stream is closed when this happens. You get a message if the stream was a file:

EXAMPLE 4.4.2

```
*** End of Input File 2
:   File closed, GOING UP!
```

But if Genstat meets the end of a text structure used as an input stream, then it returns automatically, and you do not get any message.

4.4.3 The OUTPUT directive

OUTPUT

Defines where output is to be stored or displayed.

Options

PRINT	= strings	Additions to output (dots, page, unchanged); default d,p
DIAGNOSTIC	= strings	What diagnostic printing is required (faults, warnings, extra); default f,w
WIDTH	= scalar	Limit on number of characters per record; default width of output file

Parameter

scalar	Channel number of output file

An OUTPUT statement defines where output is to go for all subsequent statements in a program, unless a statement contains a CHANNEL option. Thus

    OUTPUT 2
    PRINT X
    & [CHANNEL=3] Y
    ANOVA X

sends the values of X, and the analysis of X by the ANOVA statement, to the file on the second output channel, and the values of Y to the file on the third. The channel must have been opened (4.3); and if it has already been used for output since it was opened, subsequent output will be appended.

The PRINT option controls two aspects of the output produced by some directives: whether a line of dots is printed at the start, and whether a new page is started. This can also be controlled by the OUTPRINT option of the SET directive: see 2.2.1 for details.

Similarly, the DIAGNOSTIC option is precisely as in SET (2.2.1).

The WIDTH option specifies the maximum width to be used when producing output. The default value is the width specified when the file was opened (4.3.1), but you can

4.5 Storing and retrieving structures

subsequently decrease it. You cannot use OUTPUT to set the width to a greater value than that specified when the file was opened.

4.5 Storing and retrieving structures

Frequently you will want to save information that you have put into a data structure. This section shows you how to transfer information to various other storage media on the computer, so that you can access the information easily later on.

Note the difference between storing and merely printing. When you give the statement

    PRINT [CHANNEL=2] X

you put only the identifier and the values of X into the character file attached to output channel 2. But if you give the statement

    STORE [CHANNEL=2] X

you put the whole structure X into the binary file attached to backing-store channel 2. That is, all the attributes of X are stored there too; for example, what kind of structure it is, how long it is, and so on.

There are other ramifications of STORE, that we describe in this section. Subsection 4.5.1 describes the simplest use of storage and retrieval; this may be enough for most of your needs. Subsection 4.5.2 describes subfiles, userfiles, and workfiles. Subsections 4.5.3 to 4.5.6 describe the four directives available: STORE, RETRIEVE, CATALOGUE, and MERGE.

4.5.1 Simple storing

Here is an example to illustrate the simplest way of storing data, and then retrieving it. First, storing:

    OPEN 'EXAMPLE'; CHANNEL=1; FILETYPE=backingstore
    SCALAR A; VALUE=2
    VARIATE [VALUES=1...4] B
    "Store structures A and B"
    STORE [CHANNEL=1] A,B

This stores the information from scalar A and variate B in the file named EXAMPLE which is opened on backing-store channel number 1 (see 4.3.1). There is actually an invisible intermediate stage here: A and B are first stored in a *subfile* by the STORE statement; this subfile is then stored in the file EXAMPLE. The default name for the subfile is SUBFILE.

Now here is a job in which one of these structures is retrieved:

EXAMPLE 4.5.1a

```
  1  OPEN 'EXAMPLE'; CHANNEL=1; FILETYPE=backingstore
  2  "Retrieve structure A only"
```

```
      3   RETRIEVE [CHANNEL=1] A
      4   PRINT A

                   A
               2.000
```

So far, the file consists of one subfile: you can add more if you want. To do this, you must give a subfile name:

```
          OPEN 'EXAMPLE'; CHANNEL = 1; FILETYPE = backingstore
          TEXT [VALUES = 'STORING MORE DATA','ON BACKING STORE'] T
          "Add new subfile called Newset to file"
          STORE [CHANNEL = 1; SUBFILE = Newset] T
```

There are now two subfiles in the file, called SUBFILE and Newset.

Next, the text structure T is retrieved:

EXAMPLE 4.5.1b

```
      1   OPEN 'EXAMPLE'; CHANNEL=1; FILETYPE=backingstore
      2   "Retrieve T and print it"
      3   RETRIEVE [CHANNEL=1; SUBFILE=Newset] T
      4   PRINT T

                     T
        STORING MORE DATA
         ON BACKING STORE
```

You can add as many new subfiles as you want, in exactly the same way as above. But you must keep the subfile names distinct within each file.

4.5.2 Subfiles, userfiles and workfiles

Before going any further, you need to know how structures are stored.

A subfile is itself merely a portion of the backing-store file. Each subfile starts with a *catalogue*, recording which structures it stores. Then come the attributes (Chapter 3) and values of each structure stored. Note that a subfile cannot be changed by subsequent statements, for example to include further structures, though it can be completely replaced; for an example, see 4.5.5.

There are two types of subfiles. *Ordinary subfiles* can hold all types of structures except procedures; *procedure subfiles* hold only procedures (and their dependent structures).

4.5 Storing and retrieving structures

Whenever you store a structure in a subfile, Genstat automatically stores also all the structures to which it points. If these latter also point to further structures, then they are stored too, and so on. Some of the structures may be unnamed (1.4.3) and some structures may be system structures (2.5.1).

For example

```
TEXT [VALUES=A,B,C] T
FACTOR [LABELS=T; VALUE=1...3] F
STORE F
```

creates a subfile containing factor F. The complete definition of factor F depends on text T to supply level names. So T is stored too. In turn, text T depends on a system structure, which is therefore added to the store. Hence to save factor F, Genstat has actually had to save three structures.

When you store a structure with a suffixed identifier, Genstat may have to set up a series of pointer structures if they are not already present (1.3.3 and 3.3.4). An example is:

```
VARIATE [VALUES=1,2] V[1,2]
STORE [PRINT=catalogue] V[1]
```

The first line sets up a pointer structure V, pointing to V[1] and V[2]. To store variate V[1], a pointer structure V has to be set up in the subfile, pointing to V[1] only. Thus two structures are saved on backing store, namely V and V[1]. The original pointer V in the program is left unchanged. (If the example had stored the whole of V, no such complications would have arisen.)

You can retrieve any pointer structure that you have set up in this way; you could then use it in the same way as any other pointer. But when a pointer has to be set up only so that something else can be included in a subfile, it never gets textual suffixes. So, to store textual suffixes, you must have defined them and stored the pointer explicitly.

A backing-store file then consists of several subfiles; in fact the file can exist even if it is empty. However, if a file that contains anything not stored by a Genstat backing-store statement is attached as a backing-store file, it will be rejected.

A file that can be read by another job is called a *userfile*; it is permanent, in the sense that it will continue to exist after you have finished the job that created it. The userfile starts with a catalogue of subfiles followed by the contents of each subfile. How many userfiles you can have open at any one time will depend on your local computer: you can find this out by putting:

```
HELP environment,channel
```

Each job can have one temporary file called the *backing-store workfile* which also consists of a set of subfiles. The workfile's catalogue is deleted at the end of each job. The workfile itself may be overwritten in a later job in the same Genstat program, and on most computers it will automatically be deleted by STOP. But if you use the computer's device for abruptly terminating a run of Genstat, then the workfile may survive.

A subfile name can be either an unsuffixed identifier or a suffixed identifier (1.3.3) with a numerical suffix. The identifiers of subfiles are kept in a separate catalogue to the identifiers of data structures, so you do not have to worry about using distinct identifiers for a data structure and a subfile.

A minor technicality that you will almost never have to worry about is that suffixed identifiers are used automatically to define the pointer to subfiles. So a dummy subfile is set up, which acts like a pointer; this subfile contains no data. Thus if you create a backing-store file which contains a subfile with identifier Sub[1][2][3], the system itself will set up subfiles with identifiers Sub, Sub[1] and Sub[1][2]: thus you cannot use these names as subfile names.

4.5.3 The STORE directive

STORE

To store structures in a subfile of a backing-store file.

Options

PRINT	= string	What to print (catalogue); default *
CHANNEL	= scalar	Channel number of the backing-store file where the subfile is to be stored; default 0, i.e. the workfile
SUBFILE	= identifier	Identifier of the subfile; default SUBFILE
LIST	= string	How to interpret the list of structures (inclusive, exclusive, all); default i
METHOD	= string	How to append the subfile to the file (add, overwrite, replace); default a, i.e. clashes in subfile identifiers cause a fault (note: replace overwrites the complete file)
PASSWORD	= text	Password to be stored with the file; default *
PROCEDURE	= string	Whether subfile contains procedures only (no, yes); default n

Parameters

IDENTIFIER	= identifiers	Identifiers of the structures to be stored

---

4.5 Storing and retrieving structures

| STOREDIDENTIFIER | = identifiers | Identifier to be used for each structure when it is stored |

The structures that you specify by the IDENTIFIER parameter are put into the backing-store file indicated by the CHANNEL option; you can modify how the IDENTIFIER list is interpreted by using the LIST option. The structures that get stored in the subfile are merely copies of the structures in the job, which therefore remain available for you to use again.

Sometimes you may want a structure to have a different name in a subfile: you do this by using the STOREDIDENTIFIER parameter. For example,

    VARIATE [VALUES = 1,2] V
    STORE V; STOREDIDENTIFIER = Var

stores a structure with identifier V as a structure with identifier Var in backing store.

All structures in a subfile must have distinct identifiers: you will get a fault reported if you try to give two the same name. You should be particularly careful in this regard if you are storing structures inside a procedure. For example, an identifier may be used for one structure within the procedure, and for another outside; but they cannot both be stored in the same subfile.

Procedures that you have retrieved automatically from libraries (6.3) cannot be stored by STORE.

You might want to rename only some of the structures in the IDENTIFIER list. You can do this in two ways, each using the STOREDIDENTIFIER parameter. Either give the same identifier as in the IDENTIFIER parameter, or enter * in the correct position in the STOREDIDENTIFIER list. For example, you could store X and Y, renaming only Y as Yy, by

    STORE X,Y; STOREDIDENTIFIER = X,Yy

or by

    STORE X,Y; STOREDIDENTIFIER = *,Yy

You can give an unnamed structure in the list of either parameter. But you will not be able to retrieve structures that are stored as unnamed structures.

The PRINT option controls the printing of catalogues. You get the catalogues of subfiles in the backing-store file, and the catalogue of the subfile created by the directive STORE itself (more details in 4.5.5).

The CHANNEL option selects the backing-store channel on which you want to store the subfile (4.3.1). If you set the value to zero, then the backing-store file is the workfile; otherwise it is a userfile. You must have opened a userfile in advance (4.3.1).

The SUBFILE option gives a name to the subfile being created.

The LIST option modifies the way the IDENTIFIER parameter is dealt with. You use the 'inclusive' setting to store the structures that you have specified by the parameters; this is the default.

You use 'all' setting to store all those structures in the current job to which you have given an identifier and for which you have declared a type. If the statement is inside a procedure, then only structures defined within the procedure are stored (6.3). If you are storing procedures, then the setting 'all' will store all procedures that you have created in this job by PROCEDURE or RETRIEVE statements.

You use the 'exclusive' setting to store everything that you have not included in the IDENTIFIER parameter: that is, all the other named structures currently accessible, or all the other procedures created in this job. Note, though, that some of the structures in the IDENTIFIER list would be stored if they were needed to complete a set of structures on backing store. If you use this setting, the STOREDIDENTIFIER parameter is ignored.

An example is:

    TEXT [VALUES=a,b] T
    FACTOR [LABELS=T] F
    TEXT [VALUES='variate text'] Vt
    VARIATE V; EXTRA=Vt

This creates four named structures, T, F, V and Vt. Then

    STORE [LIST=inclusive] T

means store structure T;

    STORE [LIST=all]

means store all the four structures that have identifiers;

    STORE [LIST=exclusive] F,T

means store Vt and V; and

    STORE [LIST=exclusive] Vt,T

results in all four structures being saved, because V points to Vt, and F points to T.

If the subfile name already exists on the backing-store file, the storing operation will not normally work. But you can set the option METHOD=overwrite to overwrite the old subfile: that is, the old subfile will be replaced by the new subfile.

You can even get rid of the entire backing-store file, by putting METHOD=replace. It removes the contents of the backing-store file and creates a new file with the new subfile only.

To make your files secure, you can use a password. You specify this by the PASSWORD option; spaces, case, and newlines are significant in the password. Once you have done this, you must include the same password in any future use of STORE or MERGE with this same userfile (4.5.6). You cannot change the password in a userfile once you have set it, but you can use the MERGE directive to create a new userfile with no password or with a new password. If you set the password to be a text that you have restricted, then the restriction will be ignored.

The PROCEDURE option tells Genstat what type of subfile should be created. The 'no' setting causes an ordinary subfile to be formed and the 'yes' setting causes a procedure subfile to be formed.

4.5.4 The RETRIEVE directive

RETRIEVE
Retrieves structures from a subfile.

Options

CHANNEL	= scalar	Channel number of the backing-store file containing the subfile; default 0, i.e. workfile
SUBFILE	= identifier	Identifier of the subfile; default SUBFILE
LIST	= string	How to interpret the list of structures (inclusive, exclusive, all); default i
MERGE	= string	Whether to merge structures with those already in the job (no, yes); default n, i.e. a structure whose identifier is already in the job overwrites the existing one, unless it has a different type

Parameters

IDENTIFIER	= identifiers	Identifiers to be used for the structures after they have been retrieved
STOREDIDENTIFIER	= identifiers	Identifier under which each structure was stored

---

You get information back from a subfile of a backing-store file using the RETRIEVE directive; you specify which backing-store file by the CHANNEL option. A copy of the subfile is left unchanged in the file. Retrieving one structure might bring with it a chain of associated structures: that is, all the structures to which it points, and the structures to which they point, and so on.

For example, suppose you store the three structures with identifiers T, V, and F, along with an unnamed structure storing information about T, in a subfile called SUBFILE in backing-store file FILE1:

        OPEN 'FILE1'; CHANNEL = 1; FILETYPE = backingstore

```
TEXT [VALUES = a,b,c] T
VARIATE V; EXTRA = T
FACTOR [LABELS = T] F
STORE [CHANNEL = 1] T,V,F
```

Then you can get V back by the statement

```
RETRIEVE [CHANNEL = 1] V
```

You also get T and the unnamed structure, since V depends on T, and T on the unnamed structure. We call V, T, and the unnamed structure, a *complete set* from the subfile.

Normally you specify the structures to be retrieved in the IDENTIFIER parameter. But sometimes you may want a structure to have a different identifier in the current job from its identifier in the backing-store subfile. You can do this by specifying the identifier that you want it to have in the job using the IDENTIFIER parameter, and the identifier it has in the subfile using the STOREDIDENTIFIER parameter. For example, suppose a subfile in the workfile contains the structure Old and that you want to call it New in the job. Then you would put:

```
RETRIEVE IDENTIFIER = New; STOREDIDENTIFIER = Old
```

If you retrieve structures that were stored in procedures, there can be problems if the same identifier is used twice, as for the STORE directive (4.5.3).

You are not allowed to give identical identifiers to two retrieved structures. Moreover, you are not allowed to have the same identifier referring to a structure of one type in a subfile and to a structure of a different type in your job.

You can rename some retrieved structures while leaving others with the names they have in the subfile. For a structure that you do not want to change, you specify in STOREDIDENTIFIER either the same name as you have given it in the IDENTIFIER parameter, or else a * in the correct position in the list: this works in the same way as TOREDIDENTIFIER in STORE (4.5.3).

Genstat knows whether you are retrieving a procedure by the type of the subfile that you have told it you are accessing. You are not allowed to rename a procedure as a suffixed identifier or a directive name.

You can even rename a structure so that it is unnamed in the job. Suppose, for example, that a structure T already exists in the job and that you want to retrieve V stored in file FILE1 as above. Then, as we have seen, the structure T will also be retrieved. You can avoid the structure T in the job being overwritten by making the retrieved version of the stored structure T unnamed:

```
OPEN 'FILE1'; CHANNEL = 1; FILETYPE = backingstore
RETRIEVE [CHANNEL = 1] V,!T(a); STOREDIDENTIFIER = V,T
```

Note that the value, *a*, of the unnamed text !T(a) will be replaced by the values stored for T. Alternatively you could rename T to be Tnew by changing the statement to

4.5 Storing and retrieving structures

```
RETRIEVE [CHANNEL = 1] V,TNew; STOREDIDENTIFIER = V,T
```

When you are retrieving a suffixed identifier, Genstat matches the numerical suffix only, and not the whole structure that is denoted by the identifier. For example, suppose pointer P stored in a subfile points to structures with identifiers A, B, C, and D, and that P has numerical suffixes 1 to 4 respectively. Also suppose that in your current job, you have never mentioned pointer P either directly or indirectly. Then the statement

```
RETRIEVE [CHANNEL = 1] P[2]
```

gives the identifier P[2] to the retrieved structure corresponding to the unsuffixed identifier B on backing store.

It can happen that a structure that you are retrieving from a subfile overwrites the values of an existing structure in your program. So you might lose a structure with suffixed identifiers when you are retrieving a pointer or a compound structure. For example, suppose there exists a userfile called FILE2 containing a pointer P, pointing to structures A, B, C, and D. We set up three variates P[5,6,7], and then retrieve the pointer P:

```
OPEN 'FILE2'; CHANNEL = 1; FILETYPE = backingstore
VARIATE [VALUES = 1...6] P[5,6,7]
RETRIEVE [CHANNEL = 1] P
```

After this, P points to A, B, C, and D. The reference of pointer P to the structures called P[5,6,7] is lost. For more details about pointers, see 3.3.4.

The CHANNEL option is the same as in the STORE directive.

The SUBFILE option specifies the identifier of the subfile from which you want to retrieve structures.

The LIST option is similar to its namesake in the STORE directive, but it now refers to the named structures in the subfile.

Normally when you retrieve a complete subset of structures, Genstat overwrites all structures in the job that have the same identifier (after any renaming). As a result, some other structures already in the job may become inconsistent and will be destroyed (2.4.2). You can avoid this happening by setting the MERGE option to 'yes'. Genstat then does not overwrite structures with the same name and type. But a consequence of this might be that some of the retrieved structures will now become inconsistent and will be destroyed in the program (though they will remain in the subfile, of course).

4.5.5 The CATALOGUE directive

CATALOGUE

Displays the contents of a backing-store file.

Options

PRINT	= strings	What to print (subfiles, structures); default s,st
CHANNEL	= scalar	Channel number of the backing-store file; default 0, i.e. the workfile
LIST	= string	How to interpret the list of subfiles (inclusive, exclusive, all); default i
SAVESUBFILE	= text	To save the subfile identifiers; default *

Parameters

SUBFILE	= identifiers	Identifiers of subfiles in the file to be catalogued
SAVESTRUCTURE	= texts	To save the identifiers of the structures in each subfile

You use this directive to print the catalogues that have been set up by the STORE or MERGE directives, or to store information from the catalogues in a text. There are three types of catalogue: catalogues of subfiles in a backing-store file, catalogues of structures in an ordinary subfile, and catalogues of procedures in a procedure subfile.

You use the SUBFILE parameter to specify the subfiles that you want catalogued.

The SAVESTRUCTURE parameter allows you to set up texts, one for each subfile in the SUBFILE parameter. Each text contains the identifiers of all structures with an unsuffixed identifier in the subfile. Each identifier is put on a separate line, and the characters ,\ are appended to all but the last line. You would normally use these texts as a macro; the ,\ makes them useable as lists of identifiers. If the text is used as a macro, it is subject to the restriction on the length of statements (1.6).

The PRINT option signifies which catalogues are to be printed: these are either of subfiles, or of structures within subfiles. The 'subfiles' setting is for printing the catalogue of subfiles that are in the backing-store file that is attached to the channel specified by the CHANNEL option. The 'structures' setting prints the catalogue of structures or procedures that are in the subfiles that you have specified by the SUBFILE parameter.

The CHANNEL option is the same as in the STORE directive.

The LIST option is similar to its namesake in the STORE directive, but it now refers to the named structures in the subfile.

You may want to put the catalogue of subfiles into a text. You do this with the SAVESUBFILE option. It sets up a text to contain the identifiers of all the subfiles in a backing-store file. Each identifier is on a separate line with the characters ,\ appended

4.5   Storing and retrieving structures   131

to all but the last line.

For example, suppose there exists a userfile called FILE3 containing three subfiles: two ordinary subfiles (called Sub[1] and Sub[2]), and a procedure subfile called Sub3 containing a procedure called PROCS. Then you add a fourth subfile called Sub4 by:

>   TEXT [VALUES = 'variate heading'] T
>   POINTER [SUFFICES = !(1,3); NVALUES = !T('vector-1','vector-3')] P
>   VARIATE P[],V; EXTRA = T
>   OPEN 'FILE3'; CHANNEL = 1; FILETYPE = backingstore
>   STORE [CHANNEL = 1; SUBFILE = Sub4] P,V

So the subfile contains five named structures with identifiers P, V, P[1], P[2] and T, and also two unnamed structures associated with P and T.

You can get the catalogue, and create a text to hold it, by the statement

>   CATALOGUE [CHANNEL = 1; PRINT = structures] Sub4;
>   SAVESTRUCTURE = Tsub4

This prints:

EXAMPLE 4.5.5a

```
***** catalogue *****

catalogue of structures in the subfile Sub4

      entry   identifier      type           points to

        1       P             pointer        3
                                             pointer values
                                             unit            entry     labels
                                             1               4         vector_1
                                             3               5         vector_3
        2       V             variate        6
        3                     text           7
        4       P[1]          variate        6
        5       P[3]          variate        6
        6       T             text           8
        7                     system
        8                     system
```

If you now put

>   PRINT Tsub4

you get the following output:

EXAMPLE 4.5.5b

```
       Tsub4
          P,\
          V,\
            T
```

The catalogue lists the structures in the subfile. The identifiers of structures are printed if they exist; you also get details of any dependencies among structures, referenced by the number in the entry column. The identifier column gives the numerical suffix, and the labels column gives textual suffixes.

You get more information with a pointer structure or a compound structure (for example, LRV, 3.6.1); you may find this helpful with complicated pointer-trees, so that you can work out the different ways of referring to the same structure. (For an example of a complicated pointer-tree, see 3.3.4).

The text, Tsub4 above, gives the names of all the structures in the subfile that you can access directly. You will find these texts especially useful when you use them as macros. For example, here is a job illustrating how you can effectively add a variate Vnew to a subfile Sub4 on backing-store file FILE3.

```
OPEN 'FILE3'; CHANNEL = 1; FILETYPE = backingstore
CATALOGUE [CHANNEL = 1; PRINT = *] Sub4; SAVESTRUCTURE = Tsub4
VARIATE [VALUES = 1,2] Vnew
RETRIEVE [CHANNEL = 1; SUBFILE = Sub4] ##Tsub4
STORE [CHANNEL = 1; SUBFILE = Sub4; METHOD = overwrite] ##Tsub4,Vnew
```

You get one type of catalogue for procedure subfiles, as in

```
CATALOGUE [CHANNEL = 1] Sub3; SAVESTRUCTURE = Tsub3
```

which produces

EXAMPLE 4.5.5c

```
***** catalogue *****

catalogue of procedures in the procedure subfile Sub3

    entry   identifier

    1       PROCS
```

This kind of catalogue lists only the identifiers of procedures stored in the subfile. You never get the system structures associated with the procedure; there may be many of them.

You get another kind of catalogue for ordinary subfiles, or whole userfiles, as in

```
CATALOGUE [CHANNEL = 1; PRINT = subfiles; SAVESUBFILE = Tuser]
```

This produces

4.5 Storing and retrieving structures

EXAMPLE 4.5.5d

```
***** catalogue *****

catalogue of subfiles in the userfile   1
    entry   identifier      type
    1       Sub4            ordinary
    2       Sub3            procedure
    3       Sub             system
    4       Sub[2]          ordinary
    5       Súb[1]          ordinary
```
---

and a text Tuser with the values 'Sub4,\','Sub3,\','Sub[2],\','Sub[1]'.

Because the statement set the PRINT option, you get all the subfiles in the backing-store file. Included among them are system-type subfiles, which can remind you that you cannot include any subfile called Sub in this backing-store file. Sub does not appear in text Tsub4, since its inclusion would make the text difficult to use as a macro.

4.5.6 The MERGE directive

MERGE

Copies subfiles from backing-store files into a single file.

Options

PRINT	= string	What to print (catalogue); default *
OUTCHANNEL	= scalar	Channel number of the backing-store file where the subfiles are to be stored; default 0, i.e. the work file
METHOD	= string	How to append subfiles to the OUT file (add, overwrite, replace); default a, i.e. clashes in subfile identifiers cause a fault (note: replace overwrites the complete file)
PASSWORD	= text	Password to be checked against that stored with the file; default *

Parameters

SUBFILE	= identifiers	Identifiers of the subfiles
INCHANNEL	= scalars	Channel number of the backing-store file containing each subfile
NEWSUBFILE	= identifiers	Identifier to be used for each subfile in the new file

You use the MERGE directive to transfer subfiles between backing-store files. You can either add the subfiles to another backing-store file, or put them into a new backing-store file. The *target file* is where you are sending the subfiles. The set of subfiles that you are transferring is called the *merge subfile-set*.

You use the SUBFILE parameter to specify the subfiles that you want added into the merge subfile-set. You get these subfiles from a backing-store file that you have previously opened on the channel number given by the INCHANNEL parameter. If you do not specify the INCHANNEL parameter, Genstat assumes that the subfiles are coming from the workfile (channel 0). You can give a new identifier to a subfile by using the NEWSUBFILE parameter.

You cannot put the channel number of the target file in the INCHANNEL parameter. Also, you cannot give the same name to two subfiles in the merge subfile-set; to avoid doing this, rename subfiles by the NEWSUBFILE parameter. Where subfiles have suffixed names, you should take note of any dummy system-subfiles that are created.

Sometimes you will want to add all the subfiles in a backing-store file to the merge subfile-set; you can do this by entering * in the SUBFILE parameter. You can also rename a subfile. Suppose, for example, that the workfile has four subfiles called S1, S2, S3 and Sub. You want all four subfiles in the merge subfile-set, but you want to rename Sub as S4:

    MERGE *,Sub; INCHANNEL = 0; NEWSUBFILE = *,S4

You can leave some subfile names unchanged while you are renaming others. For a subfile that you do not want to change, either specify in the NEWSUBFILE parameter the same identifier as in the SUBFILE parameter, or else enter * in the correct position in the list.

Subfiles are merged in a fixed order. First come the subfiles from the backing-store file with the lowest channel number; these are added in the order in which they are stored in that backing-store file. Then come the subfiles from the file with the next highest channel number, ordered similarly. And so on.

The PRINT option controls the printing of the catalogues of subfiles in the new backing-store file (4.5.5).

The OUTCHANNEL option selects the backing-store channel for the target file: that is, for the new merged file. If you set its value to be zero, then the workfile is used; otherwise it is a userfile. You must already have opened the file if it is a userfile.

Normally the statement will fail if you name a subfile with an identifier that already exists in the target file. But you can change that by the METHOD option. The 'overwrite' setting indicates that when clashes occur in a subfile name, the old subfile is to be replaced by the new subfile. The 'replace' setting means remove the entire contents of the target file, and create a new file containing the merged subfile-set only. The final order of subfiles in the new merged file is determined as follows. If the target file was the workfile, the subfiles that you want to keep from the workfile are followed by the subfiles in the merged subfile-set. But if the target file is a userfile, things are the other way round: the merged subfile-set is followed by the subfiles that you want to keep.

The PASSWORD option is the same as in the STORE directive.

4.6 Storing and retrieving data and programs in unformatted files

Suppose you want to break off in the middle of working interactively. To avoid losing what you have done, or leaving the computer idle (the break may be longer than you intend), you can record the current state of your job, then abandon it, and later re-enter it at the point where you left off. The RECORD and RESUME directives allow you to do this. But before using them, you must first have opened an unformatted file (4.3.1). You can also store values of individual data structures in unformatted files; this is particularly useful for communicating with other programs.

4.6.1 The RECORD directive

RECORD

Dumps a job so that it can later be restarted by a RESUME statement.

Option

| CHANNEL | = scalar | Channel number of the unformatted file where information is to be dumped; default 1 |

No parameters

This directive sends information about your current job to the unformatted file specified by the CHANNEL option. The information includes the attributes and values of all your data structures, and the current graphics settings (7.2). Note, however, that no details are kept of external files: that is, of any files that you have opened.

After you use RECORD, you can continue normally (you can postpone your break). If you use RECORD with the same channel number again, the earlier state of your job is lost.

4.6.2 The RESUME directive

RESUME

Restarts a recorded job.

Option

CHANNEL = scalar Channel number of the unformatted file where the information was dumped; default 1

No parameters

RESUME recovers the information stored by a previous RECORD statement. This information need not be from an earlier session: it can be from earlier in the same one. When you use RESUME, you lose all the data structures of the current job: they are replaced by the structures from the recorded job. External files that you had attached before you used the RESUME directive remain in the same state, but the graphics settings will now be from the recorded job.

If you used the RECORD directive in a procedure or a FOR loop, then the job is not resumed at that point. Instead, it restarts at the statement after the procedure call, or after the outermost ENDFOR statement.

The amount of space available for data in the current job need not be the same as that in the recorded job. But you will get a fault if the available space is too small: that is, if the space needed by the recorded job is greater than space available in the current job.

Here is an example that uses the RECORD and RESUME directives:

EXAMPLE 4.6.2

```
  1   OPEN 'Dump'; CHANNEL=1; FILETYPE=unformatted
  2   VARIATE [VALUES=1,2] A
  3   RECORD
  4   PRINT A

              A
          1.000
          2.000

  5   STOP


  1   OPEN 'Dump'; CHANNEL=1; FILETYPE=unformatted
  2   RESUME
  3   CALCULATE A = A+1
  4   PRINT A
```

4.6 Storing and retrieving data and programs in unformatted files

```
         A
      2.000
      3.000

  5   STOP
```

4.6.3 Storing and reading data with unformatted files

You can store and retrieve individual structures in unformatted files using READ (4.1) and PRINT (4.2). Here are three reasons why you might want to store data in unformatted files.

(a) Temporarily to free some space for numerical data.

(b) To read large data sets that you have to input several times. Input from character files is slow. So after vetting a large data set, to read it later on will be more efficient if you transfer its contents to an unformatted file. Alternatively you could use backing store, but then the stored data are indexed, and access to them will take longer.

(c) To read numerical data written by another Fortran program.

You cannot transfer text structures to or from unformatted files produced by Genstat or other Fortran programs.

When you invoke Genstat, it automatically creates an *unformatted workfile* to which unformatted output is sent by default, and from which unformatted input is taken by default. On most computers, this file will be automatically deleted by the STOP statement.

You can use all the options of PRINT, but some are irrelevant. The only options of READ that are relevant are CHANNEL, REWIND, UNFORMATTED, and SERIAL.

You will mostly want to read and write structures serially, because it is quicker. Also the values of the structures transferred in parallel must all be of the same *mode*: that is, they must all be real numbers or all be integers. In particular, values of factors are stored as integers, and so cannot be stored in parallel with values of scalars, variates, matrices or tables all of which are stored as real numbers.

Now an example. First we open a file, and declare some variates, matrices, and factors:

```
OPEN 'BIO'; CHANNEL = 3; FILETYPE = unformatted
VARIATE X,Y,Z; VALUES = !(11...19),!(21...29),!(31...39)
MATRIX [ROWS = 2; COLUMNS = 3; VALUES = 11,12,13,21,22,23] M
FACTOR [LEVELS = 3; VALUES = 1,3,2,3,1,2,2,2,1,3] F
```

The next two statements store data for M and F on the file named BIO and data for X, Y and Z (in parallel) on the workfile:

```
PRINT [CHANNEL = 3; SERIAL = yes; UNFORMATTED = yes] M,F
PRINT [UNFORMATTED = yes] X,Y,Z
```

You can now free the space for numerical data for other purposes, by putting:

		DELETE X,Y,Z,F,M

The data is safe in the files.

If you now change the lengths of structures, be careful to reset them to their original values before you use unformatted READ to recover the data values from the file.

You must rewind files before you can read the data written at the beginning of them:

		READ [UNFORMATTED = yes; REWIND = yes] X,Y,Z
		READ [CHANNEL = 3; SERIAL = yes; UNFORMATTED = yes; REWIND = yes] M,F

You can also re-use the external file BIO in a later job.

4.6.4 Communicating with other programs

Your computer may impose constraints on the extent to which you can read data written by other programs (perhaps in languages other than Fortran). Firstly, the computer that created the data file must be compatible with the one that you are using. Secondly, the style of reading and writing of the data used by the programs must be compatible.

With two compatible computers, you can communicate between Genstat and another program either with character files or with binary files. To use binary files, the other program must be able to deal with files created by Fortran WRITE statements; to understand this, you need some knowledge of the Fortran which underlies Genstat.

Genstat puts data in unformatted files by a simple Fortran WRITE statement of the form

		WRITE (UNIT) (ARRAY(K),K = 1,N)

ARRAY here is either REAL or INTEGER. The value of N will not exceed a certain limit which on most computers is 1024. Structures with a larger number of values than N are split up into several lines, all but (perhaps) the last of which is of size N.

When Genstat reads unformatted files, an analogous Fortran READ statement is used, with the same assumptions on record length. If you wish to read a file that has already been created, you must know how it is arranged. If it does not conform to these Genstat conventions, you will have to use a Fortran program to re-arrange it suitably.

There is more information about communicating with other programs in 12.3 and 12.5.

P.K.L.
H.R.S.
A.D.T.

5 Data handling

Genstat has many facilities for doing calculations and for manipulating data. You may need to use these before one of Genstat's more powerful directives; for example, you may want to transform your data before doing a regression analysis or an analysis of variance. However you may also find these facilities useful even if you have no intention of doing any statistical analysis.

The CALCULATE directive (5.1) allows you to do straightforward arithmetic operations on any numerical data structure. It also enables you to make logical tests on data: for example, you may want to check whether two variates have the same values; you can do these checks on factors, texts, and pointers as well. You can use CALCULATE for matrix operations: for example, matrix multiplication, inversion, and Choleski decompositions (5.1.3 and 5.2.4). Other directives let you form eigenvalues (latent roots) and singular values (5.7). CALCULATE can do calculations with tables, and these need not have identical sets of classifying factors (5.1.4). You can also form marginal summaries over any of the dimensions of a table (5.2.5 and 5.8.2). Other directives allow you to combine "slices" in any dimension (5.3.4) and to tabulate values, for example, to summarize data from surveys (5.8.1).

The functions that you can use in the expressions are described in 5.2. These allow you to transform data and do various other standard operations, relevant to the different types of structure.

There are several directives for transferring and sorting values (5.3). You can transfer data from one set of structures into another: for example you may want to copy the columns of a matrix into a list of variates (5.3.1). You can re-order the units in vectors according to numerical or textual "keys" (5.3.2), or you can put them into random order (5.3.3). Randomization is of course particularly useful for the allocation of treatment factors in designed experiments. Factor values can also be generated in systematic order (5.5.1).

Genstat has several directives for manipulating textual data. You can omit complete lines, or append one text onto the end of another (5.3.1). You can form a text each of whose lines is made up from sections of lines from several texts concatenated together (5.4.1). For less standard manipulations, you can use the general text editor (5.4.2).

5.1 Numerical calculations

The main directive that you use for calculations in Genstat is CALCULATE, which is described in the first part of this section. The particular calculation is defined by a Genstat expression. The rules for these are given in 1.5.2, but below we give examples to show exactly how to use them. Expressions also occur in RESTRICT (2.4.1), the directives for program control (6.2), and FITNONLINEAR (8.6); so this section is relevant also to several other areas of Genstat.

Subsection 5.1.1 contains general information about CALCULATE, describing its options and illustrating the operators that can occur in expressions. Information about calculations with particular data structures is given in 5.1.2 (scalars, factors, variates, and texts), 5.1.3 (matrices) and 5.1.4 (tables). The rules for implicit declarations in CALCULATE are given in 5.1.5. How to define subsets of vectors or matrices using qualified identifiers is described in 5.1.6. The functions that can be used in expressions are described in 5.2.

The other directive described in this section is INTERPOLATE (5.1.7), which allows you to interpolate values at intermediate points of an observed sequence.

5.1.1 The CALCULATE directive

CALCULATE

Calculates numerical values for data structures.

Options

PRINT	= string	Printed output required (summary); default * i.e. no printing
ZDZ	= string	Value to be given to zero divided by zero (missing, zero); default m
TOLERANCE	= scalar	If the scalar is non missing, this defines the smallest non-zero number; otherwise it accesses the default value, which is defined automatically for the computer concerned

Parameter

	expression	Expression defining the calculations to be performed

---

The parameter of CALCULATE is unnamed, and is an expression. An expression (1.5.2) consists of identifier lists, operators and functions. However, for an expression to be valid in CALCULATE, it must include the assignment operator (=). For example

    CALCULATE 5,6

will fail, with an error message, even though the list 5,6 is an expression. The simplest form of the expression in CALCULATE is to assign values from one structure to another of the same type. For example:

    VARIATE [VALUES=1...4] Va
    CALCULATE Vb = Va

The values of the variate Va are given to the structure Vb; since Vb has not been declared previously Genstat defines it implicitly, here as a variate. The rules for implicit declarations in CALCULATE are described in 5.1.5, but you may prefer to declare everything explicitly until you are confident in using CALCULATE.

A complete list of the operators available for expressions is given in 1.3.6. Most of the operators in this list act element-by-element on the values of data structures of the same type, the exceptions being the compound operator *+ (matrix multiplication) and the four relational operators, .IS., .ISNT., .IN. and .NI. The assignment operator (=) has been demonstrated above; the next example shows the arithmetic operators +, −, *, /, and ** operating element-by-element on variates, X and Y:

EXAMPLE 5.1.1a

```
 1  VARIATE [VALUES=10,12,14,16,*,20] X
 2  VARIATE [VALUES=4,3,2,1,0,-1] Y
 3  CALCULATE Vadd = X + Y
 4  & Vsub = X - Y
 5  & Vmult = X * Y
 6  & Vdiv = X / Y
 7  & Vexp = X ** Y
 8  PRINT X,Y,Vadd,Vsub,Vmult,Vdiv,Vexp; FIELDWIDTH=9; DECIMALS=2

        X        Y     Vadd     Vsub    Vmult     Vdiv      Vexp
    10.00     4.00    14.00     6.00    40.00     2.50  10000.00
    12.00     3.00    15.00     9.00    36.00     4.00   1728.00
    14.00     2.00    16.00    12.00    28.00     7.00    196.00
    16.00     1.00    17.00    15.00    16.00    16.00     16.00
        *     0.00        *        *        *        *         *
    20.00    -1.00    19.00    21.00   -20.00   -20.00      0.05
```

A missing value in either or both of the variates gives a missing value in the resulting variate.

You can use the operator minus (−) in two ways: either as *dyadic* minus, to subtract one operand from another, as shown above; or as *monadic* minus, to change the sign of a single operand. Genstat gives monadic minus high precedence, which means that when it appears in an expression, it is one of the first operations to be done. Thus you must take care when using monadic minus to change the sign of the result of an expression. In particular, these two CALCULATE statements will give the same values to Vb and Vc:

    CALCULATE Vb = Va**2
    CALCULATE Vc = −Va**2

This is because the operator − appears as monadic minus, and so the sign of the values of Va are changed before being squared; what you need to put is

    CALCULATE Vc = −(Va**2)

In logical and relational expressions, Genstat uses the value 0 to represent false, and the value 1 to represent true. In fact any non-zero non-missing value is taken as representing a true value.

For numerical structures, Genstat has the relational operators, .EQ., .NE., .LT., .LE., .GT., and .GE. with their symbolic equivalents. Note that the symbolic equivalent for .EQ. is the compound operator ==. When using this compound operator, be sure to enter two equals signs; giving a single equals sign by mistake will lead to unexpected results, since this is the assignment operator. For text structures, the appropriate relational operators are .EQS. and .NES. The two texts must have the same number of units (or lines).

EXAMPLE 5.1.1b

```
 1   VARIATE [VALUES=1,2,3,4,5,*,*,1] X
 2   & [VALUES=5,4,3,2,1,1,*,*] Y
 3   TEXT [VALUES=a,b,c,d,e,'','',a] Tx
 4   & [VALUES=a,x,c,d,y,a,'',''] Ty
 5   CALCULATE Veq = X.EQ.Y
 6   & Vne = X.NE.Y
 7   & Vlt = X.LT.Y
 8   & Vle = X.LE.Y
 9   & Vgt = X.GT.Y
10   & Vge = X.GE.Y
11   & Veqs = Tx.EQS.Ty
12   & Vnes = Tx.NES.Ty
13   PRINT X,Y,Veq,Vne,Vlt,Vle,Vgt,Vge,Tx,Ty,Veqs,Vnes; \
14       FIELDWIDTH=5; DECIMALS=0

   X    Y  Veq  Vne  Vlt  Vle  Vgt  Vge   Tx   Ty Veqs Vnes
   1    5    0    1    1    1    0    0    a    a    1    0
   2    4    0    1    1    1    0    0    b    x    0    1
   3    3    1    0    0    1    0    1    c    c    1    0
   4    2    0    1    0    0    1    1    d    d    1    0
   5    1    0    1    0    0    1    1    e    y    0    1
   *    1    0    1    *    *    *    *         a    0    1
   *    *    1    0    *    *    *    *              1    0
   1    *    0    1    *    *    *    *    a         0    1
```

With most of the relational operators, a missing value in either operand, or in both, gives a missing result. The exceptions are .EQ. and .NE., .EQS. and .NES. When both operands are missing, .EQ. gives a true result and .NE. gives a false result. A missing value for a text is a null string, which is treated in the same way by .EQS. and .NES. as missing numbers are by .EQ. and .NE.

To test whether or not a dummy points to a particular identifier, you use the relational operators .IS. and .ISNT. For example, to store in Sca the result of a test to check whether dummy D points to Va, you would put

    CALCULATE Sca = D.IS.Va

or to test that D does not point to Vb, you would put

5.1 Numerical calculations

   CALCULATE Sca = D.ISNT.Vb

The final pair of relational operators, .IN. and .NI., represent logical inclusion and non-inclusion. These two operators are distinct from the other relational operators in that each value in the structure on the left-hand side is compared in turn with every value in the structure on the right-hand side.

For .IN., the result is true if the value on the left-hand side is included in the set of values in the right-hand structure; otherwise the result is false. The .NI. operator forms the complement of .IN. The length of the result is taken from the length of the left-hand structure, since it is the values of the left-hand structure that are being tested. For a very simple example, suppose that the variate X contains the values 1,2,1,1,3,5,1,2,1,4, and that the variate Evens contains 0,2,4. Then the statement

   CALCULATE S = X.IN.Evens

records in S the position of the even-valued elements in X: that is, S is given the values 0,1,0,0,0,0,0,1,0,1.

With a factor on the left-hand side of .IN. or .NI. and a variate on the right-hand side, Genstat checks the levels of the factor against the values in the variate. Alternatively, if the factor has a labels vector, you can specify a text against which Genstat will compare the labels. In the next example, the variate Large records which elements of the factor Size have values that lie in the set {4.8, 6}, and the variate Notab records which elements of the factor Type have values that lie outside the set {a, b}:

EXAMPLE 5.1.1c

```
  1  FACTOR [LEVELS=!(1.2,2.4,3.6,4.8,6)] Size; \
  2     VALUES=!(1.2,4.8,6,2.4,3.6,2.4,1.2,6)
  3  FACTOR [LABELS=!T(a,b,c,d)] Type; VALUES=!T(2(a,b,c,d))
  4  CALCULATE Large = Size .IN. !(4.8,6)
  5  & Notab = Type .NI. !T(a,b)
  6  PRINT Size,Large,Type,Notab; FIELDWIDTH=6; DECIMALS=1,0,0,0

  Size Large  Type Notab
   1.2     0     a     0
   4.8     1     a     0
   6.0     1     b     0
   2.4     0     b     0
   3.6     0     c     1
   2.4     0     c     1
   1.2     0     d     1
   6.0     1     d     1
```

You can use the logical operators .AND., .NOT., .OR., and .EOR. to combine the results of the relational operators, and form a single logical result: .NOT. reverses true and false results; .OR. gives a true result only if one or both operands are true; .AND. gives a true result if both operands are true; and .EOR. gives a true result if one of the operands is true, but a false result if both are true or both false. This example shows the results of the four logical operators. Notice that a missing value in either operand gives a missing value in the result.

EXAMPLE 5.1.1d

```
  1  VARIATE [VALUES=3(0,1,2),1,*] X
  2  &        [VALUES=(0,1,2)3,*,*] Y
  3  CALCULATE Vnot = .NOT. Y
  4  &         Vor  = X .OR. Y
  5  &         Vand = X .AND. Y
  6  &         Veor = X .EOR. Y
  7  PRINT X,Y,Vnot,Vor,Vand,Veor; FIELDWIDTH=5; DECIMALS=0
```

```
    X    Y Vnot  Vor Vand Veor
    0    0    1    0    0    0
    0    1    0    1    0    1
    0    2    0    1    0    1
    1    0    1    1    0    1
    1    1    0    1    1    0
    1    2    0    1    1    0
    2    0    1    1    0    1
    2    1    0    1    1    0
    2    2    0    1    1    0
    1    *    *    *    *    *
    *    *    *    *    *    *
```

If the expression contains lists, Genstat does several calculations. For example,

    CALCULATE A,B,C = X,Y,Z + 1,2,3

is equivalent to the three CALCULATE statements:

    CALCULATE A = X + 1
    CALCULATE B = Y + 2
    CALCULATE C = Z + 3

Genstat takes the items in the lists in parallel, and recycles any lists that are shorter than the list of primary arguments. In CALCULATE, the primary arguments are the identifiers on the left-hand side of the assignment operator (=). In the above example, each list had three identifiers, and so no recycling was done; but in the statement

    CALCULATE A,B,C = X,Y + 1,2,3

the second list is of length only two, and so is recycled to give the calculations:

    CALCULATE A = X + 1
    CALCULATE B = Y + 2
    CALCULATE C = X + 3

If the longest list is not on the left-hand side of the assignment operator, CALCULATE gives a diagnostic.

There may be occasions when you want to combine several expressions into a single CALCULATE statement to make your program look neater. But this has to be done with care. Suppose you have two variates X and Y; X is to be multiplied by 10, and Y is to be divided by 180. Naively, you might write down the statement:

CALCULATE X,Y = X*10,Y/180

The CALCULATE statement is syntactically correct, but it does not do what you want: the statement corresponds to the pair

CALCULATE X = X*10/180
CALCULATE Y = X*Y/180

which is quite different from the intention. Genstat interprets the elements of the list on the right-hand side as 10 and Y, not as "X*10" and "Y/180". If you really want to combine these two operations together, you should put

CALCULATE X,Y = X,Y * 10,1 / 1,180

Then the three lists on the right-hand side are taken in parallel: X, 10, and 1 together, and Y, 1, and 180 together.

CALCULATE has three options: PRINT, ZDZ, and TOLERANCE. If you set the PRINT option to 'summary', then you get a summary of the values of the structure whenever an assignment is made. Genstat does this for all structures except scalars and factors. The output has the same form as in the READ directive (4.1.1): identifier, minimum value, mean value, maximum value, number of values, number of missing values, and whether or not the set of values is skew. In this example two assignments are made, and summaries printed for the variates B and C.

EXAMPLE 5.1.1e

```
  1   VARIATE [VALUES=1,4,*,7,10] A
  2   CALCULATE [PRINT=summary] C = (B = 2*A) + 1
      Identifier    Minimum       Mean    Maximum     Values    Missing
               B       2.00      11.00      20.00          5          1
               C       3.00      12.00      21.00          5          1
```

You get a warning if you try to use CALCULATE to do something invalid; for example, take a logarithm or square root of a negative number. Genstat then inserts a missing value. The one exception is dividing zero by zero: for this Genstat does not print a warning, and you can use option ZDZ to specify whether the result should be a missing value (ZDZ=missing) or zero (ZDZ=zero); the default is 'missing'. In this example, the variate %dm is formed with zeroes in the positions corresponding to where Fresh_wt and Dry_wt both have zeroes.

EXAMPLE 5.1.1f

```
  1  VARIATE [VALUES=15.74,88.61,48.70,0,49.37] Fresh_wt
  2  & [VALUES=3.21,11.3,7.83,0,7.23] Dry_wt
  3  CALCULATE [ZDZ=zero] %dm = 100*Dry_wt / Fresh_wt
  4  PRINT Dry_wt,Fresh_wt,%dm; FIELDWIDTH=9; DECIMALS=2

   Dry_wt  Fresh_wt       %dm
     3.21     15.74     20.39
    11.30     88.61     12.75
     7.83     48.70     16.08
     0.00      0.00      0.00
     7.23     49.37     14.64
```

Arithmetic operations with real numbers can suffer from rounding errors. Genstat uses real arithmetic for all its operations in CALCULATE, and so makes allowance for cases where rounding error may cause problems: in other words, very small numbers are taken to be zero. But sometimes you may want to do calculations with numbers that are genuinely very small, and you may then need to change the value that Genstat uses to assess the rounding-off. You do this using the TOLERANCE option.

5.1.2 Expressions with scalars and vectors

Examples of addition and multiplication with scalars were shown in Chapter 1. The next example shows a calculation involving both scalars and variates. Several scalars are used to transform the variate Mpg (miles per gallon) into its metric counterpart Lp100k (litres per 100 kilometres). The scalar values are applied to every unit of the variate. You will see that the zero value in Mpg causes Genstat to print the warning "Attempt to divide by zero"; a missing value is given to the corresponding position in Lp100k. This warning is printed only once per operation; so subsequent zero values in Mpg do not trigger it again.

EXAMPLE 5.1.2a

```
  1  VARIATE [VALUE=0,10,20,30,32...40,0,50] Mpg
  2  VARIATE Lp100km
  3  SCALAR Lpt,Cmin,Ydm,Inyd,Mkm; VALUE=0.568,2.54,1760,36,1000
  4  CALCULATE Lp100km = 8 * Lpt * 100 * Mkm * 100 / \
  5     ( Mpg * Ydm * Inyd * Cmin )

******* Warning (Code CA 18). Statement 1 on Line 5
Command: CALCULATE Lp100km = 8 * Lpt * 100 * Mkm * 100 / ( Mpg * Ydm * Inyd * Cmin )

Attempt to divide by zero
Attempt to divide by zero occurs at unit 1
```

5.1 Numerical calculations

```
  6  PRINT Lp100km,Mpg; FIELDWIDTH=8; DECIMALS=2

   Lp100km       Mpg
         *      0.00
     28.24     10.00
     14.12     20.00
      9.41     30.00
      8.82     32.00
      8.30     34.00
      7.84     36.00
      7.43     38.00
      7.06     40.00
         *      0.00
      5.65     50.00
```

If you use CALCULATE with variates that are restricted, Genstat applies the same restriction to all the variates. Thus any variates in a CALCULATE statement that are restricted must all be restricted in the same way; Genstat will apply this restriction also to any that are not explicitly restricted. In this example, the variate Fresh_wt is restricted to those of its values that correspond to the first level of the factor Block. Only these units are involved in the calculation; the other units are left unchanged. Here the variate %dm had no values, and so the units in Block 2 are given missing values.

EXAMPLE 5.1.2b

```
  1  VARIATE[VALUES=15.74,88.61,48.70,49.37,18.96,12.13,23.38,48.16] Fresh_wt
  2  & [VALUES=3.21,11.3,7.83,7.23,3.55,2.6,4.0,6.43] Dry_wt
  3  VARIATE [NVALUES=8] %dm
  4  FACTOR [LEVELS=2; VALUES=4(1,2)] Block
  5  RESTRICT Fresh_wt; CONDITION=Block.EQ.1
  6  CALCULATE %dm = Dry_wt*100/Fresh_wt
  7  PRINT %dm

       %dm
     20.39
     12.75
     16.08
     14.64
         *
         *
         *
         *
```

When Genstat implicitly declares a structure during a CALCULATE operation, it also by default sets its attributes to match those of the structures in the calculation: for details, see 5.1.5. We now do the same calculation, but leave Genstat to declare the structure %dry_m as a variate. In particular, the length of %dry_m will be the same as that of Fresh_wt and %dry_m will become restricted in the same way as Fresh_wt. Thus, the PRINT statement shows only the values of

%dry_m corresponding to Block 1; in fact, all the other values of %dry_m are missing. Genstat would also have carried the restriction across if we had declared %dry_m as a variate but had left the CALCULATE statement to set its number of values.

EXAMPLE 5.1.2c

```
  8  CALCULATE %dry_m = Dry_wt*100/Fresh_wt
  9  PRINT %dry_m

     %dry_m
      20.39
      12.75
      16.08
      14.64
```

If you put a factor in a calculation, Genstat will use its levels, as in the example below where V takes its values from the factor F. The function NEWLEVELS (5.2.1) allows you to specify an alternative levels variate to be used instead in the calculation. The example uses the values 3.5 and 6.4, instead of the values 2 and 4 in the levels variate of the factor F, when forming the values of the variate Vn.

EXAMPLE 5.1.2d

```
  1  FACTOR [LEVELS=!(2,4)] F; VALUES=!((2,4)4)
  2  CALCULATE V = F
  3  & Vn = NEWLEVELS(F; !(3.5,6.4))
  4  PRINT V,Vn

         V            Vn
     2.000         3.500
     4.000         6.400
     2.000         3.500
     4.000         6.400
     2.000         3.500
     4.000         6.400
     2.000         3.500
     4.000         6.400
```

If the factor is on the left-hand side of the equals sign, Genstat checks that each of the results of the calculation is an acceptable level. This allows you to define the values of a factor from a variate, or from another factor. However you must already have declared the factor, with its levels and labels vectors; factors cannot be declared implicitly. The next example first sets the values of a factor Rate from the variate Setting; it then uses the NEWLEVELS function to form the values of the factor Amount, whose first level corresponds to levels 1 and 2 of the factor Rate

5.1 Numerical calculations

and whose second level corresponds to levels 3 and 4.

EXAMPLE 5.1.2e

```
  1  VARIATE [VALUES=1,3,2,1,4,3,1,2] Setting
  2  FACTOR [LEVELS=!(1.25,2.5,3.75,5)] Rate
  3  CALCULATE Rate = Setting*1.25
  4  FACTOR [LABELS=!T(lower,higher)] Amount
  5  CALCULATE Amount = NEWLEVEL(Rate; !(1,1,2,2))
  6  PRINT Setting,Rate,Amount; FIELDWIDTH=8; DECIMALS=2

 Setting     Rate   Amount
    1.00     1.25    lower
    3.00     3.75   higher
    2.00     2.50    lower
    1.00     1.25    lower
    4.00     5.00   higher
    3.00     3.75   higher
    1.00     1.25    lower
    2.00     2.50    lower
```

Text structures can occur only with the relational operators, .EQS., .NES., .IN., and .NI. described in 5.1.1. The result of any expression is a number, so you cannot create a text with CALCULATE, even if the structures on which the operations are being done are texts.

5.1.3 Expressions with matrices

All the arithmetic, relational, and logical operators that we have now seen being used with variates can also be used with rectangular matrices, symmetric matrices, and diagonal matrices. The basic rule for working with different matrix structures is that their dimensions must conform. This means that, for each pair of matrices, row dimension must match row dimension and column dimension must match column dimension. Consider the matrices Mx, My, and Mz, and the symmetric matrix Smz declared here:

```
MATRIX [ROWS=3; COLUMNS=4] Mz,My
MATRIX [ROWS=3; COLUMNS=3] Mx
SYMMETRICMATRIX [ROWS=3] Smz
```

The dimensions of Mz and My conform; but the dimensions of Mx and Mz do not, since Mx and Mz have different numbers of columns, three and four respectively. Similarly the dimensions of the symmetric matrix Smz and the matrix Mx conform; but the dimensions of Smz and Mz do not.

For simplicity, our examples mostly involve addition; but remember that you can replace the operator + with any of the other arithmetic, logical, or relational operators. Matrix multiplication is described towards the end of this subsection.

In the first example, two rectangular matrices, Ma and Mb (each with four rows and three columns) are added together to form Mc. Notice that Genstat operates in turn on each element of these two matrices, and that the new structure Mc is a matrix also with four rows and three columns.

EXAMPLE 5.1.3a

```
 1    MATRIX [ROWS=4; COLUMNS=3] Ma,Mb; \
 2      VALUES=!(1...12),!(5,4,6,12,10,11,7,9,8,3,1,2)
 3    & Mc
 4    CALCULATE Mc = Ma + Mb
 5    PRINT Mc
```

```
           Mc
            1          2          3

    1     6.00       6.00       9.00
    2    16.00      15.00      17.00
    3    14.00      17.00      17.00
    4    13.00      12.00      14.00
```

When you calculate with diagonal matrices, each one must have the same number of rows. Similarly, with symmetric matrices the row dimensions must match. When you add, subtract, multiply, divide, or exponentiate a symmetric matrix, only those elements that are stored by Genstat are operated on. Here the two symmetric matrices Sma and Smb are added together to form another symmetric matrix Smc; this is done element by element:

EXAMPLE 5.1.3b

```
 6    SYMMETRICMATRIX [ROWS=4] Sma,Smb; \
 7      VALUES=!(1...10),!(7,8,4,9,5,2,10,6,3,1)
 8    & Smc
 9    CALCULATE Smc = Sma + Smb
10    PRINT Smc
```

```
          Smc

    1    8.00
    2   10.00     7.00
    3   13.00    10.00     8.00
    4   17.00    14.00    12.00    11.00
            1         2        3        4
```

If you use a symmetric matrix in a calculation together with a matrix, it will be extended to include the values above the diagonal, before the calculation is done. Similarly, diagonal matrices are extended for calculations with matrices or symmetric matrices. This example adds the diagonal matrix Da to the symmetric matrix Sma and puts the results in the matrix Md:

EXAMPLE 5.1.3c

```
 11   DIAGONALMATRIX [ROWS=4; VALUES=3,2,4,1] Da
 12   MATRIX [ROWS=4; COLUMNS=4] Md
 13   CALCULATE Md = Sma + Da
 14   PRINT Md
```

```
              Md
               1         2         3         4
     1     4.000     2.000     4.000     7.000
     2     2.000     5.000     5.000     8.000
     3     4.000     5.000    10.000     9.000
     4     7.000     8.000     9.000    11.000
```

You can also use variates along with matrices, provided their dimensions conform. Genstat treats variates as column matrices: that is, with n rows and 1 column. This example adds the variate Va to the four-by-one matrix Me.

EXAMPLE 5.1.3d

```
 15   VARIATE [NVALUE=4; VALUES=4,2,1,3] Va
 16   MATRIX [ROWS=4; COLUMNS=1; VALUES=10,4,7,2] Me
 17   CALCULATE Me = Me + Va
 18   PRINT Me
```

```
              Me
               1
     1    14.000
     2     6.000
     3     8.000
     4     5.000
```

You can use a scalar with any of the matrix structures; the scalar is applied to every element of the matrix, in exactly the same way as when scalars and variates occur together in a calculation (5.1.2). Here the scalar Sca is added to every element of the symmetric matrix Sma.

EXAMPLE 5.1.3e

```
 19   SCALAR Sca; VALUE=3
 20   CALCULATE Sma = Sma + Sca
 21   PRINT Sma
```

```
            Sma
   1      4.000
   2      5.000    6.000
   3      7.000    8.000    9.000
   4     10.000   11.000   12.000   13.000
             1        2        3        4
```

The multiplication operator (*) means element-by-element multiplication for the two matrices, not matrix multiplication. For example:

EXAMPLE 5.1.3f

```
  22   MATRIX [ROWS=4; COLUMNS=3] Mf
  23   CALCULATE Mf = Ma * Mb
  24   PRINT Mf

                Mf
                 1          2          3
       1       5.00       8.00      18.00
       2      48.00      50.00      66.00
       3      49.00      72.00      72.00
       4      30.00      11.00      24.00
```

For matrix multiplication you use the compound operator *+, or the function PRODUCT (5.2.4). You must be careful that the column dimension of the first matrix matches the row dimension of the second. In this example, the four-by-four matrix Mh is formed from the matrix product of Ma with Mg, a matrix with 3 rows and 4 columns.

EXAMPLE 5.1.3g

```
  25   MATRIX [ROWS=3; COLUMNS=4; VALUES=1,4,7,10,2,5,8,11,3,6,9,12] Mg
  26   MATRIX [ROWS=4; COLUMNS=4] Mh
  27   CALCULATE Mh = Ma *+ Mg
  28   PRINT Mh

                Mh
                 1          2          3          4
       1       14.0       32.0       50.0       68.0
       2       32.0       77.0      122.0      167.0
       3       50.0      122.0      194.0      266.0
       4       68.0      167.0      266.0      365.0
```

Be careful not to get confused between * and *+.

5.1 Numerical calculations

The rules for implicit declarations when combining matrices are in 5.1.5. The rules for qualified identifiers of matrices are in 5.1.6. Genstat provides several special matrix functions, including the INVERSE function, that can be included in CALCULATE statements: see 5.2.4.

5.1.4 Expressions with tables

You can use tables in expressions in much the same way as you would any other numerical structure. Arithmetic, relational, and logical operators act element-by-element, as do the general functions (5.2.1).

Tables in expressions must be either all without margins or all with margins. If you try to mix tables with and without margins, Genstat will report an error.

Calculating with tables is very straightforward when they have the same factors in their classifying sets. In this example, two tables are added together:

EXAMPLE 5.1.4a

```
  1  FACTOR [LEVELS=2; LABELS=!T(Woburn,Rothamsted)] Soil
  2  &      [LEVELS=2; LABELS=!T(low,medium)] Acidity
  3  TABLE  [CLASSIFICATION=Soil,Acidity] Ta,Tb; \
  4         VALUES=!(6.91,4.98,4.86,*),!(6.38,4.68,6.49,*)
  5  & Tc
  6  CALCULATE [PRINT=summary] Tc = Ta + Tb

     Identifier    Minimum       Mean    Maximum     Values    Missing
             Tc       9.66      11.43      13.29          4          1

  7  PRINT Tc

                        Tc
       Acidity         low     medium
          Soil
        Woburn       13.29       9.66
    Rothamsted       11.35          *
```

When tables have different classifying sets, there are two cases to consider. We illustrate them with the assignment operator, but the rules apply to any operation. The first case is when the table on the left-hand side has a factor in its classifying set that is not in the classifying set of the table on the right-hand side. In this case, all levels of that factor, including margins, get the same value. In the next example the values of the table Tb are repeated over the levels of the factor Block, which is the factor additional in the table Td. In other words the table Tb has been extended by an additional factor: perhaps the easiest way of thinking about what happens is that each level of the extra factor contains a whole copy of the table on the right-hand side.

EXAMPLE 5.1.4b

```
  8  FACTOR [LEVELS=2] Block
  9  TABLE [CLASSIFICATION=Soil,Acidity,Block] Td
 10  CALCULATE Td = Tb
 11  PRINT Td
```

		Td	
	Block	1	2
Soil	Acidity		
Woburn	low	6.380	6.380
	medium	4.680	4.680
Rothamsted	low	6.490	6.490
	medium	*	*

The second case is when the table on the right-hand side has a factor in its classifying set that is not in the classifying set of the table on the left-hand side. Now the values in the margin over that factor are taken for the left-hand table. If the table has no margins, they must be calculated first. By default Genstat forms marginal totals, but you can use the special table functions (5.2.5) to form other types of margin. In this example, marginal totals are calculated for table Td over the factor Block, and the results are placed in the previously declared table Tc.

EXAMPLE 5.1.4c

```
 12  CALCULATE Tc = Td
 13  PRINT Tc
```

	Tc	
Acidity	low	medium
Soil		
Woburn	12.76	9.36
Rothamsted	12.98	*

The classifying set of a table has two forms – one taken from the sequence in which the factors appeared in the CLASSIFICATION option of the TABLE declaration, the other determined by the order in which the identifiers of the factors are stored within Genstat. The second of these is called the *ordered classifying set*, and is the one used by CALCULATE for all operations on tables. CALCULATE permutes the values of tables so that they correspond to the ordered classifying set.

There are then two consequences. The first is that if a fault occurs while an operation on a table is being done, its values may have been permuted, and so may no longer be in the order corresponding to the classifying set specified in the CLASSIFICATION option of the TABLE declaration. But this problem arises only if there has been a fault, since CALCULATE does not permanently permute the values.

The second consequence concerns implicit declarations. Tables declared implicitly have their classifying set and ordered classifying set the same. In the first example, adding the two tables with the same classifying set but not declaring the resultant table Tc, made its classifying set the same as its ordered classifying set. So, if you want your tables printed with the factors in a particular order, you must declare the table before its values are assigned in CALCULATE, or else use the PERMUTE option in PRINT (4.2).

5.1.5 Rules for implicit declarations

Undeclared structures on the left-hand side of an assignment (=) in an expression are declared automatically: this will be referred to as *implicit declaration*. We now describe the rules that determine the type of the declared structure. You do not need to know about these rules unless you intend to let Genstat do these declarations for you.

The assignment operator (=) can appear anywhere in an expression, and so you must understand the order of evaluation. For example, in the CALCULATE statement

    CALCULATE Vc = Va*Vb

the result of Va*Vb is not placed directly in Vc: CALCULATE forms an intermediate structure whose values in this case are the results of Va*Vb; then the values of the intermediate structure are assigned to Vc. On assignment, the resultant structure is given the values of the intermediate structure, also its type and other attributes if they have not been previously declared (implicitly or explicitly) for the resultant.

When structures of the same type are combined, the rule is that the intermediate structure will be of the same type; the same rule applies to tables with identical classifying sets. When structures of different types are combined, you need to know what form the intermediate structure takes.

In the rules listed and explained below, .OP. refers to any arithmetic, logical, or relational operator, except .IS., .ISNT., .IN., .NI., .EQS., and .NES. which have their own rules described earlier (5.1.1). The dimensions of operands must conform in any operation involving matrices and variates.

Combination	Type of intermediate structure	Assignment
Scalar.OP.Scalar	Scalar	Any
Variate.OP.Scalar	Variate	Variate,Factor
Variate.OP.Variate	Variate	Variate,Factor
Factor.OP.Scalar	Variate	Variate,Factor
Factor.OP.Variate	Variate	Variate,Factor
Factor.OP.Factor	Variate	Variate,Factor
Diagonal.OP.Scalar	Diagonal	Diagonal,Symmetric
Diagonal.OP.Variate	*invalid*	

Diagonal.OP.Factor	*invalid*	–
Diagonal.OP.Diagonal	Diagonal	Diagonal,Symmetric,Matrix
Symmetric.OP.Scalar	Symmetric	Diagonal,Symmetric,Matrix
Symmetric.OP.Variate	*invalid*	–
Symmetric.OP.Factor	*invalid*	–
Symmetric.OP.Diagonal	Symmetric	Diagonal,Symmetric,Matrix
Symmetric.OP.Symmetric	Symmetric	Diagonal,Symmetric,Matrix
Matrix.OP.Scalar	Matrix	Matrix
Matrix.OP.Variate	Matrix	Matrix,Variate
Matrix.OP.Factor	Matrix	Matrix,Variate
Matrix.OP.Diagonal	Matrix	Diagonal,Symmetric,Matrix
Matrix.OP.Symmetric	Matrix	Diagonal,Symmetric,Matrix
Matrix.OP.Matrix	Matrix	Matrix
Table.OP.Scalar	Table	Table
Table.OP.Variate	*invalid*	–
Table.OP.Factor	*invalid*	–
Table.OP.Diagonal	*invalid*	–
Table.OP.Symmetric	*invalid*	–
Table.OP.Matrix	*invalid*	–
Table.OP.Table	Table	Table

The second column indicates the type of structure resulting from the operation. In the last rule, which is Table.OP.Table, the classifying set of the intermediate table is the union of the two classifying sets. For example:

```
FACTOR [LEVELS=2] Fa,Fb,Fc
TABLE [CLASSIFICATION=Fa,Fb] Ta
TABLE [CLASSIFICATION=Fa,Fc] Tb
CALCULATE Tc = Ta+Tb
```

In any operation that involves both Ta and Tb, the intermediate table will have the classifying set Fa, Fb, and Fc. The classifying set of a table has two forms, described above at the end of 5.1.4. All tables in CALCULATE have their values permuted according to the ordered classifying set. On assignment, the ordered classifying set is transferred to the new table, which Genstat declares implicitly. So the classifying set and ordered classifying set are the same for tables declared implicitly: again see 5.1.4.

The third column of the above table, headed "Assignment", lists the types of structure to which you can assign values from the intermediate structure. Thus there is a fair amount of flexibility. Notice that all the intermediate structures listed above are real-valued; this is the reason why factors cannot be declared implicitly. However, you can assign a variate to a factor, so long as the values of the variate correspond to the values that you defined earlier for the levels variate of the factor (5.1.2).

Most functions produce a result with the same type as their first argument, but there are some exceptions; see 5.2.

5.1 Numerical calculations

5.1.6 Rules for qualified identifiers

Qualified identifiers were introduced in 1.5.2, together with the rules for expanding them into lists. The rules for their use are similar to the rules for the arguments of the ELEMENTS function (5.2.7). How many qualifiers a structure can have is determined by its dimensionality. The dimensionality of scalars is defined to be zero, and so they cannot be qualified. Tables have varying numbers of dimensions, up to nine, but in Release 1 of Genstat 5 cannot be qualified. The dimensionalities of the stuctures that can be qualified are as follows.

    1: variate, text, factor, diagonal matrix, and symmetric matrix.
    2: matrix and symmetric matrix.

Notice that a symmetric matrix has a dimensionality of either one or two, and so can be qualified in two ways; these are described below.

When you use vector structures in qualified identifiers, you can have different subsets from the contributing vectors; but for the calculation to work, the number of values contributed from each must be the same: see lines 5 to 6 of the next example. Genstat then ignores any restrictions on the vectors; in fact qualified identifiers provide an alternative way of specifying subsets of vectors. The example illustrates how to use qualifications with variates, texts, and a factor. In each case the qualified vector is a vector with fewer values than the original, but of the same type: for example, Ta$[!(1,3,5)] is a text with three values instead of six.

EXAMPLE 5.1.6a

```
  1  VARIATE [NVALUES=5] Va; VALUES=!(1...5)
  2  TEXT [NVALUES=6] Ta,Tb; VALUES=!T(a,b,c,d,e,f),!T(a,a,c,c,f,f)
  3  FACTOR [NVALUES=8; LEVELS=3] Fa; VALUES=!(1,3,2,3,1,2,3,1)
  4  VARIATE [VALUES=12(0)] Vb
  5  CALCULATE Vb$[!(3,6,10)] = Va$[!(1,2,5)] * \
  6     (Ta$[!(1,3,5)] .EQS. Tb$[!(2,4,6)]) + Fa$[!(5,7,2)]
  7  PRINT Vb; DECIMALS=0

        Vb
         0
         0
         2
         0
         0
         5
         0
         0
         0
         3
         0
         0
```

When you have a qualified diagonal matrix, the subset of values is itself a diagonal matrix. Similarly a symmetric matrix, qualified by a single list, is also a symmetric matrix: the qualifier indicates which rows and columns are to be included.

For example:

EXAMPLE 5.1.6b

```
 1  SYMMETRICMATRIX [ROWS=4] Sma; VALUES=!(1...10)
 2  & [ROWS=3] Smb
 3  CALCULATE Smb = Sma$[!(1,4,2)]
 4  PRINT Sma,Smb; FIELDWIDTH=6; DECIMALS=0
```

```
        Sma

   1     1
   2     2     3
   3     4     5     6
   4     7     8     9    10

         1     2     3     4


        Smb

   1     1
   2     7    10
   3     2     8     3

         1     2     3
```

Symmetric matrices can have two qualifiers, in which case Genstat treats the result as a rectangular matrix. Rectangular matrices must have two qualifiers. In this example, the values of the rectangular matrix Mb are formed from the addition of the values in rows 1 and 4, and columns 2 and 3, of the symmetric matrix Sma to the values in rows 1 and 4, and columns 3 and 4, of the matrix Ma.

EXAMPLE 5.1.6c

```
 1  MATRIX [ROWS=4; COLUMNS=5] Ma; VALUES=!(1...20)
 2  & [ROWS=2; COLUMNS=2] Mb
 3  SYMMETRICMATRIX [ROWS=4] Sma; VALUES=!(1...10)
 4  CALCULATE Mb = Sma$[!(1,4);!(2,3)] + Ma$[!(1,4);!(3,4)]
 5  PRINT Ma,Mb; FIELDWIDTH=6; DECIMALS=0
```

```
             Ma
              1     2     3     4     5

        1     1     2     3     4     5
        2     6     7     8     9    10
        3    11    12    13    14    15
        4    16    17    18    19    20


             Mb
              1     2

        1     5     8
        2    26    28
```

5.1 Numerical calculations

All the examples above show how to form vectors and matrices that have fewer values than the original: that is, the vectors and matrices take their values from subsets of the source structures. You can form larger vectors and matrices by using repeated values in the qualifier set. Here the matrix Mc, with four rows and three columns is formed from the two-by-two matrix Mb.

EXAMPLE 5.1.6d

```
  1  MATRIX [ROWS=2; COLUMNS=2; VALUES=5,7,6,2] Mb
  2  MATRIX [ROWS=4; COLUMNS=3] Mc
  3  VARIATE [NVALUES=4; VALUES=1,2,2,1] Va
  4  & [NVALUES=3; VALUES=1,1,2] Vb
  5  CALCULATE Mc = Mb$[Va; Vb]
  6  PRINT Mc; FIELDWIDTH=6; DECIMALS=0

             Mc
              1     2     3

       1      5     5     7
       2      6     6     2
       3      6     6     2
       4      5     5     7
```

Instead of using variates to qualify the structures, you can use any numerical structure, and these structures can be qualified too. Genstat treats any structure used as a qualifier as a one-dimensional list of values. So you can build very complicated qualifications in this way. The only limitation is that the set of values of the qualifiers must form a valid address list for the parent structure. In this example, the complicated qualification reduces to assigning the value 3 to the element in row 4 and column 3 of the matrix Ma.

EXAMPLE 5.1.6e

```
  1  VARIATE [NVALUES=6] Va; VALUES=!(1,4,3,2,4,3)
  2  MATRIX [ROWS=4; COLUMNS=6] Ma; VALUES=!(1...24)
  3  CALCULATE Ma$[Va$[2]; Ma$[1; 3]] = 3
  4  PRINT Ma; FIELDWIDTH=6; DECIMALS=0

             Ma
              1     2     3     4     5     6

       1      1     2     3     4     5     6
       2      7     8     9    10    11    12
       3     13    14    15    16    17    18
       4     19    20     3    22    23    24
```

You can use text to qualify structures, since it can label the rows and columns of matrices and the units of vectors. In this example the matrix Mb is formed with numbers of rows and columns equal to the number of values (that is lines) of the texts Tsa and Tsb.

EXAMPLE 5.1.6f

```
 1   TEXT [NVALUES=6] Ta; VALUES=!T(a,b,c,d,e,f)
 2   & [NVALUES=4] Tb; VALUES=!T(g,h,i,j)
 3   & [NVALUES=3] Tsa; VALUES=!T(d,a,f)
 4   & Tsb; VALUES=!T(i,h,j)
 5   MATRIX [ROWS=Ta; COLUMNS=Tb] Ma; VALUES=!(1...24)
 6   CALCULATE Mb = Ma$[Tsa; Tsb]
 7   PRINT Ma,Mb; FIELDWIDTH=6; DECIMALS=0
```

	Ma			
Tb	g	h	i	j
Ta				
a	1	2	3	4
b	5	6	7	8
c	9	10	11	12
d	13	14	15	16
e	17	18	19	20
f	21	22	23	24

	Mb		
	1	2	3
1	15	14	16
2	3	2	4
3	23	22	24

You can put in a missing identifier (*) to mean the complete set of elements from the dimension concerned. The next example shows how to transfer the values from columns 1 and 2 of the matrix Ma into the variates Vc1 and Vc2 respectively. Using qualified identifiers for transferring rows and columns of matrices to and from variates is more straightforward than using the EQUATE directive (5.3.1). The missing identifier (*) in the first qualifier for Ma indicates that Genstat is to take all the rows.

EXAMPLE 5.1.6g

```
 1   MATRIX [ROWS=5; COLUMNS=4] Ma; VALUES=!(1...20)
 2   VARIATE [NVALUES=5] Vc1,Vc2
 3   CALCULATE Vc1,Vc2 = Ma$[*; 1,2]
 4   PRINT Ma; FIELDWIDTH=6; DECIMALS=0
```

	Ma			
	1	2	3	4
1	1	2	3	4
2	5	6	7	8
3	9	10	11	12
4	13	14	15	16
5	17	18	19	20

```
  5   & Vc1,Vc2; FIELDWIDTH=6; DECIMALS=0
      Vc1   Vc2
        1     2
        5     6
        9    10
       13    14
       17    18
```

Single values from qualified identifiers have the same type as their parent. They are not treated as scalars in a CALCULATE statement. If you want them to be used as scalars, you can include an embedded assignment in the expression. For example, to multiply the variate Va by the value in row 2 and column 1 of the matrix Ma, you should put:

SCALAR Sca
CALCULATE Vb = Va * (Sca = Ma$[2;1])

If you tried to use the expression Va*Ma$[2;1], you would get an error message, since Genstat would object to multiplying the variate Va by the one-by-one matrix Ma$[2;1].

5.1.7 The INTERPOLATE directive

INTERPOLATE

Interpolates values at intermediate points.

Options

CURVE	= string	Type of curve to be fitted to calculate the interpolated value (linear, cubic); default l
METHOD	= string	Type of interpolation required (interval, value, missing): for METHOD=v, values are interpolated for each point in the NEWINTERVAL variate and stored in the NEWVALUE variate; for METHOD=i, points are estimated in the NEWINTERVAL variate for the observations in the NEWVALUE variate; while for METHOD=m, the NEWVALUE and NEWINTERVAL lists are irrelevant, INTERPOLATE now interpolates for missing values in the OLDVALUE and

		OLDINTERVAL variates (except those missing in both variates). Default i
Parameters		
OLDVALUES	= variates	Observations from which interpolation is to be done
NEWVALUES	= variates	Results of each interpolation
OLDINTERVALS	= variates	Points at which each set of OLDVALUES was observed
NEWINTERVALS	= variates	Points for each set of NEWVALUES

If you have a set of pairs of observations (x, y), you can use interpolation to estimate either a value y for a value x that need not be in the set, or a value x for a value y that likewise need not be in the set. The simplest way to do it is by joining successive pairs of observations by straight lines and reading off the appropriate values in between: then the two cases are called *linear interpolation* (getting y from x) and *inverse linear interpolation* (getting x from y). Genstat can alternatively join the points by cubic functions instead of straight lines. Genstat uses the term *values* to describe the set of y-values and *intervals* for the set of x-values, no matter whether you are doing direct or inverse interpolation.

Genstat does the interpolation for each parallel set of variates in the parameter lists. Each variate in the OLDINTERVALS list specifies the x-values of a set of observed points; the corresponding variate in the OLDVALUES list specifies the corresponding y-values. The variates in the NEWINTERVALS and NEWVALUES lists are for the x-values and y-values of the interpolated points. The values of the variates in the OLDVALUES and OLDINTERVALS lists must be monotonically increasing or decreasing; if they are not, Genstat reports an error.

If you set METHOD = value, Genstat does ordinary interpolation. You use the NEWINTERVALS variate to specify the x-values for which you require interpolated y-values. Genstat calculates the y-values and stores them in the corresponding NEWVALUES variate; this variate will be declared implicitly if you have not declared it already. In the next example, wheat plants have been sampled on five occasions and their growth stage (Zadoks) assessed. The program interpolates values, which it stores in variate Nzad, to estimate the growth stage that the plant has reached after 50, 100, and 150 days.

EXAMPLE 5.1.7a

```
1   VARIATE [NVALUES=6] Zadoks,Days; \
2        VALUES=!(0,15,23,35,65,95),!(0,50,84,119,147,182)
3   & [NVALUES=3] Nzadoks,Ndays; VALUES=!(25,50,75),!(50,100,150)
4   INTERPOLATE [METHOD=value] Zadoks; NEWVALUES=Nzad; \
```

5.1 Numerical calculations

```
  5    OLDINTERVALS=Days; NEWINTERVALS=Ndays
  6  PRINT Ndays,Nzad; FIELDWIDTH=8; DECIMALS=2

   Ndays    Nzad
   50.00   15.00
  100.00   28.49
  150.00   67.57
```

Similarly, if you set METHOD=interval, Genstat does inverse interpolation. You must then specify the y-values in the NEWVALUES variate. Genstat calculates the x-values and stores them in the corresponding NEWINTERVALS variate which, again, will be declared implicitly if necessary. Inverse interpolation is the default. This example uses the same data as above, but does inverse linear interpolation to estimate how long after planting we have to wait for the plant to reach growth stages 25, 50, and 75 Zadoks.

EXAMPLE 5.1.7b

```
  7  INTERPOLATE [METHOD=interval] Zadoks; NEWVALUES=Nzadoks; \
  8      OLDINTERVALS=Days; NEWINTERVALS=Nd
  9  PRINT Nzadoks,Nd; FIELDWIDTH=8; DECIMALS=2

 Nzadoks      Nd
   25.00   89.83
   50.00  133.00
   75.00  158.67
```

If you set METHOD=missing, Genstat ignores the NEWVALUES and NEWINTERVALS parameters; it estimates values for x or y when the other is missing, placing the results in the previously missing position of the OLDVALUES or the OLDINTERVALS variates. Ordinary interpolation is used when the missing value is in y, and inverse interpolation when it is in x. If both the x-value and the y-value are missing for a particular unit, no values can be interpolated for it, and it remains missing. To do linear interpolation requires that both the x-value and the y-value should be non-missing for the point on each side of the unit with the missing value. For cubic interpolation, there must be two non-missing points on each side of the unit. In the next example the missing value in Yval at unit 2 is replaced with the interpolated value 2.85, while the one at unit 4 remains missing because the x-value is missing there too. The missing value at unit 9 of Xint is replaced by 5.96, while the one at unit 4 again stays missing. Notice also that Genstat ignores the NEWINTERVALS setting Xnewint.

EXAMPLE 5.1.7c

```
  1   VARIATE [NVALUES=9] Yval,Xint; \
  2       VALUES=!(2.5,*,3.2,*,4.3,4.8,7.2,7.3,8.7),!(1,2,3,*,4,5,*,6,7)
  3   PRINT Xint,Yval; FIELDWIDTH=8; DECIMALS=2

    Xint    Yval
    1.00    2.50
    2.00       *
    3.00    3.20
       *       *
    4.00    4.30
    5.00    4.80
       *    7.20
    6.00    7.30
    7.00    8.70

  4   INTERPOLATE [METHOD=missing] OLDVALUE=Yval; OLDINTERVAL=Xint ;\
  5       NEWINTERVAL=Xnewint
  6   PRINT Xint,Yval; FIELDWIDTH=8; DECIMALS=2

    Xint    Yval
    1.00    2.50
    2.00    2.85
    3.00    3.20
       *       *
    4.00    4.30
    5.00    4.80
    5.96    7.20
    6.00    7.30
    7.00    8.70
```

The CURVE option has two settings, 'linear' and 'cubic'. By default, CURVE = linear, and successive pairs of observations are connected by straight-line segments for linear, or inverse-linear, interpolation. For cubic interpolation you set CURVE = cubic; there must then be at least four values in each of the OLDVALUES and OLDINTERVALS variates.

5.2 Functions for use in expressions

This section lists and describes the functions that you can use in expressions. The general form is:

CALCULATE y = LOG10(x)

Here LOG10 is the name of a function, and the identifier enclosed in brackets is its argument. Throughout this section we use lower case for identifiers that are arguments or results of functions, such as x and y above, to contrast with the upper-case conventionally used in this manual for function names, such as LOG10.

The argument of a function can be a list of identifiers, or even an expression. Some functions may need two arguments, in which case the arguments are separated by a semicolon (;). For example:

CALCULATE w = SORT(x; y + z)

5.2 Functions for use in expressions

(For an explanation of SORT, see below.) Genstat checks that you have given the correct number of arguments. With some functions, you do not need to set the second and subsequent arguments; in that case, you should omit the semicolons that would follow the last argument that you do use.

The functions here are divided into classes as follows: general and mathematical functions (5.2.1), scalar functions (5.2.2), variate functions (5.2.3), matrix functions (5.2.4), table functions (5.2.5), dummy functions (5.2.6), elements of structures (5.2.7), and statistical functions (5.2.8). They are described in alphabetical order within each subsection. At the beginning of each class we set out the valid types of argument for each function, and the type of the result. We give synonyms, and abbreviations for the function names where these have fewer than four letters: for example, the matrix function INVERSE has the two abbreviations INV and I. You can abbreviate any function to four letters (1.5.1): for example, LOG10 could be written as LOG1 (but that would be unwise, since your program would lose clarity).

5.2.1 General and mathematical functions

In this subsection, x and y represent identifiers, or lists of identifiers, of any structures containing numerical data: that is, scalars, variates, factors, tables, matrices, diagonal matrices or symmetric matrices; s represents a scalar, f a factor and v a variate. Where x and y occur together as arguments they must be of the same type. Apart from NEWLEVELS, which produces a variate from a factor, the result of any of these functions has the same type as that of the first argument.

ABS(x) gives the absolute value of x: |x|.

ARCCOS(x) gives the inverse cosine of x ($-1 \leqslant x \leqslant 1$); the result is in radians.

ARCSIN(x) gives the inverse sine of x ($-1 \leqslant x \leqslant 1$); the result is in radians.

CIRCULATE(x; s) treats x as a circular stack and shifts the values of x round the stack according to the value and sign of s. For example, if x contains 1,2,3,4,5, and s is -2, then the result is 3,4,5,1,2; if s were 2, the result would be 4,5,1,2,3. If you omit the second operand, CIRCULATE moves the values by one place to the right: that is, $s=1$.

COS(x) gives the cosine of x, for x in radians.

CUMULATE(x) or CUM(x) forms the cumulative sum of the values of x: for example, the result from x with values 1,5,4 is 1,6,10. If the operand is a scalar, the result is the value of the scalar.

DIFFERENCE(x; s) forms the differences between consecutive elements of x: that is, the ith element of the result is $x_i - x_{i-s}$ If you omit the second operand, first differences are formed ($s=1$). If $i-s<1$ or $i-s>n$, where n is the number of values of x, the ith element is set to missing.

EXP(x) gives the exponential function of x: e^x.

INTEGER(x) or INT(x) gives the integer part of x: [x].

LOG(x) gives the natural logarithm of x (x>0).

LOG10(x) gives the logarithm to base 10 of x (x>0).

MVREPLACE(x; y) replaces missing values in x with corresponding values from y. Elements with missing values in both x and y produce a warning message.

NEWLEVELS(f; x) forms a variate from the factor f; the variate x contains values to correspond to the levels, and should be of the same length as the number of levels of the factor. The result of this function is a variate of the same length as f. For an example see 5.1.2.

REVERSE(x) reverses the values of x: for example, the result from x with values 1,2,3 is 3,2,1.

ROUND(x) rounds to nearest integer.

SHIFT(x; s) shifts the values of x by s places (to the right or left according to the sign of s). This is not a circular shift, and so some positions lose values; these are replaced with missing values. That is, the ith element of the result is the value that was in element $i-s$ unless $i-s \leqslant 0$.

SIN(x) gives the sine of x, for x in radians.

SORT(x; y) sorts the elements of x into the order that would put the values of y into ascending order; the values of y are left unchanged. If the second argument is omitted, the values of x are sorted into ascending order. x can be the same structure as y. See below for an example.

SQRT(x) gives the square root of x (x≥0).

The next example illustrates the functions DIFFERENCE, INTEGER, ROUND, MVREPLACE, SIN, and SORT. In the example of SORT, Genstat sorts the missing values in variate Va to the beginning of the array; the sorted order within the missing values is completely arbitrary. Tied values, too, are sorted arbitrarily, although in this example the tied values are by chance listed in their order of occurrence in the variate Va.

EXAMPLE 5.2.1

```
  1  VARIATE [VALUES=-0.4,4.1,8.4,*,-1.6,5.7,-2.3] Va
  2  CALCULATE Vb = DIFFERENCE(Va; 2)
  3  PRINT Va,Vb; FIELDWIDTH=6; DECIMALS=1

     Va     Vb
   -0.4      *
    4.1      *
    8.4    8.8
      *      *
   -1.6  -10.0
    5.7      *
   -2.3   -0.7

  4  CALCULATE Iva = INTEGER(Va)
  5  &         Rva = ROUND(Va)
  6  PRINT Va,Iva,Rva; FIELDWIDTH=6; DECIMALS=1
```

```
      Va    Iva    Rva
     -0.4   0.0    0.0
      4.1   4.0    4.0
      8.4   8.0    8.0
        *     *      *
     -1.6  -1.0   -2.0
      5.7   5.0    6.0
     -2.3  -2.0   -2.0
   7  VARIATE [VALUES=1,2,3,27.3,5,6,7] Vb
   8  CALCULATE Vc = MVREPLACE(Va; Vb)
   9  PRINT Vc; DECIMALS=2

           Vc
        -0.40
         4.10
         8.40
        27.30
        -1.60
         5.70
        -2.30

  10  CALCULATE Ve = SIN(Vc)
  11  PRINT Ve; FIELDWIDTH=8; DECIMALS=3

      Ve
  -0.389
  -0.818
   0.855
   0.827
  -1.000
  -0.551
  -0.746

  12  VARIATE [VALUES=3,1,*,*,1,4,7,4,*] Vsa
  13  &       [VALUES=1...9] Vsb
  14  CALCULATE Vsc = SORT(Vsb; Vsa)
  15  PRINT Vsc; FIELDWIDTH=6; DECIMALS=0

 Vsc
   9
   4
   3
   2
   5
   1
   6
   8
   7
```

5.2.2 Scalar functions

The scalar functions let you do two things. You can find an attribute of the structure in the argument: for example, NVALUES gives the number of values. Or you can form a summary value from the values of the structure: for example, MEAN gives the mean of the non-missing values of the structure.

In this subsection, x can be any numerical structure (scalar, variate, factor, rectangular matrix, symmetric matrix, diagonal matrix, or table), f is a factor, and m is either a rectangular matrix, a symmetric matrix, or a diagonal matrix. All the functions produce a scalar result from each structure in the argument list; all except NMV and NVALUES exclude missing values in the structure. Thus, the function MEAN is equivalent to SUM divided by NOBSERVATIONS, and the function NOBSERVATIONS is equivalent to NVALUES minus NMV.

MAXIMUM(x) or MAX(x) finds the maximum of the values of x.

MEAN(x) gives the mean of the values of x.

MEDIAN(x) or MED(x) finds the median of the values of x.

MINIMUM(x) or MIN(x) finds the minimum of the values of x.

NCOLUMNS(m) gives the number of columns of matrix m.

NLEVELS(f) gives the number of levels of factor f.

NMV(x) counts the number of missing values in x.

NOBSERVATIONS(x) counts the number of observations (non-missing values) in x.

NROWS(m) gives the number of rows of matrix m.

NVALUES(x) gives the number of values, including missing values, of x (the length of x).

SUM(x) or TOTAL(x) gives the sum of the values in x.

VARIANCE(x) or VAR(x) gives the variance of the values in x (the divisor being the number of non-missing values in x, minus 1).

For example:

EXAMPLE 5.2.2

```
  1   VARIATE [VALUES=8,2,16,4,1,10,*,30] Va
  2   " Med, Mn, Tot, Obs, and Nv are declared implicitly (as scalars). "
  3   CALCULATE Med = MEDIAN(Va)
  4   &  Mn  = MEAN(Va)
  5   &  Tot = SUM(Va)
  6   &  Obs = NOBSERVATIONS(Va)
  7   &  Nv  = NVALUES(Va)
  8   PRINT Med,Mn,Tot,Obs,Nv; FIELDWIDTH=8; DECIMALS=2

     Med       Mn      Tot      Obs       Nv
    8.00    10.14    71.00     7.00     8.00

  9   FACTOR [LEVELS=!(1,2,4,8)] Ff
 10   CALCULATE Nl = NLEVELS(Ff)
 11   PRINT Nl; FIELDWIDTH=6; DECIMALS=1

     Nl
    4.0
```

5.2.3 Variate functions

Variate functions produce summaries across a set of variates. They each have a single argument, which is a pointer to the set of variates to be summarized. The variates

5.2 Functions for use in expressions

must all be of the same length. If any of them is restricted, that restriction is applied to all of them; if several are restricted, each restriction must be to the same set of units. The result of each function is a variate of length equal to the length of the variates in the pointer. For example, if p points to the variates X1, X2, and X3, each of length n, VMEANS(p) produces a variate of length n, whose ith unit contains the mean of the values in the unit i of X1, X2, and X3.

All the functions except VNMV and VNVALUES exclude missing values. Thus, the function VMEANS is equivalent to VSUMS divided by VNOBSERVATIONS, and the function VNOBSERVATIONS is equivalent to VNVALUES minus VNMV.

VMAXIMA(p) finds the maximum of the values in each unit over the variates in pointer p.

VMEDIANS(p) finds the median of the values in each unit over the variates in pointer p.

VMINIMA(p) finds the minimum of the values in each unit over the variates in pointer p.

VNMV(p) counts the number of missing values in each unit over the variates in pointer p.

VNOBSERVATIONS(p) counts the number of observations (non-missing values) in each unit over the variates in pointer p.

VNVALUES(p) gives the total number of values in each unit over the variates in pointer p: that is the number of variates in p.

VSUMS(p) or VTOTAL(p) gives the sum of the non-missing values in each unit over the variates in pointer p.

VVARIANCES(p) gives the variance of the non-missing values in each unit over the variates in pointer p.

EXAMPLE 5.2.3

```
 1   VARIATE [NVALUES=6] X,Y,Z; \
 2       VALUES=!(28,*,18,26,*,17),!(12,27,*,34,*,15),!(17,25,3(*),20)
 3     & Min,Mean,Max,Obs,Nval,Tot
 4   POINTER [VALUES=X,Y,Z] P
 5   CALCULATE Min  = VMINIMA(P)
 6         & Mean = VMEANS(P)
 7         & Max  = VMAXIMA(P)
 8         & Obs  = VNOBSERVATIONS(P)
 9         & Nval = VNVALUES(P)
10         & Tot  = VTOTALS(P)
11   PRINT X,Y,Z,Min,Mean,Max,Obs,Nval,Tot; FIELDWIDTH=8; DECIMALS=1
```

X	Y	Z	Min	Mean	Max	Obs	Nval	Tot
28.0	12.0	17.0	12.0	19.0	28.0	3.0	3.0	57.0
*	27.0	25.0	25.0	26.0	27.0	2.0	3.0	52.0
18.0	*	*	18.0	18.0	18.0	1.0	3.0	18.0

```
26.0      34.0       *       26.0     30.0     34.0      2.0      3.0     60.0
  *         *         *         *        *        *      0.0      3.0       *
17.0      15.0     20.0      15.0     17.3     20.0      3.0      3.0     52.0
```

5.2.4 Matrix functions

These functions operate on the various types of matrix available in Genstat. There is no general rule defining the type of the resulting structure. For some of the functions you can specify a variate, which is treated as a rectangular matrix with one column: any restriction on the variate is then ignored. (Remember that matrices cannot be restricted.) A *matrix* is a rectangular, symmetric or diagonal matrix structure; a *square matrix* is a rectangular matrix with the same number of rows as of columns.

CORRMAT(x) forms a correlation matrix from a symmetric matrix x that contains sums of squares and products: the values of the resulting symmetric matrix c are formed by $c_{ij} = x_{ij} / \sqrt{(x_{ii} \times x_{jj})}$.

CHOLESKI(x) forms the Choleski decomposition of a symmetric matrix x; this produces a square matrix L such that x = LL' and such that upper off-diagonal elements are zero. The symmetric matrix x must be positive semi-definite.

DETERMINANT(x) or DET(x) or D(x) forms the determinant of a symmetric matrix or a square matrix; the result is a scalar. Genstat uses the decomposition x = LU, and the determinant is defined to be $\prod \{l_{ii} \times u_{ii}\}$.

INVERSE(x) or INV(x) or I(x) forms the inverse of a non-singular symmetric matrix, or a square matrix; the result is a square matrix or a symmetric matrix, according to the type of x. For a square matrix, Genstat uses Crout's method by forming the lower and upper triangular decomposition of the matrix, x = LU, and inverting L and U separately. Genstat uses the equivalent decomposition (Choleski) for symmetric matrices, which must be positive semi-definite. To form the generalized inverses of singular rectangular matrices, you should use the directives SVD (5.7.1) or FLRV (5.7.2). If you want the generalized inverse of a diagonal matrix, D, you can put:

CALCULATE [ZDZ=zero] Dinv = D/D/D

LTPRODUCT(x; y) forms the left transposed product of x and y: that is, the matrix product of the transpose of x with y, which can also be written T(x)*+y. The structures x and y can be matrices or variates. The number of rows of x must equal the number of rows of y. The result is a rectangular matrix with number of rows equal to the number of columns of x and number of columns equal to the number of columns of y; but if both x and y are diagonal matrices, the result is also a diagonal matrix.

PRODUCT(x; y) forms the matrix product of x and y; this can also be written x*+y using the operator *+. The structures x and y can be matrices or variates. The number of columns of x must equal the number of rows of y. The result is a rectangular matrix with number of rows equal to the number of rows of x and number of columns equal to the number of columns of y; but if both x and y are diagonal matrices, the result is also a diagonal matrix.

QPRODUCT(x; y) forms the quadratic product of x and y; it can thus be written as x*+y*+ T(x) but the use of QPRODUCT is more efficient. x is a rectangular matrix or a variate, and y is a symmetric matrix or a diagonal matrix or a scalar. The number of columns of x must be the same as the number of rows of y. The result is a symmetric matrix with number of rows equal to the number of rows of x.

RTPRODUCT(x; y) forms the right transposed product of x and y: that is, the matrix product of x with the transpose of y, which can also be written x*+T(y). The structures x and y can be matrices or variates. The number of columns of x must equal the number of columns of y. The result is a rectangular matrix with number of rows equal to the number of rows of x and number of columns equal to the number of rows of y; but if both x and y are diagonal matrices, the result is also a diagonal matrix.

SOLUTION(x:y) solves a set of simultaneous linear equations x*+b=y: that is

$$x_{11} \times b_1 + x_{12} \times b_2 + \ldots + x_{1n} \times b_n = y_1$$
$$\vdots$$
$$x_{n1} \times b_1 + x_{n2} \times b_2 + \ldots + x_{nn} \times b_n = y_n$$

The function thus finds b, as in the alternative expression

   CALCULATE b = PRODUCT(INVERSE(x); y)

but the use of SOLUTION is more efficient and numerically stable than using PRODUCT and INVERSE: x is a square matrix and y is a rectangular matrix or a variate. The number of rows of x must be the same as the number of rows of y. The result is a rectangular matrix with numbers of rows and columns the same as y.

SUBMAT(x) forms sub-triangles or sub-rectangles of a rectangular or symmetric matrix x, whose dimensions must be labelled by pointers. The structure to receive the values must have been declared already, as a rectangular or symmetric matrix according to the type of x, and have each of its dimensions also labelled by a pointer whose values are included in the pointer of the corresponding dimension of x. The correspondence between the values of the pointers that label the resulting matrix and those labelling x determines which rows and columns of x appear in the result. The same effect can be obtained by using the function ELEMENTS with a single list or expression for symmetric matrices, and with two lists for rectangular matrices. Just as with the ELEMENTS function, the resulting matrix can be made larger than x, by specifying repeated identifiers in its pointers.

TRACE(x) forms the trace of matrix x: that is, the sum of its diagonal elements. x can be a square matrix, a diagonal matrix or a symmetric matrix. The result is a scalar.

TRANSPOSE(x) or T(x) forms the transpose of x, where x is a rectangular matrix or a variate. The result is a rectangular matrix.

EXAMPLE 5.2.4

```
 1   SYMMETRICMATRIX [ROWS=4] Sma; \
 2      VALUES=!(36,40,64,65,90,144,80,110,175,225)
 3   MATRIX [ROWS=4; COLUMNS=4] Chsma
 4   CALCULATE   Chsma = CHOLESKI(Sma)
 5   PRINT Chsma; FIELDWIDTH=8; DECIMALS=3
```

```
              Chsma
                  1        2        3        4

          1   6.000    0.000    0.000    0.000
          2   6.667    4.422    0.000    0.000
          3  10.833    4.020    3.237    0.000
          4  13.333    4.774    3.511    3.479
```

```
 6   MATRIX [ROWS=3; COLUMNS=3] Ma; VALUES=!(1,1,2,3,4,5,1,4,2)
 7   & Mainv
 8   CALCULATE Mainv = INVERSE(Ma)
 9   PRINT Ma; FIELDWIDTH=8; DECIMALS=3
```

```
              Ma
                  1        2        3

          1   1.000    1.000    2.000
          2   3.000    4.000    5.000
          3   1.000    4.000    2.000
```

```
10   & Mainv; FIELDWIDTH=8; DECIMALS=3
```

```
              Mainv
                  1        2        3

          1  -4.000    2.000   -1.000
          2  -0.333    0.000    0.333
          3   2.667   -1.000    0.333
```

```
11   MATRIX [ROWS=3; COLUMNS=3] Mx; VALUES=!(1,1,2,3,4,5,1,4,2)
12   & [ROWS=3; COLUMNS=1] My; VALUES=!(4,5,6)
13   & Bxy
14   CALCULATE Bxy = SOLUTION(Mx; My)
15   PRINT Bxy; FIELDWIDTH=8; DECIMALS=3
```

```
              Bxy
                  1

          1 -12.000
          2   0.667
          3   7.667
```

```
16   VARIATE Va,Vb,Vc,Vd,Ve,Vf,Vg,Vh,Vi,Vj
17   POINTER Pa,Pb,Pc,Pd; VALUES=!P(Va,Vb,Vc,Vd,Ve,Vf),!P(Vg,Vh,Vi,Vj), \
18      !P(Vc,Va,Vf,Ve),!P(Vi,Vh,Vg)
19   MATRIX [ROWS=Pa; COLUMNS=Pb] Ma ; VALUES=!(1...24)
20   & [ROWS=Pc; COLUMNS=Pd] Mb
21   CALCULATE Mb = SUBMAT(Ma)
22   PRINT Ma; FIELDWIDTH=8; DECIMALS=1
```

```
                    Ma
              Pb    Vg       Vh       Vi       Vj
              Pa
              Va   1.0      2.0      3.0      4.0
              Vb   5.0      6.0      7.0      8.0
              Vc   9.0     10.0     11.0     12.0
              Vd  13.0     14.0     15.0     16.0
```

```
             Ve    17.0    18.0    19.0    20.0
             Vf    21.0    22.0    23.0    24.0
 23   & Mb; FIELDWIDTH=8; DECIMALS=1
                  Mb
            Pd    Vi      Vh       Vg
            Pc
            Vc    11.0    10.0     9.0
            Va     3.0     2.0     1.0
            Vf    23.0    22.0    21.0
            Ve    19.0    18.0    17.0
```

5.2.5 Table functions

The table functions operate on tables to produce new values for extended or summarized tables; for example,

CALCULATE tr = TMEANS(ta)

takes means of certain of the cells in table ta and puts them in the table tr. If the resulting table, tr above, has already been declared, it must have the same status for margins as the corresponding argument table: ta above. But if tr is left to be declared implicitly, it will be given margins whether or not they occur in ta. Factors occurring in ta but not in tr are collapsed (using the function TMEANS in this example) for all combinations of the levels of the factors common to both tables, and the resulting values are put into tr. Factors occurring in tr but not in ta are treated as described in 5.1.4; that is, all levels of such factors get the same value. Finally, if tr has margins, these are filled in according to the function specified. For example, if tr is classified by factors A and B but ta is classified by A, B and C,

CALCULATE tr = TMEANS(ta)

will put, in each cell of tr, means over the levels of factor C. A further example is given below, after the descriptions of the functions.

TMAXIMA(t) forms margins of maxima for table t.

TMEDIANS(t) forms margins of medians for table t.

TMEANS(t) forms margins of means for table t.

TMINIMA(t) forms margins of minima for table t.

TNOBSERVATIONS(t) forms margins counting the numbers of observations (non-missing values) in table t.

TNMV(t) forms margins counting the numbers of missing values in table t.

TNVALUES(t) forms margins counting the numbers of values, missing or non-missing, in table t.

TSUMS(t) or TTOTALS(t) forms margins of totals for table t.

TVARIANCES(t) forms margins of between-cell variances for table t.

EXAMPLE 5.2.5

```
  1  FACTOR [LEVEL=2] Fa,Fb,Fc
  2  TABLE [CLASSIFICATION=Fa,Fb,Fc; MARGIN=yes] Tabin ; !(1...27)
  3  & [CLASSIFICATION=Fa,Fb; MARGIN=yes] Tabmean
  4  CALCULATE Tabmean = TMEANS(Tabin)
  5  PRINT Tabmean; FIELDWIDTH=6; DECIMALS=1

           Tabmean
      Fb       1      2 Margin
      Fa
      1      1.5    4.5    3.0
      2     10.5   13.5   12.0

   Margin    6.0    9.0    7.5
```

5.2.6 Dummy functions

The function UNSET allows you to check whether a dummy is set; this is useful particularly in procedures (6.3.1) and FOR loops (6.2.1).

UNSET(d) gives a scalar logical value (0 or 1) indicating whether or not the dummy d is set: that is, whether or not d points to another structure.

5.2.7 Elements of structures

The ELEMENTS function has a similar role to qualified identifiers (5.1.6). Two functions, EXPAND and RESTRICTION, are available to derive sets of values from the results of a RESTRICT statement (2.4.1).

ELEMENTS(x; e1; e2) specifies a set of elements of x; e1 and e2 are expressions. As with qualified identifiers, you cannot specify elements of scalars or tables. You cannot use a text in any of the arguments of ELEMENTS. However the ability to specify expressions in the second and third arguments, instead of merely structures, is one way in which the use of ELEMENTS is more powerful that the use of qualified identifiers.

The rules of dimensionality, and the specification of the expressions e1 and e2, which identify the elements in each dimension, are similar to those for qualified identifiers (5.1.6). If x is a variate, a factor, or a diagonal matrix, you do not specify the third argument e2; the type of the result is the same as that of x. You can also omit the third argument if x is a symmetric matrix, in which case the result is also a symmetric matrix; or you can specify both expressions, in which case the result is a rectangular matrix. For rectangular matrices, both e1 and e2 must be specified, and the result is a rectangular matrix. Genstat evaluates each expression and treats the result as a one-dimensional list of values. In the next example the values of the symmetric matrix Smb are taken from the rows of the symmetric matrix Sma indicated by the variate Va.

EXAMPLE 5.2.7a

```
  1  SYMMETRICMATRIX [ROWS=5] Sma; VALUES=!(15...1)
  2  & [ROWS=3] Smb
  3  VARIATE Va; VALUES=!(5,4,2)
  4  CALCULATE Smb = ELEMENTS(Sma; Va)
  5  PRINT Sma,Smb; FIELDWIDTH=5; DECIMALS=0
```

```
     Sma

  1   15
  2   14   13
  3   12   11   10
  4    9    8    7    6
  5    5    4    3    2    1

        1    2    3    4    5

     Smb

  1    1
  2    2    6
  3    4    8   13

        1    2    3
```

ELEMENTS is the only function that you are allowed to put on the left-hand side of an assignment. In the next example the values of the matrix Ma are assigned to the symmetric matrix Sma in the rows and columns indicated by the variates Va and Vb. Note that, since Sma is symmetric, values that Va and Vb point to above the main diagonal are transposed to their corresponding position below the main diagonal.

EXAMPLE 5.2.7b

```
  6  VARIATE Va,Vb; VALUES=!(5,3,1),!(2,5,4)
  7  MATRIX [ROWS=3; COLUMNS=3; VALUES=1...9] Ma
  8  CALCULATE ELEMENTS(Sma; Va; Vb-1) = Ma
  9  PRINT Sma; FIELDWIDTH=4; DECIMALS=0
```

```
     Sma

  1    7
  2   14   13
  3    9   11    6
  4    8    8    5    6
  5    1    4    3    2    1

        1    2    3    4    5
```

EXPAND(x; s) forms a variate of zeroes and ones from the values of x, which Genstat takes to be a list of unit numbers; usually you will have formed x as the save structure from a RESTRICT statement. The second argument, s, is a scalar defining the length of the result; if you omit it, and EXPAND cannot determine the length of the result from its context, Genstat will set up a variate with the length of the units structure (2.2.3).

RESTRICTION(x) forms a variate with ones in the positions corresponding to the set of units to which x is currently restricted; the other units of the result are left unchanged. If this variate is declared implicitly here, it will be restricted in the same way and have the same number of values as x. If you put the RESTRICTION function by itself into the CONDITION parameter of the RESTRICT directive (2.4.1), the restriction on x is passed to all the vectors that you have listed in first parameter list of RESTRICT.

EXAMPLE 5.2.7c

```
   1  VARIATE [VALUES=35,24,27,26,42,57] Age
   2  RESTRICT Age; CONDITION=Age>30; SAVESET=Va
   3  CALCULATE Vb = EXPAND(Va; 8)
   4  PRINT Va,Vb; FIELDWIDTH=6; DECIMALS=0

           Va     1     5     6

           Vb     1     0     0     0     1     1     0     0

   5  VARIATE [VALUES=6(-1)] Rest
   6  CALCULATE Rest = RESTRICTION(Age)
   7  " Cancel the restriction on Age. "
   8  RESTRICT Age
   9  PRINT Age,Rest; FIELDWIDTH=6; DECIMALS=0

  Age  Rest
   35    1
   24   -1
   27   -1
   26   -1
   42    1
   57    1
```

5.2.8 Statistical functions

The statistical functions are the angular transformation (ANGULAR), probabilities from the Normal, Chi-square, and F distributions (NORMAL, CHISQ, FRATIO, FPROB), and the inverses of these probability functions (NED, CED, FED). You can also form values for log-likelihoods of samples from the four distributions binomial, gamma, Normal and Poisson (LLBINOMIAL, LLGAMMA, LLNORMAL, LLPOISSON). And you can generate sets of uniform pseudo-random numbers (URAND). In the descriptions below, %p, p and x are identifiers of any numerical data structure. Any constraints on the possible values of %p, p and x are given with each description. Except for the four log-likelihood functions and the URAND function, the result is a

5.2 Functions for use in expressions

structure of the same type, dimension, and number of values as the structure in the first argument.

The first argument for the log-likelihood functions must be a variate. The second and third arguments can be scalars or variates; if they are variates, they must be of the same length as the variate in the first argument. The meaning of the second and third arguments is given with each description, as well as the form of the expression used to calculate the log-likelihood. For these functions, the result is a scalar.

ANGULAR(%p) or ANG(%p) provides the angular transformation: %p is a percentage with $0 < \%p < 100$. The function forms

$$x = (180/\pi) \times \arcsin(\sqrt{(\%p/100)})$$

and so the result x is in degrees $0 < x < 90$.

CED(p; s) gives the chi-square equivalent deviate for probability p $(0 < p < 1)$, with number of degrees of freedom s: that is, the positive real number x that leaves a proportion p to the left of it under the chi-square curve.

CHISQ(x; s) gives the probability that a random variable with the chi-square distribution, with number of degrees of freedom s, is less than x.

FED(p; s1; s2) gives the F-distribution equivalent deviate for probability p $(0 < p < 1)$, with numbers of degrees of freedom s1 and s2 for the numerator and denominator of a variance ratio respectively: that is, the positive real number x that leaves a proportion p to the left of it under the F-distribution curve.

FRATIO(x; s1; s2) or FPROBABILITY(x; s1; s2) gives the probability that a random variable with the F-distribution, with numbers of degrees of freedom s1 and s2, is less than x.

LLBINOMIAL(x; n; p) or LLB(x; n; p) provides the log-likelihood function for the binomial distribution with sample size n and mean proportion p (n and p are scalars or variates):

$$\sum \{x \times \text{LOG}(n \times p/x) + (n-x) \times \text{LOG}(n \times (1-p)/(n-x))\}$$

LLGAMMA(x; m; d) or LLG(x; m; d) provides the log-likelihood function for the gamma distribution with mean m and index d (m and d are scalars or variates):

$$\sum \{d \times (\text{LOG}(d \times x/m) - x/m) - \text{LOGGAMMA}(d)\}$$

LLNORMAL(x; m; v) or LLN(x; m; v) provides the log-likelihood function for the Normal distribution with mean m and variance v (m and v are scalars or variates):

$$-\tfrac{1}{2} \sum \{\text{LOG}(v) + (x-m) \times (x-m)/v\}$$

LLPOISSON(x; m) or LLP(x; m) provides the log-likelihood function for the Poisson distribution with sample size m (m is a scalar or a variate):

$$\sum \{x \times \text{LOG}(m/x) + x - m\}$$

NED(p) gives the Normal equivalent deviate for probability p $(0 < p < 1)$: that is, the

178   5 Data handling

real number x that leaves a proportion p to the left of it under the standard Normal curve.

NORMAL(x) provides the normal probability integral; that is, the probability that a random variable with the standard Normal distribution is less than x.

URAND(s1; s2) provides a uniform pseudo-random number generator, giving values in the range (0,1): the algorithm is a modified version of that in Wichmann and Hill (1982). The same algorithm is used by RANDOMIZE (5.3.3). s1 is a scalar which specifies the seed for the random numbers. The seed must have a non-zero value on the first occasion that you use URAND in a job; subsequently you can give a zero value to continue the sequence of random numbers. s2 is also a scalar; if you set this, the result is a variate of length equal to the value of the scalar. If you omit s2, the type of the result of URAND is determined from the context of the expression: that is from the type of the structure that is to receive the values that are generated; if the receiving structure has not been declared already, it will be declared implicitly as a variate with the length of the units structure (2.2.3).

This example illustrates the functions LLNORMAL, NED, and URAND:

EXAMPLE 5.2.8

```
  1   VARIATE [VALUES=4.0,-3.5,-1.3,-2.8,1.9,2.5,0.3,-0.8,1.2,0.9] Va
  2   SCALAR Mean,Variance,Loglik; VALUE=0.6,1.9,*
  3   CALCULATE Loglik = LLNORMAL(Va; Mean; Variance)
  4   PRINT Loglik; FIELDWIDTH=9; DECIMALS=4

    Loglik
  -16.7198

  5   VARIATE Vb; VALUES=!(0.1,0.45,*,0.2,0.83,-0.3,0.95)
  6   & Vc
  7   " There is an invalid value in unit 6; this is
 -8     given a missing value and a warning is printed. "
  9   CALCULATE Vc = NED(Vb)

******* Warning (Code CA 7). Statement 1 on Line 9
Command: CALCULATE Vc = NED(Vb)

Invalid value for argument of function
The first argument of the NED    function in unit 6 has the value    -0.3000

 10   PRINT Vb,Vc; FIELDWIDTH=8,10; DECIMALS=2,3

            Vb        Vc
          0.10    -1.282
          0.45    -0.126
             *         *
          0.20    -0.842
          0.83     0.954
         -0.30         *
          0.95     1.645

 11   SCALAR Seed ; 849153
 12   CALCULATE Vva = URAND(Seed; 6) * 10
 13   & Vvb = URAND(0; 4)
```

```
14  PRINT Vva,Vvb; FIELDWIDTH=8; DECIMALS=3

    Vva     2.349   9.668   0.560   1.379   6.632   3.453
    Vvb     0.477   0.042   0.532   0.050
```

5.3 Transferring and manipulating values

Four things are described in this section: how to transfer values from one set of data structures to another (5.3.1), how to re-order the units of vectors according to the values of an index vector (5.3.2), how to re-order randomly (5.3.3), and how to combine together slices of multi-way structures (5.3.4).

5.3.1 The EQUATE directive

EQUATE lets you copy values from one set of data structures to another. For example, you may wish to copy the values from a one-way table into a variate, or from a matrix into a set of variates (one variate for each row, or for each column), or the other way round, from variates into a matrix.

EQUATE

Transfers data between structures of different sizes or types (but the same modes i.e. numerical or text) or where transfer is not from single structure to single structure.

Options

OLDFORMAT	= variate	Format for values of OLDSTRUCTURES; within the variate, a positive value N means take N values, $-N$ means skip N values and a missing value means skip to the next structure; default * i.e. take all the values in turn
NEWFORMAT	= variate	Format for values of NEWSTRUCTURES; within the variate, a positive value N means fill the next N positions, $-N$ means skip N positions and a missing value means skip to the next structure; default * i.e. fill all the positions in turn

180   5   Data handling

Parameters		
OLDSTRUCTURES	= identifiers	Structures whose values are to be transferred; if values of several structures are to be transferred to one item in the NEWSTRUCTURES list, they must be placed in a pointer
NEWSTRUCTURES	= identifiers	Structures to take each set of transferred values; if several structures are to receive values from one item in the OLDSTRUCTURES list, they must be placed in a pointer

The general idea is that the values in the structures in the OLDSTRUCTURES list are copied into those in the NEWSTRUCTURES list. Each item in the OLDSTRUCTURES list specifies a single data structure, or a single set of data structures, containing the values to be transferred. A single structure can be a factor, or a text, or any one of the structures that contain numbers (scalar, variate, rectangular matrix, diagonal matrix, symmetric matrix, or table). If you want to give a set of structures you must put them into a pointer. All the structures in the set must contain the same kind of values: that is, they must all be texts, or all factors, or must all contain numbers (but they need not all be the same kinds of numerical structure: they could, for example, be a mixture of variates and matrices).

The corresponding entry in the NEWSTRUCTURES list indicates where the transferred data are to be placed. It is either a single structure or a pointer to a set of structures; the structures must be of a type suitable to store the values to be transferred.

Now an example: information about the employees of a firm has been typed in series in two separate sections, and the statement in lines 19 and 20 copies them into one; for each employee there are three pieces of information – name, grade, and hours.

EXAMPLE 5.3.1a

```
  1   OPEN 'EMPLOYEE.DAT'; CHANNEL=2
  2   " Read values for the first 6 employees,
 -3      in series, into Name1, Grade1 and Hours1."
  4   TEXT [NVALUES=6] Name1
  5   FACTOR [NVALUES=6; LEVELS=3] Grade1
  6   VARIATE [NVALUES=6] Hours1
  7   READ [PRINT=data,errors; CHANNEL=2; SERIAL=yes] Name1,Grade1,Hours1

      1   Clarke Innes Adams Jones Day Grey :
      2   2 1 2 1 1 3 :
      3   45 51 40 46 44 40 :

  8   " Read values for the final 4 employees,
 -9      in series, into Name2, Grade2 and Hours2."
```

5.3 Transferring and manipulating values

```
 10    TEXT [NVALUES=4] Name2
 11    FACTOR [NVALUES=4; LEVELS=3] Grade2
 12    VARIATE [NVALUES=4] Hours2
 13    READ [PRINT=data,errors; CHANNEL=2; SERIAL=yes] Name2,Grade2,Hours2

   4    Edwards Baker Hill Foster :
   5    2 2 3 1 :
   6    47 42 40 41 :

 14    " Use EQUATE to put information about all the employees
-15      into single vectors Name, Grade and Hours."
 16    TEXT [NVALUES=10] Name
 17    FACTOR [NVALUES=10; LEVELS=3] Grade
 18    VARIATE [NVALUES=10] Hours
 19    EQUATE !P(Name1,Name2),!P(Grade1,Grade2),!P(Hours1,Hours2); \
 20       NEWSTRUCTURES=Name,Grade,Hours
 21    PRINT Name,Grade,Hours

        Name       Grade       Hours
       Clarke        2         45.00
        Innes        1         51.00
        Adams        2         40.00
        Jones        1         46.00
          Day        1         44.00
         Grey        3         40.00
      Edwards        2         47.00
        Baker        2         42.00
         Hill        3         40.00
       Foster        1         41.00
```

Except with a format (see below) Genstat ignores where each structure within a set from the OLDSTRUCTURES list ends and another one begins: that is, it treats the set as being a concatenated list of values. Similarly, it treats the structures in each NEWSTRUCTURES set as an unstructured list of positions that are to receive values. The old values are repeated as often as is necessary to traverse all the new positions. The next example forms a matrix M with repeated and alternating rows taken from variates R1 and R2.

EXAMPLE 5.3.1b

```
  1   VARIATE [VALUES=1...6] R1
  2      & [VALUES=101...106] R2
  3   " Form a matrix M whose rows are R1, R2, R1 and R2."
  4   MATRIX [ROWS=4; COLUMNS=6] M
  5   EQUATE !P(R1,R2); NEWSTRUCTURES=M
  6   PRINT M; FIELDWIDTH=6; DECIMALS=0

                    M
                    1     2     3     4     5     6

             1      1     2     3     4     5     6
             2    101   102   103   104   105   106
             3      1     2     3     4     5     6
             4    101   102   103   104   105   106
```

You can use the OLDFORMAT and NEWFORMAT options to control how the old values and new positions are traversed. The setting for each of these is a variate whose values are interpreted as follows:

(a) a positive integer n means take the next n values (OLDFORMAT) or fill the next n positions (NEWFORMAT);

(b) a negative integer $-n$ means skip the next n values or positions;

(c) a missing value ∗ means skip to the end of the structure.

As usual, Genstat recycles when it runs out of values. That is, if the contents of one of the variates is exhausted before all the NEWSTRUCTURES positions have either been filled or skipped, then that variate is repeated. For example:

EXAMPLE 5.3.1c

```
  7  " Form variates C[1...6] containing the values in the columns of M."
  8  VARIATE [NVALUES=4] C[1...6]
  9  EQUATE [OLDFORMAT=!((1,-5)4,-1)] M; NEWSTRUCTURES=C
 10  PRINT C[1...6]; FIELDWIDTH=6; DECIMALS=0

  C[1]   C[2]   C[3]   C[4]   C[5]   C[6]
     1      2      3      4      5      6
   101    102    103    104    105    106
     1      2      3      4      5      6
   101    102    103    104    105    106
```

This gives the variates C[1...6] the values in the columns of M. It does it by taking one column at a time from M, skipping the values in the other columns. (Remember that the values of M are held row-by-row.) In detail, what happens is this. For C[1], the format !((1, −5)4, −1) in line 9 takes the value in row 1 column 1, then skips the elements in the remaining five columns of row 1 before taking the value from column 1 of row 2. For C[1] this continues for each row of M, until the final element of the format, −1, skips column 1 of row 1, so that C[2] is given the values in column 2, and so on.

Notice that, as pointer C is automatically available to refer to C[1...6] (see 3.3.4), there is no need to put, for example, !P(C[1...6]).

The final part of the example shows how to form a matrix from a set of variates that contain the values for the columns.

EXAMPLE 5.3.1d

```
 11  " Reform values of M so that its columns are C[1...6] in reverse order."
 12  EQUATE [OLDFORMAT=!((1,-3)6,-1)] !P(C[6...1]); NEWSTRUCTURES=M
 13  PRINT M; FIELDWIDTH=6; DECIMALS=0
```

```
          M
          1     2     3     4     5     6
   1      6     5     4     3     2     1
   2    106   105   104   103   102   101
   3      6     5     4     3     2     1
   4    106   105   104   103   102   101
```

If you are transferring values between factors, Genstat will check that each value to be transferred is valid for the factor in the NEWSTRUCTURES list. If both the factors have been declared with labels, Genstat checks that each transferred value has a label that is one of the labels of the new factor; otherwise it checks that the level is a valid level for the new factor.

5.3.2 The SORT directive

SORT

Sorts units of vectors according to an index vector or defines factors to represent groupings of the units.

Options

INDEX	= vector	Variate or text whose values are to define the ordering; default is to use the first vector in the OLDVECTOR list
GROUPS	= factor	To save (unsorted) groupings of the units; default *
DIRECTION	= string	Order in which to sort (ascending, descending); default a
DECIMALS	= scalar	Number of decimal places to which to round before sorting numbers; default * i.e. no rounding
LIMITS	= variate or text	Upper group limits for forming groups; default *
NGROUPS	= scalar	Number of groups to form when LIMITS is not specified; if NGROUPS is also unspecified, each distinct value (allowing for rounding) defines a group; default *
LEVELS	= variate	Variate to become the levels vector of the GROUP factor and store mid-rank values of the groups; default *

LABELS	= text	Text to become the labels vector of the GROUP factor and store mid-rank values of the groups; default *

Parameters

OLDVECTOR	= vectors or pointers	Factors, pointers, texts, or variates whose values are to be sorted
NEWVECTOR	= vectors or pointers	Structure to receive each set of sorted values; if any are omitted, the values are placed in the corresponding OLDVECTOR

SORT has two separate purposes: the reordering of values of vectors according to an "index" vector, and the sorting of the units of the index vector into groupings which are then stored in a factor.

The index vector can be either a text or a variate. You can specify it explicitly by the INDEX option. If you omit the INDEX option, Genstat uses the first vector in the OLDVECTOR list. The DECIMALS option allows you to define the number of decimal places that are taken into account for an index variate: for example DECIMALS = 0 would round each value to the nearest integer. If you do not set this, there is no rounding.

Vectors or pointers whose values are to be sorted are listed by the OLDVECTOR parameter. The units of each structure are permuted in exactly the same way, into an ordering determined from the index vector. For example, suppose that the variates Age and Income, the text Name and the factor Sex have the values shown here:

EXAMPLE 5.3.2a

```
  1  VARIATE [VALUES=18,50,24,49,61,29,32,42,36,40] Age
  2  & [VALUES=3000,17500,5000,20000,7000,4500, \
  3     12000,18000,15500,17500] Income
  4  TEXT [VALUES=Clarke,Innes,Adams,Jones,Day, \
  5     Grey,Edwards,Baker,Hill,Foster] Name
  6  FACTOR [LABELS=!T(male,female); VALUES=2,1,1,1,2,2,1,1,2,1] Sex
  7  PRINT Age,Income,Name,Sex

        Age      Income        Name         Sex
      18.00        3000       Clarke      female
      50.00       17500        Innes        male
      24.00        5000        Adams        male
      49.00       20000        Jones        male
      61.00        7000          Day      female
      29.00        4500         Grey      female
      32.00       12000      Edwards        male
      42.00       18000        Baker        male
      36.00       15500         Hill      female
      40.00       17500       Foster        male
```

5.3 Transferring and manipulating values

We now show how to sort the units of these vectors so that the names are in alphabetical order.

EXAMPLE 5.3.2b

```
  8    SORT [INDEX=Name] Age,Income,Name,Sex
  9    PRINT Age,Income,Name,Sex

         Age       Income         Name        Sex
       24.00         5000        Adams       male
       42.00        18000        Baker       male
       18.00         3000       Clarke     female
       61.00         7000          Day     female
       32.00        12000      Edwards       male
       40.00        17500       Foster       male
       29.00         4500         Grey     female
       36.00        15500         Hill     female
       50.00        17500        Innes       male
       49.00        20000        Jones       male
```

Here the index vector Name is also one of the vectors being sorted; indeed if you list it first, then you can omit the INDEX option:

SORT Name,Age,Income,Sex

However it need not be among the vectors sorted. Moreover, you can specify new vectors to contain the sorted values, and thus keep the unsorted values in the original vectors. For example

SORT [INDEX=Name] Age,Income,Name,Sex; NEWVECTOR=A,*,N,S

would place the sorted values of Age, Name, and Sex into A, N, and S; as there is a null entry (*) corresponding to Income in the NEWVECTOR list, the sorted incomes would replace the original values of Income. Any undeclared vector in the NEWVECTOR list is declared implicitly to match the corresponding OLDVECTOR.

The next section of the example sorts the units into order of descending age.

EXAMPLE 5.3.2c

```
 10    SORT [DIRECTION=descending] Age,Income,Name,Sex
 11    PRINT Age,Income,Name,Sex

         Age       Income         Name        Sex
       61.00         7000          Day     female
       50.00        17500        Innes       male
       49.00        20000        Jones       male
       42.00        18000        Baker       male
       40.00        17500       Foster       male
       36.00        15500         Hill     female
       32.00        12000      Edwards       male
       29.00         4500         Grey     female
       24.00         5000        Adams       male
       18.00         3000       Clarke     female
```

Here the index vector is a variate. But the DIRECTION option can apply to textual index vectors as well, when ascending order is interpreted as being alphabetical order. The default of DIRECTION is to sort into ascending order.

The second use of SORT is to define values of a factor from the index vector, which again can be either a variate or a text. In the simplest way of doing this, the factor is defined to have a level for each distinct value of the index vector. For this, all you do is specify the identifier of the factor, using the GROUPS option. You need not have declared the factor already; it will be declared automatically if necessary. For example, we can set up a factor Inclev with a level for each of the recorded incomes as follows:

EXAMPLE 5.3.2d

```
  12   SORT [INDEX=Income; GROUPS=Inclev]
  13   PRINT Income,Inclev

       Income      Inclev
         7000           4
        17500           7
        20000           9
        18000           8
        17500           7
        15500           6
        12000           5
         4500           2
         5000           3
         3000           1
```

Notice that the values of Inclev are not sorted, but match the original order of the index vector Income. To sort the values you would need to include Inclev in the OLDVECTOR list.

Alternatively, you can divide the index values into groups to be represented by the factor. You can use the LIMITS option to specify the range of values for each group. The limits vector is a text or a variate, depending on the type of the index vector; its values specify end limits for the intervals corresponding to the groups. The interpretation of the limits depends on whether DIRECTION has been set to 'ascending' or 'descending'. For example, to divide the ages into classes 0–19, 20–29, 30–39, 40–49, 50–59, and over 59, you give the limits vector the five values 19, 29, 39, 49, and 59. Each group includes any values at the limit, and there is a final group to include values beyond the final limit.

EXAMPLE 5.3.2e

```
  14   SORT [INDEX=Age; GROUPS=Ageclass; LIMITS=!(19,29,39,49,59)]
  15   PRINT Age,Ageclass
```

```
      Age  Ageclass
    61.00         6
    50.00         5
    49.00         4
    42.00         4
    40.00         4
    36.00         3
    32.00         3
    29.00         2
    24.00         2
    18.00         1
```

If you set the DIRECTION option to descending, the values of the limits vector should also be in descending order: for example

SORT [INDEX = Age; GROUPS = Ageclass; DIRECTION = descending; \
LIMITS = !(60,21)]

would divide the ages into classes: 60 and above, 59 down to (and including) 21, and under 21.

You can also ask SORT itself to set limits that will partition the units into groups of nearly equal size. For this you specify the NGROUPS option instead of LIMITS. (If you give both LIMITS and NGROUPS, NGROUPS is ignored.) For example, to form three groups according to Income:

EXAMPLE 5.3.2f

```
  16  SORT [INDEX=Income; GROUP=Incgroup; NGROUP=3]
  17  PRINT Income,Incgroup

      Income  Incgroup
        7000         2
       17500         3
       20000         3
       18000         3
       17500         3
       15500         2
       12000         2
        4500         1
        5000         1
        3000         1
```

If the index vector is a variate, you can use the LEVELS option to specify a variate in which Genstat will store the median values of the units in each group; this variate becomes the levels vector of the factor. Similarly, for an index text, option LABELS can specify a text for median alphabetic values, and to become the labels vector of the factor. These two options can be used however the factor is formed: that is, whether it has a level for each value of the index vector or whether it represents ranges of values. For Inclev, in the example above, the corresponding levels variate would simply have the Income values 3000, 4500, 5000, 7000, 12000,

15500, 17500, 18000, and 20000; for Incgroup the variate would have values 5000, 12000, and 17500.

SORT takes account of any restrictions (2.4.1) on any of its vectors: the index vector, the grouping factor, or any of the old or new vectors. If more than one vector is restricted, then each such restriction must be the same. When Genstat is sorting units of vectors, the units that you have restricted out are left in their original positions. When Genstat is forming a factor in SORT, it gives missing values to the excluded units.

5.3.3 The RANDOMIZE directive

RANDOMIZE

Randomizes the units of a designed experiment or the elements of a factor or variate.

Options

BLOCKSTRUCTURE	= formula	Block model according to which the randomization is to be carried out; default * i.e. as a completely-randomized design
EXCLUDE	= factors	(Block) factors whose levels are not to be randomized
SEED	= scalar	Seed for the random-number generator; default 12345

Parameter

	factors or variates	Structures whose units are to be randomized according to the defined block model

---

In its simplest form, RANDOMIZE allows you to permute the values of a list of factors or variates into a random order. You list these structures with the parameter of RANDOMIZE; Genstat gives them all exactly the same randomization, which is produced by a set of random numbers generated from the SEED option. For example

    RANDOMIZE [SEED = 144556] X,Y

puts the values of X and Y into an identical random order. The seed can be any positive integer, but only the last six digits of its integer part are used. Thus the seeds 2144556 and 7144556.3 are both equivalent to the seed 144556.

If you have restricted any of the structures in the parameter list (2.4.1), then all will

5.3 Transferring and manipulating values

be treated as though they were restricted; moreover, all the restricted structures must be restricted in exactly the same way.

Another use of RANDOMIZE is to randomize the allocation of treatments to units in a designed experiment.

In the analysis of designed experiments (Chapter 9), the blocking structure of an experiment is defined by a block formula. How to construct and interpret these formulae is described in detail in 9.2. Provided the only operators in a block formula are the nesting (/) and crossing (*) operators, it also specifies the correct randomization of the experiment.

The nesting operator specifies that one factor is to be randomized within another one. The simplest example is the randomized block design: its block formula is Blocks/Plots; a separate randomization of plots is done for each block. Another example is a split-plot design, the formula for which is Blocks/Wplots/Subplots; this means randomize first the levels of Blocks, then the levels of Wplots within levels of Blocks, and finally the levels of Subplots within the levels of Blocks and Wplots. In other words, there is a separate randomization of Wplots for each Block, and a separate randomization of Subplots for each Wplot. A similar formula and randomization would apply to a resolvable incomplete-block design.

The crossing operator specifies that the factors are to be randomized independently of each other. For example the formula Rows*Cols means randomize the levels of Rows and Cols separately. Thus the same randomization of Cols appears within each Row. This is the block formula associated with a row and column design, for example a Latin square.

You specify the block formula by the BLOCKSTRUCTURE option, which thus defines the way in which the randomization is to be carried out. Genstat does not randomize the factors in the block structure themselves, unless you put them into the parameter list. This is because the original order of the block-factor levels often describes actual positions in the experiment; for example, in a field. So you will be interested in keeping these values, rather than the random ordering of them that is used to allocate treatments.

For example, to produce a randomization of a randomized block design, you could put:

EXAMPLE 5.3.3a

```
  1  UNITS [NVALUES=16]
  2  FACTOR [LEVELS=4; VALUES=4(1...4)] Blocks,Rows
  3  &      [VALUES=(1...4)4] Plots,Cols
  4  &      [LABELS=!T(A,B,C,D)] Dose
  5  PRINT Blocks,Plots,Dose

     Blocks          Plots         Dose
          1              1            A
          1              2            B
          1              3            C
          1              4            D
```

190   5   *Data handling*

```
         2             1          A
         2             2          B
         2             3          C
         2             4          D
         3             1          A
         3             2          B
         3             3          C
         3             4          D
         4             1          A
         4             2          B
         4             3          C
         4             4          D

  6   RANDOMIZE [BLOCKSTRUCTURE=Blocks/Plots; SEED=556743] Dose
  7   PRINT Blocks,Plots,Dose

    Blocks        Plots        Dose
      1             1            C
      1             2            B
      1             3            D
      1             4            A
      2             1            C
      2             2            B
      2             3            A
      2             4            D
      3             1            B
      3             2            C
      3             3            D
      3             4            A
      4             1            A
      4             2            C
      4             3            D
      4             4            B
```

Notice that the values of the Blocks and Plots factors have not been randomized because they did not appear in the parameter list. Note also that the block formula for this design is Blocks/Plots and not just Blocks. This is because the formula must define each experimental unit by a unique combination of the block factor levels, for example block 1, plot 3. To put a block formula of just Blocks would not give Genstat any information about what to do with the elements of the blocks.

If you replaced this RANDOMIZE statement with

    RANDOMIZE [BLOCKSTRUCTURE=Rows∗Cols] Dose

the randomization would be as shown here:

EXAMPLE 5.3.3b

```
  8   FACTOR [LABELS=!T(A,B,C,D)] Dose; VALUES=!((1...4)4)
  9   RANDOMIZE [BLOCKSTRUCTURE=Rows*Cols; SEED=432546] Dose
 10   PRINT Rows,Cols,Dose
```

```
      Rows    Cols    Dose
        1       1      B
        1       2      C
        1       3      D
        1       4      A
        2       1      B
        2       2      C
        2       3      D
        2       4      A
        3       1      B
        3       2      C
        3       3      D
        3       4      A
        4       1      B
        4       2      C
        4       3      D
        4       4      A
```

Since Rows and Cols are randomized separately, the same randomization of Dose appears within each Row. If you had input the values of the factor Dose in the correct order for a Latin square design, this RANDOMIZE statement would have randomized it appropriately.

You should use the EXCLUDE option if you want to restrict the randomization so that one or more of the factors in the block formula is not randomized. The most common instance where this is required is when one of the treatment factors is time-order, which cannot be randomized. For example, suppose the main plot treatments in a split-plot experiment were lengths of time between two chemicals being mixed together, and that the analysis is of the amount of gas produced. If all the jars of chemicals needed to be mixed up at the beginning of the day, and the analyses were performed after the appropriate time lapse, the standing times would have to be in the same order in each replicate. A suitable randomization could be produced by

EXAMPLE 5.3.3c

```
  1  UNITS [NVALUES=18]
  2  FACTOR [LEVELS=3; VALUES=6(1,2,3)] Block
  3   & [LABELS=!T(A,B,C); VALUES=(1...3)6] Method
  4   & [LEVELS=2; LABELS=!T('2 hours','4 hours'); VALUES=3(1,2)3] Time
  5   & [LABELS=*] Mplot
  6  PRINT Block,Time,Method

     Block         Time       Method
       1         2 hours         A
       1         2 hours         B
       1         2 hours         C
       1         4 hours         A
       1         4 hours         B
       1         4 hours         C
       2         2 hours         A
       2         2 hours         B
       2         2 hours         C
       2         4 hours         A
```

```
                  2         4 hours          B
                  2         4 hours          C
                  3         2 hours          A
                  3         2 hours          B
                  3         2 hours          C
                  3         4 hours          A
                  3         4 hours          B
                  3         4 hours          C
  7   RANDOMIZE [BLOCKSTRUCTURE=Block/Mplot/Method; EXCLUDE=Mplot; \
  8       SEED=888667] Time,Method
  9   PRINT Block,Time,Method

              Block        Time       Method
                1        2 hours          C
                1        2 hours          A
                1        2 hours          B
                1        4 hours          A
                1        4 hours          B
                1        4 hours          C
                2        2 hours          C
                2        2 hours          B
                2        2 hours          A
                2        4 hours          C
                2        4 hours          B
                2        4 hours          A
                3        2 hours          C
                3        2 hours          B
                3        2 hours          A
                3        4 hours          B
                3        4 hours          A
                3        4 hours          C
```

In this example we have also used a simplification of the terminology for the block structure: we have used a treatment factor, Method, to specify what is actually a term in the block formula. The strict specification of the structure has a block factor that is synonymous with Method; but having to specify such duplicate structures can be wasteful, and may not conform to the way in which such experiments are described colloquially. In fact the RANDOMIZE statement above could be modified further to remove the Mplot factor:

> RANDOMIZE [BLOCKSTRUCTURE = Block/Time/Method; EXCLUDE = Time; \
>   SEED = 888667] Method

The SEED option determines which randomization Genstat gives. If you use the same seed, you will get the same random numbers, and hence the same randomization (provided the block formula and the block factors are the same as before). You should always, therefore, use the SEED option to make sure that you are not repeatedly using the same random numbers – unless you specifically want to.

This repeatability can, however, be useful, since it means that you need not store the randomization that you have used for an experiment: you can reproduce it afterwards. For example, in the randomized block design above, you could store just the number of blocks (four), the number of plots per block (also four), and the seed (556743).

5.3.4 The COMBINE directive

COMBINE

Combines or omits "slices" of a multi-way data structure (table, matrix or variate).

Options

OLDSTRUCTURE	= identifier	Structure whose values are to be combined; no default i.e. this option must be set
NEWSTRUCTURE	= identifier	Structure to contain the combined values; no default i.e. this option must be set

Parameters

OLDDIMENSION	= factors or scalars	Dimension number or factor indicating a dimension of the OLDSTRUCTURE
NEWDIMENSION	= factors or scalars	Dimension number or factor indicating the corresponding dimension of the NEWSTRUCTURE; this can be omitted if the dimensions are in numerical order
OLDPOSITIONS	= pointers, texts or variates	These define positions in each OLDDIMENSION: pointers are appropriate for matrices whose rows or columns are indexed by a pointer; texts are for matrices indexed by a text, variates with a textual labels vector, or tables whose OLDDIMENSION factor has labels; and variates either refer to levels of table factors or numerical labels of matrices or variates, if these are present, otherwise they give the (ordinal) number of the position. If omitted, the positions are assumed to be in (ordinal) numerical order. Margins of tables are indicated by

		missing values
NEWPOSITIONS	= pointers, texts or variates	These define positions in each NEWDIMENSION, specified similarly to OLDPOSITIONS; these indicate where the values from the corresponding OLDDIMENSION positions are to be entered (or added to any already entered there)
WEIGHTS	= variates	Define weights by which the values from each OLDDIMENSION coordinate are to be multiplied before they are entered in the NEWDIMENSION

Sometimes you may wish to reclassify a table to have factors different from those that you used in its declaration. Provided the changes consist either of omitting levels of the classifying factors, or of combining several levels together, you can use COMBINE to do this. You specify the original table using the OLDSTRUCTURE option, and a table to contain the reclassified values using the NEWSTRUCTURE option; if you have not already declared the new table, it will be declared implicitly. You must specify both of these options.

You can modify several of the classifying factors at a time. You list the factors of the original table with the OLDDIMENSION parameter, and the equivalent factors of the new table with NEWDIMENSION. An alternative way of doing this is to give a dimension number, specifying the position of the factor in the classifying set of the table (3.5); for the NEWDIMENSION list, this requires that you have already declared the new table. You can even omit the list of dimensions if they would be in ascending numerical order.

In this example, the table Sales contains the number of items of some product sold by a retailer with shops in nine towns, in the years 1979 to 1984. Lines 20 to 23 declare and form a table Csales in which the sales are classified by the country where the sale was made, instead of the town; so there is one OLDDIMENSION, the factor Town, and a corresponding NEWDIMENSION, Country.

EXAMPLE 5.3.4a

```
  1  TEXT  [VALUES= Aberdeen,Birmingham,Cardiff,Dundee,Edinburgh, \
  2        Liverpool,Manchester,Sheffield,Swansea] Townname
  3  VARIATE [VALUES=1979,1980,1981,1982,1983,1984] Yearnum
  4  FACTOR [LABELS=Townname] Town
  5  FACTOR [LEVELS=Yearnum]  Year; DECIMALS=0
  6  TABLE  [CLASSIFICATION=Town,Year] Sales
  7  READ Sales
```

```
   Identifier    Minimum      Mean   Maximum     Values    Missing
        Sales      343.0     676.3    1158.0         54          0

17  PRINT Sales; FIELDWIDTH=8; DECIMALS=0

                Sales
        Year     1979    1980    1981    1982    1983    1984
        Town
    Aberdeen      608     635     672     692     685     723
  Birmingham      618     601     784     720     863     921
     Cardiff      757     743     785     816     783     737
      Dundee      343     391     358     366     418     470
   Edinburgh      714     751     710     763     788     830
   Liverpool      816     859     820     938    1007    1158
  Manchester      662     632     758     721     893     837
   Sheffield      531     569     615     624     607     593
     Swansea      416     461     478     462     497     520

18  FACTOR [LABELS=!T(England,Wales,Scotland)] Country
19  " Form a table Csales, classified by country instead of town."
20  COMBINE [OLDSTRUCTURE=Sales; NEWSTRUCTURE=Csales] \
21     OLDDIMENSION=Town; NEWDIMENSION=Country; \
22     OLDPOSITIONS=!(2,6,7,8,1,4,5,3,9); \
23     NEWPOSITIONS=!T(4(England),3(Scotland),2(Wales))
24  PRINT Csales; FIELDWIDTH=8; DECIMALS=0

                Csales
        Year     1979    1980    1981    1982    1983    1984
     Country
     England     2627    2661    2977    3003    3370    3509
       Wales     1173    1204    1263    1278    1280    1257
    Scotland     1665    1777    1740    1821    1891    2023
```

Each of the levels of Country is a combination of several levels of Town. You use the OLDPOSITIONS and NEWPOSITIONS parameters to specify how this combining is to be done. These parameters specify a pair of vectors for each pair of old and new dimensions, listing positions within the old dimension and the corresponding positions to which they are mapped in the new dimension. The positions can be defined in terms of either the levels or the labels of the factor that classifies the dimension. In the example, the vector for the old dimension Town is an unnamed variate !(2,6,7,8,1,4,5,3,9) whose values refer to the levels (1 to 9); the vector for Country is an unnamed text !T(4(England),3(Scotland),2(Wales)) whose values are labels of Country. The correspondence between the two sets of values is:

Town level	Town label	Country label	Country level
2	Birmingham	England	1
6	Liverpool	England	1
7	Manchester	England	1
8	Sheffield	England	1
1	Aberdeen	Scotland	2
4	Dundee	Scotland	2
5	Edinburgh	Scotland	2
3	Cardiff	Wales	3
9	Swansea	Wales	3

Thus, as you can see, the values in the original table for the English towns (Birmingham, Liverpool, Manchester, and Sheffield) are allocated to Country England in the new table, the Scottish towns (Aberdeen, Dundee, and Edinburgh) are allocated to Scotland, and Cardiff and Swansea are allocated to Wales.

If you omit the vector for one of the dimensions, it is assumed to contain each value once only, taken in the order in which they occur in the levels vector of the factor. Thus the OLDPOSITIONS variate could be omitted in

COMBINE [OLDSTRUCTURE = Sales; NEWSTRUCTURE = Csales] \
    OLDDIMENSION = Town; NEWDIMENSION = Country;
    OLDPOSITIONS = !(1...9);
    NEWPOSITIONS = !T(Scotland,England,Wales,2(Scotland),3(England),Wales)

You indicate a margin of the table by a missing value in a variate, or by a null string in a text.

Values in the original table can be allocated to more than one place. Also, as we have mentioned already, you can modify more than one dimension at a time. In the next section of the example, the Years dimension is modified as well as the Town dimension: years 1979 and 1980 are omitted, while the other years are allocated to two summary lines as well as to themselves in the new dimension Ysummary. Thus the new table Salesum has lines giving sales for the individual years, interspersed with bi-annual totals. Note that the interspersing of the summary lines is ensured by the order in which the FACTOR declaration specifies the labels of the factor Yearsum.

EXAMPLE 5.3.4b

```
25    " Form a table classified by country and year including biannual totals."
26    FACTOR [LABELS=!T('1981','1982','1981-2','1983','1984','1983-4')] Yearsum
27    TABLE   [CLASSIFICATION=Yearsum,Country] Salesum
28    COMBINE [OLDSTRUCTURE=Sales; NEWSTRUCTURE=Salesum] \
29       OLDDIMENSION=Town,Year; NEWDIMENSION=Country,Yearsum; \
30       OLDPOSITIONS=!(2,6,7,8,1,4,5,3,9),!V((1981...1984)2); \
31       NEWPOSITIONS=!T(4(England),3(Scotland),2(Wales)), \
32       !T('1981','1982','1983','1984',2('1981-2','1983-4'))
33    PRINT Salesum; FIELDWIDTH=8; DECIMALS=0
```

	Salesum		
Country	England	Wales	Scotland
Yearsum			
1981	2977	1263	1740
1982	3003	1278	1821
1981-2	5980	2541	3561
1983	3370	1280	1891
1984	3509	1257	2023
1983-4	6879	2537	3914

In parallel with the vectors of positions, you can also specify a variate of weights by which the values are multiplied before being entered into the new table. Thus, for example, forming summary lines of means instead of totals would require an extra parameter list

    WEIGHTS = *,!(1,1,1,1,0.5,0.5,0.5,0.5)

Although the main way in which you will use COMBINE is likely to be for tables, you can also use it on rectangular matrices and even variates. For these, the dimensions can only be numbers: number 1 refers to the rows of a matrix, and 2 to the columns; number 1 refers to the rows (or units) of a variate. The position vectors refer to the labels vectors of matrices (3.4.1), which can be variates, texts, or pointers; or they refer to the unit labels of a variate (3.3.1), which can be held in either a variate or a text. If a dimension has no labels vector, you use a variate to specify its positions; then each value of the variate gives the number of a row, column, or unit. You can do the same also if the labels vector is something other than a variate: that is, a text or a pointer.

5.4 Operations on text

A text structure (3.3.2) is a vector each line of which contains a string of characters. So you can use it to label the units of other vectors, or to contain a complete piece of description.

    The first part of this section describes the CONCATENATE directive (5.4.1) which allows you to concatenate several texts together side by side so that each line of the new text is formed by joining together a series of lines, one from each of the original texts. You can omit characters at the beginning and end of the component lines; so this also gives you a way of truncating the lines of a text. An alternative form of concatenation places whole texts one after another, possibly omitting some of their lines: you can do this with the EQUATE directive (5.3.1).

    The remaining parts of the section describe the EDIT directive (5.4.2). This is a subsystem within Genstat; it has its own command syntax (5.4.3), allowing you to delete and insert series of characters, or to substitute one series for another, or to delete and insert complete lines, and so on.

    Some general directives, described elsewhere, are also useful for manipulating text. The SORT directive allows you to sort the units of a text into alphabetical order or to form a factor from a text (5.3.2). You can test for equality and inequality of the lines of texts in the expressions that occur in CALCULATE (5.1), in RESTRICT (2.4.1), and in the directives for program control (6.2). You can direct output from the PRINT directive (4.2) into a text, which has the added advantage of letting you place numerical values into a text, and READ can take its input from a text (4.1.5).

5.4.1 The CONCATENATE directive

CONCATENATE

Concatenates and truncates lines (units) of text structures.

Option

NEWTEXT	= text	Text to hold the concatenated/truncated lines; default is the first OLDTEXT vector

Parameters

OLDTEXT	= texts	Texts to be concatenated
WIDTH	= scalars or variates	Number of characters to take from the lines of each text; if * or omitted, all the (unskipped) characters are taken
SKIP	= scalars or variates	Number of characters to skip at the left-hand side of the lines of each text; if * or omitted, none is skipped

---

CONCATENATE joins lines of several texts together, side by side, to form a new text. You can specify the identifier of this text by the NEWTEXT option, in which case it need not already have been declared as a text. If you do not specify NEWTEXT, Genstat places the new textual values into the first text in the OLDTEXT parameter list (replacing its existing values).

The texts to be concatenated are specified by OLDTEXT; they should all contain the same number of lines, unless you want to insert an identical series of characters into every line of the new text: a series of characters that is to be duplicated within every line can be specified either as a string, or in a single-valued text. In the next example, the string ', ' inserts a comma and a space into every line of the NEWTEXT Fullname.

EXAMPLE 5.4.1a

```
  1  TEXT [VALUES='1. Adams','2. Baker','3. Clarke','4. Day','5. Edwards', \
  2    '6. Field','7. Good','8. Hall','9. Irving','10. Jones'] Name
  3  TEXT [VALUES='B.J.','J.S.','K.R.','A.T.','R.S.', \
  4    'T.W.','S.I.','D.M.','H.M.','C.C.'] Initials
  5  "Form text Fullname containing the number, name, and initials."
  6  CONCATENATE [NEWTEXT=Fullname] OLDTEXT=Name,', ',Initials
  7  PRINT Fullname; JUSTIFICATION=left

Fullname
1. Adams, B.J.
2. Baker, J.S.
```

```
     3. Clarke, K.R.
     4. Day, A.T.
     5. Edwards, R.S.
     6. Field, T.W.
     7. Good, S.I.
     8. Hall, D.M.
     9. Irving, H.M.
    10. Jones, C.C.
```

The WIDTH and SKIP parameters enable you to omit characters at the start and end of the lines. For example:

EXAMPLE 5.4.1b

```
  8   " Now reform Fullname to contain just the first initial and the name."
  9   CONCATENATE [NEWTEXT=Fullname] OLDTEXT=Initials,Name; \
 10      WIDTH=2,*; SKIP=*,!(9(2),3)
 11   PRINT Fullname; JUSTIFICATION=left

Fullname
B. Adams
J. Baker
K. Clarke
A. Day
R. Edwards
T. Field
S. Good
D. Hall
H. Irving
C. Jones
```

If you give a variate in the SKIP list, then it must contain a value for each line of the text in the OLDTEXT list; the value indicates the number of characters to be omitted at the beginning of that line. Alternatively, you can give a scalar if the same number of characters is to be omitted at the start of every line. In the example, the null entry for Initials (indicated by *) specifies that no characters are to be omitted.

Similarly the WIDTH parameter specifies how many characters are to be taken, after omitting any initial characters as specified by SKIP. In the example WIDTH has a scalar setting of 2 for Initials, so that only the first initial followed by a dot is taken for each name.

CONCATENATE takes account of restrictions (2.4.1) on any of the vectors that occur in the statement. If more than one vector is restricted, then each such restriction must be the same. The values of the units that are excluded by the restriction are left unchanged.

5.4.2 The EDIT directive

```
EDIT
```

Edits text vectors.

Options

CHANNEL	= scalar or text	Text structure containing editor commands, or a scalar giving the number of a channel from which they are to be read; default is the current input channel
END	= text	Character(s) to indicate the end of the commands read from an input channel; default is the character colon (:)
WIDTH	= scalar	Limit on the line width of the text; default *
SAVE	= text	Text to save the editor commands for future use; default *

Parameters

OLDTEXT	= texts	Texts to be edited
NEWTEXT	= texts	Text to store each edited text; if any of these is omitted, the corresponding OLDTEXT is used

The EDIT directive edits each text in the OLDTEXT list, storing the results in the corresponding structure in the NEWTEXT list. It both edits and stores each text before moving on to the next. If you have not already declared any of the texts in the NEWTEXT list, it will be declared implicitly. If you do not specify a NEWTEXT, the edited version simply replaces the values of the original: thus the old text will be overwritten by the new text. You can also omit a text from the OLDTEXT list, which you might do if you wanted to form the values of the new text entirely from within the editor. If any of the old texts are restricted, they must all be restricted to exactly the same set of units. Then only those units will be involved in the edit. When a restriction is in force, you cannot add or delete any units (or lines).

The CHANNEL option tells Genstat where to find the editing commands. A scalar specifies the number of an input channel from which the commands are to be read. Alternatively, you can specify a text structure containing the commands. In either case the commands should be terminated by the string specified by the END option.

The end string can be more than one character; the default is the single character colon (:). Genstat gives a warning if you have forgotten to specify the end string in a text of commands. The default for the CHANNEL option is to take input from the current input channel.

The WIDTH option specifies the maximum line length for vectors of commands and of text, the default being 80 and the maximum being 255.

The SAVE option allows you to specify a text structure to store the edit commands, so that you can save them for future EDIT statements.

5.4.3 Commands for the EDIT directive

The commands that you can use to edit text are described in this subsection. You can give commands to the editor in upper or lower case. You can put as many commands as you like on a line, subject only to the width restriction set by the WIDTH option. Commands must be separated by at least one space. You cannot put spaces into the middle of a command, unless they are part of a character string (or part of a sequence of commands).

The character that separates the parts of a command is written here as /, but you can use any character for this other than a space or a digit.

Genstat puts the lines from the old text into an internal *buffer*, where they are modified according to the commands that you specify. While you are editing, Genstat moves a notional *marker* around the buffer. The marker can be moved backwards or forwards along a line or between lines. So you can move around the text and modify the lines in any order. Some commands move the marker automatically, as explained in the definitions below. If the marker is before the first line of text it is at the [start] position; if it is after the last line of text it is at the [end] position. The line that currently contains the marker is called the *current line*. Genstat does not write anything to the new text until the edit has been completed.

Some commands allow you to specify a number: for example D*n* deletes the next *n* lines. Genstat gives a warning message if this number is zero or is not an integer.

The command definitions are as follows.

	A	Insert the next line of text from the buffer, immediately after the marker within the current line.
	B	Break the current line at the marker position. Text before the marker is written as a new line to the internal buffer and text after the marker becomes the new current line with the marker at character position 1.
	C	Cancel edits performed on the current line by restoring it to the form in which it was most recently read from the buffer. Note that if you have previously edited the line and then moved to some other line, it is the previously edited form that will be given, not the form as originally in the old text; also, if you have

given any A or B commands during your modification of the current line, their effects are not undone, so for example any lines that have been inserted into the current line by A will be lost.

D	Delete the current line, and make the next line the current line with the marker at character position 1.
Dn	Delete the next n lines (including the current line), making the next line after that the current line with the marker at position 1.
D+n	Synonymous with Dn.
D∗	Delete from the current line to end of text. The current line is then [end].
D+∗	Synonymous with D∗.
D−	Delete the current line, making the previous line the current line with the marker at character position 1.
D−n	Delete the current and previous n lines, making the line before that the current line with the marker at character position 1.
D−∗	Delete the current line and all previous lines, the current line is then [start].
D/s/	Delete from the current line to the next line containing the character string s. The marker is placed immediately before the character string s in the located line; anything before s in that line is deleted.
D−/s/	The same as D/s/, except that it moves backwards through the text, deleting all lines from and including the current one until the first occurrence of a line containing the character string s. The marker is placed immediatley before the located character string s; anything on that line after s is deleted.
F/i/	Inserts the contents of the text structure with identifier i immediately before the current line. The marker is not moved.
I/s/	Inserts the string s as a new line immediately before the current line. The marker is not moved.
L	Moves the marker to the start of the next line, which can be [end].
Ln	Moves the marker to the start of the nth line after the current line. So L1 gives the next line.
L+n	Is synonymous with Ln.
L+	Is synonymous with L or L+1.
L+∗	Moves the marker to [end].

L−n	Moves the marker to the start of the nth line before the current line, which can be [start]. L−1 gives the line immediatley before the current line.
L−	Is synonymous with L−1.
L−*	Moves the marker to [start].
L+/s/	Moves the marker to the position immediately before the next occurrence of the character string s after the current marker position; this occurrence need not be on the current line. If the string s is not found, the marker will be located at [end].
L−/s/	Moves the marker to the position immediately before the first occurrence of the string s before the current marker position; this occurrence need not be on the current line. If the string s is not found, the marker will be located at [start].
P+n	Moves the marker n characters to the right of the current position within the current line. You cannot move the marker beyond the maximum line length (which will vary between computers, but is normally the same as the width of your local line-printer).
P+*	Moves the marker to the position immediately after the last non-blank character in the current line. This can be to the left of the current marker position.
P−n	Moves the marker n characters to the left of the current position within the current line. The marker cannot be moved to the left of character position 1.
P−*	Moves the marker to the position immediately before the first non-blank character after character position 1. This can be to the right of the current marker position.
Pn	Moves the marker to the character position n within the current line, counting from the left and starting at 1. The maximum value of n varies between computers but is normally the same as the width of your local line-printer.
Q	Abandons the current edit, leaving the original text unaltered.
S/s/t/	Substitutes the string t for the next occurrence of string s after the marker within the current line. The marker is moved to he character position immediately after the last character in t. If s is null (when the command is S//t/) then t is inserted immediately after the marker. If t is null (when the command is S/s//), then s is deleted from the line.

V	Turns on the verification mode. Then, if you are working interactively, the current line will be displayed each time that Genstat prompts you for commands. By default the marker is indicated by the character > but you can change this by the Vc or V+c command.
Vc	Turns on the verification mode (see V), and changes the marker character to c.
V+c	Is synonymous with Vc.
V−	Turns verification mode off (see V).
($cseq$)n	Repeats the command sequence, $cseq$, n times. The command sequence $cseq$ can be any valid combination of editing commands, each separated by at least one space. The complete sequence, including brackets and repeat count, must all be on a single line. You can nest sequences up to a depth of 10.
($cseq$)∗	Repeats the command sequence $cseq$ until [end] or [start] is encountered. In all other respects ($cseq$)∗ behaves exactly as ($cseq$)n; so it would be equivalent to putting n equal to some very large number.

EXAMPLE 5.4.3

```
>  " An interactive run using the editor:
>    Genstat prompts for the statement with the string '> ';
>    within the editor, the prompt is 'EDIT> '."
>  TEXT [NVALUES=10] Name
>  OPEN 'NAMES.DAT'; CHANNEL=2
>  READ [CHANNEL=2] Name

>  EDIT Name
>B.J. Adams
EDIT> S//Mr. /
Mr. >B.J. Adams
EDIT> L
>J.S. Baker
EDIT> (S//Dr. / L)4
>T.W. Field
EDIT> L-2
>Dr. A.T. Day
EDIT> S/D/M/
M>r. A.T. Day
EDIT> L+2
>T.W. Field
EDIT> S//Ms. /
Ms. >T.W. Field
EDIT> L
>S.I. Good
EDIT> S//Mr. /
Mr. >S.I. Good
```

```
EDIT> L S//Miss. /
Miss. >D.M. Hall
EDIT> (L S//Dr. /)2
Dr. >C.C. Jones
EDIT> :
> PRINT Name; JUSTIFICATION=left

Name
Mr.   B.J. Adams
Dr.   J.S. Baker
Dr.   K.R. Clarke
Mr.   A.T. Day
Dr.   R.S. Edwards
Ms.   T.W. Field
Mr.   S.I. Good
Miss. D.M. Hall
Dr.   H.M. Irving
Dr.   C.C. Jones
```

5.5 Operations on factors

You use factors in Genstat to indicate groupings of the units of vectors. You would need to do this, for example, in the analysis of designed experiments (Chapter 9), or when forming tabular summaries of group totals, means, maxima, minima, and so on (5.8). This section describes the GENERATE directive which allows you to define factor values in a systematic order. You can also use it to define values for the pseudo-factors required to specify partially balanced experimental designs (9.7.3).

There are other directives that allow you to manipulate factor values; these are described elsewhere in this chapter. Once you have generated factor values, you can use the RANDOMIZE directive to randomize the order of their units; this is particularly useful for treatment factors in designed experiments (5.3.3). The use of factors within expressions, and in CALCULATE in particular, is described in 5.1.1 and 5.1.2. This allows you to form a variate from a factor, either taking its declared levels or by taking an alternative set of levels using the NEWLEVELS function. CALCULATE also allows you to assign values of a variate to a factor, provided you have already declared the factor with levels including all the values taken by the variate. But a more satisfactory method is to use the GROUPS option of SORT (5.3.2), which will declare a factor automatically while forming its values from either a variate or a text. These levels can cover every distinct value of the variate or text, or groups of values; you can specify these groups yourself, or have them defined automatically.

5.5.1 The GENERATE directive

> GENERATE
>
> Generates factor values for designed experiments: with no options set, factor values are generated in standard order; the options allow pseudo-factors to be generated describing confounding in partially balanced experimental designs.

Options

TREATMENTS	= formula	Model term for which pseudo-factors are to be generated; default *
REPLICATES	= formula	Factors defining replicates of the design; default *
BLOCKS	= formula	Block term; default *

Parameter

	factors	Factors whose values are to be generated

GENERATE is invaluable when you have a set of data that is to be read in a systematic order: for example, you may want to take all the observations within one group, then the same number of observations within the next group, and so on until an equal number of observations has been read for every group. You can then define values of the grouping factor or factors by GENERATE; so the only values that you need to read are the observed data. Designed experiments are the obvious instance where the data are structured in this way: for example, you might have all the data from the first block, then all those from the second block, and so on (see Chapter 9).

The best way to understand GENERATE is to look at some examples. The values of a set of factors that you have defined by GENERATE are said to be in *standard order*: that is their units are arranged so that the levels of the first factor occur in the same order as in its levels vector then, within each level of the first factor, the levels of the second factor are arranged similarly, and so on. For example

```
FACTOR [NVALUES=24; LEVELS=2] A
& [LEVELS=!(4,1,2)] B
& [LEVELS=4] C
GENERATE A,B,C
```

gives A, B, and C the values

```
A: 1 1 1 1 1 1 1 1 1 1 1 1 2 2 2 2 2 2 2 2 2 2 2 2
B: 4 4 4 4 1 1 1 1 2 2 2 2 4 4 4 4 1 1 1 1 2 2 2 2
C: 1 2 3 4 1 2 3 4 1 2 3 4 1 2 3 4 1 2 3 4 1 2 3 4
```

Placing a number or a scalar in the parameter list has the same effect as if a factor with that number of levels had been listed. Thus to generate values only for A and C, all that you require is

```
GENERATE A,3,C
```

To generate values for just B and C is even simpler since the cycling process is itself recycled until all the units have been covered. Omitting A therefore causes all combinations of a level of B with a level of C to be used twice, in the same pattern as displayed above; so you need specify only

    GENERATE B,C

You get a warning if one of the cycles is incomplete, as would happen for example if B and C had 18 values instead of 24.

An example describing how to use the options of GENERATE to form the values of pseudo-factors is given in 9.7.3.

5.6 Operations on dummies and pointers

You use dummies when you want a series of statements to operate on one of several different data structures, but you do not want to specify exactly which one (3.2.2). By referring to a dummy instead of any specific structure, you can make the statements apply to whichever structure you want. The commonest use of dummies is in loops (6.2.1), and in procedures (6.3.1). In this section we describe an alternative way of specifying a value for a dummy, by using the ASSIGN directive.

ASSIGN also enables you to change the values of elements of pointers, which are used mainly to specify collections of data structures for directives such as EQUATE (5.3.1), or as a convenient way of specifying lists of structures (1.4.4 and 3.3.4). You can do tests on the values of dummies and pointers by the .IS. and .ISNT. operators (5.1.1).

5.6.1 The ASSIGN directive

ASSIGN

Sets elements of pointers and dummies.

No options

Parameters

STRUCTURE	= identifiers	Values for the dummies or pointer elements
POINTER	= dummies or pointers	Structure that is to point to each of those in the STRUCTURE list
ELEMENT	= scalars or texts	Unit or unit label indicating which pointer element is to be set; if omitted, the first element is assumed

---

ASSIGN allows you to set individual elements of pointers, or to assign a value to a dummy. The parameter POINTER lists the pointers or dummies whose values you want to set; the values that you want to give them are listed by the STRUCTURE parameter. You pick out the individual elements of pointers by the ELEMENT parameter; a scalar identifies the element by its suffix number, while a text identifies it by its label. This example sets the dummy Yvar to point to the variate Height, and elements 1 and 2 of the pointer Xvars to Protein and Vitamins, respectively.

```
VARIATE Height,Protein,Vitamins
POINTER [NVALUES=2] Xvars
DUMMY Yvar
ASSIGN Height,Protein,Vitamins; POINTER=Yvar,2(Xvars); ELEMENT=1,1,2
```

Element 1 is assumed unless you specify otherwise; so to set just Yvar we need only put

```
ASSIGN Height; POINTER=Yvar
```

5.7 Operations on matrices and compound structures

The CALCULATE directive (5.1.1 and 5.1.3) allows you to do arithmetic operations on matrices element by element: addition, subtraction, multiplication, division, and exponentiation, as well as logical operations of testing for equality and inequality, and so on; you can also do matrix multiplication. There are several functions for standard operations on matrices, such as taking inverses, described in 5.2.4. You can combine and omit rows or columns of a rectangular matrix using the COMBINE directive (5.3.4). EQUATE allows you to transfer values to matrices from another structure, and vice versa (5.3.1), or you can select sub-matrices with CALCULATE, using qualified identifiers (5.1.6).

You cannot do calculations directly with a complete compound structure like an LRV or an SSPM, but you can do calculations with the individual elements. For example, to take the diagonal matrix of latent roots from an LRV structure, L, and divide it by the trace, you could put

```
CALCULATE L['Roots'] = L['Roots'] / L['Trace']
```

This section describes the directives SVD and FLRV, which allow you to form singular value and eigenvalue decompositions; it also describes the FSSPM directive, which calculates sums of squares and products and all the associated information stored in an SSPM structure. These operations form the basis of many common statistical methods. The FTSM directive, which forms preliminary values of a time-series model in a TSM structure, is described in 11.7.1.

5.7.1 The SVD directive

SVD

Calculates singular value decompositions of matrices
i.e. (LEFT *+ SINGULAR *+ TRANSPOSE(RIGHT)).

Option

PRINT	= strings	Printed output required (left, singular, right); default * i.e. no printing

Parameters

INMATRIX	= matrices	Matrices to be decomposed
LEFT	= matrices	Left-hand matrix of each decomposition
SINGULAR	= diagmats	Singular values (middle) matrix
RIGHT	= matrices	Right-hand matrix of each decomposition

Suppose that we have a rectangular matrix A with m rows and n columns, and that p is the minimum of m and n. The singular value decomposition can be defined as

$$_mA_n = {_mU_p} *+ {_pS_p} *+ {_pV_n}$$

The diagonal matrix S contains the p singular values of A, ordered such that

$$s_1 \geqslant s_2 \geqslant ... \geqslant s_p \geqslant 0$$

The matrices U and V contain the left and right singular vectors of A, and are orthonormal:

$$U'U = V'V = I_p$$

The smaller of U and V will be orthogonal. So, if A has more rows than columns, $m > n$, $p = n$ and $VV' = I_p$.

The INMATRIX parameter specifies the matrices to be decomposed. The algorithm uses Householder transformations to reduce A to bi-diagonal form, followed by a QR algorithm to find the singular values of the bi-diagonal matrix (Golub and Reinsch 1971). The other parameters allow you to save the component parts of the decomposition: LEFT, SINGULAR, and RIGHT for U, S, and V respectively.

The PRINT option allows you to print any of the components of the decomposition; by default, nothing is printed. All p columns of the matrices are printed, even if you are storing only the first r columns (see below). In this example all the component parts are printed.

EXAMPLE 5.7.1a

```
  1   MATRIX [ROWS=6; COLUMNS=4] A; VALUES=!(15,5,9,16,3,20,7,12,22,17,10,11, \
  2     13,8,1,23,2,4,6,14,18,21,24,19)
  3   SVD [PRINT=LEFT,SINGULAR,RIGHT] A
```

3...

***** Singular value decomposition *****

***   Singular Values   ***

	1	2	3	4
	65.30	17.75	14.29	10.82

***   Left Singular Vectors   ***

	1	2	3	4
1	0.35066	−0.33717	−0.30338	0.26324
2	0.32642	0.30654	0.69925	−0.39495
3	0.45861	0.18847	−0.51086	−0.52922
4	0.37075	−0.71706	0.15091	−0.29662
5	0.20711	−0.27069	0.36641	0.39876
6	0.61629	0.41157	−0.03151	0.49765

***   Right Singular Vectors   ***

	1	2	3	4
1	0.50011	−0.13783	−0.80932	−0.27549
2	0.50254	0.53368	0.40553	−0.54607
3	0.40479	0.48070	−0.09456	0.77210
4	0.57749	−0.68199	0.41425	0.17258

The least-squares approximation of rank r to A can be formed as

$$A_r = U_r S_r V_r'$$

where U_r and V_r are the first r columns of U and V, and S_r contains the first r singular values of A (Eckart and Young 1936). This value r is analogous to the dimensionality in multivariate analysis (10.1.1). Thus, when you want to save the singular values and the corresponding matrices of vectors, the rules for the sizes of output structures from multivariate directives apply: Genstat decides how many columns to store according to the number of columns that you have defined for the matrices in the LEFT, SINGULAR, and RIGHT lists. So you must have declared these structures already, and they must all have the same number of columns. If you save the results for r singular values, and $r<p$, then the first r singular values will be saved, along with the corresponding columns of singular vectors.

EXAMPLE 5.7.1b

```
  4   MATRIX [ROWS=6; COLUMNS=2] Ua,Uua
  5   & [ROWS=4; COLUMNS=2] Va,Vva
  6   DIAGONALMATRIX [ROWS=2] Sa
  7   SVD A; LEFT=Ua; SINGULAR=Sa; RIGHT=Va
```

5.7 Operations on matrices and compound structures

```
  8   CALCULATE A2 = Ua *+ Sa *+ TRANSPOSE(Va)
  9   PRINT [RLWIDTH=6] A; FIELDWIDTH=9; DECIMALS=3
```

```
            A
            1         2         3         4

     1   15.000     5.000     9.000    16.000
     2    3.000    20.000     7.000    12.000
     3   22.000    17.000    10.000    11.000
     4   13.000     8.000     1.000    23.000
     5    2.000     4.000     6.000    14.000
     6   18.000    21.000    24.000    19.000
```

```
 10   & A2; FIELDWIDTH=9; DECIMALS=3
```

```
            A2
            1         2         3         4

     1   12.276     8.313     6.392    17.304
     2    9.910    13.615    11.243     8.598
     3   14.515    16.834    13.730    15.012
     4   13.861     5.373     3.681    22.660
     5    7.426     4.232     3.165    11.087
     6   19.119    24.122    19.801    18.257
```

The diagonal matrix Sa saves the first two singular values, while the first two left singular vectors are stored in the matrix Ua. A2 is a least-squares approximation to A, based on $r=2$ singular values (an Eckart-Young approximation, in other words, of rank 2).

One practical application of the singular value decomposition is to form generalized inverses of matrices. If you use the singular value decomposition you get the Moore-Penrose generalized inverse, which is sometimes called the *pseudo-inverse*.

This example verifies that the necessary properties of the Moore-Penrose inverse are satisfied. You need to set the ZDZ option of CALCULATE to 'zero' when calculating Splus, the generalized inverse of the diagonal matrix of singular values, in case any of the singular values is zero. The default for ZDZ would set the corresponding elements of Splus to be missing (5.1.1).

EXAMPLE 5.7.1c

```
 11   MATRIX [ROWS=6; COLUMNS=4] Uda
 12   & [ROWS=4; COLUMNS=4] Vda
 13   DIAGONALMATRIX [ROWS=4] Sda
 14   SVD A; LEFT=Uda; SINGULAR=Sda; RIGHT=Vda
 15   CALCULATE [ZDZ=zero] Splus = Sda / Sda / Sda
 16   & Aplus = Vda *+ Splus *+ TRANSPOSE(Uda)
 17   & Aa,Aap = A,Aplus *+ Aplus,A *+ A,Aplus
 18   PRINT A; FIELDWIDTH=9; DECIMALS=3
```

212   5 *Data handling*

```
                A
                1         2         3         4
        1   15.000     5.000     9.000    16.000
        2    3.000    20.000     7.000    12.000
        3   22.000    17.000    10.000    11.000
        4   13.000     8.000     1.000    23.000
        5    2.000     4.000     6.000    14.000
        6   18.000    21.000    24.000    19.000

  19   & Aa; FIELDWIDTH=9; DECIMALS=3

                Aa
                1         2         3         4
        1   15.000     5.000     9.000    16.000
        2    3.000    20.000     7.000    12.000
        3   22.000    17.000    10.000    11.000
        4   13.000     8.000     1.000    23.000
        5    2.000     4.000     6.000    14.000
        6   18.000    21.000    24.000    19.000

  20   & Aplus; FIELDWIDTH=9; DECIMALS=3

              Aplus
                1         2         3         4         5         6
        1    0.016    -0.029     0.044     0.007    -0.027    -0.009
        2   -0.029     0.052     0.021     0.001    -0.016    -0.009
        3    0.014    -0.022    -0.026    -0.039     0.020     0.051
        4    0.011     0.005    -0.026     0.030     0.029    -0.003

  21   & Aap; FIELDWIDTH=9; DECIMALS=3

               Aap
                1         2         3         4         5         6
        1    0.016    -0.029     0.044     0.007    -0.027    -0.009
        2   -0.029     0.052     0.021     0.001    -0.016    -0.009
        3    0.014    -0.022    -0.026    -0.039     0.020     0.051
        4    0.011     0.005    -0.026     0.030     0.029    -0.003

  22   CALCULATE Asa,Aspa = A,Aplus *+ Aplus,A
  23   PRINT Asa; FIELDWIDTH=9; DECIMALS=3

               Asa
                1         2         3         4         5         6
        1    0.398    -0.305     0.113     0.248     0.158     0.218
        2   -0.305     0.845     0.059     0.124     0.083     0.109
        3    0.113     0.059     0.787     0.115    -0.354     0.113
        4    0.248     0.124     0.115     0.762     0.208    -0.219
        5    0.158     0.083    -0.354     0.208     0.409     0.203
        6    0.218     0.109     0.113    -0.219     0.203     0.798

  24   & Aspa; FIELDWIDTH=9; DECIMALS=3

              Aspa
                1         2         3         4
        1    1.000     0.000     0.000     0.000
        2    0.000     1.000     0.000     0.000
        3    0.000     0.000     1.000     0.000
        4    0.000     0.000     0.000     1.000
```

5.7.2 The FLRV directive

FLRV

Forms the values of LRV structures.

Options

PRINT	= strings	Printed output required (roots, vectors); default * i.e. no printing
NROOTS	= scalar	Number of roots or vectors to print; default * i.e. print them all
SMALLEST	= string	Whether to print the smallest roots instead of the largest (no, yes); default n

Parameters

INMATRIX	= matrices	Matrices whose latent roots and vectors are to be calculated
LRV	= LRVs	LRV to store the latent roots and vectors from each INMATRIX
WMATRIX	= matrices	(Generalized) within-group sums of squares and products matrix used in forming the "two-matrix decomposition"; if any of these is omitted, it is taken to be the identity matrix, giving the usual spectral decomposition

---

This directive solves the two related eigenvalue problems

1) $AX = XL$
2) $AX = WXL$

of which the first gives the spectral decomposition of the $n \times n$ matrix A. Here A and W are both symmetric matrices, L is a diagonal matrix, and X is a square matrix. The structures X and L will usually be the first two elements of an LRV structure.

In problem 1, the *one-matrix problem*, XLX is the spectral decomposition of the symmetric matrix A. Here L is a diagonal matrix containing the n latent roots, or eigenvalues, of A ordered such that

$$l_1 \geq l_2 \geq \ldots \geq l_n$$

The columns of the $n \times n$ matrix X are the corresponding latent vectors, or eigenvectors. The matrix X is orthogonal:

$$X'X = XX' = I_n$$

The spectral decomposition is the basis for several multivariate methods (Chapter 10).

In problem 2, the *two-matrix problem*, both A and W must have the same number of rows, n, and W must be positive semi-definite. Now the latent roots are the n elements of the diagonal matrix L and are the successive maxima of

$$l = (x'Ax) / (x'Wx)$$

where x is the corresponding column of the $n \times n$ matrix X, normalized so that $X'WX = I$. The two-matrix decomposition is particularly relevant for canonical variate analysis (10.2.2).

For either problem, the sum of the latent roots is stored in the element of the LRV labelled 'Trace'. In the one-matrix problem, this is also the trace of the original matrix A; but for the two-matrix problem, it is the trace of $W^{-1}A$. Latent roots are often expressed as percentages of the trace (see Chapter 10).

The method used for the spectral decomposition first reduces the matrix to tri-diagonal form using Householder transformations (Martin, Reinsch, and Wilkinson 1968); this is followed by a QL algorithm for finding the eigenvalues and eigenvectors (Bowdler, Martin, Reinsch, and Wilkinson 1968). The two-matrix problem is solved using two spectral decompositions, each computed as for the first problem.

The three options of FLRV control the printing of the results. You use the PRINT option to specify whether you want the roots or vectors to be printed. If you request the roots to be printed, the trace will be printed as well. By default nothing is printed. The NROOTS option governs how many of the roots and vectors are printed, while the SMALLEST option determines whether the largest or smallest roots, and corresponding vectors, are printed.

The INMATRIX parameter lists the matrices for which latent roots and vectors are to be calculated. You can use the LRV parameter to save the latent roots and vectors, and the trace; you must previously have declared these structures to be LRVs, with the correct number of rows.

This example forms the LRV structure, Ulrv, from a matrix of sums of squares and products (5.7.3), and prints the two smallest roots in order of descending magnitude. The trace is printed together with the latent roots, and the latent roots are printed as percentages of the trace.

EXAMPLE 5.7.2a

```
1   VARIATE [NVALUE=13] U[1...7]
2   OPEN 'HARVF.DAT'; CHANNEL=2
3   READ [CHANNEL=2] U[]
```

Identifier	Minimum	Mean	Maximum	Values	Missing
U[1]	7.91	10.62	12.71	13	0
U[2]	7.71	10.25	12.15	13	0
U[3]	8.32	10.46	13.16	13	0

5.7 Operations on matrices and compound structures

```
            U[4]      9.19    10.86    13.06    13      0
            U[5]      7.72    10.46    13.08    13      0
            U[6]      8.69    10.53    12.82    13      0
            U[7]      8.81    10.31    11.99    13      0
  4   SSPM [TERMS=U[]] Us; SSP=Ussp
  5   FSSPM Us
  6   LRV [ROWS=7; COLUMNS=6] Ulrv
  7   FLRV [PRINT=roots,vectors; NROOTS=2; SMALLEST=yes] Ussp; LRV=Ulrv

7...........................................................
```

***** Spectral decomposition *****

*** Latent Roots ***

              6.00   7.00
              6.881   1.122

*** Percentage variation ***

              6.00   7.00
              4.25   0.69

*** Trace ***

     161.7

*** Latent vectors ***

                  6.00   7.00
        1   0.4770   -0.4040
        2   -0.4345   -0.1617
        3   -0.0525   0.1007
        4   0.1123   0.4762
        5   0.1425   -0.0629
        6   0.4178   -0.5527
        7   -0.6111   -0.5141

You can save a subset of the latent roots and vectors by supplying an LRV structure with fewer columns than rows. However this saves only the largest roots and the corresponding vectors. You cannot save the smallest roots directly, as the SMALLEST option applies only to printing. If you want to save the smallest roots, then you must save the complete set of roots and vectors, and extract the last columns of the matrix, for example using qualified identifiers (5.1.6). These rules are the same as those applied in the directives for multivariate analysis (see 10.1.1).

For the two-matrix problem, you specify the matrix W using the WMATRIX parameter. As an example we take W to be the diagonal of the matrix A. In this case, the solution is equivalent to the spectral decomposition of the correlation matrix derived from A, although the normalization of the latent vectors will be different. This example shows the equivalence of the two analyses.

EXAMPLE 5.7.2b

```
  8  CALCULATE Usspcor = CORRMAT(Ussp)
  9  PRINT Usspcor; FIELDWIDTH=7; DECIMALS=3
```

```
   Usspcor

 1   1.000
 2   0.215   1.000
 3   0.179   0.113   1.000
 4   0.294   0.439  -0.002   1.000
 5  -0.137   0.100  -0.049   0.345   1.000
 6  -0.754  -0.013  -0.014   0.065   0.062   1.000
 7   0.177  -0.112   0.021   0.419   0.258  -0.359   1.000
         1       2       3       4       5       6       7
```

```
 10  DIAGONALMATRIX Dusp
 11  SYMMETRICMATRIX Sdusp
 12  CALCULATE Sdusp = (Dusp = Ussp)
 13  LRV [ROWS=7; COLUMNS=7] Uclrv,Usclrv
 14  FLRV [PRINT=roots,vectors; NROOTS=4] Usspcor; LRV=Uclrv
```

14..

***** Spectral decomposition *****

*** Latent Roots ***

```
                1        2        3        4
             2.114    1.593    1.244    0.936
```

*** Percentage variation ***

```
                1        2        3        4
             30.20    22.76    17.78    13.37
```

*** Trace ***

```
     7.000
```

*** Latent vectors ***

```
                    1         2         3         4
          1    0.5558   -0.3518   -0.1574    0.1238
          2    0.2649    0.2787   -0.6400    0.2819
          3    0.1227   -0.1055   -0.4073   -0.8902
          4    0.4249    0.5042   -0.1067    0.0873
          5    0.1510    0.5429    0.2671   -0.1397
          6   -0.4809    0.4648   -0.1942   -0.1236
          7    0.4138    0.1497    0.5284   -0.2651
```

```
 15  & Ussp; LRV=Uclrv; WMATRIX=Sdusp
```

15..

***** Two-matrix latent decomposition *****

*** Latent Roots ***

```
                1        2        3        4
             2.114    1.593    1.244    0.936
```

5.7 Operations on matrices and compound structures 217

```
*** Percentage variation ***
                    1         2         3         4
                30.20     22.76     17.78     13.37

*** Trace ***

       7.000

*** Latent vectors ***
                    1         2         3         4
           1    0.09616   0.06086  -0.02723  -0.02142
           2    0.06482  -0.06821  -0.15664  -0.06899
           3    0.02494   0.02144  -0.08279   0.18093
           4    0.09889  -0.11735  -0.02484  -0.02033
           5    0.02429  -0.08736   0.04298   0.02248
           6   -0.10395  -0.10046  -0.04198   0.02672
           7    0.13848  -0.05011   0.17682   0.08871
```

A similar use of the two-matrix problem is when W is obtained from previous samples of the same set of variables as those in A.

For a symmetric matrix A, you can use FLRV to form an inverse of A in much the same way as the singular value decomposition. If A is singular, you form the Moore-Penrose inverse (pseudo inverse). This example follows the lines of the SVD example for the generalized inverse of a matrix (5.7.1).

EXAMPLE 5.7.2c

```
 16   SYMMETRICMATRIX [ROWS=3; VALUES=10,13,17,17,22,29] Smx
 17   LRV [ROWS=3; COLUMNS=3] Lsmx; VECTORS=Vsmx; ROOTS=Rts
 18   FLRV [PRINT=roots,vectors] Smx; LRV=Lsmx

18...........................................................

***** Spectral decomposition *****

*** Latent Roots ***
             Rts
               1         2         3
           55.80      0.20      0.00

*** Percentage variation ***
             Rts
               1         2         3
           99.65      0.35      0.00

*** Trace ***

Lsmx['Trace']
       56.00

*** Latent vectors ***
```

```
            Vsmx
              1          2          3
    1      0.4233    -0.0512     0.9045
    2      0.5500    -0.7788    -0.3015
    3      0.7199     0.6251    -0.3015

 19   " The value 1.E-6 is to check for roots which,
-20     but for numerical round-off, would be zero.
-21     This might need to be changed in another example. "
 22   CALCULATE [ZDZ=zero] Irts = (Rts > 1.E-6) / Rts
 23   CALCULATE Ismx = Vsmx *+ Irts *+ TRANSPOSE(Vsmx)
 24   PRINT Ismx; FIELDWIDTH=8; DECIMALS=2

            Ismx
              1          2          3
    1       0.02       0.21      -0.16
    2       0.21       3.08      -2.46
    3      -0.16      -2.46       1.99
```

The relationship between the singular value decomposition of a rectangular matrix A and the spectral decompositions of $A'A$ and AA' is as follows. If $A = USV'$ is the singular value decomposition for A, then $A'A = VSU'USV' = VS^2V'$ and $AA' = USV'VSU' = US^2U'$, since $U'U = V'V = I$. The rank of matrix A is q and $q \leq \min(m,n)$, which is p in our earlier notation (5.7.1); q corresponds to the number of non-zero singular values, and the diagonal matrix S consists of the q non-zero singular values followed by $(p-q)$ zero values. This shows that the squares of the q singular values of A are equivalent to the non-zero latent roots of the two symmetric matrices, $A'A$, and AA', derived from A. It also shows that the matrices U and V contain the first p latent vectors of AA' and $A'A$, respectively. For further details, see Rao (1973, Chapter 1) or Digby and Kempton (1987, Appendix A.8).

5.7.3 The FSSPM directive

FSSPM

Forms the values of SSPM structures.

Options

PRINT	= strings	Printed output required (correlations, wmeans, SSPM); default * i.e. no printing
WEIGHTS	= variate	Weightings for the units; default * i.e. all units with weight one
SEQUENTIAL	= scalar	Used for sequential formation of SSPMs; a positive value indicates that

		formation is not yet complete (see READ directive); default * i.e. not sequential
SAVE	= identifier	Regression work structure (see TERMS directive); default *

Parameter

	SSPMs	Structures to be formed

FSSPM forms the values for the component parts of SSPM structures, based on the information supplied when the SSPM structures were declared (3.6.2). You can use an SSPM as input to the regression directive TERMS (8.2.2), or the multivariate directives, PCP and CVA (10.2). The method used to form the SSPM is based on the updating formula for the means and corresponding corrected sums of squares and cross products (Herraman 1968).

FSSPM has one parameter which lists the SSPM structures whose values are to be formed. Genstat takes account of restrictions on any of the variates or factors forming the terms of the SSPM, or on the weights variate or grouping factor if you have specified them. If any of these vectors has a missing value, the corresponding unit is excluded from all the means and all the sums of squares and products. You can also exclude units by setting their weights to zero.

In this example, units 1, 5, and 7 are omitted. Notice that the 'wmeans' setting of the PRINT option is ignored, as the GROUPS option of the SSPM directive has not been set.

EXAMPLE 5.7.3a

```
  1  VARIATE [NVALUE=10] Va[1...6]
  2  OPEN 'HARVFB.DAT'; CHANNEL=2
  3  READ [CHANNEL=2] Va[]

    Identifier    Minimum        Mean     Maximum      Values    Missing
         Va[1]      15.70       36.86       47.10          10          0
         Va[2]      32.30       37.91       55.60          10          0    Skew
         Va[3]      29.40       37.47       53.00          10          0
         Va[4]      26.20       33.66       44.00          10          0
         Va[5]      13.20       38.06       51.90          10          0
         Va[6]      12.70       36.74       54.60          10          0

  4  VARIATE Weight; VALUES=!(0,1,1,1,0,1,0,1,1,1)
  5  SSPM [TERMS=Va[]] Ssva
  6  FSSPM [PRINT=wmean,correlation,sspm; WEIGHT=Weight] Ssva

*** Degrees of freedom ***

Sums of squares:   6
Sums of products:  5
Correlations:      5
```

```
*** Sums of squares and products ***

    Va[1]       1       482.54
    Va[2]       2        91.37      88.11
    Va[3]       3       248.40     141.25     559.24
    Va[4]       4       -82.84    -105.96     -75.79     270.99
    Va[5]       5      -305.37     -52.51    -142.92     248.19
    Va[6]       6       122.76     -30.11     248.49      43.17

                          1          2          3          4

    Va[5]       5       983.05
    Va[6]       6      -593.52     799.23

                          5          6

*** Means ***

    Va[1]              33.54
    Va[2]              35.39
    Va[3]              38.34
    Va[4]              34.67
    Va[5]              34.17
    Va[6]              34.27

*** Sum of weights ***

            7.000

*** Correlation matrix ***

    Va[1]       1   1.000
    Va[2]       2   0.443   1.000
    Va[3]       3   0.478   0.636   1.000
    Va[4]       4  -0.229  -0.686  -0.195   1.000
    Va[5]       5  -0.443  -0.178  -0.193   0.481   1.000
    Va[6]       6   0.198  -0.113   0.372   0.093  -0.670   1.000

                      1       2       3       4       5       6
```

When you have very many units, you may not be able to store them all at the same time within Genstat. You can then use the SEQUENTIAL option of READ (4.1.6) to read the data in conveniently sized blocks, and the SEQUENTIAL option of FSSPM to control the accumulation of the sums of squares and products. The SSPM is updated for each block of data in turn until the end of data is found.

EXAMPLE 5.7.3b

```
  7   OPEN 'HARV.DAT'; CHANNEL=3
  8   SCALAR Sseq; 0
  9   VARIATE [NVALUE=10] V[1...5]
 10   SSPM [TERMS=V[]] Vssp
 11   FOR [NTIMES=999]
 12      READ [CHANNEL=3; SEQUENTIAL=Sseq] V[]
 13      FSSPM [SEQUENTIAL=Sseq; PRINT=SSPM] Vssp
 14      EXIT Sseq <= 0
 15   ENDFOR
```

5.7 Operations on matrices and compound structures

```
    Identifier    Minimum       Mean    Maximum    Values    Missing
         V[1]        8.52      10.13      11.75        10          0
         V[2]       8.910      9.829     10.800        10          0
         V[3]        8.69      10.95      13.08        10          0
         V[4]        7.71      10.01      11.65        10          0
         V[5]        9.29      10.50      12.34        10          0

    Identifier    Minimum       Mean    Maximum    Values    Missing
         V[1]        8.32      10.29      12.71        10          0
         V[2]        7.72      11.00      13.16        10          0
         V[3]        8.93      10.79      12.66        10          0
         V[4]        7.91      11.02      13.06        10          0
         V[5]        8.81      10.90      13.07        10          0


*** Degrees of freedom ***

Sums of squares:   20
Sums of products:  19


*** Sums of squares and products ***

V[1]        1        27.75
V[2]        2         4.72      39.99
V[3]        3        -4.55      -3.45     34.06
V[4]        4         4.01      16.81      7.59     55.32
V[5]        5         6.76      24.19     10.79     16.81

                      1           2         3         4

V[5]        5        34.71

                      5


*** Means ***

V[1]              10.23
V[2]              10.50
V[3]              10.86
V[4]              10.64
V[5]              10.80

*** Number of units used ***

         21
```

Notice that the PRINT option has no effect until the last set of values is processed, when READ sets the scalar indicator to a negative value (4.1.6).

If you use an SSPM as input to the TERMS directive (8.2.2), Genstat will copy the information into a special structure called a DSSP. This uses extra precision for storage on computers where real numbers are not represented by enough bits to guard against unacceptable inaccuracies of round-off during the regression calculations. The extra precision is also used while the information is calculated by FSSPM, but it will be lost when the values are placed in the SSPM. The SAVE option allows you to save the DSSP formed internally by FSSPM, so that you can put it straight into TERMS.

5.8 Operations on tables

A table is a structure that stores numerical summaries of data that are classified into groups. The TABULATE directive lets you form tables from a variate, given also factors to define the groups; you can form and print tables containing counts, means, totals, minima, maxima, or variances of the observations in each group (5.8.1). You can also use tables to save means, effects, and numbers of replications from an analysis of variance (9.6.1), or predictions from regression and generalized linear models (8.3.4 and 8.4.4).

You can do numerical calculations on the values in tables, using the CALCULATE directive (5.1.1, 5.1.4, and 5.2.5). You can re-form a table to omit or combine levels of any of the classifying factors (5.3.4); and you can include, omit, or recalculate margins (5.8.2).

5.8.1 The TABULATE directive

TABULATE

Forms summary tables of variate values.

Options

PRINT	= strings	Printed output required (counts, totals, nobservations, means, minima, maxima, variances); default * i.e. no printing
CLASSIFICATION	= factors	Factors classifying the tables; default * i.e. these are taken from the tables in the parameter lists
COUNTS	= table	Saves a table counting the number of units with each factor combination; default *
SEQUENTIAL	= scalar	Used for sequential formation of tables; a positive value indicates that formation is not yet complete (see READ); default *
MARGINS	= string	Whether the tables should be given margins if not already declared (no, yes); default n
IPRINT	= string	Whether to print the identifier of the table or the identifier of the (associated) variate that was used to

form it (identifier,
associatedidentifier); default i

Parameters

DATA	= variates	Data values to be tabulated
TOTALS	= tables	Tables to contain totals
NOBSERVATIONS	= tables	Tables containing the numbers of non-missing values in each cell
MEANS	= tables	Tables of means
MINIMA	= tables	Tables of minimum values in each cell
MAXIMA	= tables	Tables of maximum values in each cell
VARIANCES	= tables	Tables of cell variances

TABULATE allows you to produce the various types of tabular summary listed in the settings of its PRINT option. If you want to save the summaries in tables, for manipulating or for printing later on, you should list identifiers of the tables in the appropriate parameter list: for example, you would save the totals in a table T by including T in the list for the TOTALS parameter. The other parameters similarly give the other kinds of summary. If you merely want to print the summaries, you do not usually need to list any tables: you need only specify the PRINT option; the only exception to this is with sequential tabulation, described at the end of this subsection. The variates whose values are to be summarized are listed with the DATA parameter.

Any table that you have not declared in advance will be declared implicitly. If you have not declared any of the tables, the classifying factors are taken from the CLASSIFICATION option, which in that case you must have set; likewise, the MARGINS option determines whether or not the tables will have margins. Otherwise these two options are ignored, and the undeclared tables are defined to have the same classifying factors and status for margins as the tables that you have declared previously; all these previously declared tables must have the same set of classifying factors, and must be all with margins or all without margins.

The next example concerns goods of two different types dispatched to four different towns. In the print of the data you will notice that the book-keeping has been rather slack. There is one consignment (in line 6) where the type has not been recorded. With such observations, Genstat cannot find out what the group should be because one of the factor values is missing; so they are ascribed to the *unknown* cell associated with the table (3.5). In the declaration in line 9, the scalar that stores this value has been named so that it can be referred to in later calculations. After the tabulation (line 10), table Totdisp stores the total number of items of each type dispatched to each town, and the scalar Udisp summarizes the observations with unknown type or destination.

EXAMPLE 5.8.1a

```
  1  VARIATE [NVALUES=15] Quantity,Charge
  2  FACTOR [NVALUES=15; LABELS=!T(A,B)] Type
  3   & [LABELS=!T(London,Manchester,Birmingham,Bristol)] Town
  4  READ [PRINT=data,errors] Town,Quantity,Type; FREPRESENTATION=labels

  5       London 10 A    Manchester   5 B  Birmingham  10 B     Bristol 25 A
  6   Manchester 10 *    Birmingham 100 B      London 200 B  Manchester 25 A
  7       Bristol 50 A   Birmingham  25 A     Bristol  25 B      London 25 A
  8        London 50 B   Manchester  25 B      London  50 A   :

  9  TABLE [CLASSIFICATION=Town,Type] Totdisp; UNKNOWN=Udisp
 10  TABULATE Quantity; TOTALS=Totdisp
 11  PRINT Totdisp; DECIMALS=0

                 Totdisp
        Type        A            B
        Town
      London       85          250
  Manchester       25           30
  Birmingham       25          110
     Bristol       75           25

Unknown     Totdisp        10
```

The second part of the example illustrates what happens when a value of the data variate is missing. Variate Charge stores the charge to be made for the transport of each consignment, and you will see that three of the values are missing (because these invoices have not yet been prepared). In the tables listed with the parameters, missing data values are ignored. For example, the table Invoices is declared automatically by the NOBSERVATIONS parameter to hold the number of invoices sent to each destination; it excludes the observations where Charge has a missing value. Similarly Payment contains the total charge to be paid on behalf of each destination, ignoring the missing values. You can however obtain a count of the numbers of units that would have contributed to each group if no values had been missing: you use the COUNTS option if you want to save the table, or put PRINT=counts if you want to print it. So table Nconsign contains the total number of consignments made to each destination (regardless of whether the corresponding charge is missing or not). The data variates are irrelevant for counts, and so you need not list any if counts are all that you require.

EXAMPLE 5.8.1b

```
 12  READ [PRINT=data,errors] Charge

 13  10 20 15 15 * 60 80 30 25 15 25 15 40 * * :
```

```
14  TABULATE [CLASSIFICATION=Town; COUNTS=Nconsign] DATA=Charge; \
15    TOTALS=Payment; NOBSERVATIONS=Invoices
16  PRINT Nconsign,Invoices,Payment; DECIMALS=0,0,2
```

```
              Nconsign     Invoices      Payment
       Town
     London          5            4       145.00
 Manchester          4            2        50.00
 Birmingham          3            3        90.00
     Bristol         3            3        65.00
```

If there are no observations in one of the groups, the corresponding cell will be zero in a table of numbers of observations or counts; in a table of totals, means, minima, maxima, or variances, the cell will contain a missing value.

If you have many observations to summarize, there may be insufficient space within Genstat for you to read them all and then form the tables. To cater for such situations, Genstat allows you to process the data in sections, using the SEQUENTIAL option of TABULATE in conjunction with the SEQUENTIAL option of READ (4.1.6). After READ, the absolute value of the option indicates the number of units that have been read in this particular section; the value is positive during interim sections and negative or zero once the terminator at the end of the data is reached. TABULATE will not print any tables until the final section has been processed. If you want to see the intermediate tables, you can include a PRINT statement after the TABULATE statement. With sequential tabulation, you must supply tables in the parameter lists for all the tables that TABULATE is to print. Moreover any tables that Genstat needs for calculating these tables must also be supplied: tables of means need tables of numbers of observations; tables of totals also need tables of numbers of observations (in order to identify empty cells); and tables of variances require tables of means and tables of numbers of observations.

All this is illustrated in the final part of the example, which also shows how to use the IPRINT option to print the identifier of the variate from which the table was formed, instead of the identifier of the table. Also notice that this time the table formed has a margin (3.5). As there is only one type of table being printed, Genstat has labelled the margin appropriately (as "Mean"). If several types of table were printed, Genstat would label the margins as "Margin".

EXAMPLE 5.8.1c

```
1  UNITS [NVALUES=20]
2  FACTOR [LABELS=!T(UK,EEC,other)] National
3  FACTOR [LABELS=!T(male,female)] Sex
4  VARIATE Baggage
5  VARIATE Excess; EXTRA=' baggage per person in Kilograms'; DECIMALS=3
6  OPEN 'FLIGHT.DAT'; CHANNEL=2
7  SCALAR S
8  FOR [NTIMES=999]
```

```
 9  READ [CHANNEL=2; SEQUENTIAL=S] Baggage,Sex,National; \
10      FREPRESENTATION=labels
11  CALCULATE Excess=(Baggage>20)*(Baggage-20)
12  TABULATE [PRINT=mean; CLASSIFICATION=Sex,National; SEQUENTIAL=S; \
13      MARGINS=yes; IPRINT=associatedidentifier] \
14      Excess; NOBSERVATIONS=Ntemp; MEANS=Mtemp
15  EXIT S <= 0
16  ENDFOR
```

```
    Identifier    Minimum        Mean     Maximum      Values     Missing
    Baggage         15.00       20.50       28.00          20           0


    Identifier    Minimum        Mean     Maximum      Values     Missing
    Baggage         17.00       22.45       35.00          20           0


    Identifier    Minimum        Mean     Maximum      Values     Missing
    Baggage         15.00       20.65       30.00          20           0


    Identifier    Minimum        Mean     Maximum      Values     Missing
    Baggage         15.00       20.35       28.00          20           3


              Excess baggage per person in Kilograms
    National            UK         EEC       other          Mean
    Sex
      male          1.292       2.364       3.857         2.265
      female        1.200       1.400       1.250         1.250

    Mean            1.256       2.063       2.909         1.896
```

The printed table summarizes the amount of excess baggage per person for the passengers on a particular flight. There are 77 passengers. The factors and variates are declared to have length 20, so the data are read in three sections of size 20 and a final section of size 17. The setting of the SEQUENTIAL option is the scalar S: it has the value 20 for the first three times that the loop is executed, and −17 on the final time. Notice that the variate Baggage is given missing values in units 18 to 20 in the final section: the value −17 in S tells Genstat that these units are not to be included in the tabulation. The loop construct FOR-ENDFOR is described in 6.2.1, and the EXIT directive in 6.2.4.

5.8.2 The MARGIN directive

MARGIN

Forms and calculates marginal values for tables.

Option

CLASSIFICATION	= factors	Factors classifying the margins to be formed; default * requests all margins to be formed

Parameters

OLDTABLE	= tables	Tables from which the margins are to be taken or calculated
NEWTABLE	= tables	New tables formed with margins
METHOD	= strings	Way in which the margins are to be formed for each table (totals, means, minima, maxima, variances, medians, deletion, or a null string to indicate that the marginal values are all to be set to the missing value); if unspecified, t is assumed

You can use MARGIN to extend a table to contain marginal values, or to change the marginal values of a table that already has margins, or to delete the margins from a table. The tables whose margins are to be changed are specified by the OLDTABLES parameter. If you specify only this parameter, the new values replace those of the original tables. For example, in 3.5, the statement

    MARGIN Classnum,Schoolnm

formed margins of totals over all the classifying factors for the tables, Classnum and Schoolnm; the new values, including the margins, replaced the original values of Classnum and Schoolnm.

However, if you want to retain the original values, you can specify new tables to contain the amended values, using the NEWTABLES list. These tables will be declared automatically, if you have not declared them already.

This example creates the new tables, Classt and Schoolt, with margins of totals, using the values in the tables Classnum and Schoolnm.

EXAMPLE 5.8.2a

```
  1  FACTOR [LABELS=!T(boy,girl)] Sex
  2  FACTOR [LEVELS=5] Class
  3  FACTOR [LEVELS=2] School
  4  TABLE [CLASSIFICATION=Class,Sex; \
  5     VALUES=15,17,29,31,34,30,33,35,28,27] Classnum
  6  TABLE [CLASSIFICATION=School,Class,Sex; VALUES=15,17,29,31,34, \
  7     30,33,35,28,27,18,16,33,31,35,36,34,33,31,32] Schoolnm
  8  MARGIN Classnum,Schoolnm; NEWTABLE=Classt,Schoolt
  9  PRINT Classt,Schoolt; DECIMALS=0

                Classt
      Sex        boy          girl         Margin
    Class
        1         15           17            32
        2         29           31            60
        3         34           30            64
```

```
                      4          33          35          68
                      5          28          27          55

                   Margin       139         140         279

                               Schoolt
                  Sex           boy        girl       Margin
      School    Class
        1          1            15          17          32
                   2            29          31          60
                   3            34          30          64
                   4            33          35          68
                   5            28          27          55

                 Margin        139         140         279

        2          1            18          16          34
                   2            33          31          64
                   3            35          36          71
                   4            34          33          67
                   5            31          32          63

                 Margin        151         148         299

      Margin       1            33          33          66
                   2            62          62         124
                   3            69          66         135
                   4            67          68         135
                   5            59          59         118

                 Margin        290         288         578
```

You can form other types of margin by setting the METHOD parameter. The next example forms the tables Classno and Schoolno with margins of means and maxima respectively.

EXAMPLE 5.8.2b

```
   10   MARGIN Classnum,Schoolnm; NEWTABLE=Classno,Schoolno; \
   11       METHOD=means,maxima : PRINT Classno,Schoolno; DECIMALS=0

                   Classno
          Sex       boy        girl       Margin
         Class
           1        15          17          16
           2        29          31          30
           3        34          30          32
           4        33          35          34
           5        28          27          28

         Margin     28          28          28

                   Schoolno
                   Sex         boy        girl       Margin
        School   Class
          1        1            15          17          17
                   2            29          31          31
                   3            34          30          34
                   4            33          35          35
```

5.8 Operations on tables

```
              5        28          27         28
         Margin       34          35         35
 2            1        18          16         18
              2        33          31         33
              3        35          36         36
              4        34          33         34
              5        31          32         32
         Margin       35          36         36
Margin        1        18          17         18
              2        33          31         33
              3        35          36         36
              4        34          35         35
              5        31          32         32
         Margin       35          36         36
```

All the examples so far have been of adding margins. But you can delete them too: if you set METHOD=deletion, all the margins of the tables are deleted but the body of the table is retained.

The CLASSIFICATION option specifies the list of factors for which you want to form marginal values. This example forms a margin of totals for the factor Class in the table Classnum.

EXAMPLE 5.8.2c

```
 12   MARGIN [CLASSIFICATION=Sex] Classnum
 13   PRINT Classnum; DECIMALS=0

                  Classnum
         Sex         boy        girl      Margin
       Class
           1         15          17          32
           2         29          31          60
           3         34          30          64
           4         33          35          68
           5         28          27          55
      Margin          0           0           0
```

Genstat puts missing values in the margins that are excluded if the METHOD parameter is set to maxima or minima; for other settings of METHOD, Genstat puts in zeroes. The classifying sets for each table can be different, but all the factors in the CLASSIFICATION option must be in the classifying sets of each OLDTABLE. So, for example,

    MARGIN [CLASSIFICATION = Sex,School] Classnum, Schoolnm

would fail because the factor School is not in the classifying set of Classnum.

A.E.A.
P.K.L.
R.W.P.
P.J.V.
R.P.W.

6 Job control

In this chapter we tell you more about the structure of a Genstat program (6.1). We also describe the directives that allow you to loop over sequences of statements, or to choose between alternative sets of statements (6.2). These give you the flexibility to write general programs, where the exact analysis to be performed depends on information derived from the data when the program is run. Programs that are frequently required can be stored in procedures, as described in 6.3. This not only makes them simpler for you to use; it also means that you can make them easily available to other users, by means of libraries or by articles in the Genstat Newsletter (see 12.1).

6.1 Genstat programs

A Genstat program is a sequence of statements involving either standard Genstat directives or procedures (1.1). You may often wish to examine several different sets of data within the same program: for example, you may want to be able to collect several analyses in one batch run (1.1.11), or to be able to end one analysis and start a different one, with different data, when you are running Genstat interactively (1.1.10).

The JOB and ENDJOB directives allow you to partition a Genstat program into separate jobs. Each of these is self-contained. Many of the settings that define the environment of your program carry over from job to job, unless you explicitly modify them by options of JOB, and the status of any files attached to Genstat also remain unchanged. However, values and identifiers of all procedures and data structures are deleted at the end of a job. Any setting defined by a UNITS statement (2.2.3) is deleted, as are the special structures set up by analyses like regression and analysis of variance (2.2.1). The graphics environment is also reset to the initial default (7.2).

6.1.1 The JOB directive

JOB

Starts a Genstat job.

Options

INPRINT	= strings	Printing of input as in PRINT option of INPUT (statements, macros,

OUTPRINT	= strings	procedures, unchanged); default u Additions to output as in PRINT option of OUTPUT (dots, page, unchanged); default u
DIAGNOSTIC	= strings	Output to be printed for a Genstat diagnostic (warnings, faults, extra); default f,w
ERRORS	= scalar	Limit on number of error diagnostics that may occur before the job is abandoned; default * i.e. no limit
Parameter		
	text	Name to identify the job

You use the JOB directive to start a new job. Its options also allow you to modify the Genstat environment (2.2). These all have the default 'unchanged', which keeps any settings that you may have had in the previous job, or which have been defined by the initial defaults (2.2). You use option INPRINT to select those parts of the input that are to appear in the output file. The OUTPRINT option specifies any additional output that is to be generated by certain directives (see 2.2.1): 'dots' gives a line of dots and the statement number at the start of the output from each of these directives; 'page' causes the output also to start on a new page. The DIAGNOSTICS option specifies which types of diagnostic message should be printed (2.1.2). There are two classes of message: 'warnings' and 'faults'. If DIAGNOSTICS=warnings, then faults will be printed as well, since these are more serious than warnings. The setting 'extra' is used to generate a dump (2.5.1) after a diagnostic has been printed; this will also produce warnings and faults. You can inhibit output of both warnings and faults by putting DIAGNOSTICS=*. The ERRORS option puts a limit on the number of error messages that are printed after a fatal error has occurred.

All four options of JOB have equivalents in SET (2.2.1), and INPRINT and OUTPRINT correspond also to the PRINT options of INPUT (4.4.1) and OUTPUT (4.4.3) respectively.

JOB also has a parameter which you can set to a text to identify the job; this is printed as part of the message at the end of the job.

You do not need to give a JOB statement at the start of a program, unless you wish to define an identifying text or to modify the environment. Similarly JOB can be omitted after an ENDJOB statement.

6.1.2 The ENDJOB directive

ENDJOB

Ends a Genstat job.

No options or parameters

The ENDJOB directive terminates a job, printing a message summarizing how much workspace has been used. For example:

EXAMPLE 6.1.2

```
  1  JOB 'Example of ENDJOB message'
  2  PRINT 'This job just prints this message.'

This job just prints this message.

  3  ENDJOB

******** End of Example of ENDJOB message.  Maximum of 820 data units used at
line 2 (41918 left)
```

You do not need to give an ENDJOB statement before a JOB statement, as JOB will automatically end any existing job before it starts another. Thus you can begin a new job by specifying either JOB or ENDJOB (or both).

6.1.3 The STOP directive

STOP

Ends a Genstat program.

No options or parameters

The STOP directive terminates a sequence of jobs. It tells the computer that you have finished using Genstat. It also ends the existing job, so there is no need to give an ENDJOB statement beforehand.

Genstat will ignore any input following a STOP statement.

6.2 Program control in Genstat

Usually the statements in a Genstat job are executed in sequence, until either ENDJOB or STOP is reached. But, as with most programming languages, you may sometimes want to control the order in which the statements are executed.

If you have several sets of data that are all to be analysed in the same way, you may want to repeat the necessary series of statements for each set. You can do this by preceding the series with a FOR statement, and ending it with an ENDFOR statement. The FOR directive also allows you to specify dummy structures (3.2.2) which point in turn to the data structures of the successive sets.

To be able to write general programs, you may need to be able to choose between alternative sets of statements, according to the exact form of a particular set of data. There are two ways in which you can do this. The directives IF, ELSIF, ELSE, and ENDIF allow you to define *block-if* structures (6.2.2). Alternatively, the directives CASE, OR, ELSE, and ENDCASE allow you to choose between sets of statements according to an integer value (6.2.3).

The directive EXIT (6.2.4) allows you to abandon any of these control structures while the program is being executed. The exit can be dependent on a condition, for example on an invalid data value or even on a Genstat diagnostic. The Genstat language is designed in accordance with the principles of structured programming: there is no way of "labelling" a statement and no equivalent of the Fortran "GO TO" construct.

6.2.1 FOR loops

FOR

Introduces a loop; subsequent statements define the contents of the loop, which is terminated by the directive ENDFOR.

Options

NTIMES	= scalar	Number of times to execute the loop; default is to execute as many times as the length of the first parameter list or once if the first list is null
COMPILE	= string	Whether to execute each statement as it is compiled or to compile all statements before execution as a block (each, all); default a

Parameters

	dummies	are set up implicitly by the statement; each dummy appears to be a parameter

---

> **ENDFOR**
>
> Indicates the end of the contents of a loop.
>
> **No options or parameters**

The FOR loop is a series of statements, or a *block*, that is repeated several times. The FOR directive introduces the loop and indicates how many times it is to be executed. In its simplest form, FOR has no parameters. You set the number of times by the NTIMES option, giving either a single number or a scalar. Thus the iterative calculation of a square root in the example in 1.7.2 can be specified in a FOR loop like this:

```
FOR [NTIMES=3]
  CALCULATE Previous = Root
  & Root = (X/Previous + Previous)/2
  PRINT Root,Previous; DECIMALS=4
ENDFOR
```

The sequence of CALCULATE and PRINT statements is repeated three times (exactly as in 1.7.2).

The parameters of FOR allow you to write a loop whose contents apply to different data structures each time it is executed. Unlike other directives, the parameter names of FOR are not fixed for you by Genstat: you can put any valid identifier before each equals sign. Each of these then refers to a Genstat dummy structure, as described in 3.2.2; so you must not have declared them already as any other type of structure. The first time that the loop is executed, they each point to the first data structure in their respective lists, next time it is the second structure, and so on. The list of the first parameter must be the longest; other lists are recycled as necessary. You can specify as many parameters as you need. For example

```
FOR Ind=Age,Name,Salary; Dir='descending','ascending'
  SORT [INDEX=Ind; DIRECTION=#Dir] Name,Age,Salary
  PRINT Name,Age,Salary
ENDFOR
```

is equivalent to the sequence of statements

```
SORT [INDEX=Age; DIRECTION='descending'] Name,Age,Salary
PRINT Name,Age,Salary
SORT [INDEX=Name; DIRECTION='ascending'] Name,Age,Salary
PRINT Name,Age,Salary
SORT [INDEX=Salary; DIRECTION='descending'] Name,Age,Salary
PRINT Name,Age,Salary
```

printing the units of the text Name, and variates Age and Salary, first in order of descending ages, then in alphabetic order of names, and finally in order of descending salaries.

You can put other control structures inside the loop. So you can have loops within loops, for example.

When you are running Genstat interactively (1.1.10), you may wish to execute the first pass through the loop as you type it. If you specify option COMPILE = each, each statement will be compiled and executed by Genstat immediately after it has been read. The default of COMPILE = all causes Genstat to read the entire contents of the loop before it executes any statements. When you are using loops interactively, you may find it helpful to specify the PAUSE option of SET; this requests Genstat to pause after every so many lines of output (2.2.1). Another useful directive is BREAK, which specifies an explicit break in the execution of the loop (6.4.1).

6.2.2 Block-if structures

The component parts of a *block-if* structure are delimited by IF, ELSIF, ELSE, and ENDIF statements.

IF

Introduces a block-if control structure.

No options

Parameter

expression	Logical expression, indicating whether or not to execute the first set of statements.

ELSIF

Introduces a set of alternative statements in a block-if control structure.

No options

Parameter

expression	Logical expression to indicate whether or not the set of statements is to be executed.

ELSE

Introduces the default set of statements in block-if or in multiple-selection control structures.

No options or parameters

ENDIF

Indicates the end of a block-if control structure.

No options or parameters

A *block-if* structure consists of one or more alternative sets of statements. The first of these is introduced by an IF statement. There may then be further sets introduced by ELSIF statements. Then you can have a final set introduced by an ELSE statement, and the whole structure is terminated by an ENDIF statement. Thus the general form is: first

    IF expression
        statements

then either none, one, or several blocks of statements of the form

    ELSIF expression
        statements

then, if required, a block of the form

    ELSE
        statements

and finally the statement

    ENDIF

Each expression must evaluate to a single number, which is treated as a logical value: a zero value is treated as false and non-zero as true (5.1.1). Genstat executes the block of statements following the first true expression. If none of the expressions is true, the block of statements following ELSE (if present) is executed.

You can thus use these directives to built constructs of increasing complexity. The simplest form would be to have just an IF statement, then some statements to execute, and then an ENDIF. For example:

    IF MINIMUM(Sales) < 0
        PRINT 'Incorrect value recorded for Sales.'
    ENDIF

If the variate Sales contains a negative value, the PRINT statement will be executed. Otherwise Genstat goes straight to the statement after ENDIF.

To specify two alternative sets of statements, you can include an ELSE block. For example

```
IF Age < 20
   CALCULATE Pay = Hours*1.75
ELSE
   CALCULATE Pay = Hours*2.5
ENDIF
```

calculates Pay using two different rates: 1.75 for Age less than 20, and 2.5 otherwise.

Finally, to have several alternative sets, you can include further sets introduced by ELSIF statements. Suppose that we want to assign values to X according to the rules:

$X = 1$ if $Y = 1$
$X = 2$ if $Y \neq 1$ and $Z = 1$
$X = 3$ if $Y \neq 1$ and $Z = 2$
$X = 4$ if $Y \neq 1$ and $Z \neq 1$ or 2

This can be coded in Genstat as follows:

```
IF Y == 1
   CALCULATE X = 1
ELSIF Z == 1
   CALCULATE X = 2
ELSIF Z == 2
   CALCULATE X = 3
ELSE
   CALCULATE X = 4
ENDIF
```

If Y is equal to 1, the first CALCULATE statement is executed to set X to 1. If Y is not equal to 1, Genstat does the tests in the ELSIF statements, in turn, until it finds a true condition; if none of the conditions is true, the CALCULATE statement after ELSE is executed to set X to 4. Thus, for $Y = 99$ and $Z = 1$, Genstat will find that the condition in the IF statement is false. It will then test the condition in the first ELSIF statement; this produces a true result, so X is set to 2. Genstat then continues with whatever statement follows the ENDIF statement.

Block-if structures can be nested to any depth, to give conditional constructs of even greater flexibility.

6.2.3 The multiple-selection control structure

The directives CASE, OR, ELSE, and ENDCASE allow you to specify alternative blocks of statements, to be selected according to the value of an expression yielding a single integer value.

CASE

Introduces a "multiple-selection" control structure.

No options

Parameter

        expression   Expression which is evaluated to an integer, indicating which set of statements to execute

---

OR

Introduces a set of alternative statements in a "multiple-selection" control structure.

No options or parameters

ENDCASE

Indicates the end of a "multiple-selection" control structure.

No options or parameters

A *multiple-selection* control structure consists of several alternative blocks of statements. The first of these is introduced by a CASE statement. This has a single parameter, which is an expression that must yield a single number. Subsequent blocks are each introduced by an OR statement. There can then be a final block, introduced by an ELSE statement, as in the block-if structure (6.2.2). The whole structure is terminated by an ENDCASE statement. Thus the general form is: first

    CASE expression
        statements

then either none, one, or several blocks of statements of the form

    OR
        statements

then, if required, a block of the form

    ELSE
        statements

and finally the statement

    ENDCASE

Genstat rounds the expression in the CASE expression to the nearest integer, k say, and then executes the kth block of statements. If there is no kth block present (as for example if k is negative) the block of statements following the ELSE statement is executed, if there is such a block; otherwise an error diagnostic is given. The next example prints the salient details about each day in the song *The twelve days of Christmas*. The scalar Day indicates which day it is.

```
CASE Day
   PRINT 'a partridge in a pear tree'
OR
   PRINT 'two turtle doves and a partridge in a pear tree'
OR
   PRINT 'three French hens, two turtle doves \
and a partridge in a pear tree'
OR
   PRINT 'four calling birds, three French hens ...'
OR
   PRINT 'five gold rings ...'
OR
   PRINT 'six geese a-laying ...'
OR
   PRINT 'seven swans a-swimming ...'
OR
   PRINT 'eight maids a-milking ...'
OR
   PRINT 'nine drummers drumming ...'
OR
   PRINT 'ten pipers piping ...'
OR
   PRINT 'eleven ladies dancing ...'
OR
   PRINT 'twelve lords a-leaping ...'
ELSE
   PRINT 'sorry, no delivery today'
ENDCASE
```

CASE statements can be nested to any depth.

6.2.4 Exit from control structures

Sometimes you may want simply to abandon part of a program: you may be unable to do any further calculations or analyses. For example, if you are examining several subsets of the units, you would wish to abandon the analysis of any subset that turned out to contain no observations. Another example would be if you wanted to abandon the execution of a procedure whenever an error diagnostic has appeared. The EXIT directive allows you to exit from any control structure.

EXIT

Exits from a control structure.

Options

NTIMES	= scalar	Number of control structures, N, to exit; default 1. If N exceeds the number of control structures of the specified type that are currently active, the exit is to the end of the outer one; while for N negative, the exit is to the end of the $-N'$th structure (in order of execution)
CONTROL	= string	Type of control structure to exit (job, for, if, case, procedure); default f
REPEAT	= string	Whether to go to the next set of parameters on exit from a FOR loop (no, yes); default n

Parameter

	expression	Logical expression controlling whether or not an exit takes place.

In its simplest form EXIT has no parameter setting, and the exit is unconditional: Genstat will always exit from the control structure or structures concerned. You are most likely to use this as part of an ELSE block of a block-if or multiple-selection structure. For example

```
SCALAR Sprint,Strace
FOR Setting=PRINT,TRACE; Sset=Sprint,Strace
  IF Setting .EQS. 'yes'
    CALCULATE Sset = 1
  ELSIF Setting .EQS. 'no'
```

```
        CALCULATE Sset = 0
      ELSE
        EXIT [CONTROL = procedure]
      ENDIF
    ENDFOR
```

checks that PRINT and TRACE have valid settings (either 'yes' or 'no') and exits from a procedure if either of them has some other value. In the example in 6.3.1 the procedure is Squarert, and PRINT and TRACE are its two options.

The CONTROL option specifies the type of control structure from which to exit. The default setting is 'for', causing an exit from a FOR loop (6.2.1). For the other settings: 'if' causes an exit from a block-if structure (6.2.2), 'case' exits from a multiple-selection structure (6.2.3), 'procedure' exits from a procedure (6.3), and 'job' causes the entire job to be abandoned. Sometimes, to exit from one type of control structure, others must be left too. To exit from the procedure in the above example, requires Genstat to exit also from the FOR loop and the block-if structure. Generally, Genstat does these nested exits automatically, as required. However, inside a procedure, you can exit only from FOR loops and block-if or multiple-selection structures that are within the procedure. You cannot put, for example,

```
    EXIT [CONTROL = if]
```

within a part of the procedure where there is no block-if in operation, and then expect Genstat to exit both from the procedure and from a block-if structure in the outer program from which the procedure was called. Genstat regards a procedure as a self-contained piece of program.

The NTIMES option indicates how many control structures of the specified type to exit from. If you ask Genstat to exit from more structures than are currently in operation in your program, it will exit from as many as it can and then print a warning. If NTIMES is set to zero or to missing value no exit takes place. If NTIMES is set to a negative value, say $-n$, the exit is to the end of the nth structure of the specified type, counting them in the order in which their execution began. Consider this example:

```
    FOR I = A[1...3]
      FOR J = B[1...3]
        FOR K = C[1...3]
          FOR L = D[1...3]
            "contents of the inner loop, including:"
            EXIT [NTIMES = Nexit]
            "amongst other statements"
          ENDFOR "end of the loop over D[]"
        ENDFOR "end of the loop over C[]"
      ENDFOR "end of the loop over B[]"
    ENDFOR "end of the loop over A[]"
```

If the scalar Nexit has the value 2, the exit is to the end of the loop over C[ ]; so the exit is from the loop over D[ ] and the loop over C[ ]. But if NEXIT has the value -2 the exit is to the end of the loop over B[ ]; this is the second loop to have been started.

A further possibility when EXIT is used within a FOR loop is that you can choose either to go right out of the loop and continue by executing the statement immediately after the ENDFOR statement, or to go to ENDFOR and then repeat the loop with the next set of arguments. To repeat the loop, you need to set option REPEAT = yes. For example, suppose that variates Height and Weight contain information about children of various ages, ranging from five to 11. The RESTRICT statement causes the subsequent GRAPH statement to plot only those units of Height and Weight where the variate Age equals Ageval (2.4.1). The EXIT statement ensures that the graph is not plotted if there are no units of a particular age; the program then continues with Ageval taking the next value in the list.

```
FOR Ageval = 5,6,7,8,9,10,11
  RESTRICT Height,Weight; CONDITION = Age.EQ.Ageval
  EXIT [REPEAT = yes] NVALUE(Height).EQ.0
  GRAPH Height; X = Weight
ENDFOR
```

This example also illustrates the use of the parameter of EXIT, to make its effect conditional. The parameter is an expression which must evaluate to a single number which Genstat interprets as a logical value. If the value is zero, the condition is false and no exit takes place; for other values the condition is true and the exit takes effect as specified. This is particularly useful for controlling the convergence of iterative processes: the use of EXIT to avoid unnecessary iterations in the square-root calculation in 6.2.1 is shown in 6.3.1.

6.3 Procedures

Once you start to write programs for complicated tasks, you may wish to keep them to use again in future. The most convenient way of doing this is to form them into procedures. You may also wish to use procedures written by other people.

Using a Genstat procedure looks exactly the same as using one of the standard Genstat directives. You give the name of the procedure, in full, and then specify options and parameters as required. When Genstat meets a statement with a name that it does not recognize as one of the standard Genstat directives, it first looks to see whether you have a procedure of that name already stored in your program. Then it looks in any procedure library that you may have attached explicitly to your program, taking these in order of their channel number (6.3.2). The people who manage your computer can define a special *site* library and arrange for this to be attached to Genstat automatically when it is run. If they have done so, this library will be examined next. Finally Genstat looks in the official Genstat procedure library

(12.1.1), which is also attached automatically to your program.

Thus the official library allows new facilities to be offered to all users. Or your computer manager can make procedures available that cover the special needs of the users at your site, and these will over-ride any procedures of the same name in the official library. Or you can form your own libaries of the procedures that you find particularly useful, and these will always be taken in preference to procedures in the site or the official library. Note however that a procedure cannot have the same name as any of the Genstat directives (6.3.1).

Information is transferred to and from a procedure only by means of its options and parameters. Otherwise the procedure is completely self-contained. Anyone who uses it does not need to know how the program inside operates, what data structures it contains, nor what directives it uses. The data structures inside the procedure are local to the procedure and cannot be accessed from outside.

The first part of this section describes how to write procedures. Later we describe how to access libraries of procedures, and how to form libraries of your own.

6.3.1 Forming a procedure

You start the definition of a Genstat procedure by a PROCEDURE statement. This has no options, and a single parameter which defines the name of the procedure.

PROCEDURE

Introduces a Genstat procedure.

No options

Parameter

	text	Name of the procedure

The name of the procedure can be up to eight characters with the same rules as for the identifiers of data structures: the first character must be a letter, the second to the eighth can be either letters or digits, and characters beyond the eighth are ignored. However the name cannot be suffixed, neither must it be the same as the name of any of the standard Genstat directives, nor any of their valid abbreviations. Thus you could have a procedure with the name CALCULUS but not CALC or CALCUL as these are abbreviations of the directive CALCULATE. When you use the procedure, you must give the name in full. After the procedure statement, you must define what options and parameters the procedure is to have; this is done by the directives OPTION and PARAMETER respectively. Only one of each of these should be given, and they must appear immediately after the PROCEDURE statement, but it does not matter which of the two you give first. They have very similar syntaxes, except that OPTION

6.3 Procedures 245

has an extra parameter which allows you to define default values to be taken when a user of the procedure does not specify a particular option. If you do not wish to define options or parameters for a procedure you can simply omit these directives; alternatively you can use OPTION or PARAMETER but with none of their parameters set, which has precisely the same effect.

OPTION

Defines the options of a Genstat procedure.

No options

Parameters

NAME	= texts	Names of the options
MODE	= strings	Mode of each option (e, f, p, t, v, as for unnamed structures); default p
DEFAULT	= identifiers	Default values for each option

PARAMETER

Defines the parameters of a Genstat procedure.

No options

Parameters

NAME	= texts	Names of the parameters
MODE	= strings	Mode of each parameter (e, f, p, t, v, as for unnamed structures); default p

---

The NAMES parameter of each directive defines the identifiers that you use, within the procedure itself, to refer to the data structures that are to take information into and out of the procedure. Genstat defines a dummy data structure (3.2.2) for each of these identifiers, and these point to the structures that the user supplies when the procedure is called. In the example below, the PARAMETER statement (line 4) defines two parameters for the procedure Squarert, with names X and ROOTX. In line 39, the procedure Squarert is invoked with parameter ROOTX = Rx, and so the dummy ROOTX within the procedure points to the scalar Rx in the outer program. When you call the procedure (as in line 39), you have the choice of putting the name ROOTX in capital letters, or small letters, or any mixture of the two; this corresponds to the rules for the names of options and parameters of directives. To avoid ambiguity, Genstat automatically converts the identifiers of the dummies to be all in capital letters, and it is in this form that you must use them in the statements within the procedure.

The MODE parameter tells Genstat whether the setting of each option or parameter of the procedure is to be a number (v), or an identifier of a data structure (p), or a string (t), or an expression (e), or a formula (f). These codes are exactly the same as those that indicate the mode of the values to appear within the brackets containing an unnamed structure (1.4.3).

You also use the OPTION and PARAMETER directives to extend the Genstat language; the syntax of the OPTION directive is then slightly different (12.4.1).

After these two statements, you then list the statements that are to be executed when the procedure is called: these statements are the sub-program that makes up the procedure. Any data structures defined within the procedure are local to the procedure and cannot be accessed from outside. So you can use any identifiers for the structures, without having to worry about whether they may also be used outside by someone who may later use the procedure. You end these statements making up the procedure by an ENDPROCEDURE statement.

ENDPROCEDURE

Indicates the end of the contents of a Genstat procedure.

No options or parameters

Once you have defined a procedure, its subsequent use is very easy. This example shows how to form into a procedure the calculation of a square root, described in 6.2.1. At line 39, Squarert is used to calculate the square root of 48 and save the result in scalar Rx. In this call, the options PRINT and TRACE take their default values of 'no'. In line 41, the PRINT option requests Squarert to print the result; while the call in line 42 also illustrates the TRACE option, which allows you to follow the convergence of the process.

EXAMPLE 6.3.1

```
   1   PROCEDURE 'SQUARERT'
   2     " Define the options & parameters of the procedure "
   3     OPTION NAME='PRINT','TRACE'; MODE=t; DEFAULT='no'
   4     PARAMETER NAME='X','ROOTX'; MODE=p
   5     " Check that the option settings are valid and set scalars
  -6       Sprint and Strace to indicate whether they are set to yes "
   7     SCALAR Sprint,Strace
   8     FOR Setting=PRINT,TRACE; Sset=Sprint,Strace
   9       IF Setting .EQS. 'yes'
  10         CALCULATE Sset = 1
  11       ELSIF Setting .EQS. 'no'
  12         CALCULATE Sset = 0
  13       ELSE
  14         EXIT [CONTROL=p]
```

6.3 Procedures

```
15        ENDIF
16      ENDFOR
17   " Check for invalid setting of parameter X ( i.e. < 0 ) "
18   IF X < 0
19      PRINT 'X < 0 :  square root cannot be calculated'
20   ELSE
21   " Calculate convergence limit & initialize "
22      CALCULATE Clim = X/10000   & ROOTX = X
23   " Loop until convergence "
24      FOR [NTIMES=20]
25         CALCULATE Previous = ROOTX
26         & ROOTX = (X/Previous + Previous)/2
27         IF Strace
28            PRINT [IPRINT=*] ROOTX
29         ENDIF
30         EXIT ABS(Previous-ROOTX) < Clim
31      ENDFOR
32      IF Sprint
33         PRINT [IPRINT=*] 'square root of ',X,' is ',ROOTX; \
34            JUSTIFICATION=left
35      ENDIF
36   ENDIF
37 ENDPROCEDURE
38 SCALAR Rx
39 SQUARERT X=48; ROOTX=Rx
40 PRINT Rx

       Rx
    6.928

41 SQUARERT [PRINT=yes] X=48; ROOTX=Rx

square root of  48.00        is        6.928

42 SQUARERT [PRINT=yes; TRACE=yes] 81; ROOTX=Rx

    41.00

    21.49

    12.63

     9.521

     9.014

     9.000

     9.000

square root of  81.00        is        9.000
```

A very much better way of calculating square roots within Genstat is to use the function SQRT in CALCULATE (5.1.1): this example is simply to illustrate the basic rules of procedures. However there are many, more complicated, procedures within the Genstat procedure library (12.1.1), making sophisticated techniques available without requiring you to know any of the details of how they are programmed.

There are several functions that you may find useful when writing procedures. You might use these either in CALCULATE (5.1.1), or in the program control directives (6.2). Some of the functions enable you to access information about the structures that have been supplied in the options or parameters of the procedure. For example: the function NVALUES allows you to find out the length of a structure; NROWS enables you to find out the number of rows of a matrix; and so on (5.2.2). Alternatively you can use the GETATTRIBUTE directive (2.5.2). You might want to use this information to check that the supplied structures are suitable for the operations that the procedure is to carry out; or you might use it in the definition of the local structures required within the procedure.

You can use the UNSET function (5.2.6) to check whether the user has set a particular option or parameter. If this option or parameter is necessary for the procedure to be executed, you can use the EXIT directive to abandon execution. Otherwise, you could use the ASSIGN directive (5.6.1) to set the corresponding dummy to some default structure within the procedure. For example, the following statements could be inserted at the start of the procedure Squarert to abandon the procedure if X is not set, and to set ROOTX to Root if ROOTX has not been set.

```
IF UNSET(X)
   PRINT 'Parameter X not given.'
   EXIT [CONTROL = procedure]
ENDIF
  IF UNSET(ROOTX)
  SCALAR Root
  ASSIGN Root; POINTER = ROOTX
ENDIF
```

You may change some aspects of the Genstat environment within a procedure. This may be the intended purpose of the procedure; but if it is an unwanted side effect, you should reset them at the end of the procedure. To do this you should use GET (2.3.1) to store information about the current settings on entry to the procedure, and then use SET (2.2.1) on exit from the procedure to reset any that you have changed. SET and GET also allow you to control the printing of diagnostics; you can use them in conjunction with DISPLAY (2.1.3) to allow a procedure to perform some of its own error handling.

You can use other procedures from within a procedure; in fact you can even call the procedure itself, allowing you to write recursive programs. However, these auxiliary procedures must already exist when the procedure is defined: they must either have been defined already within your program or be available within one of the libraries attached to your job; you cannot define a procedure within another procedure or within any other control structure.

You are allowed to redefine an existing procedure if you wish to change any of the statements that it contains. To do this you specify the PROCEDURE statement, as usual, followed by the statements making up the new version of the procedure, and then an ENDPROCEDURE statement. This is exactly the same as the method used for defining a new procedure, except that you cannot change the option or parameter specifications: if either OPTION or PARAMETER is given again, Genstat will give an error diagnostic.

6.3.2 Using a procedure library

A *procedure library* is a particular kind of backing-store file that is used to store procedures. It can be used like any other backing store file: you can store procedures in the file, then retrieve them later for further use, using the methods described in 4.5. However you will usually find a library more convenient to use when it is attached to one of the input channels reserved just for procedure libraries. You can then only read procedures from the file and you cannot add new procedures; but the procedures are retrieved from the library automatically, as described at the start of this section.

Several libraries can be attached to a Genstat job. The standard Genstat procedure library is attached automatically, and you may have a local site library that is also attached automatically; your local documentation should give details of this. To attach your own libraries, you can use the OPEN directive (4.3.1). For example:

OPEN 'graphicslib'; CHANNEL = 2; FILETYPE = procedurelibrary

Alternatively, you may be able to open the file in the command that you use to run Genstat; this should also be described in your local documentation.

6.3.3 Forming a procedure library

Procedure libraries are backing-store files that are formed using the normal backing-store directives, as described in full in 4.5. Individual procedures are best stored in separate subfiles; to make constructing and maintaining these libraries easier, you should give the subfiles the same names as the procedures. Using the same name also means that the automatic retrieval works more efficiently. To store procedures you use the STORE directive, for example:

STORE [CHANNEL = 1; SUBFILE = Jacknife ; PROCEDURE = yes] Jacknife

Some procedures may contain references to auxiliary procedures for performing particular parts of an analysis; in this case the searching of the library is more efficient if the additional procedures are contained in the same subfile as the main procedure: for example

STORE [CHANNEL = 1; SUBFILE = Plot ; PROCEDURE = yes] Plot,Scalex,Scaley

While you are developing a procedure library you will need to use it like any other backing-store file, retrieving any procedures that are required by using RETRIEVE. To edit a procedure library you can use either the STORE or MERGE directives. You can display the contents of a library and subfiles using CATALOGUE.

6.4 Debugging Genstat programs

If you are writing a general program in the Genstat language (as in any other high-level language) you may often find that your program is syntactically correct and can be executed by Genstat, but nevertheless produces the wrong answers: somewhere in the logic of your program you have made a mistake. To allow such errors to be identified and corrected, Genstat has two directives, BREAK and DEBUG, that allow you to interrupt the execution of your program. You can then execute other statements, for example to examine the contents of data structures or modify their values, or even to exit from a control structure. This is particularly useful inside a procedure: the data structures used by the procedure are local and cannot normally be accessed from outside; during a break you remain within the procedure and so all the local data structures can be accessed. The BREAK directive allows you to insert breakpoints explicitly; so you must plan its use in advance when you are writing the code. Alternatively you can use DEBUG to insert breakpoints implicitly. This allows you for example to debug an existing procedure without having to edit and redefine it.

6.4.1 Breaking into the execution of a program

BREAK

Suspends execution of the statements in the current channel or control structure and takes subsequent statements from the channel specified.

Option

CHANNEL = scalar   Channel number; default current input channel

No parameters

A BREAK statement has the effect of temporarily halting execution of the current set of statements so that you can execute some other statements. The CHANNEL option determines where these statements are to be found. Usually (and by default) they are in the current input channel. The statements are read and executed, one at a time, until an ENDBREAK statement is reached, at which point control returns to the statements originally being executed.

> **ENDBREAK**
>
> Returns to the original channel or control structure and continues execution.
>
> **No options or parameters**

BREAK also provides a convenient way of interrupting a loop or a procedure so that you can read one set of output before the next is produced. For example:

EXAMPLE 6.4.1

```
  1  VARIATE [NVALUES=13] X,Y,LogY
  2  READ X,Y

     Identifier    Minimum      Mean    Maximum    Values    Missing
              X       6.00     30.00      60.00        13          0
              Y      72.50     95.42     115.90        13          0

 16  CALCULATE LogY = LOG(Y)
 17  FOR Dum=Y,LogY
 18      MODEL Dum
 19      TERMS X
 20      FIT [PRINT=summary] X
 21      BREAK
 22      RDISPLAY [PRINT=estimates]
 23      BREAK
 24  ENDFOR
```

24...

```
***** Regression Analysis *****

*** Summary of analysis ***

               d.f.         s.s.         m.s.
Regression        1       1831.9      1831.90
Residual         11        883.9        80.35
Total            12       2715.8       226.31

Change           -1      -1831.9      1831.90

Percentage variance accounted for 64.5

* MESSAGE: The following units have large residuals:
                    10                 2.04

* MESSAGE: The following units have high leverage:
                     1                 0.34

***** break at statement 5 in For Loop
 25  ENDBREAK
```

25...

```
***** Regression Analysis *****

*** Estimates of regression coefficients ***
```

```
              estimate           s.e.                 t
Constant       117.57           5.26              22.34
X              -0.738           0.155             -4.77
```

***** break at statement 7 in For Loop
   26 ENDBREAK

26..

***** Regression Analysis *****

*** Summary of analysis ***

```
              d.f.       s.s.         m.s.
Regression      1      0.21732      0.217322
Residual       11      0.09931      0.009028
Total          12      0.31663      0.026386

Change         -1     -0.21732      0.217322
```

Percentage variance accounted for 65.8

* MESSAGE: The following units have high leverage:
                         1   0.34

***** break at statement 5 in For Loop
   27 ENDBREAK

27..

***** Regression Analysis *****

*** Estimates of regression coefficients ***

```
              estimate           s.e.                 t
Constant        4.7876          0.0558             85.83
X              -0.00804         0.00164            -4.91
```

***** break at statement 7 in For Loop
   28 ENDBREAK

6.4.2 Putting automatic breaks into a program

DEBUG

Puts an implicit BREAK statement after the current statement and after every NSTATEMENTS subsequent statements, until an ENDDEBUG is reached.

Options

CHANNEL	= scalar	Channel number; default current input channel
NSTATEMENTS	= scalar	Number of statements between breaks; default 1

6.4 Debugging Genstat programs

No parameters

A DEBUG statement causes an immediate break, and then further breaks at regular intervals until you issue an ENDDEBUG statement. The interval between breaks is specified by the NSTATEMENTS option; by default, breaks take place after every statement. During the break, Genstat takes statements from the channel specified by the CHANNEL option; by default they are taken from the current input channel. Each individual break is terminated by an ENDBREAK, exactly like a break invoked explicitly by the BREAK directive (6.4.1).

ENDDEBUG

Cancels a DEBUG statement.

No options or parameters

For example:

EXAMPLE 6.4.2

```
   1   PROCEDURE 'POLAR'
   2      PARAMETER 'X','Y','R','THETA'
   3      " Takes (x,y) and returns (r,theta) "
   4      CALCULATE R = SQRT(X*X + Y*Y)
   5      CALCULATE THETA = ARCCOS(X/R)
   6      CALCULATE THETA = THETA + 2*(3.14159 - THETA)*(Y < 0)
   7   ENDPROCEDURE
   8   SCALAR Xpos,Ypos; VALUE=3,4
   9   DEBUG
  10   POLAR Xpos; Y=Ypos; R=Radius; THETA=Angle
***** break at statement 1 in Procedure POLAR
" CALCULATE R = SQRT(X*X + Y*Y)"
  11   ENDBREAK
***** break at statement 2 in Procedure POLAR
" CALCULATE THETA = ARCCOS(X/R)"
  12   PRINT R

       Radius
        5.000

  13   ENDBREAK
***** break at statement 3 in Procedure POLAR
" CALCULATE THETA = THETA + 2*(3.14159 - THETA)*(Y < 0)"
  14   PRINT THETA

       Angle
       0.9273

  15   ENDBREAK
***** break at statement 4 in Procedure POLAR
"ENDPROCEDURE"
```

```
16    CALCULATE Deg = THETA*180/3.14159
17    PRINT Deg

         Deg
       53.13

18    ENDDEBUG
19    PRINT Xpos,Ypos,Radius,Angle

         Xpos       Ypos      Radius       Angle
        3.000      4.000       5.000      0.9273
```

S.A.H.
H.R.S.

7 Graphical display

The basic type of graphical output from Genstat uses just the normal characters that are available on a line printer. You would use it to make rough checks on data, perhaps using a v.d.u. rather than a printer; whatever the device, though, we shall refer to this style as *line-printer* output. For example, you could quickly check that the residuals from a linear regression are distributed in the way they are supposed to be (Chapter 8). Histograms, graphs and contour plots are available in this basic style.

In addition to line-printer output there is a separate set of directives for producing *high-quality* output on graphical devices such as plotters and graphics monitors. These have a lot in common with the line-printer output, but contain additional options and parameters to take advantage of the more sophisticated facilities available on graphical devices. Obviously the quality and range of output is very dependent on the devices that you use; line-printer output, on the other hand, should be similar on all devices.

7.1 Line-printer graphics

There are three directives that you can use to get line-printer output: HISTOGRAM, GRAPH, and CONTOUR. You can use options and parameters to modify the annotation, the symbols used, the size of plot, and so on. Several options apply generally to all three directives: these are described in this section. Others are more specific and are left until the descriptions of the relevant directives.

Normally, output goes to the current output channel, but you can use the CHANNEL option to direct it to another. (See Chapter 4 for details of output channels.) For example, when you are working interactively, you might send a graph to a secondary output file so that you can print it later.

The TITLE option lets you supply one or more lines of text as an overall title for the output. You specify titles for the axes by the YTITLE and XTITLE options; these are not relevant to histograms. You can supply the text settings of these options directly, or give them as the identifier of a text structure (3.3.2). For example:

    GRAPH [XTITLE='Nitrogen Applied (kg/ha)'] Y; X

or

    TEXT Title
    READ [CHANNEL=2; SERIAL=yes; SETNVALUES=yes] Title,Data
    HISTOGRAM [TITLE=Title] Data

Genstat prints the y-axis title as a column of characters down the left-hand side of a graph or contour map. New lines are ignored, so that strings within the text are concatenated. Genstat truncates the title if necessary: the maximum number of characters is the number of rows of the frame plus four. The x-axis title is printed as ordinary text below the graph, though centred. Again, strings are truncated if necessary: the maximum number of characters is the number of columns of the frame plus four.

7.1.1 The HISTOGRAM directive

HISTOGRAM

Produces histograms of data on the terminal or line printer.

Options

CHANNEL	= scalar	Channel number of output file; default is the current output file
TITLE	= text	General title; default *
LIMITS	= variate	Variates of group limits for classifying variates into groups; default *
NGROUPS	= scalar	When LIMITS is not specified, this defines the number of groups into which a data variate is to be classified; default is the integer value nearest to the square root of the number of values in the variate
LABELS	= text	Group labels
SCALE	= scalar	Number of units represented by each asterisk; default 1

Parameters

DATA	= identifiers	Data for the histograms; these can be either a factor indicating the group to which each unit belongs, a variate whose values are to be grouped, or a one-way table giving the number of units in each group
NOBSERVATIONS	= tables	One-way table to save numbers in the groups
GROUPS	= factors	Factor to save groups defined from a variate

7.1 Line-printer graphics

SYMBOLS	= texts	Characters to be used to represent the bars of each histogram
DESCRIPTION	= texts	Annotation for key

Histograms provide quick and simple visual summaries of data values. The data are divided into several groups; the histogram that is output consists of a line of asterisks for each group, the length of each line being proportional to the number of units in the group; you also get the number in each group at the left-hand end of each line. For example:

EXAMPLE 7.1.1a

```
  1  VARIATE [NVALUES=25] Data
  2  READ Data

     Identifier    Minimum       Mean    Maximum     Values    Missing
           Data      0.000      3.960      9.000         25          0

  4  HISTOGRAM Data

Histogram of Data

               -    2   9 *********
           2 - -    4   7 *******
           4 - -    6   5 *****
           6 - -    8   2 **
           8 -      2 **

Scale:  1 asterisk represents 1 unit.
```

You can form histograms for data stored in variates, factors, or tables. If a histogram is to be formed from a variate, Genstat sorts its values into groups that are defined by upper and lower bounds. Alternatively you can specify this grouping using a factor. Moreover, you can use one-way tables of counts to supply the group counts directly. If you give a list of variates, you get a parallel histogram. For each group there is one row of asterisks printed for each identifier, labelled by the corresponding name. For example:

EXAMPLE 7.1.1b

```
  5  VARIATE [NVALUES=30] X,Y
  6  TEXT [VALUES=small,medium,large] Size
  7  READ X,Y

     Identifier    Minimum       Mean    Maximum     Values    Missing
              X      1.000      3.667      9.000         30          0
              Y      0.000      3.100      7.000         30          0

 10  HISTOGRAM [LABELS=Size] Y,X
```

```
Histogram of Y and X

  small    Y 18 ******************
           X 15 ***************
 medium    Y 10 **********
           X 12 ************
  large    Y  2 **
           X  3 ***

Scale:  1 asterisk represents 1 unit.
```

Here the two variates are sorted into the same groups. You can also draw parallel histograms of variates that have different numbers of values.

If parallel histograms are to be formed from several factors, they must all have the same number of levels and the labels or levels of the first factor will be used to identify the groups. Likewise, if you are forming parallel histograms from several tables, they must all have the same number of values and the classifying factor of the first table will define the labelling of the histogram.

You can use the DESCRIPTION parameter to provide texts that will label the histogram in place of the variate, factor or table identifiers.

The SYMBOLS parameter lets you put some other character in place of the asterisk. For example:

    HISTOGRAM Variate; SYMBOLS='+'

If you specify more than one character in the string, Genstat uses them in order, recycled as necessary, until each histogram bar is of the correct length:

EXAMPLE 7.1.1c

```
   11   HISTOGRAM Data; SYMBOLS='X-O-'

Histogram of Data

              -   2   9  X-O-X-O-X
          2 -     4   7  X-O-X-O
          4 -     6   5  X-O-X
          6 -     8   2  X-
          8 -         2  X-

Scale:  1 character represents 1 unit.
```

You can save the group counts using the NOBSERVATIONS parameter, which forms a one-way table of counts of appropriate size. The missing-value cell of this table will contain a count of the number of units that were missing and were therefore unclassified. Moreover, when you are forming the histogram of a variate, you can record which group each unit went into by using the GROUPS parameter.

You can set the LABELS option to provide explicit labels for the groups of the histogram. Otherwise Genstat uses labels associated with the structures in the DATA parameter: from the labels or levels of a factor that has either been specified explicitly or classifies a one-way table.

You can define explicitly the groups into which a variate is sorted by using either the LIMITS or NGROUPS option. The LIMITS option needs a variate giving the group limits; for example:

EXAMPLE 7.1.1d

```
 12   VARIATE [VALUES=1,2,3,5,7,8,10] Uplimits
 13   HISTOGRAM [LIMITS=Uplimits] Data; NOBSERVATIONS=Counts
Histogram of Data grouped by Uplimits

              -    1.00   5 *****
       1.00  -    2.00   4 ****
       2.00  -    3.00   2 **
       3.00  -    5.00   7 *******
       5.00  -    7.00   4 ****
       7.00  -    8.00   1 *
       8.00  -   10.00   2 **
      10.00  -           0

Scale:  1 asterisk represents 1 unit.
```

Here, Limits is a variate with seven values producing a histogram of the data split into eight groups; ≤1, 1–2, 2–3, 3–5, 5–7, 7–8, 8–10, >10. (The group 3–5, for example, contains values that are greater than 3 and less than or equal to 5: this rule for boundaries is always followed.) Genstat sorts the values of the variate of limits into ascending order.

Alternatively you can specify the number of groups by using the NGROUPS option; this causes Genstat to work out appropriate limits to form groups of equal width. For example:

    HISTOGRAM [NGROUPS=5] Data

If you specify neither NGROUPS nor LIMITS, Genstat calculates the default number of groups to be the integer value nearest to the square root of the number of values (as in the first example which sorts 25 values into five groups).

When Genstat plots the histogram of a factor, the number of groups of the histogram is the number of levels of the factor. The value for each group is the number of units of the factor occurring at the corresponding level. For example:

EXAMPLE 7.1.1e

```
 14  TEXT [VALUES=apple,banana,peach,cherry,pear,orange] Name
 15  FACTOR [LEVELS=6; LABELS=Name; NVALUES=32] Fruit
 16  READ Fruit

 18  HISTOGRAM Fruit
```

```
Histogram of Fruit

 apple   3 ***
banana   2 **
 peach   8 ********
cherry   5 *****
  pear   8 ********
orange   6 ******

Scale:  1 asterisk represents 1 unit.
```

When Genstat plots the histogram of a one-way table, the number of groups is the number of levels of the factor classifying the table.

EXAMPLE 7.1.1f

```
 19  PRINT [NDOWN=0] Counts; FIELDWIDTH=8

       Counts
           1       2       3       4       5       6       7       8
           5       4       2       7       4       1       2       0

 20  HISTOGRAM Counts
```

```
Histogram of Counts

            1 5 *****
            2 4 ****
            3 2 **
            4 7 *******
            5 4 ****
            6 1 *
            7 2 **
            8 0

Scale:  1 asterisk represents 1 unit.
```

Normally one asterisk will represent one unit. However, if there is a lot of data so that group counts become large, Genstat may not be able to fit enough asterisks into one row. In this case it alters the scaling so that one asterisk represents more units. You can set the scaling explicitly by the SCALE option; the value you give is rounded to the nearest integer, and determines how many units should be represented by each asterisk.

7.1.2 The GRAPH directive

GRAPH

Produces scatter and line graphs on the terminal or line printer.

Options

CHANNEL	= scalar	Channel number of output file; default is current output file
TITLE	= text	General title; default *
YTITLE	= text	Title for y-axis; default *
XTITLE	= text	Title for x-axis; default *
YLOWER	= scalar	Lower bound for y-axis; default *
YUPPER	= scalar	Upper bound for y-axis; default *
XLOWER	= scalar	Lower bound for x-axis; default *
XUPPER	= scalar	Upper bound for x-axis; default *
MULTIPLE	= variate	Numbers of plots per frame; default * i.e. all plots are on a single frame
JOIN	= string	Order in which to join points (ascending, given); default a
EQUAL	= string	Whether/how to make bounds equal (no, scale, lower, upper); default n
NROWS	= scalar	Number of rows in the frame; default * i.e. determined automatically
NCOLUMNS	= scalar	Number of columns in the frame; default * i.e. determined automatically
YINTEGER	= string	Whether y-labels integral (no, yes); default n
XINTEGER	= string	Whether x-labels integral (no, yes); default n

Parameters

Y	= identifiers	Y coordinates
X	= identifiers	X coordinates
METHOD	= strings	Type of each graph (point, line, curve, text); if unspecified, p is assumed
SYMBOLS	= factors or texts	For factor SYMBOLS, the labels (if defined), or else the levels, define plotting symbols for each unit,

		whereas a text defines textual information to be placed within the frame for METHOD = text or the symbol to be used for each plot for other METHOD settings; if unspecified, * is used for points, with integers 1-9 to indicate coincident points, ' and . are used for lines and curves
DESCRIPTION	= texts	Annotation for key

The simplest form of the GRAPH directive produces a point plot; this is sometimes called a scatter plot:

EXAMPLE 7.1.2a

```
  1  VARIATE [VALUES=-16,-7,9,16,7,-8,-12,-5,0,10,4,-4,-3,3,16] X
  2  &       [VALUES=0,-14,-12.5,0,14,0,12,0,-10,-9,5,6,-6,-1.5,16] Y
  3  GRAPH Y; X
```

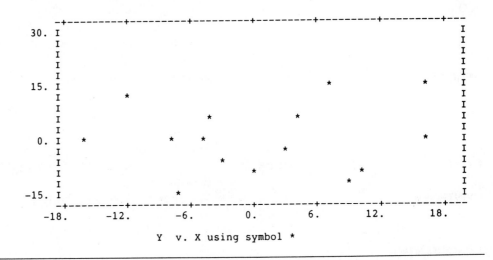

The identifiers Y and X here are of variates, with equal length; Genstat uses their values pairwise to give the coordinates of the points to be plotted. If you give lists of identifiers as parameters, Genstat plots the variates within one frame, a pair at a time; for example,

    GRAPH Y[1...3]; X[1,2]

superimposes plots of Y[1] v X[1], Y[2] v X[2] and Y[3] v X[1]. The usual rules governing the parallel expansion of lists apply here: the Y parameter determines the

7.1 Line-printer graphics

the number of plots within the frame, and Genstat recycles the X parameter if necessary.

The identifiers in the Y and X parameters need not be of variates, but can contain any numerical structures: scalars, variates, tables or matrices. The only constraints are that corresponding structures must have the same numbers of values, and that tables must not have margins.

There are four types of graph available, which you select by the METHOD parameter: 'point' (the default), 'line', 'curve' and 'text'.

A line plot is one in which each point is joined to the next by a straight line.

If you specify the 'curve' method, then Genstat uses a smoothed curve (fitted by a cubic spline). By default, Genstat sorts the data so that the x-values are in ascending order before the points are joined. But you can modify this by the JOIN option, which allows the order of points to be just as given in the variate. If the data contain a missing value, and you use the setting JOIN=given, there will be a corresponding break in the line being drawn.

Lines drawn using methods 'line' or 'curve' do not include marks for the data points; you should plot these separately if you want them. For example:

EXAMPLE 7.1.2b

```
  4  VARIATE [VALUES=-0.1,0.1...0.9] V
  5  & [VALUES=5.5,9.9,8.7,2.3,1.3,5.5] W
  6  GRAPH [TITLE='Point and curve plot'; NROWS=16; NCOLUMNS=61] W,W; V; \
  7    METHOD=curve,point; SYMBOLS=*,'X'; DESCRIPTION='Fitted curve    ...',*
```

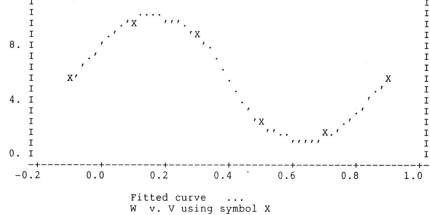

Here W is plotted against V twice, first with the 'curve' method and then with the 'point' method.

The fourth plotting method is 'text'. You can use this for placing an item of text within a graph as extra annotation. For example:

```
SCALAR Xt,Yt; VALUE = 20,10
TEXT [VALUES = 'Y = aX + b'] T
GRAPH Y,Yt; X,Xt; METHOD = line,text; SYMBOLS = *,T
```

This plots a line, defined by the variates Y and X. In addition, the text T is printed within the frame formed by the axes. The scalars Xt and Yt give the coordinates of the point at which the text is to start.

The plotting symbol is the character or characters used to mark either points or lines on a graph. The default symbol for points is the asterisk, and for lines is a combination of dots and single quotes: you can see these in the above examples. If several points coincide, Genstat replaces the asterisk by a digit between 2 and 9, representing the number of coincidences, with 9 meaning nine or more. You can specify alternative symbols using the SYMBOLS parameter, which you can set to either a text or a factor for point plots. If you specify a text with a single string, the string is used to label each point; otherwise, the text must have one string for each point.

By default, Genstat automatically calculates the extents of the axes from the data to be plotted, in such a way that all data can be contained within the frame. You can set one or more of the bounds of the axes by options YLOWER, YUPPER, XLOWER and XUPPER. By setting the upper bound of an axis to a value that is less than the lower bound, you can reverse the usual convention for plotting in which the y-values increase upwards and the x-values increase to the right.

The EQUAL option allows you to place constraints on the bounds for the axes. The default setting 'no' (meaning no constraint) uses the boundary values as set by the options or calculated from the data. The settings 'lower' and 'upper' constrain the lower or upper bounds of the two axes to be equal: for example, you may want to plot the line $y = x$ along with the data, and you may want to make it go through the bottom left-hand corner of the frame. The 'scale' setting adjusts the y-bounds and x-bounds so that the physical distance on one axis corresponds as closely as possible to physical distance on the other: for example, one centimetre represents the same number of units on each axis.

Normally each GRAPH statement produces one frame, and Genstat sets the size so that it will fill one v.d.u. screen or line-printer page. The size is defined by the number of characters in each row and the number of rows in the frame, a row being one line of output. You can adjust the size of the frame by using the NROWS and NCOLUMNS options; the minimum allowed is three rows and three columns, and the maximum number of columns is 16 characters less than the width of the current output channel (4.4.3). There is no maximum on the number of rows allowed. The number of rows that you specify does not include two rows at the top and bottom which are used for the frame and the x-axis. You can redefine the size of a "page" when you open an output file (4.3.1), for example, if you want to avoid having to specify the same numbers of rows and columns for several graphs.

The MULTIPLE option lets you generate several frames with one statement, by specifying a list of scalars. The number of values in this list defines the number of frames to be output; each value gives the number of structures to plot in each frame. For example:

GRAPH [MULTIPLE = !(2, 1, 2)] A,B,C,D,E; X[1...3]

will produce three frames; the first containing A v X[1] and B v X[2], the second containing C v X[3] and the third containing D v X[1] and E v X[2]. The sum of the values in the MULTIPLE list gives the total number of structures required to form the plots, which must therefore be equal to the length of the Y parameter list. The X list will be recycled if necessary, as here.

If you set the NCOLUMNS option to be suitably small, Genstat may be able to fit more than one frame across a page. In that case you will automatically get several graphs printed side by side when you use the MULTIPLE option.

Setting the options YINTEGER and XINTEGER constrains the axis markings to be integral.

You can annotate the graph by using the TITLE, XTITLE, and YTITLE options described at the beginning of this section. If you set none of these, a key will be produced below the graph: in the examples above you can see that GRAPH lists the identifiers and plotting symbols for each pair. You can change this by the DESCRIPTION parameter, which prints a line of text for each plot; for example:

GRAPH Y; X; DESCRIPTION = 'Plot of Y versus X'

7.1.3 The CONTOUR directive

CONTOUR

Produces contour maps of two-way arrays of numbers (on the terminal/printer).

Options

CHANNEL	= scalar	Channel number of output file; default is current output file
INTERVAL	= scalar or variate	Contour interval for scaling (scalar) positions of the contours (variate); default * i.e. determined automatically
TITLE	= text	General title; default *
YTITLE	= text	Title for y-axis; default *

XTITLE	= text	Title for x-axis; default *
YLOWER	= scalar	Lower bound for y-axis; default 0
YUPPER	= scalar	Upper bound for y-axis; default 1
XLOWER	= scalar	Lower bound for x-axis; default 0
XUPPER	= scalar	Upper bound for x-axis; default 1
YINTEGER	= string	Whether y-labels integral (no, yes); default n
XINTEGER	= string	Whether x-labels integral (no, yes); default n
LOWERCUTOFF	= scalar	Lower cut-off for array values; default *
UPPERCUTOFF	= scalar	Upper cut-off for array values; default *
Parameters		
GRID	= identifiers	Pointers (of variates representing the columns of a data matrix), matrices or two-way tables specifying values on a regular grid
DESCRIPTION	= texts	Annotation for key

A contour map lets you see three-dimensional patterns in two dimensions. For example, you may have data sampled over an area on a regular grid, or you may want to represent a function of two variables. Examples are the concentration of some chemical in plants over an experiment, or a likelihood surface as a function of two parameters (8.6). You can for example use contours to find rough estimates of the values of the parameters that maximize the likelihood. The CONTOUR directive produces output for a line printer by using cubic interpolation between the grid points. You can best understand the output by looking at the next example, below. CONTOUR scales the values on the surface according to the interval between contour lines, which you set by the INTERVAL option. The scaled grid values are taken modulo 10, discarding the fractional part, and then truncated to integers. In the example below, the first scaled contour value is -3.869; this becomes 6.131 when taken modulo 10 and then 6 after truncation. Genstat prints only even digits. Thus, taking all these rules into account, you can see the contours as the boundaries between the blank areas and the printed digits.

7.1 Line-printer graphics

The GRID parameter can be one of three types of structure; but to use it, you need to understand how the layout of the values is interpreted by CONTOUR. Suppose you have a matrix, M, with four rows and five columns. When Genstat prints M using the PRINT directive (4.2.1), the layout is

$$\begin{matrix} m_{11} & \cdots & \cdots & m_{15} \\ \vdots & & & \vdots \\ \vdots & & & \vdots \\ m_{41} & \cdots & \cdots & m_{45} \end{matrix}$$

In normal printing, the origin of the rows and columns is at the top left-hand corner. The same applies to a table classified by two factors, and to pointers of variates where each variate is a column. The order of rows is reversed in contour plots: thus with the matrix M, the first row in the above display occurs at the bottom of the grid. Additionally, CONTOUR prints an array of the scaled values at grid points in the rows-reversed order. The DCONTOUR directive (7.3.3) also reverses the rows to give the origin in the conventional place for contour plots.
***

EXAMPLE 7.1.3a

```
  1  MATRIX [ROWS=5; COLUMNS=7] Xval,Yval; VALUES=!((1...7)5),!(7(1...5))
  2  CALCULATE Zval = (Xval-2.5)*(Xval-6)*Xval - 10*(Yval-3)*(Yval-3)
  3  TEXT [VALUES='   Z(x,y) = x*(x-2.5)*(x-6) - 10*(y-3)**2   '] Top
  4  TEXT [VALUES='X values'] Bottom
  5  TEXT [VALUES='Y values'] Side
  6  CONTOUR [TITLE=Top; YTITLE=Side; XTITLE=Bottom] Zval

Contour plot of Zval at intervals of     8.400

** Scaled values at grid points **
   -3.8690    -4.2857    -5.2976    -6.1905    -6.2500    -4.7619    -1.0119
   -0.2976    -0.7143    -1.7262    -2.6190    -2.6786    -1.1905     2.5595
    0.8929     0.4762    -0.5357    -1.4286    -1.4881     0.0000     3.7500
   -0.2976    -0.7143    -1.7262    -2.6190    -2.6786    -1.1905     2.5595
   -3.8690    -4.2857    -5.2976    -6.1905    -6.2500    -4.7619    -1.0119
```

```
                  Z(x,y) = x*(x-2.5)*(x-6) - 10*(y-3)**2

                 0.0000    0.1667    0.3333    0.5000    0.6667    0.8333    1.0000
                    '         '         '         '         '         '         '
            1.000-66666              4444444444              4444444    666    888-
                  666666666666666          4444444444444444444444       666     88
                      6666666666                                      66666      88    0
                8888888              6666666666                    666666      888    00
                88888888888888          6666666666666666666666        888       00
                   88888888888             666666666666             8888        00     2
            0.750-       888888888                                    8888       000    22-
                  0000              8888888888                       888888      000     22
         Y         0000000000              88888888888                 8888888      000      22
                   0000000000000             8888888888888888888888888       000     222
         v         00000000000000              888888888888888888888       000      22
         a         000000000000000               88888888888888888888       000      22
         l  0.500-0000000000000000000               88888888888888888888       0000     22    -
         u         000000000000000               88888888888888888888       000      22
         e         00000000000000              888888888888888888888       000      22
         s         0000000000000             8888888888888888888888888       000     222
                   0000000000              88888888888                 8888888      000      22
                  0000              8888888888                       888888      000     22
            0.250-       888888888                                    8888       000    22-
                      88888888888             666666666666             8888        00     2
                88888888888888          6666666666666666666666        888         00
                8888888              6666666666                    666666      888    00
                      6666666666                                      66666      88    0
                  666666666666666          4444444444444444444444       666     88
            0.000-66666              4444444444              4444444    666    888-
                    '         '         '         '         '         '         '

                                            X values
```

In this example, a function of two variables is calculated, and the shape of the function is displayed with the CONTOUR directive. Titles have been given to the x-axis and the y-axis, and there is an overall title which is the algebraic form of the function. Default values for the upper and lower limits to the axes are used here, as well as a default interval for the contour.

Eight of the thirteen options control the labelling of the axes; we now describe these together.

7.1 Line-printer graphics

You give titles for the y-axis and x-axis as texts in YTITLE and XTITLE. Genstat uses the first line of the text, and centralizes it if it is too short. If the text is too long, Genstat prints only its beginning. You get a warning if a text contains more than one line.

Unlike GRAPH (7.1.1), you cannot leave CONTOUR to label the axes using the grid values. The YUPPER and YLOWER options set values for the end points of the y-axis: at what values the intermediate tick marks occur is calculated from these. If you do not set these, Genstat uses the default values 1 and 0. If you set the YINTEGER option to 'yes', then the values at the tick marks are printed as integers. The options XUPPER, XLOWER and XINTEGER similarly control the labelling of the x-axis.

The INTERVAL option allows you to set the interval at which the boundaries between blanks and digits appears. If your grid values ranged from 17 to 72 then boundaries would be seen at grid values of 20, 25 ... 70 when the contour interval is set to 5. You should take some care in setting the INTERVAL option, since it can clarify the output. You can see the effect of INTERVAL (and the other options we have discussed) in the next example.

EXAMPLE 7.1.3b

```
   7   FACTOR [LEVELS=5] Fy
   8   & [LEVELS=7] Fx
   9   TABLE [CLASSIFICATION=Fy,Fx] Tzval
  10   EQUATE OLD=Zval; NEW=Tzval
  11   CONTOUR [INTERVAL=5; TITLE=Top; YTITLE=Side; XTITLE=Bottom; \
  12       YINTEGER=yes; XINTEGER=yes; YUPPER=5; YLOWER=1; XUPPER=7; XLOWER=1] \
  13       Tzval

Contour plot of Tzval at intervals of      5.000

** Scaled values at grid points **
   -6.5000    -7.2000    -8.9000   -10.4000   -10.5000    -8.0000    -1.7000
   -0.5000    -1.2000    -2.9000    -4.4000    -4.5000    -2.0000     4.3000
    1.5000     0.8000    -0.9000    -2.4000    -2.5000     0.0000     6.3000
   -0.5000    -1.2000    -2.9000    -4.4000    -4.5000    -2.0000     4.3000
   -6.5000    -7.2000    -8.9000   -10.4000   -10.5000    -8.0000    -1.7000
```

```
              Z(x,y) = x*(x-2.5)*(x-6) - 10*(y-3)**2

                 1          2          3          4          5          6          7
                 ,          ,          ,          ,          ,          ,          ,
              5-        222222     000000                        0000   222 44 66 8-
                 44444444444     222222     0000000000000        222    44   6 88
                        444444      2222222            22222      444   6  8   0
                 666666666666    444444       22222222       444     66 88 00
                         666666      4444444             44444    666   8   0   2
                 88888888888     666666       4444444444444      666   88   0 22
              4-        8888888    666666                       6666   888 00 22 4-
                 0000          888888     66666666        6666666      888  00  2 44
     Y           0000000000      888888       666666666666666    888      00   2   4
                 000000000000     888888          66666666       888     00  22  44
     v                 00000000        888888                    888      00   2   4  6
     a                 •  0000000       888888                  8888      00  22   4  6
     l           3-       0000000       888888                  8888     000  22  44 6-
     u                    0000000       888888                  8888      00  22   4  6
     e                    00000000      888888                   888      00   2   4  6
     s           000000000000     888888          66666666       888     00  22  44
                 0000000000      888888        666666666666666   888      00   2   4
                 0000          888888      66666666        6666666     88   00  2  44
              2-        8888888    666666                       6666   888 00 22 4-
                 88888888888     666666       4444444444444      666   88   0 22
                         666666      4444444             44444    666    8   0   2
                 666666666666    444444       22222222       444     66 88 00
                        444444      2222222            22222      444   6  8   0
                 44444444444     222222     0000000000000        222    44   6 88
              1-        222222     000000                        0000   222 44 66 8-
                 ,          ,          ,          ,          ,          ,          ,

                                         X values
```

Here the values in the matrix Zval are put into the table Tzval. The upper and lower limits to the axes are set, as are the YINTEGER and XINTEGER options (thus giving whole numbers at the tick marks).

You use the UPPERCUTOFF and LOWERCUTOFF options to set a window for the grid values that will form the contours. All values above or below these are printed as "X".

EXAMPLE 7.1.3c

```
 14   MARGIN Tzval; METHOD=medians
 15   CONTOUR [TITLE=Top; YTITLE=Side; XTITLE=Bottom; YINTEGER=yes;
 16     XINTEGER=yes; YUPPER=5; YLOWER=1; XUPPER=7; XLOWER=1; UPPERCUTOFF=
 17     Tzval
```

```
Contour plot of Tzval at intervals of    8.400

** Scaled values at grid points **
   -3.8690   -4.2857   -5.2976   -6.1905   -6.2500   -4.7619   -1.0119
   -0.2976   -0.7143   -1.7262   -2.6190   -2.6786   -1.1905    2.5595
    0.8929    0.4762   -0.5357   -1.4286   -1.4881    0.0000    3.7500
   -0.2976   -0.7143   -1.7262   -2.6190   -2.6786   -1.1905    2.5595
   -3.8690   -4.2857   -5.2976   -6.1905   -6.2500   -4.7619   -1.0119

              Z(x,y) = x*(x-2.5)*(x-6) - 10*(y-3)**2
                1         2         3         4         5         6         7
                ,         ,         ,         ,         ,         ,         ,
             5-66666                4444444444                 4444444   666   888-
                666666666666666             444444444444444444444      666    88
                               6666666666                           66666      88    0
                 8888888               6666666666                  666666    888   00
                 88888888888888            666666666666666666666      888      00   X
                         88888888888                666666666666       8888     00XXX
           4-                888888888                               8888      000XXXX-
                    0000                  8888888888                    888888     000  XXXX
   Y              0000000000                  88888888888             8888888    000 XXXXX
                 0000000000000              88888888888888888888888888        000XXXXXX
   v             00000000000000             888888888888888888888888           000  XXXXXX
   a             000000000000000             88888888888888888888888           000XXXXXXX
   l           3-0000000000000000             888888888888888888888            0000XXXXXXX-
   u             00000000000000                888888888888888888888            000XXXXXXX
   e             00000000000000             888888888888888888888888           000  XXXXXX
   s             0000000000000              888888888888888888888888888        000XXXXXX
                  0000000000                 88888888888              8888888    000 XXXXX
                     0000                     8888888888                 888888      000  XXXX
           2-                888888888                                8888      000XXXX-
                         88888888888                666666666666       8888      00XXX
                 88888888888888            6666666666666666666666      888      00  X
                  8888888                6666666666                666666    888      00
                               6666666666                           66666      88    0
                666666666666666             444444444444444444444     666      88
             1-66666                4444444444                 4444444   666   888-
                ,         ,         ,         ,         ,         ,         ,

                                      X values
```

The effect here of setting the UPPERCUTOFF and LOWERCUTOFF options is to change the default contour interval. This default gives you approximately 10 contours between the values of the options. If you leave either unset, then Genstat uses the maximum or minimum grid value respectively.

Note in this example also that the table Tzval has been given margins which are ignored by the CONTOUR statement.

The text of the TITLE option appears at the top of the output from each CONTOUR statement, and so you should use it to give a general description of what the contour lines represent. The DESCRIPTION parameter is specific to each contour line: it is printed at the bottom of the picture.

EXAMPLE 7.1.3d

```
 18    VARIATE [NVALUES=5] Core[1...5]
 19    READ Core[]

       Identifier    Minimum       Mean    Maximum    Values    Missing
         Core[1]       5.000      8.200     11.000         5          0
         Core[2]        6.00      67.60     195.00         5          0
         Core[3]       129.0      940.6     2315.0         5          0
         Core[4]       10.00      36.00      77.00         5          0
         Core[5]       7.000      9.400     15.000         5          0

 25    "
-26       Data are core samples taken from a wetland rice experiment to examine
-27       the leaching of ammonium nitrate.  Cores were taken centrally and
-28       5 and 10 cm either side of the central core.  Concentration of
-29       ammonium nitrate measured at depths of 4,8...20 cm.
-30    "
 31    TEXT [VALUES=' Samples taken 40 days after placement ', \
 32         ' of 2 grams supergranule urea. '] Coredesc
 33    CALCULATE Core[] = LOG10(REVERSE(Core[]))
 34    CONTOUR [YTITLE='Soil depth in cm'; XTITLE='Distance from central core'; \
 35         YINTEGER=yes; XINTEGER=yes; YUPPER=4; YLOWER=20; \
 36         XUPPER=10; XLOWER=-10] Core; DESCRIPTION=Coredesc

Contour plot of Core at intervals of     0.267

** Scaled values at grid points **
     3.1704     2.9193     7.9179     3.7515     3.5799
     3.3880     7.1797    12.6222     6.4687     3.1704
     2.6222     8.5911    11.3557     7.0772     3.1704
     3.7515     5.7926    11.3091     5.5415     3.5799
     3.9068     4.8808     8.2790     3.7515     4.4121
```

```
                     -10         -5          0          5         10
                      '          '          '          '          '
               4- 2222222222    4   66         66   44                -
                     222      44 66   88 888   66 444
                          444    6  88    0    8  66 4444
                          444   66 88   00 00   88 66     444
s                        4444   66 88 00        00  8   6     444
o                         444   66 88 00     2   00   8  66   4444
i              8-   444    66   8   0   2222   00  8  66     444     -
l                   444    66  88 00    2222   00 88  66     44
                     44  66  88  00     2222   00  88 666    444
d              2  44   66  88 000        222   00  88   66   444
e              2  44    6  88  000            000 88   66    44
p              2  44 66  88    000             00  88   66   44
t             12-2  44  66  88   000           000  88  66    44    -
h              2  44   66  88   0000           000  88 666    444
               2  44  666  88    000            00  88   66   44
i                 44   666  88   00             00  88  66    4444
n                 444   666 88   000            00  88 66     444
                  4444    66 88   00            00   8   6     4444
c             16-    4444     6   8  00      000 88  66       4444   -
m                  44444   66 88   00  00    8    6          4444
                   44444    6  88   000       88 66  4444
                   44444    66 88    0         88  6    444
                   444444  66  88              88   66  444           4
                   444444    66   888888       66   44                4
              20-  4444444   666     88        66   44        44-
                      '          '          '          '          '
                                Distance from central core

                       Samples taken 40 days after placement
                             of 2 grams supergranule urea.
```

7.2 The environment for high-quality graphics

The quality of the graphics that you can obtain depends on the devices that your computer has available. In particular, the output from your graphical devices may not look the same as that reproduced in Figure 7.2.

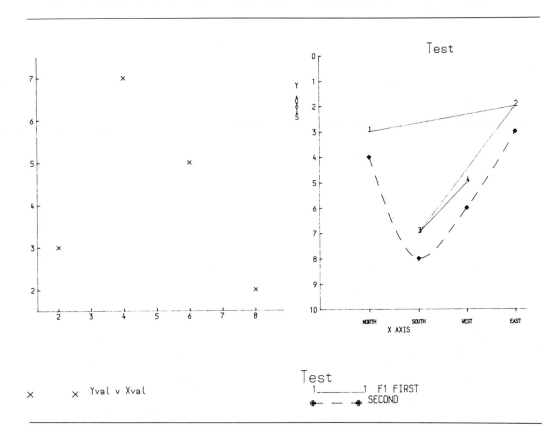

The first of these graphs was produced by the statements:

    VARIATE [VALUES = 3,2,7,5] Yval
    VARIATE [VALUES = 2,8,4,6] Xval
    "Note that the values of Xval are not in ascending order."
    DGRAPH Yval; X = Xval

This simple statement uses the default settings of the environment for high-quality graphics. By judicious changes to that environment, you can get essentially the same three Genstat statements to produce the second graph; we describe how to do this later (end of 7.3.2).

Genstat can produce four types of pictures:

> histograms using DHISTOGRAM
> graphs using DGRAPH
> contours using DCONTOUR
> pie charts using DPIE

All of these, except pie charts, have analogous line-printer forms in HISTOGRAM, GRAPH, and CONTOUR respectively. They share certain options and parameters, and for these the settings have the same meanings.

When you use these directives, the pictures are drawn on a *graphical device*, in a *graphical window*, using a *graphical pen*. You can give a *key* to the picture to provide additional information. These terms are explained below.

A *graphical device* can either produce graphical output directly allowing you to switch between looking at character output or graphical output, or can be used to form a picture from an output file called a *metafile*. A metafile is a file of coded instructions, using the conventions of an underlying graphics package: three of the commonest are GHOST, GINO and GKS. To find out which is used on your computer, put

> HELP environment,pictures,possible

You then get graphical output by processing the metafile with the relevant graphical software provided on your computer for the graphical package. You name the metafile using the OPEN directive (4.3.1):

> OPEN NAME='filename'; CHANNEL=1; FILETYPE=graphics

There is only one channel available for graphics files, and so you must close one metafile before opening another (4.3.4). If you send output to a metafile without naming it, a default name may be assumed: see your local documentation.

On a particular computer there may be more than one device that you can use for displaying Genstat pictures. You may be able to use some of these interactively; also, they may have different maximum window sizes (7.2.4). The HELP statement above will tell you what devices you can use; read your local documentation to find out their attributes, such as the maximum window size. You can direct output to only one device at a time: you control it by the DEVICE directive which is described in 7.2.3.

You will get strange results if you try to use a non-graphical terminal as a graphical terminal, or if you try to use a graphical device that is not available with Genstat.

A *graphical window* is a part of the page or screen on which a picture will appear: that is, it is a rectangular portion of the total available area on a graphical device. Such a rectangle is defined in special coordinates that define the position and size of a window within the screen; these are called *normalized device coordinates*. They are independent of the values of the data that are to be plotted. For most devices these coordinates are in the range 0 to 1 in the X and Y directions with (0, 0) in the bottom left corner; what physical distance is represented by a

7.2 The environment for high-quality graphics

coordinate difference of, for example, 0.1 will vary between devices.

You can define up to eight graphical windows, and they are referred to by the numbers 1 to 8. They are used for the pictures described above, and for the keys associated with them which contain additional information to help in interpreting the pictures. You change the size and position of these windows by the FRAME directive (7.2.4).

Associated with a window are the attributes of its axes. These influence how axes are drawn in the DGRAPH statement. They can also influence the graphical presentation of histograms (DHISTOGRAM) and contours (DCONTOUR). You change them by the AXES directive (7.2.1).

You can specify that certain parts of a picture are to be drawn by a particular *graphical pen*. How those parts are drawn is determined by the attributes of that pen. Obvious attributes are those of colour and linestyle (full or dashed). A less obvious attribute is the *brush*; this specifies how parts of a histogram, or pie chart, are filled in. There are eight pens identified by the numbers 1 to 8, and you change their attributes by the PEN directive (7.2.2).

The directives that define the environment change only the parameters that you specify; unspecified parameters retain their previous values, or their default values (7.3).

7.2.1 The AXES directive

AXES

Defines the axes in each window of high-quality graphs and contours.

Option

EQUAL	= string	Whether/how to make axes equal (no, scale, lower, upper); default n

Parameters

WINDOW	= scalars	Numbers of the windows
YTITLE	= texts	Title for the y-axis in each window
XTITLE	= texts	Title for the x-axis in each window
YLOWER	= scalars	Lower bound for y-axis
YUPPER	= scalars	Upper bound for y-axis
XLOWER	= scalars	Lower bound for x-axis
XUPPER	= scalars	Upper bound for x-axis
YINTEGER	= strings	Whether y-labels integral (no, yes)

XINTEGER	= strings	Whether x-labels integral (no, yes)
YMARKS	= scalars or variates	Distance between each tick mark on y-axis (scalar) or positions of the marks (variate)
XMARKS	= scalars or variates	Distance between each tick mark on x-axis (scalar) or positions of the marks (variate)
YLABELS	= texts	Labels at each mark on y-axis
XLABELS	= texts	Labels at each mark on x-axis
YORIGIN	= scalars	Position on y-axis at which x-axis is drawn
XORIGIN	= scalars	Position on x-axis at which y-axis is drawn
STYLE	= strings	Style of axes (none, x, y, xy, box, grid)

The WINDOW parameter specifies the graphical window that is to be modified according to the settings of the other parameters. The YLOWER and YUPPER parameters specify the lower and upper bounds that Genstat uses in constructing the y-axis; by default, Genstat calculates the bounds from the data. You can set the lower bound to be numerically greater than the upper, but they must not be equal. The XLOWER and XUPPER parameters set bounds for the x-axis in a similar way. The values specified in these parameters are on the scale of the data that are plotted, and are independent of the normalized device coordinates used in FRAME (7.2.4).

The YORIGIN parameter determines the value on the y-axis through which the x-axis passes. If its value is outside the bounds, then they are suitably modified; this applies whether you have explicitly set the bounds or have left Genstat to calculate them from the data. By default, the x-axis will be drawn through the zero of the y-axis if possible, otherwise through the lower bound. The XORIGIN parameter sets the origin for the x-axis in a similar way.

You use the YTITLE and XTITLE parameters to give titles to the axes. The title is limited to a single line of up to 30 characters.

You use the YINTEGER parameter to force the tick marks on the y-axis to be integers. The parameter is ignored if the range of the y-axis is less than 1. The XINTEGER parameter similarly forces the tick marks on the x-axis to be integers.

You use the YMARKS parameter to over-ride the default annotation for the axes. If you give a single value, it becomes the increment between the tick marks; it must be positive, irrespective of whether the lower bound is greater than the upper. For example, YMARKS = 1.5 with bounds 2 and 10 causes tick marks to appear at 2, 3.5,

5, 6.5, 8, and 9.5. If the setting is a variate with more than one value, then the values are taken to be the specific positions of the tick marks on the y-axis, provided that they are within the bounds of the y-axis. For example, if the bounds were 2 and 10 and the variate Marks had values 2, 3, 5, and 9, then the setting YMARKS = Marks would cause the ticks to appear at 2, 3, 5, and 9. Note that the values may be printed in a scaled form, with a multiplier like 10^{-3} written at the top of the y-axis or at the right of the x-axis.

If you do use YMARKS to set the values for the tick marks, then the only annotation of the axis is these values, unless you set the YLABELS parameter. You can control the number of decimal places displayed by declaring a variate to store the values and setting the DECIMALS parameter of the VARIATE directive (3.1.2), and then specifying the variate in the YMARKS parameter. You can specify your own annotation for the y-axis, provided that you also set specific tick marks by YMARKS, by setting a text structure in the YLABELS parameter. For example:

```
TEXT [VALUES = Mon,Tues,Wed,Thur,Fri,Sat,Sun] Day
AXES 1; YMARKS = !(1.5,3...10.5); YLABELS = Day
```

The strings within the text are used cyclically; hence, the number of strings can be less than the number of tick marks.

The XMARKS and XLABELS parameters work in a similar way for the x-axis.

The STYLE parameter allows you to control the type of axes drawn. It must be set to one of:

none	no axes;
x	x-axis only;
y	y-axis only;
xy	both axes (default);
box	both axes, with a surrounding frame; or
grid	both axes, with an overlayed grid.

The lines of an overlayed grid are drawn to pass through the tick marks. You must specify positions for the tick marks on both the axes, or on neither.

Thus, to produce the axes in the second graph shown in Figure 7.2, the AXES statement should be:

```
AXES WINDOW = 3; YLOWER = 10; YUPPER = 0; XLOWER = 0; XUPPER = 10; \
    XMARK = Xval; XLABEL = !T(NORTH,EAST,SOUTH,WEST);   \
    YTITLE = 'Y AXIS'; XTITLE = 'X AXIS'
```

The variate Xval here has the four values 2,8,4,6, as at the beginning of this section.

The EQUAL option has the same settings as in the GRAPH directive (7.1.2).

7.2.2 The PEN directive

PEN

Defines the properties of "pens" for high-quality graphics.

No options

Parameters

NUMBER	= scalars	Numbers associated with the pens
COLOUR	= scalars	Number of the colour used with each pen
LINESTYLE	= scalars	Style for line used by each pen when joining points (zero = no line)
METHOD	= strings	Method for determining line (point, line, monotonic, closed, open)
SYMBOLS	= identifiers	Symbols for points – scalar for special symbols, texts or factors for character symbols
JOIN	= strings	Order in which points are to be joined by each pen (ascending, given)
BRUSH	= scalars	Number of the type of area filling used with each pen when drawing pie charts or histograms

---

The NUMBER parameter specifies a reference number for a style of drawing components of pictures; it must be in the range 1 to 8.

The COLOUR parameter specifies the reference number of the colour to be used; it must be in the range 1 to 4. How reference numbers are associated with actual colours depends on the graphical device. For example, many graphical plotters allow a number of pens to be clipped into holders: reference number 1 will then correspond to the pen put into holder number 1.

If points are connected by a line, then the LINESTYLE parameter specifies the linestyle, thus:

      0 no line
      1 full line
      2 dashed line, style 1
      3 dashed line, style 2
      4 dashed line, style 3

The actual dashing produced will depend on the underlying graphical package; here are the styles produced by GHOST80:

FIGURE 7.2.2a

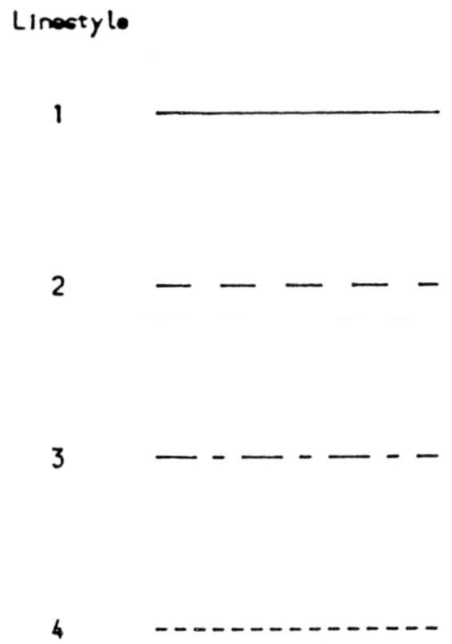

Linestyle

1 ─────────────

2 ── ── ── ── ─

3 ── · ── · ── · ·

4 - - - - - - - - - - - - - - -

The SYMBOLS parameter determines what is drawn at the points of the graph. The numbers 1 to n correspond to graphical markers, such as crosses or squares. The actual markers drawn, and the value of n, will depend on the underlying graphical package; here are the possibilities for GHOST80, GINO-F, and GKS:

FIGURE 7.2.2b

	1	2	3	4	5	6	7	8	9
GHOST80	×	⊙	✲	✳	⊡	◇	△	▽	·
GINO-F	×	○	+	×	✳	□	◇	△	▽
GKS	×	⊡	÷	✳	×	×	×	×	×

The number 0 implies that no symbol will be drawn. By giving a text in the SYMBOLS parameter, you can insert text within a graph at the associated (x, y) position. If there are several strings in the text, they are inserted cyclically. If the setting is a factor, the labels of the factor will be written on the graph (or the levels if you have not specified any labels). Note that a graphical marker is drawn so that its centre is at the specified position. Text and factor labels are drawn so that the bottom left point of the first character is at the specified position: for example, the text 'LABEL' would be placed with the foot of the first L at the position.

If you use a factor to identify individual points of a graph, it must have the same length as the structures used in the Y and X parameters of subsequent DGRAPH statements. The same is true of a text, except that a text with a single string is always allowed, giving the same annotation to all points regardless of the length of the Y and X structures.

The METHOD parameter controls how points are connected. They can be connected by curves ('monotonic', 'open', 'closed'), straight lines ('line'), or they can be unconnected ('point'). The 'monotonic' setting specifies that a smooth single-valued curve is to be drawn through the data points, and that successive x-coordinates will be non-decreasing. The data are always sorted before plotting if this setting is used; but the sorting is temporary, and leaves the order of the values unchanged in the original structure. The 'open' and 'closed' settings specify that a smooth, possibly multi-valued, curve is to be drawn through the data points, using the method of McConalogue (1970).This implies that the resulting curve is rotationally invariant, but it is not invariant under scaling. The 'closed' setting will connect the last point to the first.

Note that if the method is not 'point', then the setting LINESTYLE = 0 will cause a DGRAPH statement to draw a full line, just as if the setting were LINESTYLE = 1. A combination of SYMBOLS = 0 and METHOD = point will produce no plotting at all (and no warning).

The JOIN parameter allows you to control the order in which points are connected. 'Given' implies that the order is to be determined by the order of the data (overridden if METHOD = monotonic, see above). 'Ascending' implies that the data points are sorted (temporarily) according to the x-values. Large numbers of points can be plotted much faster if you use JOIN = ascending.

The BRUSH parameter controls how areas within histograms and pie charts are to be filled. There are 16 available patterns indicated by the integers 1 to 16. In general, the higher the number, the denser the hatching, and the longer such areas take to plot:

FIGURE 7.2.2c

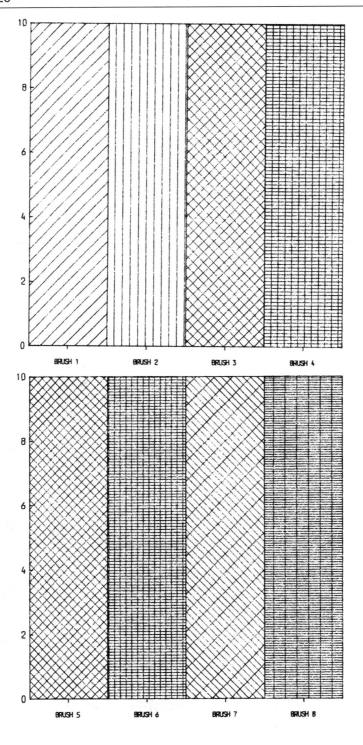

282   7 *Graphical display*

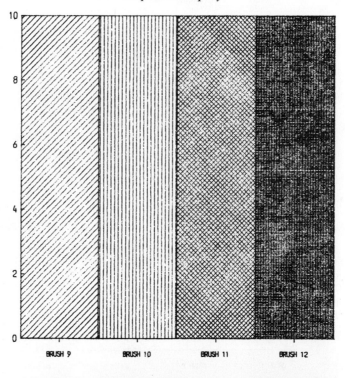

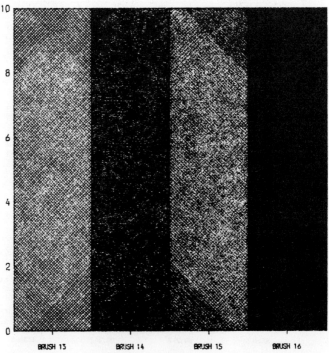

7.2 The environment for high-quality graphics

The PEN statement that was used to draw the second graph in Figure 7.2 is shown below:

```
"Set up four formal factor levels in F1"
FACTOR [LEVELS = 4; VALUES = 1...4] F1
PEN NUMBER = 1,2; COLOUR = 1; LINESTYLE = 1,2; METHOD = line,monotonic; \
     SYMBOLS = F1,3; JOIN = given,ascending
```

There are two plots. For the first, METHOD = line and LINESTYLE = 1, which cause the points to be connected with straight and solid lines; the points are identified by the integers 1 to 4 (from SYMBOLS = F1) and are connected in the order of presentation (from JOIN = given). For the second plot METHOD = monotonic and LINESTYLE = 2 cause the points to be connected by a dashed, single-valued curve, with the points indicated by the same graphical symbol (SYMBOL = 3 corresponds to a solid circle on this device). Note that JOIN = ascending is in fact redundant here, since METHOD = monotonic automatically causes the x-values to be taken in non-decreasing order.

7.2.3 The DEVICE directive

DEVICE

Switches between (high-quality) graphics devices.

No options

Parameter

| | scalar | Device number |

Sometimes there is more than one graphical device for a particular computer; this directive switches between those devices. For example,

    DEVICE 1

connects output to graphical device 1 and switches off any other graphical device. Use the HELP directive, and read your local documentation to find out the device numbers that you can use.

7.2.4 The FRAME directive

> **FRAME**
>
> Defines the positions of windows within the frame of a high-quality graph. The positions are defined in normalized device coordinates ($[0,1] \times [0,1]$).
>
> **No options**
>
> **Parameters**
>
> | | | |
> |---|---|---|
> | WINDOW | = scalars | Window numbers |
> | YLOWER | = scalars | Lower y device coordinate for each window |
> | YUPPER | = scalars | Upper y device coordinate for each window |
> | XLOWER | = scalars | Lower x device coordinate for each window |
> | XUPPER | = scalars | Upper x device coordinate for each window |

This defines the positions and sizes of windows within the total area available using normalized device coordinates. For example:

    FRAME WINDOW=1,2; YLOWER=0.25,0; YUPPER=0.75,0.2; XLOWER=0; \
        XUPPER=0.5

For window 1, this defines a rectangle whose bottom left corner is (0, 0.25) and whose top right corner is (0.5, 0.75). Note this does not define the exact size of a picture drawn in this window, as a margin is allowed for the annotation and titles.

The settings of YLOWER and XLOWER must be strictly less than those of YUPPER and XUPPER respectively; also, none of the bounds can be outside the permitted bounds for the graphical device that you have set by DEVICE. You cannot use * to reset a bound to the default value; if you try to do so, you will get an error diagnostic.

7.3 High-quality graphics

The DHISTOGRAM, DGRAPH, DCONTOUR, and DPIE directives have certain options and parameters in common. These are the TITLE, WINDOW, KEYWINDOW, and SCREEN options, and the PEN and DESCRIPTION parameters.

The TITLE option specifies the title drawn at the top of a picture. The title can be up to 80 characters in length, and it must consist of one line of text only. The default is to draw no title.

The WINDOW option can be set to an integer in the range 0 to 8. If it is 0, the picture is not drawn; otherwise the value of the setting is the number of the window that the picture is to be drawn in. The attributes associated with the window determine the characteristics of the picture; you set the attributes with the FRAME (7.2.4) and AXES (7.2.1) directives, or you can use the default attributes. The default is WINDOW = 1.

The KEYWINDOW option can also be set to an integer in the range 0 to 8. If it is 0, the key is not drawn; otherwise you get a key positioned in the window with the specified number. The key gives additional information relevant to the picture. The default is KEYWINDOW = 2.

The SCREEN option can be set to 'clear' or 'keep'. This determines whether the existing information on the screen is to be cleared or kept before the current picture is drawn. Thus, you can add to an existing picture or draw a picture in a different window. The default is to clear the screen, or, equivalently, to move to a new frame if output is to a plotter.

The PEN parameter specifies which graphical pen Genstat uses to display the values of a data structure; you can define the pens by the PEN directive (7.2.2), or use the default definitions.

The DESCRIPTION parameter specifies text to appear in the key of a picture, as in the directives for line-printer graphics. If the parameter is omitted, the identifiers of the structures being plotted will be used in the key.

Genstat sets defaults for the graphical environment so that you get reasonable output without using AXES, FRAME, or PEN. To find out the initial default, or current settings of the environment, use the HELP directive (2.1.1). The statement

> HELP environment,pictures,current

gives the following information if no AXES, FRAME, or PEN statement has been given in the current job:

EXAMPLE 7.3

```
ENVIRONMENT, PICTURES, CURRENT

CURRENT SETTINGS

The current settings of the graphics environment directives are:
                              Window
FRAME       1       2       3       4       5       6       7       8
-----------------------------------------------------------------------
YLOWER    0.250   0.000   0.000   0.000   0.000   0.000   0.000   0.000
YUPPER    1.000   0.200   1.000   1.000   1.000   1.000   1.000   1.000
XLOWER    0.000   0.000   0.000   0.000   0.000   0.000   0.000   0.000
XUPPER    0.750   1.000   1.000   1.000   1.000   1.000   1.000   1.000

-----------------------------------------------------------------------
  AXES
-----------------------------------------------------------------------
EQUAL      no      no      no      no      no      no      no      no
YTITLE   unset   unset   unset   unset   unset   unset   unset   unset
XTITLE   unset   unset   unset   unset   unset   unset   unset   unset
```

YLOWER	*	*	*	*	*	*	*	*
YUPPER	*	*	*	*	*	*	*	*
XLOWER	*	*	*	*	*	*	*	*
XUPPER	*	*	*	*	*	*	*	*
YINT	no	no	no	no	no	no	no	no
XINT	no	no	no	no	no	no	no	no
YMARKS	*	*	*	*	*	*	*	*
XMARKS	*	*	*	*	*	*	*	*
YLABELS	*	*	*	*	*	*	*	*
XLABELS	*	*	*	*	*	*	*	*
YORIGIN	*	*	*	*	*	*	*	*
XORIGIN	*	*	*	*	*	*	*	*
STYLE	xy	xy	xy	xy	xy	xy	xy	xy

Pen

PEN	1	2	3	4	5	6	7	8
COLOUR	1	2	3	4	1	2	3	4
L-STYLE	0	0	0	0	0	0	0	0
METHOD	point	point	point	point	point	point	point	point
SYMBOLS	1	2	3	4	5	6	7	8
JOIN	ascend	ascend	ascend	ascend	ascend	ascend	ascend	ascend
BRUSH	1	2	3	4	5	6	7	8

Keys are very useful for interpreting a graphical picture. The formats of the keys depend on the picture, but they have some common characteristics. The key contains a title, if you have specified one by the TITLE option, and one or more legends. The exact format of the legends is described below, when we describe the various pictures.

Sometimes the picture, without a key, would completely fill the available screen. So you might want to obtain the picture first without a key, and then obtain just the key by repeating the statement but preventing the picture from appearing. For example:

DGRAPH [WINDOW = 2; KEYWINDOW = 0 ......
DGRAPH [WINDOW = 0; KEYWINDOW = 2 ......

Obviously, you should not change any of the relevant attributes of the pens used between these DGRAPH statements.

Generally you would not want the key window to overlap the picture window. But after you have inspected a picture you can use the FRAME directive with suitable bounds in the AXES directive to insert the key inside an area of the graph that has no useful graphical information.

You get a warning message if the space in the key window is filled: there is room for 53 legends in a full-sized window. Genstat might truncate very long identifiers and descriptions, giving you a warning message. You might then want to try to extend the window size using the FRAME directive.

7.3.1 The DHISTOGRAM directive

DHISTOGRAM

Draws histograms on a plotter or graphics monitor.

Options

TITLE	= text	General title; default *
WINDOW	= scalar	Window number for the histograms; default 1
KEYWINDOW	= scalar	Window number for the key (zero for no key); default 2
LIMITS	= variate	Variates of group limits for classifying variates into groups; default *
NGROUPS	= scalar	When LIMITS is not specified, this defines the numbers of groups into which a DATA variate is to be classified; default is the integer value nearest to the square root of the number of values in the variate
LABELS	= text	Group labels; default *
APPEND	= string	Whether or not the bars of the histograms are appended together (no, yes); default n
SCREEN	= string	Whether to clear the screen before plotting or to continue plotting on the old screen (clear, keep); default c

Parameters

DATA	= identifiers	Data for the histograms; these can be either a factor indicating the group to which each unit belongs, a variate whose values are to be grouped, or a one-way table giving the number of units in each group
NOBSERVATIONS	= tables	One-way table to save numbers in the groups
GROUPS	= factors	Factor to save groups defined from a variate

PEN	= scalars	Pen number for each histogram
DESCRIPTION	= texts	Annotation for key

Nearly all the options and parameters of DHISTOGRAM are either the same as those in HISTOGRAM (7.1.1) or are common to all the directives for high-quality graphics (described above). The exception is the APPEND option, which is relevant when a number of structures are specified by the DATA parameter.

The histograms produced by DHISTOGRAM can be of two forms. By default, all the bars of a histogram are drawn parallel to each other, representing the data in each structure listed by the DATA parameter. Alternatively, if you set the APPEND option to 'yes', corresponding bars for each structure are concatenated into a single bar, so that the picture shows only one bar for each group of the data. Then the top portion of each bar corresponds to the first structure, and the bottom to the last structure.

The area representing one structure within a bar is filled according to the settings of the BRUSH and COLOUR parameters of the PEN directive for the graphical pen corresponding to that structure. In order to separate the bars when there is more than one structure, you should use more than one pen, since otherwise they will all have the same shading and colour.

The length of a bar is proportional to the number of values it represents, irrespective of the limits of the corresponding group. The width of the bar does not represent any feature of the data. Notice, therefore, that the area of the bars cannot be taken to represent relative frequencies of the groups.

You can set the YUPPER parameter of the AXES directive to determine an overall upper limit of the histogram. You can thus produce a series of histograms to the same scale by choosing a value larger than all the bars. However, Genstat ignores the setting of this parameter if the longest bar is greater than the setting; so when you use YUPPER, you must be careful to take into account the effect of the APPEND option.

The key that Genstat produces for the histogram consists of the title, if there is one, followed by a legend for each structure in the DATA parameter. The legend consists of a small rectangle that is drawn in the same colour and brush style as that used in the histogram; it is followed by the corresponding text that you set in the DESCRIPTION parameter, or, if you did not set any, the identifier of the structure in the DATA parameter.

Here is an example of a histogram of some random numbers generated by the URAND function (5.2.8); the resulting picture is in Figure 7.3.1.

```
CALCULATE Var[1...3] = URAND(1237,0,0; 30)
& Var[1...3] = 10,11,12 + NED(Var[1...3]) * 1,1.2,1.3
"Default histogram with single colour pen"
PEN 1...3; COLOUR=1
DHISTOGRAM [TITLE='Default'] Var[]; PEN=1...3
"Repeat setting the APPEND option and different brush styles"
```

7.3 High-quality graphics

```
PEN 1...3; BRUSH = 4,5,9
DHISTOGRAM [TITLE = 'Appending & new brush style'; APPEND = yes] \
   Var[]; PEN = 1...3; DESCRIPTION = 'First','Second','Last'
VARIATE [VALUES = 6,9,12,15] Limits
AXES WINDOW = 1; YUPPER = 30
DHISTOGRAM [TITLE = 'YUPPER set & limits';LIMITS = Limits]\
   DATA = Var[]; PEN = 1,2,3; DESCRIPTION = 'First','Second','Last'
STOP
```

FIGURE 7.3.1

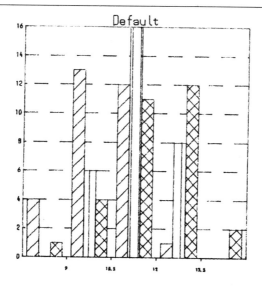

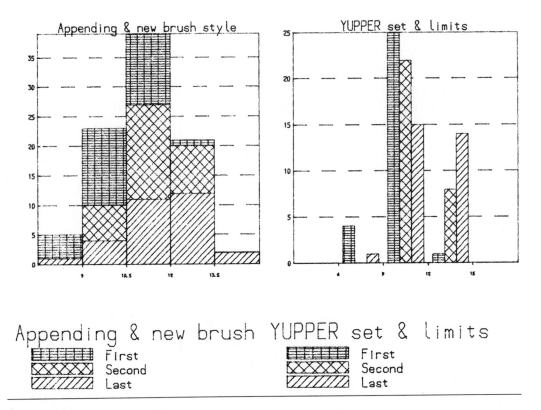

7.3.2 The DGRAPH directive

DGRAPH

Draws graphs on a plotter or graphics monitor.

Options

TITLE	= text	General title; default *
WINDOW	= scalar	Window number for the graphs; default 1
KEYWINDOW	= scalar	Window number for the key (zero for no key); default 2
SCREEN	= string	Whether to clear the screen before plotting or to continue plotting on the old screen (clear, keep); default c

Parameters

Y	= identifiers	Vertical coordinates

7.3 High-quality graphics

X	= identifiers	Horizontal coordinates
PEN	= scalars	Pen number for each graph
DESCRIPTION	= texts	Annotation for key
YLOWER	= identifiers	Lower values for vertical bars
YUPPER	= identifiers	Upper values for vertical bars
XLOWER	= identifiers	Lower values for horizontal bars
XUPPER	= identifiers	Upper values for horizontal bars

All the options of the DGRAPH directive, and the PEN and DESCRIPTION parameters, are common to all the directives for high-quality graphics, and are described at the beginning of this section.

The DGRAPH directive draws a graph on the graphical device that you have specified by the DEVICE directive. As for the GRAPH directive, corresponding structures listed by the Y and X parameters must be of equal length. See 7.1.2 for a detailed description of these parameters. If the structures contain missing values, you will get a warning in the current output file (not the graphics file) of the form:

```
                    ***  EXAMPLE 7.3.2   ***

* Message: At least one missing value in the plot of  YY v XX
```

where YY is the name of the y-variate and XX is the name of the x-variate. The type of graphical output that you get when there are missing values depends on the setting of the METHOD parameter in the latest PEN directive. If the method is 'point' then there will be no indication on the graph that any points were missing. Otherwise the connecting line will be broken at the places where a value is missing.

By default, Genstat calculates bounds on the axes that are wide enough to include all the data. You can set bounds in the AXES directive in such a way that points fall outside them. Then *clipping* occurs if you have set the METHOD parameter of PEN to anything other than 'point'. That is, lines will be drawn from points that are within bounds towards points that are out of bounds, terminating at the appropriate edge. Clipping may also occur if the method is 'monotonic', 'open', or 'closed' and you have left Genstat to set default boundaries, because these methods fit curves that may extend beyond the boundaries. If the method is 'point' then Genstat does not indicate points that are out of bounds.

Genstat uses LINESTYLE=1 if you set the LINESTYLE parameter of the PEN directive to 0 and do not also set the METHOD parameter to 'point'. But if you set the METHOD parameter to 'point' and the SYMBOLS parameter to 0, you will get no plot at all, although an entry will be made in the key if there is one. If you set the METHOD parameter to 'monotonic', the JOIN parameter will be ignored and the data will be plotted in ascending order of the x-values.

In addition to plotting points and joining the points with lines, DGRAPH can draw "error bars" parallel to either axis. You might want to use these, for example, to show confidence limits on points that have been fitted by a regression (Chapter 8). To draw error bars parallel to the y-axis you should supply variates holding the y-coordinates of the upper and lower ends of the bars; these variates should be the same length as the y-variate.

        YLOWER = Lowererr; YUPPER = Uppererr

Genstat draws a line to connect the lower to the upper error at the associated x-position. If one is missing, then a line is drawn from the non-missing one to the associated y-position. You can obtain similar error bars parallel to the x-axis using XLOWER and XUPPER.

There are only eight Genstat pens but there are four colours, four visible line styles, and up to nine graphical symbols. You can usually get distinguishable plots by redefining the pens suitably, provided the plots do not obliterate each other. You do this by giving a series of DGRAPH statements; use the same window and set the option SCREEN = keep for the second and subsequent graphs of the series. The axes are then drawn only for the first graph, and the same axis bounds are used for the subsequent ones as were determined by the first DGRAPH statement. Thus you should define axis limits that enclose all the subsequent data. Here is a program that draws sixteen graphs in a frame:

```
PEN 1...8; COLOUR = 1...4; LINESTYLE = 1; SYMBOLS = 1...8; METHOD = line
VARIATE [NVALUES = 4] Y[2...16]
& [VALUES = 1,3,2,4] Y[1]
CALCULATE Y[2...16] = Y[1...15] + 1
& Ymax = MAXIMUM(VMAXIMA(Y))
& Ymin = MINIMUM(VMINIMA(Y))
AXES 1; YLOWER = Ymin; YUPPER = Ymax
VARIATE [NVALUES = 4; VALUES = 1...4] X
FRAME 1,2; YLOWER = 0; YUPPER = 0.5,0.4; XLOWER = 0,0.5; XUPPER = 0.5,1
DGRAPH Y[1...8]; X; PEN = 1...8
PEN 1...8; LINESTYLE = 2
DGRAPH   [WINDOW = 1;   KEYWINDOW = 2;   SCREEN = keep]   Y[9...16]; \
   X; PEN = 1...8
STOP
```

Since this book is printed in black and white, all the illustrations of high-quality pictures are drawn with a black pen only. To do that here, we set COLOUR = 1 in the PEN statement above; the result is in Figure 7.3.2a.

The title may not be sufficient for your needs, so you can add extra text within the graph. If you want to print several lines, specify them as strings of a text and give plotting positions for each string. The next example shows how to add several

EXAMPLE 7.3.2a

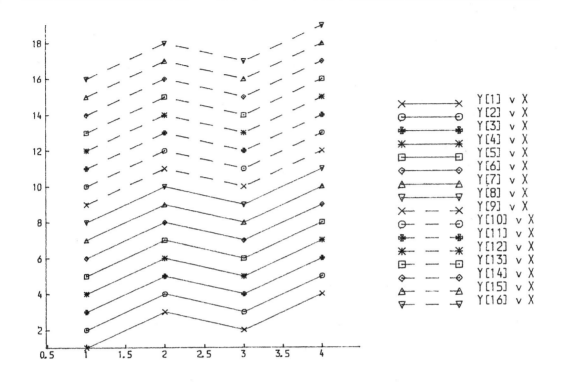

lines of text, held in the text Lines. Where they appear is determined by the variates Ypos and Xpos: thus this text is within the bounds of the axes.

```
SCALAR Yl,Yu,Xl,Xu; VALUE=0,100,0,10
AXES 1; YLOWER=Yl; YUPPER=Yu; XLOWER=Xl; XUPPER=Xu
"Set up y-positions, using the top 3% of the picture"
VARIATE [VALUES=1...4] Ypos
& [NVALUES=4] Xpos
CALCULATE Inc = (Yu − Yl) * 0.03
& Ypos = Yu − Ypos*Inc
& Xx = (Xu − Xl) * 0.5
"Ensure that the text is left-justified
TEXT [VALUES=One,Two,Three,Four] Lines
PEN 1; METHOD=point; SYMBOLS=Lines
DGRAPH [KEYWINDOW=0] Ypos; Xpos; PEN=1
STOP
```

The resulting picture is in Figure 7.3.2b.

FIGURE 7.3.2b

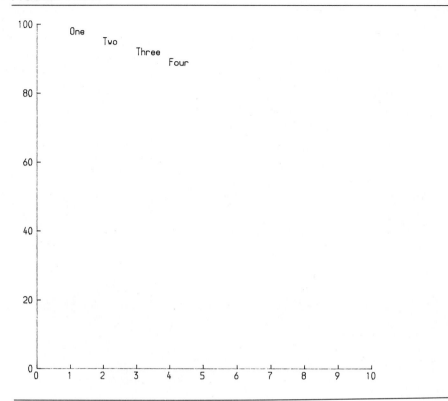

Inside the key drawn by the DGRAPH directive there is a legend for each component plot. The DESCRIPTION parameter allows you to specify your own text instead of the default, which consists simply of the identifiers used for that component of the picture. If you are plotting text you will generally not need a key. But if you add further non-textual graphs to the same diagram, you might want a key, which you can incorporate by:

DGRAPH [KEYWINDOW=2; SCREEN=keep] Y=...; X=...

For any graph, the key consists of a title (if you have specified one) and a legend for each component plot of the graph. Genstat automatically draws the legend with the same graphical pen as is used for the corresponding plot. Each legend gives three pieces of information: the setting of the SYMBOLS parameter of the PEN directive, a segment of a line drawn in the style of the relevant pen, and a description of the plot. The description is the corresponding text that you have set with the DESCRIPTION parameter; but if you have not set any, it consists of the identifiers of the y-variate and the x-variate.

7.3 High-quality graphics

If you have set the SYMBOLS parameter to be a factor, then the legend contains the identifier of the factor and its first level. If it is a text, the legend is the identifier of the text and its first string. No line will appear in the legend if you set the LINESTYLE parameter of PEN to 0 or the METHOD parameter to 'point'; otherwise it is drawn in the appropriate style. If you draw several pictures with SCREEN=keep and use the same window, the key information is appended to the key if requested and if there is room.

The complete program to produce the graphs shown in Figure 7.2 is as follows:

```
VARIATE [VALUES=3,2,7,5] Yval
VARIATE [VALUES=2,8,4,6] Xval; DECIMALS=0
FRAME 1,2; YLOWER=0.25,0; YUPPER=0.75,0.2; XLOWER=0; XUPPER=0.5
DGRAPH Yval; Xval
AXES 3; YLOWER=10; YUPPER=0; XLOWER=0; XUPPER=10; XMARK=Xval; \
   XLABEL=!T(NORTH,EAST,SOUTH,WEST);   YTITLE='Y   AXIS'; \
   XTITLE='X AXIS'
FACTOR [LEVELS=4; VALUES=1...4] F1
PEN 1,2; COLOUR=1; LINESTYLE=1,2; METHOD=line,monotonic; \
   SYMBOLS=F1,3; JOIN=given,ascending
VARIATE [VALUES=4,3,8,6] Yval2
FRAME 3,4; YLOWER=0.25,0; YUPPER=0.75,0.2; XLOWER=0.5; XUPPER=1
DGRAPH [TITLE='Test'; WINDOW=3; KEYWINDOW=4; SCREEN=keep] \
   Yval,Yval2; Xval; PEN=1,2; DESCRIPTION='FIRST','SECOND'
STOP
```

You can superimpose pictures, for example a graph on top of a histogram. To match the two pictures below it is important to supply limits for the groups in the histogram, and to make them of equal width.

```
"Generate 1000 Normal values with mean 10 and s.d. 1.5"
SCALAR Nval; VALUE=1000
CALCULATE Hval = 10 + NED(URAND(1111; Nval)) * 1.5
"Set an upper bound for the y-axis"
SCALAR Yup; VALUE=300
AXES 1; YUPPER=Yup
"Define limits and draw the histogram"
VARIATE [VALUES=3...16] Limits
DHISTOGRAM [TITLE='Histogram & graph'; LIMITS=Limits] Hval
"Calculate the mean and s.d."
CALCULATE Mu = MEAN(Hval)
  & Sigma = SQRT(VARIANCE(Hval))
"Extend the bounds for the x-axis by one increment in both directions"
SCALAR Npos,Xlow,Xup; VALUE=100,2,17
```

```
VARIATE [VALUES=0...Npos] Xval
"Calculate x and y values"
CALCULATE Xval = Xlow + (Xval * (Xup-Xlow)/Npos)
& Ycal = Nval / (SQRT(4*ARCCOS(0))* Sigma) \
  * EXP(-0.5*((Xval-Mu)/Sigma)**2)
AXES  1;  YLOWER=0;  YUPPER=Yup;  XLOWER=Xlow;  XUPPER=Xup;\
  STYLE=none
"Draw the graph"
PEN 1; COLOUR=2; METHOD=monotonic; SYMBOLS=0
DGRAPH [SCREEN=keep; KEYWINDOW=0] Ycal; Xval
STOP
```

The resulting picture is in Figure 7.3.2c.

FIGURE 7.3.2c

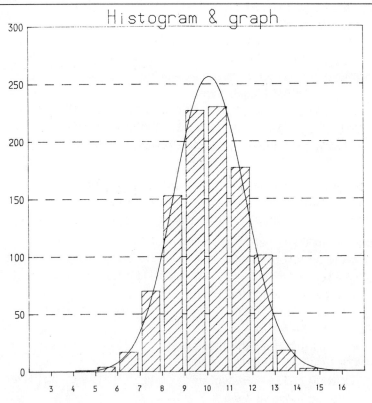

7.3.3 The DCONTOUR directive

DCONTOUR

Draws contour plots on a plotter or graphics monitor.

Options

INTERVAL	= scalar or variate	Contour interval for scaling (scalar) or positions of the contours (variate); default * i.e. determined automatically
TITLE	= text	General title; default *
WINDOW	= scalar	Window number for the plots; default 1
KEYWINDOW	= scalar	Window number for the key (zero for no key); default 2
LOWERCUTOFF	= scalar	Lower cut-off for array values; default *
UPPERCUTOFF	= scalar	Upper cut-off for array values; default *
SCREEN	= string	Whether to clear the screen before plotting or to continue plotting on the old screen (clear, keep); default c

Parameters

GRID	= identifiers	Pointers (of variates representing the columns of a data matrix), matrices or two-way tables specifying values on a regular grid
PEN	= scalars	Pen number to be used for the contours of each grid
DESCRIPTION	= texts	Annotation for key

All the options and parameters of the DCONTOUR directive are either the same as in the CONTOUR directive (7.1.3) or are common to all the directives for high-quality graphics (see the beginning of this section). The plot is also oriented in the same way as for CONTOUR.

The directive causes contour plots to be drawn to represent data in the structures specified by the GRID parameter. Each structure can be a rectangular matrix, a two-way table with or without margins, or a pointer to a set of variates. If you give a matrix, the numbers of rows and of columns must be three or more. If you give a table, the number of levels of both factors must be three or more; any marginal values will not form part of the contour plot. If you give a pointer, it must point to three or more variates of equal length, and the length must be three or more.

No missing values are allowed. For a pointer, any of the variates can be restricted, and the restriction will be applied to all the variates, but the number of values to be used must be at least three.

The levels for the contours can be produced in three ways. If you set the INTERVAL option to a variate with two or more values, they will be used as the contour levels. Alternatively, if you give a single value in the INTERVAL option, it defines the difference between each contour, whether or not you have set the LOWERCUTOFF and UPPERCUTOFF options. If the resulting number of contours is less than two, then no contours are drawn; if the number is very large, the contours may be difficult to interpret and take a long time to plot. If you do not set the INTERVAL option, Genstat will calculate a suitable interval based on the LOWERCUTOFF and UPPERCUTOFF options, or if you have not specified these, the interval will be based on the minimum and maximum values to be plotted.

You can plot values from several structures in the same window, but the resulting output may be difficult to interpret.

By default, axes will be drawn on the contour plot. You can prevent this happening by setting the STYLE parameter in the AXES directive to 'none' for the relevant window. If this parameter setting is not 'none', then by default the axis bounds will be from 1 up to the number of rows of the grid for the y-axis, and from 1 up to the number of columns for the x-axis. The settings of the parameters of the AXES directive will produce the same effect as they do in DGRAPH, but the EQUAL option has no effect.

The contour plot is a series of contour lines which Genstat labels by their level numbers. If a contour line is very short, no label will be given. The contour key gives the contour labels and the corresponding value in the scale of the data. Generally, you should define the size of the key window in the FRAME directive so that the value of the YUPPER parameter minus the value of the YLOWER parameter is large enough for all the levels that are to be used. If the settings of YLOWER and YUPPER are 0 and 1, then there is room to list 50 contour levels. The contour lines are made up of straight lines or curves according to the setting of the METHOD parameter of the corresponding pen. If the setting is 'open' or 'closed', the method of McConalogue (1970) will be used; if it is 'monotonic', the method of Butland (1980) is used. Both of these methods can produce contour lines that cross, particularly if the supplied grid of data is coarse.

Here is an example in which a matrix of function values is calculated and plotted as a contour. The columns represent a range of y-values and the rows a range of x-values.

7.3 High-quality graphics

```
"Set up interval and the number of rows and columns"
SCALAR Int,Nval[1,2]; VALUE = 20,31,31
"Set up range of values"
SCALAR Start[1,2],End[1,2];VALUE = -1.2,0,1.2,1.2
"Find intervals"
CALCULATE Inc[1,2] = (End[] - Start[])/(Nval[] - 1)
"Set up row and column matrices of ones"
MATRIX [ROWS = Nval[1]; COLUMNS = 1; VALUES = (1)#Nval[1]] One[1]
MATRIX [ROWS = 1; COLUMNS = Nval[2]; VALUES = (1)#Nval[2]] One[2]
"Calculate the matrix: the function is the Rosenbrock function
 F = 100 * (P2 - P1**2)**2 + (1 - P1)**2"
CALCULATE Vals[1,2] = Start[] - Inc[] + CUM(One[]) * Inc[]
&         Func = One[1]*+Vals[2] - (Vals[1]**2)*+One[2]
&         Func = 100 * Func**2 + (1 - Vals[1]*+One[2])**2
"Set the window sizes"
FRAME     WINDOW = 1,2;    YLOWER = 0;   YUPPER = 1;   XLOWER = 0,0.8;\
   XUPPER = 0.8,1
"Draw a plot with specified titles"
AXES WINDOW = 1; XTITLE = 'Rows'; YTITLE = 'Columns'
DCONTOUR [TITLE = 'Default'; INTERVAL = Int] Func
"Draw a plot with specified axes"
AXES WINDOW = 1; YLOWER = Start[1]; YUPPER = End[1]; \
   XLOWER = Start[2]; XUPPER = End[2]
DCONTOUR [TITLE = 'User axes'; INTERVAL = Int] Func
"Draw a plot with specified contour levels"
DCONTOUR [TITLE = 'User contour intervals';\
   INTERVAL = !(0,1,2,4,8,16,32,64,128)] Func
STOP
```

The resulting graph is in Figure 7.3.3 but with windows rearranged to fit on the page.

300   7   *Graphical display*

FIGURE 7.3.3

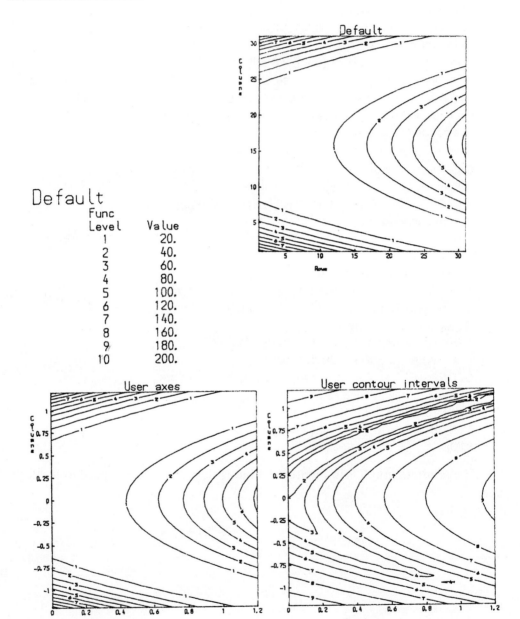

7.3 High-quality graphics

```
User axes                          User contour intervals
    Func                               Func
    Level    Value                     Level    Value
      1       20.                        1        0.
      2       40.                        2        1.
      3       60.                        3        2.
      4       80.                        4        4.
      5      100.                        5        8.
      6      120.                        6       16.
      7      140.                        7       32.
      8      160.                        8       64.
      9      180.                        9      128.
     10      200.
```

7.3.4 The DPIE directive

DPIE

Draws a pie chart on a plotter or graphics monitor.

Options

TITLE	= text	General title; default *
WINDOW	= scalar	Window number for the pie chart; default 1
KEYWINDOW	= scalar	Window number for the key (zero for no key); default 2
SCREEN	= string	Whether to clear the screen before plotting or to continue plotting on the old screen (clear, keep); default c

Parameters

SLICE	= scalars	Amounts in each of the slices (or categories)
PEN	= scalars	Pen number for each slice
DESCRIPTION	= texts	Description of each slice

All the options of the DPIE directive, and the PEN and DESCRIPTION parameters, are common to all the directives for high-quality graphics, and are described at the beginning of this section.

A pie chart is formed by taking the values of the scalars in the SLICE parameter, in order, and representing them by segments of a circle or ellipse starting at "three o'clock" and working in an anti-clockwise direction. The angle subtended by each segment is proportional to the magnitude of the value; all values must be non-negative and not missing. A segment is filled with the brush pattern associated with the relevant pen using the PEN directive. Thus, where possible, you should use as many pens as there are scalars in the SLICE parameter.

302   7   *Graphical display*

The shape of the pie chart is determined by the setting of the parameters in the FRAME directive. You get an elliptical pie chart if the window is not square: that is, if the setting of YUPPER minus that of YLOWER does not equal the setting of XUPPER minus that of XLOWER.

The key is identical to that produced by the DHISTOGRAM directive.

Figure 7.3.4 shows an example of a pie chart with four slices, drawn by the following program:

```
PEN 1...4; COLOUR = 1; BRUSH = 1,6,11,16
DPIE [TITLE = 'Pie Chart'] 1,2,4,8; PEN = 1...4
```

FIGURE 7.3.4

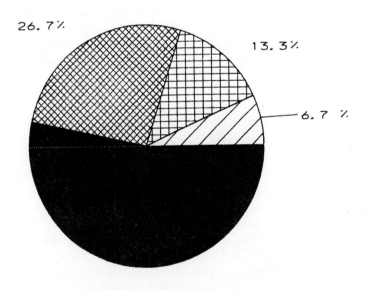

K.E.B.
S.A.H.
R.P.W.

8 Regression analysis

The simplest meaning of the word *regression* is the technique for fitting a straight line that relates one quantitative variable to another. The *response variable* is supposed to be dependent on the *explanatory variable*. How to do this simple linear regression with Genstat is described in 8.1.

In later sections we use the word regression to cover a much wider class of relationships. We look at more than two variables, at qualitative variables, and at nonlinear relationships. But the common feature is that we shall always be modelling the dependence of one variable on others.

The word linear here does not mean linear in terms of the explanatory variables, but rather linear in terms of the parameters or coefficients that have to be estimated. Thus the regression

$$y = \alpha + \beta \times x + \gamma \times x^2 + \varepsilon$$

is in fact linear: it is linear in terms of the parameters α, β, and γ, even though it is not linear in terms of the explanatory variable x.

In the model for simple linear regression, it is usually assumed that the response variable has a Normal distribution with constant variance. But other distributions can be used, and the variance need not be constant. For example, the distribution could be Poisson in which the variance is equal to the mean. These extensions are provided by generalized linear models, as described in 8.4.

In all models in this chapter, we assume that there is only one component of variation: that is, they contain only one error term like ε in the equation above. When there are more components, some results can be obtained by the methods described here: for example, you could analyse the effects of treatment factors after eliminating some blocking of the units. But for balanced designs with several components of variation you should use the methods of Chapter 9.

We assume in this chapter that you know what is the response variable and what are explanatory variables. There are more general methods of investigating relationships between variables, in which no single variable is treated as a response; see Chapter 10. We also assume that the relationship between the response variable and explanatory variables relates the mean of the response to given explanatory values. The methods of regression analysis are not applicable to law-like relationships, with values of both the response and the explanatory variables subject to error; for more details, see Sprent (1969).

Finally, we assume in this chapter that the errors in regression models are uncorrelated. For example, instances of the term ε in the equation above, for each pair of observations x and y in a set of data, are assumed to be independently distributed. When there is some correlation between the errors, the methods of Chapter 9 may be suitable, particularly if the correlation is constant within some groups of the data and zero between the groups. Alternatively, if there is a serial pattern of correlation, where the order of the observations is important, the methods of Chapter 11 may be used.

Throughout this chapter we assume that you already know about regression. We give only brief references to standard books, where you can find the theory expanded fully.

8.1 Simple linear regression

The word *simple* does not mean easy, but that there is only one explanatory variable. Suppose you have observations $\{y_i: i = 1...N\}$ of a response variable Y, and $\{x_i: i = 1...N\}$ of an explanatory variable X. Then the model for simple linear regression is:

$$y_i = \alpha + \beta \times x_i + \varepsilon_i$$

where α and β are unknown *parameters*: that is, they are numerical characteristics of the model that determine the precise nature of the relationship. The values $\{\varepsilon_i: i = 1...N\}$ are *errors* which are random variables, identically and independently distributed with a Normal distribution. For details of this model, see the books by Draper and Smith (1981) or by Weisberg (1982), or indeed any other standard statistical text.

The model can alternatively be written in matrix form:

$$y = X\beta + \varepsilon$$

where the $N \times 2$ matrix X is called the *design matrix*, and $\beta' = (\alpha, \beta)$.

Here is a Genstat program to fit a simple linear regression, along with the standard output that is produced.

EXAMPLE 8.1

```
   1   "
  -2    Simple linear relationship between % damage to apples and cropsize
  -3    Data from Snedecor and Cochran (1980) page 162.
  -4   "
   5   VARIATE [VALUE= 8, 6,11,22,14,17,18,24,19,23,26,40] Cropsize
   6   &        [VALUE=59,58,56,53,50,45,43,42,39,38,30,27] %wormy
   7   MODEL %wormy
   8   FIT Cropsize
```

8...

***** Regression Analysis *****

 Response variate: %wormy
    Fitted terms: Constant, Cropsize

```
*** Summary of analysis ***

              d.f.        s.s.         m.s.
Regression       1        948.2       948.16
Residual        10        273.8        27.38
Total           11       1222.0       111.09

Percentage variance accounted for 75.3

 * MESSAGE: The following units have large residuals:
                            4                    2.22

 * MESSAGE: The following units have high leverage:
                           12                    0.56

*** Estimates of regression coefficients ***

              estimate        s.e.            t
Constant         64.25        3.60         17.83
Cropsize         -1.013       0.172        -5.88
```

The first two statements set up variates to store the values of the two variables to be analysed; the next two statements fit the regression.

Often, more statements are needed before the regression statements are given. For example, with percentages the variances of the observations are usually not constant. Unless the range of percentages is narrow, you should correct for this by applying a transformation or better, use a generalized linear model.

For example, to replace the values for percentage damage by their logit transformation, you would insert after the VARIATE statement the line

    CALCULATE %wormy = LOG(%wormy/(100 − %wormy))

The remaining lines would be as before, to give an analysis of the logit of the percentage of damage. In this model, the variance of the transformed percentage is assumed to be constant, which corresponds to assuming that the original percentage is less variable at the extremes (near 0% or 100%) than in the middle of the range (near 50%).

Another way of coping with unequal variances is to use different distributions: see 8.4.

You can fit models to subsets of the data by using the RESTRICT directive (2.4.1). The regression directives also automatically exclude any unit that contains a missing value for either variate. But if only the response is missing, then Genstat does give you some information about the unit (8.1.2).

Most of the directives in this section are relevant also to multiple regression and to nonlinear regression. But you can understand their main features most readily by finding out first how to use them in the simplest case.

8.1.1 The MODEL directive

MODEL
Defines the response variate(s) and the type of model to be fitted for linear, generalized linear, and nonlinear regression models.

Options

DISTRIBUTION	= string	Distribution of the response variable (normal, poisson, binomial, gamma, inversenormal, multinomial); default n
LINK	= string	Link function (canonical, identity, logarithm, logit, reciprocal, power, squareroot, probit, complementaryloglog); default c (i.e. i for DIST=n, l for DIST=p, logit for DIST=b, r for DIST=g, p for DIST=i, * for DIST = m
EXPONENT	= scalar	Exponent for power link; default -2
DISPERSION	= scalar	Value of dispersion parameter in calculation of s.e.s etc; default * for DIST=n,g,i, 1 for DIST=p,b,m
WEIGHTS	= variate	Variate of weights for weighted regression; default *
OFFSET	= variate	Offset variate to be included in model; default *
GROUPS	= factor	Factor defining the groups for within-groups linear regression; default *
RMETHOD	= string	Type of residuals to form, if any, after each model is fitted (deviance, Pearson); default d
FUNCTION	= scalar	Scalar whose value is to be minimized by calculation; default *
SAVE	= identifier	To name regression save structure; default *

Parameters

Y	= variates	Response variates; only the first is used in generalized linear and nonlinear models

NBINOMIAL	= variate	Variate of binomial totals
RESIDUALS	= variates	To save residuals for each response variate after fitting a model
FITTEDVALUES	= variates	To save fitted values, and provide fitted values if no terms are given in FITNONLINEAR

In most applications, you will need only a simple form of the directive:

    MODEL identifier_of_response_variate

Notice that MODEL does not actually fit anything: it simply sets up some structures inside Genstat that are used when you give a FIT statement later on (8.1.2). So when you are doing regression, MODEL will always be accompanied by at least one other regression statement to fit a model, like FIT.

The Y parameter allows a list of variates: if you put more than one for linear regression, then you will get an analysis for each. This is a more efficient way of doing lots of linear regressions than separate pairs of MODEL and FIT statements. With generalized linear and nonlinear models (8.4 to 8.6), only the first variate will be analysed. The others will be ignored, except for taking account of missing values or of restrictions.

The NBINOMIAL parameter is relevant only for the 'binomial' setting of the DISTRIBUTION option (8.4.1).

The RESIDUALS and FITTEDVALUES parameters allow you to specify variates that contain the residuals and fitted values for each response variable. For example, above, you could put residuals into a variate R and fitted values into a variate F by changing the MODEL statement to

    MODEL %wormy; RESIDUALS = R; FITTEDVALUES = F

The residuals are the "unexplained" component of the response variable, standardized in some way according to the RMETHOD option (see below). The fitted values are the "explained" component: that is, the combination of parameters and explanatory variables fitted in the model. You can get access to these sets of values in a different way through the RKEEP directive (8.1.4).

The DISTRIBUTION, LINK, and EXPONENT options allow you to go beyond linear regression; they are described in 8.4.1.

The DISPERSION option controls how the variance of the distribution is calculated. By default, the variance is estimated from the residual mean square (8.1.2), and standard errors and standardized residuals are calculated from the estimate. If you use DISPERSION to supply a value for the variance of the Normal distribution, or for the dispersion parameter of other distributions (8.4), then standard errors and residuals are based on this given value instead.

The WEIGHT option allows you to specify a variate holding weights for each unit. Suppose, for example, you have assigned values to a weights variate W earlier in the program; then the option takes the form: WEIGHTS = W. If the weight for unit i is w_i, then regression directives will weight by w_i the contribution to the estimate of dispersion from the ith unit. In simple linear regression, the estimate of dispersion is then the weighted residual mean square:

$$\sum_{i=1}^{N} w_i \times \varepsilon_i^2 / (N-2)$$

Thus, if the variance of the response variable is not constant, and you know the relative size of the variance for each observation, you can set the weight to be proportional to the inverse of the variance of an observation.

The OFFSET option allows you to include in the regression a variable with no corresponding parameter:

$$y_i = \alpha + o_i + \beta \times x_i + \varepsilon_i$$

where o_i is the ith value of the offset variable, O say. Again, the variate O must have been given values earlier in the program. Simple linear regression including an offset is just the same as an analysis of y_i-o_i, but the offset has more sensible applications in generalized linear models (8.4.1).

The GROUPS option specifies a factor whose effects you want to eliminate before any regression is fitted. The factor should have been defined earlier. (The effects of factors on regression are discussed in 8.3.) This method of elimination is sometimes called *absorption*; you might want to use it when data from many different groups are to be modelled. GROUPS gives less information than you would get if you included the factor explicitly in the model, but it saves space inside the computer. You can use GROUPS only with linear regression.

The RMETHOD option controls how residuals are formed. By default, residuals are *deviance residuals* standardized by their estimated variance. For linear regression, these are:

$$\text{residual}_i = (y_i - f_i) \times \sqrt{(w_i/\text{variance}_i)}$$

In this, f_i is the ith fitted value, and the variance of a residual is

$$\text{variance}_i = (1 - l_i) \times (\text{estimate of dispersion})$$

The *leverage* (diagonal of the projection matrix), l_i, is defined in terms of the design matrix X and the diagonal matrix of weights W by

$$l_i = w_i \times \{X(X'WX)^{-1}X'\}_{ii}$$

The alternative *Pearson residuals* are defined in exactly the same way if the distribution is Normal. For regression models with distributions other than Normal, the two kinds of residual are different (8.4.4).

If you do not want residuals, you can set the option to asterisk (*) to save space in the computer: in that case, you will not be able to get residuals, fitted values, or leverages. If the regression is grouped, none of these quantities is available anyway.

8.1 Simple linear regression

The FUNCTION option is relevant only when you want to optimize a general function (8.6). It is ignored unless no response variates are specified by the Y parameter.

The SAVE option allows you to specify an identifier for the regression save structure. This structure stores the current state of the regression model, and can be used explicitly in the directives RDISPLAY (8.1.3), RKEEP (8.1.4), and PREDICT (8.3.4). If the identifier in SAVE is of a regression save structure that already has values, those values are deleted. You can reset the current regression save structure at any point in a program by using the SET directive (2.2.1). Then, later regression statements would use the model stored in this save structure.

8.1.2 The FIT directive

FIT
Fits a linear or generalized linear regression model.

Options

PRINT	= strings	What to print (model, summary, accumulated, estimates, correlations, fittedvalues, monitoring); default m,s,e
CONSTANT	= string	How to treat constant (estimate, omit); default e
FACTORIAL	= scalar	Limit for expansion of model terms; default 3
POOL	= string	Whether to pool ss in accumulated summary between all terms fitted in a linear model (no, yes); default n
DENOMINATOR	= string	Whether to base ratios in accumulated summary on rms from model with smallest residual ss or smallest residual ms (ss, ms); default s
NOMESSAGE	= strings	Which warning messages to suppress (dispersion, leverage, residual, aliasing, marginality); default *

Parameter

	formula	List of explanatory variates and factors, or model formula

---

A FIT statement must always be preceded by a MODEL statement, though not necessarily immediately. You can give several FIT statements after a single MODEL statement: for example, you might want to try out different explanatory variables.

The parameter of the FIT directive specifies the explanatory variables in the model. In the simple linear regression it consists of the identifier of the explanatory variate alone:

>FIT Cropsize

If you omit the parameter, Genstat fits a *null model*; that is, a model consisting of just one parameter, the overall mean:

$$y = \alpha + \varepsilon$$

The PRINT option controls output. You can give several settings at the same time, to provide reports on several aspects of the analysis.

The 'model' setting gives a description of the model, including response and explanatory variates. Here is a repeat of this aspect of the analysis for the example at the beginning of this section. (Remember that 'model' did not have to be set explicitly, because it is the default.) 'Model' gives the first lines in this output:

EXAMPLE 8.1.2a

```
  Response variate: %wormy
      Fitted terms: Constant, Cropsize

*** Summary of analysis ***

               d.f.         s.s.         m.s.
Regression       1         948.2       948.16
Residual        10         273.8        27.38
Total           11        1222.0       111.09

Percentage variance accounted for 75.3

* MESSAGE: The following units have large residuals:
                        4                    2.22

* MESSAGE: The following units have high leverage:
                       12                    0.56
```

The output from the 'summary' setting is also reproduced here: there is a summary analysis of variance, which subdivides the total sum of squares, corrected for the mean, between that explained by the regression (Regression), and that which is not explained (Residual). The table has the standard form with a column for the sums of squares (s.s.), one for the degrees of freedom (d.f.) and one for the mean squares (m.s.). The percentage variance accounted for is the *adjusted R^2 statistic*, expressed as a percentage:

>%v.a.f. = 100 × (1 − (Residual m.s.)/(Total m.s.))

8.1 Simple linear regression

However, if this statistic were to have a negative value, indicating a very poorly fitting model, then the message

---
EXAMPLE 8.1.2b

```
Residual variance exceeds variance of Y variate
```
---

would be printed instead.

Following the summary analysis of variance are two reports on apparently extreme observations in the data. First comes a list of any standardized residuals whose values are particularly large: the criterion is to list residuals greater than that value c corresponding to probability $1/d$ of being exceeded by a standard Normal deviate, where d is the number of residual degrees of freedom. However, the value $c=2.0$ is used instead of any smaller value when there are less than 20 residual degrees of freedom, and the value 4.0 is used instead of any larger value when there are more than 15,773 degrees of freedom. The second report similarly shows particularly large values of the leverage, using the criterion $c \times k/N$, where k and N are the number of parameters and number of units used in the regression model.

The reports are intended to warn you about potential "outliers" in your analysis, and about individual observations that are particularly influential in determining the parameter estimates. See Cook and Weisberg (1982) for more information about these and other model-checking techniques.

You can prevent these warnings appearing by using the NOMESSAGE option. They will not appear in any case if you have set the RMETHOD option of MODEL to *, or if you are doing a regression within groups.

The 'estimates' setting produces the last section of output in the example:

EXAMPLE 8.1.2c

```
*** Estimates of regression coefficients ***

            estimate         s.e.           t
Constant       64.25         3.60       17.83
Cropsize      -1.013        0.172       -5.88
```

The standard errors of the estimates are based here on the residual mean square. But you can supply an estimate of variance by using the DISPERSION option of MODEL; if you do this, Genstat prints a reminder about the basis of the standard errors. You can prevent this warning appearing by setting the NOMESSAGE option. The t-statistics allow you to assess whether the parameters differ significantly from zero (but be careful about the pitfalls of multiple comparisons). The number of degrees of freedom for such a test is the number of residual degrees of freedom reported in the summary analysis of variance.

The other settings you can use in the PRINT option are 'correlations', 'fitted', 'accumulated', and 'monitoring'. The first two of these produce output as shown below. There is a correlation matrix of the parameter estimates. And then there is a table of unit labels, values of response variate, fitted values, residuals, and leverages. The unit labels are those associated with the response variate using the NVALUES option of the VARIATE directive (3.3.2), if given, or the values of the unit structure (2.2.3). If neither are given, the integers 1...N are printed. If you weight the regression by setting the WEIGHTS option of the MODEL directive, the weights are also listed. The 'accumulated' and 'monitoring' settings are discussed later, in 8.2.1 and 8.4.4 respectively.

EXAMPLE 8.1.2d

```
***** Regression Analysis *****

*** Correlations ***

Constant      1   1.000
Cropsize      2  -0.908   1.000

                    1        2


*** Fitted values and residuals ***

                                  Standardized
         Unit   Response  Fitted value  residual   Leverage
           1      59.0      56.1         0.62       0.21
           2      58.0      58.2        -0.04       0.27
           3      56.0      53.1         0.60       0.15
           4      53.0      42.0         2.22       0.09
           5      50.0      50.1        -0.01       0.11
           6      45.0      47.0        -0.41       0.09
           7      43.0      46.0        -0.60       0.08
           8      42.0      39.9         0.42       0.11
           9      39.0      45.0        -1.20       0.08
          10      38.0      40.9        -0.59       0.10
          11      30.0      37.9        -1.63       0.14
          12      27.0      23.7         0.94       0.56

Mean              45.0      45.0         0.03       0.17
```

In the table, units are omitted according to any restriction in force or to any missing values of explanatory variates (8.1). Fitted values are shown, however, for units with zero weight or in which only the response variate is missing.

The CONSTANT option controls whether the constant parameter is included in the model. In simple linear regression, this parameter is the intercept, in other words the estimate of the response variable when the explanatory variable is zero. By setting CONSTANT = omit, you can prevent the constant parameter being

estimated, so that the simple linear regression becomes
$$y_i = \beta \times x_i + \varepsilon_i$$
This model is particularly useful when y_i and x_i are measurements of the same attribute of a unit, as in calibration, and when you know that they are zero together. However, you should be careful here: you must be sure that the relationship remains linear right down to zero.

When you omit the constant, the analysis of variance produced by PRINT=summary will not be corrected for the mean, so that the model will be compared with the null model $y_i=0$. (However, if the effects of factors are present in the model (8.3), the analysis will still be corrected for the mean.) The percentage variance accounted for will still be expressed as a percentage of the variance of the response variable about the mean.

The FACTORIAL option is described in 8.3.1, and the POOL and DENOMINATOR options in 8.2.1.

The NOMESSAGE option controls printing of warning messages. The 'aliasing' setting is discussed in 8.2.1 and 8.3.2, and the 'marginality' setting in 8.3.3. You use the 'leverage' and 'residual' settings to prevent warnings appearing about large leverages or residuals (those that would be produced by the 'summary' setting of the PRINT option). You use the 'dispersion' setting to prevent reminders appearing about the basis of the standard errors (as would be produced by the 'estimates' setting of the PRINT option).

8.1.3 The RDISPLAY directive

RDISPLAY
Displays the fit of a linear, generalized linear, or nonlinear model.

Options

PRINT	= strings	What to print (model, summary, accumulated, estimates, correlations, fittedvalues); default m,s,e
CHANNEL	= scalar	Channel number for output; default * i.e. current output channel
DENOMINATOR	= string	Whether to base ratios in accumulated summary on rms from model with smallest residual ss or smallest residual ms (ss, ms); default s
NOMESSAGE	= strings	Which warning messages to suppress (dispersion, leverage, residual); default *

SAVE	= identifier	Specifies save structure of model to display; default *, i.e. that of the latest model fitted

No parameters

The PRINT option has the same settings as in the FIT directive, except that no monitoring is available. The CHANNEL option selects the output channel to which the results are output, as in the PRINT directive (4.2.1). The DENOMINATOR (see 8.2.1) and NOMESSAGE options are also as in the FIT directive.

The SAVE option lets you specify the identifier of a regression save structure; the output will then relate to the most recent regression model fitted with that structure.

8.1.4 The RKEEP directive

RKEEP
Stores results from a linear, generalized linear, or nonlinear model.

Options

EXPAND	= string	Whether to put estimates in the order defined by the maximal model (no, yes); default n
SAVE	= identifier	Specifies save structure of model; default *, i.e. that of the latest model fitted

Parameters

Y	= variates	Y variates for which results are to be saved

RESIDUALS	= variates	Standardized residual for each y variate
FITTEDVALUES	= variates	Fitted values for each y variate
LEVERAGES	= variate	Leverages of the units for each y variate
ESTIMATES	= variates	Estimates of parameters for each y variate
SE	= variates	Standard errors of the estimates
INVERSE	= symmetric matrix	Inverse matrix from a linear or generalized linear model, second derivative matrix from a nonlinear model
VCOVARIANCE	= symmetric matrix	Variance-covariance matrix of parameters
DEVIANCE	= scalars	Residual ss or deviance
DF	= scalar	Residual degrees of freedom
TERMS	= pointer or formula	Fitted terms
ITERATIVEWEIGHTS	= variate	Iterative weights from a generalized linear model
LINEARPREDICTOR	= variate	Linear predictor from a generalized linear model
EXIT	= scalar	Exit status from nonlinear model
GRADIENTS	= pointer	Derivatives of fitted values with respect to parameters in a nonlinear model
GRID	= variate	Grid of function or deviance values from a nonlinear model

Structures included in the parameters need not be declared explicitly: the RKEEP directive will declare them implicitly to be of the correct type and length.

The Y parameter specifies for which response variates the results are to be saved. If you leave it out, Genstat assumes that results are to be saved for all response variates; the order in which they are dealt with is the same as you specified in the latest MODEL statement.

The RESIDUALS, FITTEDVALUES, and LEVERAGES parameters allow you to save the standardized residuals, the fitted values, and the leverages. For example, RESIDUALS = R puts the residuals in a variate R. You cannot save these values if you set the RMETHOD option to * in the MODEL statement, or if you are doing a regression within groups.

The ESTIMATES and SE parameters allow you to save the parameter estimates and their standard errors, using the same order as in the display produced by the PRINT option of FIT. However, you can change the order by setting the EXPAND option (8.2.4).

The INVERSE parameter allows you to save the inverse matrix as a symmetric matrix: that is, $(X'X)^{-1}$ where X is the design matrix. This matrix is the same for all response variates.

The VCOVARIANCE parameter allows you to save as symmetric matrices the variance-covariance matrix of the estimates for each response variate: these are formed by multiplying the inverse matrix by the relevant variance estimate based on the estimated dispersion, or on the dispersion you have supplied.

The DEVIANCE and DF parameters allow you to save as scalars the residual sum of squares and residual degrees of freedom. In models with distributions other than Normal (8.4), the first of these lets you save the deviance.

The ITERATIVEWEIGHTS and LINEARPREDICTOR parameters are discussed in 8.4.4, the EXIT and GRADIENTS parameters in 8.5.3, and the GRID parameter in 8.6.1.

8.2 Multiple linear regression

When there are several explanatory variables instead of one we have multiple linear regression. The model can be written:

$$y_i = \alpha + \beta_1 \times x_{1i} + \beta_2 \times x_{2i} + \ldots + \beta_k \times x_{ki} + \varepsilon_i$$

or in matrix form:

$$y = X\beta + \varepsilon$$

where the design matrix X has $k+1$ columns. The errors $\{\varepsilon_i\}$ will be assumed in this section to be Normally distributed, as in 8.1.

You can fit a multiple linear regression with the MODEL and FIT directives. You must now give a list of explanatory variates in FIT. Here is an example; it reads data from a file attached to the second input channel containing the values of the four explanatory variables and the dependent variable laid out in parallel columns.

EXAMPLE 8.2

```
  1   "
 -2     Multiple linear regression, relating the monthly water usage
 -3     (thousand  gallons) of a plant to four production variables:
```

```
 -4        1. Average monthly temperature (degrees F)
 -5        2. Amount of production (billion pounds)
 -6        3. Number of plant operating days in the month
 -7        4. Number of people employed
 -8      Data from Draper and Smith (1981) page 196.
 -9      "
 10   UNIT [NVALUES=17]
 11   OPEN 'WATER.DAT'; CHANNEL=2
 12   READ [CHANNEL=2] Temp,Product,Opdays,Employ,Water
```

Identifier	Minimum	Mean	Maximum	Values	Missing
Temp	39.50	64.85	81.00	17	0
Product	6.37	12.90	18.57	17	0
Opdays	19.00	21.47	25.00	17	0
Employ	129.0	181.8	206.0	17	0
Water	2.828	3.304	4.488	17	0

```
 13   MODEL Water
 14   FIT Temp,Product,Opdays,Employ


14.........................................................................


***** Regression Analysis *****

Response variate: Water
     Fitted terms: Constant, Temp, Product, Opdays, Employ

*** Summary of analysis ***
```

	d.f.	s.s.	m.s.
Regression	4	2.4488	0.61221
Residual	12	0.7438	0.06198
Total	16	3.1926	0.19954

Percentage variance accounted for 68.9

* MESSAGE: The following units have high leverage:
                        1   0.59

*** Estimates of regression coefficients ***

	estimate	s.e.	t
Constant	6.36	1.31	4.84
Temp	0.01387	0.00516	2.69
Product	0.2117	0.0455	4.65
Opdays	-0.1267	0.0480	-2.64
Employ	-0.02182	0.00728	-3.00

One thing you might want to do with multiple regression is find the subset of explanatory variables that gives the most satisfactory fit. You can search for this subset by using the ADD, DROP, SWITCH, TRY, and STEP directives. Each of these makes and reports changes to the current regression model. You must use the TERMS directive before any of these, to define a common set of units for the regression and to carry out initial calculations efficiently.

The directives described in this section let you supervise the search for suitable sets of explanatory variables. Automatic searches can be carried out by incorporating statements into a loop (6.2.1); for example, a suitable STEP statement in a loop can carry out the methods of forward selection, backward elimination, or stepwise regression (8.2.6).

8.2.1 Extensions to the FIT and RDISPLAY directives

You would usually want to divide the explained variation between explanatory variables. The summary analysis of variance from the PRINT options of FIT and RDISPLAY does not do this, so you need the further setting 'accumulated'. It divides the variation according to the order in which you listed the variables in the parameter of the FIT directive. You will find this setting useful also for summarizing changes in the regression model that you might make by the directives described later in this section.

Here is the output that you get if you set the 'accumulated' option of FIT in the example from the beginning of this section:

EXAMPLE 8.2.1

```
***** Regression Analysis *****

*** Accumulated analysis of variance ***

Change              d.f.        s.s.        m.s.        v.r.
+ Temp                 1       0.26070     0.26070      4.21
+ Product              1       1.29882     1.29882     20.95
+ Opdays               1       0.33328     0.33328      5.38
+ Employ               1       0.55603     0.55603      8.97
Residual              12       0.74380     0.06198

Total                 16       3.19263     0.19954
```

The table shows the sum of squares and degrees of freedom attributable to each individual change in the model. By default the variance ratios are obtained by dividing the mean squares by the mean square corresponding to the smallest residual sum of squares in the table; that is from the model with fewest residual degrees of freedom.

If you do not want the sum of squares and the degrees of freedom to be subdivided between changes to the explanatory variables that you make within a statement, then you should use the POOL option of FIT. There would then be just one entry in the table for each statement. The main use of POOL is with the ADD, DROP, and SWITCH directives (8.2.3). With FIT, the POOL option merely gives the same table as you would get using the 'summary' setting of the PRINT option.

The DENOMINATOR option of the FIT and RDISPLAY directives can be set to produce variance ratios in the summary based on the smallest residual mean square, rather than on the mean square corresponding to the smallest residual sum of squares. You might, for example, know in advance of doing the regression that certain variables are unlikely to have a relationship with the response variable. So you would want to be able to include the sum of squares for these variables in the residual sum of squares for the other explanatory variables. You can do that by listing the interesting variables first, and these potentially uninteresting variables last, and setting DENOMINATOR = ms.

Sometimes you will find that the effect of an explanatory variable turns out to be exactly zero. This is no problem if it happens because the correlation of the explanatory variable with the response variable is itself zero. But it is a problem if it is because the explanatory variable is a linear combination of other explanatory variables. We call this *collinearity* or *aliasing* of the explanatory variables. There is then no unique set of parameter estimates, and the method of computing information about the regression breaks down (since it involves inverting a singular matrix $X'X$). The method also becomes unstable if the explanatory variables are nearly linearly related. Therefore Genstat tests for such a linear relationship, and will not include an explanatory variable that fails the test (8.2.2). A warning message is displayed, telling you which variable is not being included; you can prevent the message appearing by the 'aliasing' setting of the NOMESSAGE option in the FIT directive.

If you then change the model, Genstat will continue to try to include this problem variable unless it is explicitly dropped. This is because the changes in the model may cause the original collinearity to disappear. If the variable is successfully included, a message is printed; again you can prevent the message appearing by the 'aliasing' setting of the NOMESSAGE option.

8.2.2 The TERMS directive

TERMS
Specifies a maximal model, containing all terms to be used in subsequent linear, generalized linear, and nonlinear models.

Options

PRINT	= strings	What to print (correlations, SSPM, wmeans); default *
FACTORIAL	= scalar	Limit for expansion of model terms; default 3
FULL	= string	Whether to assign all possible parameters to factors and interactions (no, yes); default n
SSPM	= sspm	Gives sums of squares and products on which to base calculations

TOLERANCE	= scalar	Criterion for testing for linear dependence; default * (gives 10*EPS or 10000*EPS depending on the computer's precision, where EPS is smallest real value r with 1+r greater than 1 on the computer)
Parameter		
	formula	List of explanatory variates and factors, or model formula

Note first that a TERMS statement overrules any model that has already been fitted with FIT. That is, TERMS resets the current model to be the null model.

The formula must contain all the explanatory variables you want to use in later fitting statements; for multiple regression, the formula is a simple list of variates. An example is

    TERMS Temp,Product,Opdays,Employ

This formula may include the response variates, but need not.

The TERMS directive actually fits a model: the null model containing only an intercept term (in this case a mean). It also calculates the sums of squares, the sums of products, and the means (SSPM) of all variates specified in it, including any response variates that you have included: the matrix of SSPMs is $X'X$, and is the basis of the regression calculations. The matrix is weighted if you specified weights in the MODEL statement, and the calculations are made within groups if you have specified a grouping factor. All units of the variates are used unless there are restrictions or missing values. You are not allowed to have different restrictions on the different variates. Thus you can define the set of units that Genstat uses in the calculations by putting a restriction on any one of: a response variate, an explanatory variate, the weight variate, the offset variate, or the groups factor. A missing value in any of these structures except a response variate will also exclude the corresponding unit.

The model defined by all the terms in the parameter, excluding response variates, is called the *maximal model*.

The PRINT option lets you display the calculated sums of squares, sums of products, and means, together with the degrees of freedom. You can display also the corresponding matrix of correlations between variables, and the group means if the regression is within groups.

The FACTORIAL and FULL options refer to the parameterization of factors (8.3.1 and 8.3.2).

The SSPM option lets you use values that you have already calculated for an SSPM or DSSP structure (3.6.2 and 5.7.3). You might find this especially useful when you are analysing very large sets of data: you can accumulate a DSSP sequentially to avoid storing all the data at one time (5.7.3). Later regression calculations will be based on the supplied values of the DSSP: no fitted values, residuals, or leverages will be available. The values of a supplied SSPM or DSSP are accepted without checking by the TERMS directive: Genstat simply assumes you are giving it something sensible. However, on most computers regression should not be based on sums of squares and of products stored with single-precision accuracy. All standard data structures in Genstat are in single precision (except for the DSSP structure). The TERMS directive will print a warning if you supply an SSPM structure on a computer that has less than 48 bits for single-precision storage.

The TOLERANCE option controls the detection of aliasing in subsequent model fitting. By default, a parameter in a linear or generalized linear model will be deemed to be aliased if the ratio between the original diagonal value of the SSPM ($X'X$) corresponding to this parameter and the current diagonal value of the partially inverted SSPM is less than $10 \times$ EPS. The quantity EPS depends on the computer and is defined to be the smallest number r such that the computer recognizes $1.0 + r$ as greater than 1.0 in single precision. On computers for which single and double precision are equivalent in Genstat, EPS is a much smaller quantity; so the criterion $10000 \times$ EPS is used by default. Any positive value can be supplied in the TOLERANCE option to replace this default criterion in subsequent linear regression and generalized linear regression.

8.2.3 The ADD, DROP, and SWITCH directives

ADD
Adds extra terms to a linear, generalized linear, or nonlinear model.

DROP
Drops terms from a linear, generalized linear, or nonlinear model.

SWITCH
Adds terms to, or drops them from, a linear, generalized linear, or nonlinear model.

Options
PRINT	= strings	What to print (model, summary, accumulated, estimates, correlations, fittedvalues, monitoring); default m,s,e

NONLINEAR	=	string	How to treat nonlinear parameters between groups (common, separate, unchanged); default u
CONSTANT	=	string	How to treat constant (estimate, omit, unchanged); default u
FACTORIAL	=	scalar	Limit for expansion of model terms; default 3
POOL	=	string	Whether to pool ss in accumulated summary between all terms fitted in a linear model (no, yes); default n
DENOMINATOR	=	string	Whether to base ratios in accumulated summary on rms from model with smallest residual ss or smallest residual ms (ss, ms); default s
NOMESSAGE	=	strings	Which warning messages to suppress (dispersion, leverage, residual, aliasing, marginality); default *
Parameter			
		formula	List of explanatory variates and factors, or model formula

You use the directives ADD, DROP, and SWITCH to change the current model. Broadly, ADD lets you add extra explanatory variables, DROP lets you remove variables, and SWITCH lets you simultaneously add and drop variables.

The directives have a common syntax, which is also much the same as the syntax of the FIT directive. They modify the current regression model, which may be linear, generalized linear, standard curve, or nonlinear. You must give a TERMS statement before using any of the three directives, in order to define a set of units and carry out basic calculations. If no model is fitted after the TERMS statement before an ADD, DROP, or SWITCH statement, the current model is taken to be the null model.

Here is some output that continues the example from the beginning of this section:

EXAMPLE 8.2.3

```
 16  TERMS Temp,Product,Opdays,Employ
 17  ADD [PRINT=estimates] Product,Employ,Temp
```

8.2 Multiple linear regression

```
17.........................................................................
```

***** Regression Analysis *****

*** Estimates of regression coefficients ***

	estimate	s.e.	t
Constant	3.87	1.10	3.51
Product	0.1933	0.0544	3.56
Employ	-0.01968	0.00874	-2.25
Temp	0.00804	0.00563	1.43

```
  18  DROP [PRINT=estimates] Temp

18.........................................................................
```

***** Regression Analysis *****

*** Estimates of regression coefficients ***

	estimate	s.e.	t
Constant	4.60	1.01	4.55
Product	0.2034	0.0559	3.64
Employ	-0.02157	0.00896	-2.41

```
  19  SWITCH [PRINT=estimates,accumulated] Temp,Employ

19.........................................................................
```

***** Regression Analysis *****

*** Accumulated analysis of variance ***

Change	d.f.	s.s.	m.s.	v.r.
+ Product	1	1.27017	1.27017	14.05
+ Employ	1	0.56310	0.56310	6.23
+ Temp	1	0.18417	0.18417	2.04
Residual	13	1.17519	0.09040	
- Temp	-1	-0.18417	0.18417	2.04
- Employ	-1	-0.56310	0.56310	6.23
+ Temp	1	0.28935	0.28935	3.20
Total	16	3.19263	0.19954	

*** Estimates of regression coefficients ***

	estimate	s.e.	t
Constant	1.615	0.528	3.06
Product	0.0808	0.0242	3.34
Temp	0.00996	0.00632	1.57

The formula specified in the parameter is as in the FIT directive, but you must have included all the constituent terms in the formula of the previous TERMS statement. The terms in the formula (variates in the case of multiple linear regression) are compared with those in the current regression model to form a new model.

For the ADD directive, the new model consists of all terms in the current model together with any terms in the formula: terms may appear in both the current model and the formula, in which case they will remain in the new model.

In the above example, remember first that the TERMS statement has reset the current model to be the null model. Thus the ADD statement has the same effect as

>   FIT [PRINT = estimates] Product,Employ,Temp

If the ADD statement were followed by, for example,

>   ADD Temp,Opdays

then the variate Opdays would be added to the model. The effect would be the same as for the FIT directive of the original example. If you set PRINT = estimates, then you get estimates of the coefficients for all four explanatory variables and of the constant term.

For the DROP directive, the new model consists of all terms in the current model excluding any that are in the formula: terms in the formula that are not in the current model are ignored. You can see this happening in the example. If the DROP statement had instead been

>   DROP [PRINT = estimates] Temp,Opdays

then it would have had the same effect, since Opdays does not appear in the current model as specified by the previous statements.

Terms in the formula for the SWITCH directive are dropped from the current model if they are already there, and added to it if they are not. For example, if the current model consists of R and S, the effect of

>   SWITCH S,T

is to make a new model consisting of R and T (assuming that T was included in the previous TERMS statement).

The options of the ADD, DROP, and SWITCH directives are the same as those of the FIT directive, but with the extra NONLINEAR option (8.5.4). The output you get from the 'summary' and 'accumulated' settings of the PRINT option is modified when a TERMS statement has been given, and is described in 8.2.4. The model fitted by ADD, DROP, or SWITCH will include a constant term if the previous model included one, and will not include a constant term if the previous model did not. But you can use the CONSTANT option to change this rule.

8.2.4 Extensions to output and the RKEEP directive following TERMS

Following a TERMS statement, extra output is produced by the PRINT option of the FIT, ADD, DROP, SWITCH, and RDISPLAY directives.

The summary analysis of variance produced by the 'summary' setting includes an extra line called "Change". This shows the change in the Residual line since the last model. If no previous model has been fitted, the change refers to the null model. If the current statement either only adds or only drops terms, then the change sum of squares is attributable to these changes. But if the statement both adds and drops terms, then the sum of squares is not so readily interpreted; so the corresponding mean square is replaced by *, to warn you.

The accumulated summary produced by the 'accumulated' setting of the PRINT option shows all changes made to the model since the last TERMS or FIT statement, but including any changes made by the FIT statement. You can see this after the SWITCH statement in the example: first three terms are added, then Temp is removed, and then Employ is removed and Temp reinstated.

The variance ratios from the setting PRINT = accumulated are calculated either from the smallest residual mean square, or from the residual mean square corresponding to the smallest residual sum of squares, depending on how you set the DENOMINATOR option in the statement that prints the accumulated summary. In the example, DENOMINATOR has its default value in SWITCH, and so the variance ratios are calculated from the residual mean square corresponding to the smallest residual sum of squares.

You can use the EXPAND option of the RKEEP directive to re-order the parameters of a regression model, when they are stored by RKEEP, so that they correspond to the order of the terms in the maximal model defined by the previous TERMS statement. So in the above example you could put after the TERMS statement the statement

    RKEEP [EXPAND = yes] ESTIMATE = E

This would cause E to have parameters in the following order:

    *
    Temp
    Product
    *
    Employ
    Constant

The first missing value corresponds to the response variate, which is added at the start of the list of items if not included explicitly, and the second missing value corresponds to Opdays which has not been fitted. The constant always appears last, if fitted. If you set this option of RKEEP to 'yes' all the variates and matrices that are stored by RKEEP will have information for each parameter in the maximal model. That is, for each term in the maximal model there will be one value in the variate of parameter estimates and in the variate of standard errors; and there will be one row and one column in the inverse matrix and in the variance-covariance matrix. These values or rows or columns of values will be set missing by Genstat if the corresponding parameter is not in the current model.

8.2.5 The TRY directive

> **TRY**
> Displays results of single-term changes to a linear or generalized linear model.
>
> **Options**
>
> | | | |
> |---|---|---|
> | PRINT | = strings | What to print (model, summary, accumulated, estimates, correlations, fittedvalues, monitoring); default m,s,e |
> | FACTORIAL | = scalar | Limit for expansion of model terms; default 3 |
> | POOL | = string | Whether to pool ss in accumulated summary between all terms fitted in a linear model (no, yes); default n |
> | DENOMINATOR | = string | Whether to base ratios in accumulated summary on rms from model with smallest residual ss or smallest residual ms (ss, ms); default s |
> | NOMESSAGE | = strings | Which warning messages to suppress (dispersion, leverage, residual, aliasing, marginality); default * |
>
> **Parameter**
>
> | | | |
> |---|---|---|
> | | formula | List of explanatory variates and factors, or model formula |

The essential difference between TRY and ADD is that TRY makes no permanent change to the current model. It only temporarily adds or removes some explanatory variables: after it prints the results, it causes the program to revert to the same current model as before.

The current regression model is modified by each term in the formula specified by the parameter, one term at a time, dropping terms that are in the current model and adding terms that are not. After each change, the output that you ask for by the setting of the PRINT option is displayed; then the program reverts to the original model before going to the next change. It reverts to the current model also after all the changes.

The only circumstances in which TRY does make a permanent change is when the current model includes a term that had been found to be aliased before this TRY statement was reached. If the aliased term can be fitted after dropping one of the terms in the TRY formula, then that is indeed done. The term that was dropped will be aliased thereafter.

The options are as in the FIT directive, except that there is no CONSTANT option. The 'accumulated' setting of the PRINT option will show only one change at a time. Accumulated summaries produced by later statements will not have any entries for a TRY statement.

8.2.6 The STEP directive

STEP
Selects a term to include in or exclude from a linear or generalized linear model according to the ratio of residual mean squares.

Options

PRINT	= strings	What to print (model, summary, accumulated, estimates, correlations, fittedvalues, monitoring, changes); default m,s,e,ch
FACTORIAL	= scalar	Limit for expansion of model terms; default 3
POOL	= string	Whether to pool ss in accumulated summary between all terms fitted in a linear model (no, yes); default n
DENOMINATOR	= string	Whether to base ratios in accumulated summary on rms from model with smallest residual ss or smallest residual ms (ss, ms); default s
NOMESSAGE	= strings	Which warning messages to suppress (dispersion, leverage, residual, aliasing, marginality); default *
INRATIO	= scalar	Criterion for inclusion of terms; default 1.0
OUTRATIO	= scalar	Criterion for exclusion of terms; default 1.0

Parameter

	formula	List of explanatory variates and factors, or model formula

---

Here is an example of how to use STEP.

EXAMPLE 8.2.6

```
 20  FIT [PRINT=*] Temp,Product,Employ
 21  STEP [PRINT=estimates,changes; INRATIO=4; OUTRATIO=4] \
 22     Temp,Product,Opdays,Employ
 22........................................................................
```

*** Residual mean squares ***

```
    0.06198    Adding     Opdays
    0.09040    No change
    0.09710    Dropping Temp
    0.11665    Dropping Employ
    0.16564    Dropping Product
```

***** Regression Analysis *****

*** Estimates of regression coefficients ***

```
             estimate         s.e.          t
Constant        4.60          1.01        4.55
Product       0.2034        0.0559        3.64
Employ       -0.02157       0.00896      -2.41
```

To start with, ignore the options and concentrate on the basic structure of the FIT and STEP statements. First a model with three terms is fitted. Then the STEP statement tries, one at a time, to drop Temp, Product, and Employ, and to add Opdays. After each of these it reverts to the original model. Thus far, therefore, it is like a series of TRY statements. But then STEP, unlike TRY, permanently modifies the current model according to the change that was most successful. This means (putting it loosely at the moment) that, for example, if dropping Temp produced the biggest reduction in residual mean square then Temp is permanently removed; or, when no removals are worthwhile, if adding Opdays produced the biggest reduction in residual mean square, then Opdays is permanently included. We see in fact that the former happened, and so the current model is now as displayed at the end of the example.

More rigorously, we have:

The current model is modified by each term in the formula specified by the parameter, one term at a time, as in the TRY directive. For each term, the residual sum of squares and the residual degrees of freedom are recorded; then the program reverts to the original model before trying the next term.

The current model is finally modified by the best term, according to a criterion based on the variance ratios. Suppose that the residual sum of squares and residual degrees of freedom of the current model are s_o and d_o, and of the model after making a one-term change are s_1 and d_1. If the variance ratio for any term that is dropped is greater than the value of the setting of the OUTRATIO option, then the

8.2 Multiple linear regression

term that most reduces the residual mean square is dropped. That is, a term will be dropped only if at least one term has

$$\{(s_1-s_o) / (d_1-d_o)\} / \{s_o/d_o\} > \text{OUTRATIO}$$

If the OUTRATIO option is set to *, then no term is dropped.

If no term satisfies the criterion for dropping, then the term that most reduces the residual mean square will be added to the model if its variance ratio is greater than the setting of the INRATIO option. That is, if

$$\{(s_o-s_1) / (d_o-d_1)\} / \{s_1/d_1\} > \text{INRATIO}$$

Again, if the INRATIO option is set to *, no term will be added.

If neither criterion is met, the current model is left unchanged.

The effect of the STEP directive is to make one change of a stagewise regression search. You can do the method of forward selection by repeating a STEP statement with the OUTRATIO option set to *: for example,

```
TERMS X[1...10]
FOR [NTIMES = 10]
   STEP [OUTRATIO = *] X[1...10]
ENDFOR
```

You can do the method of backward elimination similarly, setting the INRATIO option to *.

You might want to avoid the loop repeatedly making no change if some terms have little effect. To do that, cause an EXIT from the loop (6.2.4) whenever the residual d.f. does not change; you get the residual d.f. from the RKEEP directive. For example, for the method of backward elimination:

```
TERMS X[1...10]
FIT X[1...10]
RKEEP DF = D0
FOR [NTIMES = 10]
STEP [INRATIO = *; OUTRATIO = 4] X[1...10]
RKEEP DF = D1
EXIT D1.EQ.D0
CALCULATE D0 = D1
ENDFOR
```

The 'changes' setting of the PRINT option produces a list of terms with the corresponding residual mean squares and residual degrees of freedom, ordered according to the sizes of the residual mean squares. You can see this in the example. Note that this list is not available for display later by the RDISPLAY directive. The INRATIO and OUTRATIO options are explained above. The rest of the options are as in the FIT directive, except that there is no CONSTANT option.

8.3 Linear regression with grouped or qualitative data

You can incorporate in a regression the effects of grouped variables. These are sometimes called qualitative variables to distinguish them from the quantitative ones we have been looking at so far in this chapter. For example, you could fit a separate constant term for each level of some classification: you would get a series of parallel regression lines of the response variable on the quantitative variables. You might also want to fit separate slopes for the quantitative variables at each level of the classification.

More specifically, here is an example.

EXAMPLE 8.3

```
  1   "
 -2     Analysis of parallelism of linear relationship between yield of
 -3     barley plots and their brightness measured on aerial photographs;
 -4     36 plots were treated with Tridemorph fungicide against mildew.
 -5   "
  6   UNITS [NVALUES=72]
  7   FACTOR [LEVELS=!(0,1)] Tridemor; DECIMALS=0
  8   OPEN 'BARLEY.DAT'; CHANNEL=2
  9   READ [CHANNEL=2] Yield,Bright,Tridemor

      Identifier    Minimum      Mean   Maximum    Values    Missing
           Yield      3.350     4.368     5.450        72          0
           Bright     371.0     462.7     593.0        72          0

 10   MODEL Yield
 11   TERMS Bright*Tridemor
 12   FIT [PRINT=model,estimates] Bright+Tridemor
```

12...

***** Regression Analysis *****

 Response variate: Yield
      Fitted terms: Constant + Bright + Tridemor

*** Estimates of regression coefficients ***

	estimate	s.e.	t
Constant	0.918	0.627	1.46
Bright	0.00754	0.00152	4.98
Tridemor 1	-0.079	0.172	-0.46

```
 13   ADD [PRINT=model,estimates,accumulated] Bright.Tridemor
```

13...

***** Regression Analysis *****

 Response variate: Yield
      Fitted terms: Constant + Bright + Tridemor + Bright.Tridemor

8.3 Linear regression with grouped or qualitative data

*** Accumulated analysis of variance ***

Change	d.f.	s.s.	m.s.	v.r.
+ Bright	1	11.1478	11.1478	107.26
+ Tridemor	1	0.0234	0.0234	0.22
+ Bright.Tridemor	1	0.5824	0.5824	5.60
Residual	68	7.0674	0.1039	
Total	71	18.8209	0.2651	

*** Estimates of regression coefficients ***

	estimate	s.e.	t
Constant	2.611	0.938	2.78
Bright	0.00343	0.00227	1.51
Tridemor 1	−3.28	1.36	−2.41
Bright.Tridemor 1	0.00705	0.00298	2.37

Before we go into details, look at the FIT statement. Tridemor is a factor with two levels, 0 and 1. This statement fits two parallel regressions of the variate Yield on the variate Bright, one at each level of Tridemor. In other words, these two lines have the same slope but different intercepts.

Now look at the ADD statement. Here different slopes are fitted too. The joint effect of the two statements is therefore to fit two regressions, one for each level of Tridemor, which will have different slopes and different intercepts.

We now make some more formal definitions, after which we shall return to this example.

You store data from qualitative variables in factors (3.3.3). After factors have been declared and assigned values, their effects can be included in regression models. You do this by putting their identifiers in directives such as FIT and TERMS, along with the identifiers of variates storing the values of quantitative explanatory variables.

You represent the *main effect* of a factor by its identifier as a single term: a model including such a main effect has a separate intercept for each level of the factor. This corresponds to the different intercepts in the example.

Interactions between factors allow more detailed modelling of the constant term for combinations of levels of more than one factor. They are represented by terms consisting of the dot operator between factor identifiers in formulae.

Interactions between factors and variates allow modelling of the changes in the effects of the variate between combinations of levels of factors. They too are represented by terms including dot operators. This corresponds to the different slopes in the example.

"Interactions" between quantitative variables cannot be expressed in this way. However, the product of two variates, for example $\{x_i \times z_i, i = 1...N\}$ can be formed as a new variate, which can be included in a regression model along with the original variates.

8.3.1 Formulae in parameters of regression directives

Formulae are described in 1.5.3, and further details are given in 9.1.1. In regression directives you cannot use a function in a formula, nor the // operator. The basic operators are those of summation (+) and dot product (.), and if you want to you can write all formulae using just these two. The other operators provide a shorthand for representing complicated formulae. Of particular use in regression are the cross-product operator (*)

$$A*B = A + B + A.B$$

and the nesting operator (/)

$$A/B = A + A.B$$

For more complicated formulae, remember that the nesting operator is not distributive (1.5.3): for example,

$$(A + B)/C = A + B + A.B.C$$

Note also that terms can be ignored if they are put in the wrong order. For example

$$A.B + A = A.B \neq A + A.B$$

If a formula contains commas, then they are treated in the same way as + together with pairs of brackets. For example, X,Y*A is the same as (X+Y)*A, which is X+Y+A+X.A+Y.A.

The expansion of formulae into constituent terms is controlled in all regression directives by the FACTORIAL option. The default setting is 3, which excludes all interactions involving more than three identifiers. For example,

    FIT [FACTORIAL=2] A*B*C

will fit a model that includes the terms A B C A.B A.C and B.C, but excludes A.B.C.

8.3.2 How Genstat parameterizes factors

A regression model that includes the main effect of a factor contains one parameter for each level of the factor: this parameter represents the constant term for that level. If an explicit constant term is also included in the model, then some constraint must be applied to the parameters for the factors. In Genstat, the parameter corresponding to the first level of the factor is set to zero. For example, in the first model fitted at the beginning of this section, the parameter estimates are:

8.3 Linear regression with grouped or qualitative data

EXAMPLE 8.3.2a

```
*** Estimates of regression coefficients ***
                    estimate        s.e.            t
Constant              0.918        0.627         1.46
Bright              0.00754      0.00152         4.98
Tridemor 1           -0.079        0.172        -0.46
```

No parameter estimate for Tridemor 0 is shown. You can interpret the constant term here as the constant for the first level of the factor. Thus the parameter labelled Tridemor 1 is the difference between the constant for Tridemor 1 and that for Tridemor 0. If the factor had more than two levels, all other parameters associated with it would also represent differences from the first level.

This form of parameterization makes it easy to compare each level of a factor with a base level. In the example, the standard error of the estimate for Tridemor 1 shows that there is no significant difference between the constants for the levels of Tridemor.

However, you will not find this parameterization very convenient for summarizing the effect of a factor, especially when there are several factors in a model. So instead you will probably want to use the methods described in 8.4 (or indeed those described in Chapter 9).

You can get other parameterizations by modifying the terms in the model. For example, the two constants in the first model fitted can be made to be explicit parameters by using the CONSTANT option of FIT:

EXAMPLE 8.3.2b

```
  15   MODEL Yield
  16   FIT [PRINT=estimates; CONSTANT=omit] Bright,Tridemor

16......................................................................

***** Regression Analysis *****

*** Estimates of regression coefficients ***
                    estimate        s.e.            t
Bright              0.00754      0.00152         4.98
Tridemor 0            0.918        0.627         1.46
Tridemor 1            0.839        0.780         1.08
```

Since there is no constant term in this model, no constraint needs to be imposed on the parameters representing the factor.

If you want reparameterization after a TERMS statement, then you must set to 'yes' the FULL option of TERMS. This is because TERMS allocates the number of parameters for each term in the model, and automatically imposes constraints when there is over-parameterization. The setting FULL = yes specifies that a parameter is to be associated with each level of a factor, regardless of the presence of a constant term. If you also estimate a constant term, then you will find that one of the parameters representing the factor will be aliased. For example:

EXAMPLE 8.3.2c

```
  17  TERMS [FULL=yes] Bright,Tridemor
  18  FIT [PRINT=estimates] Bright,Tridemor

  18..............................................................

  ***** Regression Analysis *****

  *** Estimates of regression coefficients ***

                estimate          s.e.              t
  Constant         0.839          0.780          1.08
  Bright         0.00754        0.00152          4.98
  Tridemor 0       0.079          0.172          0.46
  Tridemor 1           0              *             *
```

The last level of the factor is aliased, and hence left as 0, since this is the last parameter to be fitted. Notice that no reports on partial aliasing of terms involving factors are given when FULL = yes, regardless of the setting of the NOMESSAGE option of the FIT directive.

Factor effects are also fully parameterized if you supplied an SSPM or a DSSP structure through the SSPM option of the TERMS directive and if the SSPM had the setting FULL = yes in the SSPM directive (3.6.2).

8.3.3 Parameterization of interaction, and marginality

The parameters representing interactions in a model are also constrained to remove over-parameterization.

For example, if A and B are factors with two and three levels respectively and the model A*B is fitted (including a constant) the parameters will be: constant, A2, B2, B3, A2.B2, A2.B3. No parameter is assigned to A1 because there is a constant, and none to B1 or A1.B1. Similarly, no parameter is assigned to A2.B1 because the main effect of A is included, and none to A1.B2 nor A1.B3 because the main effect of B is included. The terms A and B are described as being *marginal* to the term A.B. The constant term is also marginal to A and B, and to the term A.B.

In general, one term is marginal to a second if the second can be written as an interaction between the first term and a third term involving factors only; for example, A is marginal to A.B and to A.B.C.D. Whenever one term is marginal to a second, some parameters of the full set of the second term are aliased with the first term. Genstat will automatically constrain selected parameters to be zero to avoid aliasing. The automatic constraint can be removed by setting the FULL option of the TERMS directive.

In the second analysis shown at the beginning of 8.3, the fitted model is Tridemor + Bright + Tridemor.Bright. The term Tridemor.Bright is an interaction between a factor and a variate, and so represents variations in the variate between levels of the factor: that is, the regression lines of Yield on Bright are allowed to have separate slopes for the two groups of data, as well as separate intercepts.

The model is

$$y_{ij} = \alpha + \beta_i + \gamma \times x_{ij} + \delta_i \times x_{ij} + \varepsilon_{ij} \quad i = 1, 2; j = 1 \ldots 36$$

where α represents the constant term, and is the intercept for Tridemor 0. The parameters β_i represent the term Tridemor, and β_i is the difference between the intercept for the ith level of Tridemor and that for the first level (labelled Tridemor 0), so that β_1 is zero. The parameter γ represents the term Bright, and is the slope for Tridemor 0. Lastly the parameters δ_i represent the interaction term Tridemor.Bright, and δ_i is the difference between the slope for the ith level of Tridemor and that for the first level, so that δ_1 is zero. In this model, the constant is marginal to the term Tridemor, and Bright is marginal to Tridemor.Bright.

Again, you can present the results differently, either using the PREDICT directive (8.3.4), or by omitting marginal terms. The actual slopes can be made to be parameters thus:

EXAMPLE 8.3.3a

```
  19  TERMS [FULL=yes] Bright*Tridemor
  20  FIT [PRINT=estimates] Tridemor/Bright

20........................................................................

***** Regression Analysis *****

*** Estimates of regression coefficients ***
                      estimate         s.e.           t
Constant               -0.668         0.988       -0.68
Tridemor 0              3.28          1.36         2.41
Tridemor 1              0                *            *
Bright.Tridemor 0       0.00343       0.00227      1.51
Bright.Tridemor 1       0.01048       0.00192      5.45
```

Alternatively, you can restrict the form of the parameterization in the TERMS statement by omitting terms. For example:

EXAMPLE 8.3.3b

```
 21  TERMS Tridemor/Bright
 22  FIT [PRINT=estimates] Tridemor/Bright
```

22..

***** Regression Analysis *****

*** Estimates of regression coefficients ***

	estimate	s.e.	t
Constant	2.611	0.938	2.78
Tridemor 1	-3.28	1.36	-2.41
Bright.Tridemor 0	0.00343	0.00227	1.51
Bright.Tridemor 1	0.01048	0.00192	5.45

The TERMS statement here parameterizes the term Tridemor.Bright fully, even though the FULL option is not set, because the term Bright is not included in the model.

Models cannot be fitted if they exclude terms that are in the maximal model and are marginal to terms that are in the model. For example,

> TERMS Tridemor*Bright
> FIT Tridemor/Bright

produces the following warning message:

EXAMPLE 8.3.3c

```
 23  TERMS Tridemor*Bright
 24  FIT [PRINT=*] Tridemor/Bright
```

24..

```
* MESSAGE: Term Bright.Tridemor can not be added
  because term Bright is marginal to it and is not in the model
```

The term Tridemor.Bright is not fitted because Bright is marginal to it and has not been fitted. No problem occurs if you have set the FULL option of TERMS. Otherwise, you can prevent the message about marginality from appearing by using the 'marginality' setting of the NOMESSAGE option of the FIT directive.

8.3.4 The PREDICT directive

PREDICT
Forms predictions from a linear or generalized linear model.

Options

PRINT	= string	What to print (description,predictions,se); default d,p
CHANNEL	= scalar	Channel number for output; default * i.e. current output channel
COMBINATIONS	= string	Which combinations of factors in the current model to include (all, present); default a
ADJUSTMENT	= string	Type of adjustment (marginal, equal); default m
WEIGHTS	= table	Weights classified by some or all standardizing factors; default *
METHOD	= string	Method of forming margin (mean, total); default m
ALIASING	= string	How to deal with aliased parameters (fault, ignore); default f
PREDICTIONS	= table or pointer	To save tables of predictions for each y variate; default *
SE	= table or pointer	To save tables of standard errors of predictions for each y variate; default *
VCOVARIANCE	= symmetric matrix or pointer	To save variance-covariance matrices of predictions for each y variate; default *
SAVE	= identifier	Specifies save structure of model to display; default * i.e. that of the latest model fitted

Parameters

CLASSIFY	= vectors	Variates and/or factors to classify table of predictions
LEVELS	= variates or scalars	To specify values of variates, levels of factors

---

The simplest use of PREDICT is to make estimates from the regression line for specific values of the explanatory variables. For example, if you had calculated the simple linear regression of a variate Y on a variate X, then you could get the predicted value of Y at X = 3.5 (say) by

    PREDICT X; LEVELS=3.5

If you wanted the predicted values at 3.5 and 4, you would have to put these into a variate of length two. The easiest way to do that is to use an unnamed variate:

    PREDICT X; LEVELS=!(3.5,4)

Suppose now that you had a multiple regression of Y on variates X1 and X2, and that you wanted to predict the value of Y at X1 = 3.5 and 4 and X2 = 5.6, 5.7, and 5.8. Put 3.5 and 4 into one variate, and 5.6, 5.7, and 5.8 into another:

    PREDICT X1,X2; LEVELS=!(3.5,4),!(5.6,5.7,5.8)

This would give you six predicted values, one for each combination of 3.5 and 4 with 5.6, 5.7, and 5.8.

If there were a two-level factor A in the first regression, with levels 0 and 1, you might want to predict Y for X equal to 3.5 in both the presence and the absence of A. You would do that by:

    PREDICT A,X; LEVELS=!(0,1),3.5

This would produce two predicted values, classified by the levels of A.

Suppose further that there was also a second factor B in the model, along with A and X, but that you again wanted to predict Y for X equal to 3.5 in the presence and the absence of A. In other words, you wanted the predictions to be averaged over the levels of B. You do that by exactly the same statement: since it does not mention B, the predictions are automatically averaged over the levels of B.

For more complicated structures the rules are more intricate, as we shall see. But the basic ideas remain the same as in the simpler cases.

One use of the PREDICT directive is to provide summaries of the effects of explanatory variables in a regression. In simple linear regression or multiple linear regression, the parameters of the regression model may be sufficient summaries in themselves. But when you want to estimate the effects of several factors, possibly with interactions, the parameters of the regression are not usually a convenient description.

The CLASSIFY parameter specifies those variates or factors in the current regression model whose effects you want to summarize. Any variate or factor in the current model that you do not include in the parameter will be standardized in some way, according to the options. In the first and second examples above, CLASSIFY specified X, in the third X1 and X2, and in the fourth A and X.

The LEVELS parameter specifies the levels of each factor for which you want summaries, or the sets of values of each variate for which you want summaries. This is illustrated in the simple cases above. A single level or value is represented by a scalar; several levels or values must be combined into a variate (which may

8.3 Linear regression with grouped or qualitative data

of course be unnamed). A missing value in the LEVELS parameter is taken by Genstat to stand for all levels of a factor, or for the mean value of a variate. This applies only to factors or variates that you have included in the CLASSIFY parameter: others are summarized in some way, according to the options (as B in the final example above).

For example, in the program introduced at the start of this section, here is how you get a summary of the effect of factor Tridemor at the mean value of the variate Bright; the summary is based on the regression model that includes the interaction between these two terms:

EXAMPLE 8.3.4a

```
25  FIT [PRINT=*] Tridemor*Bright
26  PREDICT Tridemor,Bright; LEVEL=!(0,1),*

26........................................................

*** Predictions from regression model ***

The predictions are based on fixed values of some variates:

        Variate    Fixed value   Source of value
        Bright           462.7   Mean of variate

Response variate: Yield

        0.00           4.20
        1.00           4.18
```

The two values are estimates, based on the fitted model, of the mean yield at the value 426.7 of the explanatory variate Bright. You would get the same results by omitting the LEVELS parameter in the PREDICT statement, since all levels of Tridemor would be given by default.

You can understand how Genstat forms predictions by thinking of its calculations as consisting of two steps; shortly we give some examples that illustrate the more abstruse aspects of this process.

(a) It first calculates the full table of predictions, classified by every factor in the current model. For any variate in the model, the predictions are formed at its mean, unless you have specified some values for it in the LEVELS parameter. If you have specified such values, then they are taken as a further classification of the table of predictions.

(b) The full table of predictions is averaged over classifications that do not appear in the CLASSIFY parameter: you can control the type of averaging by the options.

Here now is some output that shows the full table of predictions constructed from

340   8   Regression analysis

the estimates of a model containing three factors. Later you will see how to use
PREDICT to get subtables of this table.

EXAMPLE 8.3.4b

```
 1    "
-2    Summary of analysis of hatching rate of H. Schachtii
-3    related to leachate and dilution rate.
-4    "
 5    UNITS [NVALUES=36]
 6    TEXT [VALUES=baresoil,emerald,emergo] Nl
 7    & [VALUES='1','1/4','1/16','1/64'] Nd
 8    FACTOR [LEVELS=3; VALUES=12(1...3)] Block
 9    FACTOR [LABELS=Nl] Leachate
10    & [LABELS=Nd] Dilution
11    OPEN 'HATCH.DAT'; CHANNEL=2
12    READ [CHANNEL=2] Leachate,Dilution,Nhatch,Nnohatch
```

Identifier	Minimum	Mean	Maximum	Values	Missing	
Nhatch	54.0	635.8	2821.0	36	3	Skew
Nnohatch	66.0	798.5	2246.0	36	0	

```
13    VARIATE Logit; EXTRA=' of percentage hatched'
14    CALCULATE %hatch = 100*Nhatch/(Nhatch+Nnohatch)
15    & [PRINT=summary] Logit = LOG(%hatch/(100-%hatch))
```

Identifier	Minimum	Mean	Maximum	Values	Missing
Logit	-2.5519	-0.4361	3.0939	36	3

```
16    MODEL Logit
17    FIT [PRINT=model,accumulated,estimates] Block+Leachate*Dilution
```

17..

***** Regression Analysis *****

Response variate: Logit of percentage hatched
  Fitted terms: Constant + Block + Leachate + Dilution + Leachate.Dilution

*** Accumulated analysis of variance ***

Change	d.f.	s.s.	m.s.	v.r.
+ Block	2	0.5256	0.2628	1.23
+ Leachate	2	38.7710	19.3855	90.61
+ Dilution	3	27.7846	9.2615	43.29
+ Leachate.Dilution	6	11.7526	1.9588	9.16
Residual	19	4.0649	0.2139	
Total	32	82.8987	2.5906	

*** Estimates of regression coefficients ***

	estimate	s.e.	t
Constant	-0.865	0.291	-2.97
Block 2	-0.125	0.204	-0.61
Block 3	-0.311	0.197	-1.58
Leachate emerald	3.856	0.378	10.21
Leachate emergo	2.342	0.546	4.29
Dilution 1/4	-0.560	0.378	-1.48

Dilution 1/16	−1.318	0.378	−3.49
Dilution 1/64	−0.491	0.378	−1.30
Leachate emerald .Dilution 1/4	−0.800	0.570	−1.41
Leachate emerald .Dilution 1/16	−0.764	0.534	−1.43
Leachate emerald .Dilution 1/64	−3.307	0.534	−6.19
Leachate emergo .Dilution 1/4	−0.112	0.664	−0.17
Leachate emergo .Dilution 1/16	−1.122	0.664	−1.69
Leachate emergo .Dilution 1/64	−2.807	0.664	−4.23

```
 18   PREDICT Block,Leachate,Dilution

18........................................................

*** Predictions from regression model ***

  Response variate: Logit of percentage hatched

            Dilution      1      1/4     1/16    1/64
    Block   Leachate
      1     baresoil    -0.87   -1.43   -2.18   -1.36
            emerald      2.99    1.63    0.91   -0.81
            emergo       1.48    0.81   -0.96   -1.82
      2     baresoil    -0.99   -1.55   -2.31   -1.48
            emerald      2.87    1.51    0.78   -0.93
            emergo       1.35    0.68   -1.09   -1.95
      3     baresoil    -1.18   -1.74   -2.49   -1.67
            emerald      2.68    1.32    0.60   -1.12
            emergo       1.17    0.49   -1.27   -2.13
```

The 'description' setting of the PRINT option corresponds to a summary of what standardization policies are used when the predictions are formed. By default, only predictions are printed with the description. Standard errors are printed if you include the 'se' setting, and the predictions will accompany them whether or not you include the 'predictions' setting. You can direct the chosen output to any channel that is open, by setting the CHANNEL option. The standard errors are relevant for predictions considered as means of those data that have been analysed, forming the means by the weighting policy defined by the option.

The word *prediction* is used because these are predictions of what the unweighted means of the data would have been if there had been some chosen kind of balance of different factor levels in the data. The standard error is not augmented by a component corresponding to the estimated variability of a new observation.

The three options COMBINATIONS, ADJUSTMENT, and WEIGHTS let you specify how the averaging is done at Step (b) above. The easiest way to understand these options is to look first at what happens by default.

By default, values in the full table from (a) are averaged with respect to all those factors that you have not included in the settings of the CLASSIFY parameter. The levels of any such factor are combined with what we call *marginal weights*: that is, by the number of occurrences of each of its levels in all the data. For instance, the occurrences of factor combinations in the most recent example can be displayed by the TABULATE directive (5.8.1):

EXAMPLE 8.3.4c

```
  19  TABULATE [CLASSIFICATION=Block,Leachate,Dilution; MARGIN=yes] \
  20     Logit; NOBSERVATIONS=Count
  21  PRINT Count; FIELDWIDTH=6; DECIMALS=0
```

		Count				
	Dilution	1	1/4	1/16	1/64	Margin
Block	Leachate					
1	baresoil	1	1	1	1	4
	emerald	1	1	1	1	4
	emergo	0	1	1	1	3
	Margin	2	3	3	3	11
2	baresoil	1	1	1	1	4
	emerald	1	0	1	1	3
	emergo	0	1	1	1	3
	Margin	2	2	3	3	10
3	baresoil	1	1	1	1	4
	emerald	1	1	1	1	4
	emergo	1	1	1	1	4
	Margin	3	3	3	3	12
Margin	baresoil	3	3	3	3	12
	emerald	3	2	3	3	11
	emergo	1	3	3	3	10
	Margin	7	8	9	9	33

The following predictions for the factor Dilution are produced by marginal weighting over the factors Block and Leachate.

EXAMPLE 8.3.4d

```
  22  PREDICT Dilution

22..............................................................

*** Predictions from regression model ***

The predictions have been standardized by averaging
fitted values over the levels of some factors:

          Factor    Weighting policy    Status of weights
          Leachate  Marginal weights    Constant over levels of other factors
```

8.3 Linear regression with grouped or qualitative data

```
      Block    Marginal weights   Constant over levels of other factors
Response variate: Logit of percentage hatched
        Dilution
           1           0.98
          1/4          0.12
         1/16         -0.93
         1/64         -1.46
```

In forming the averages for Dilution, the data from the three levels of Leachate have been combined with weights 12, 11, and 10, since these are the frequencies with which they occur in all the data. Because we are using the default settings of COMBINATIONS, ADJUSTMENT, and WEIGHTS, these weights are constant over levels of the other factors.

Now look at what these options can do. COMBINATIONS specifies which cells of the full table from (a) are to be used in the averaging. The alternative to the default of all cells being used is COMBINATIONS = present. This excludes cells for which there is no corresponding value in the data. In the example, it would exclude all the cells with 0 in the table of occurrences displayed above. This would not of course affect the numerator of the ratio for the average (since excluded cells would have had a 0 there anyway). But it will affect the denominator. For example, for Dilution 1 above, the average would be over the data corresponding to the seven non-zero cells in column 1. They would be weighted by the margins, again 12, 11, and 10. The difference from the default is that the weight 10 would enter less frequently into the denominator (in fact twice instead of four times).

COMBINATIONS overrules the LEVELS parameter. That is, when you have specified COMBINATIONS = present, any subsets of factor levels in the LEVELS parameter are ignored, and predictions will be formed for all levels that occur in the data. Likewise, the full table cannot then be classified by any sets of values of variates: only single values may be given in the LEVELS parameter.

ADJUSTMENT and WEIGHTS let you change the weights. (COMBINATIONS in a sense changes the weights too, but not under your control.)

The setting ADJUSTMENT = equal specifies that levels are to be weighted equally, when the predictions are adjusted for the effects of the standardizing factors. In the above example (with COMBINATIONS = all), the weights would then be equal instead of 12, 11, and 10. This is done by:

EXAMPLE 8.3.4e

```
  23  PREDICT [ADJUSTMENT=equal] Dilution
```

```
  *** Predictions from regression model ***
```

The predictions have been standardized by averaging
fitted values over the levels of some factors:

Factor	Weighting policy	Status of weights
Leachate	Equal weights	Constant over levels of other factors
Block	Equal weights	Constant over levels of other factors

Response variate: Logit of percentage hatched

Dilution	
1	1.06
1/4	0.19
1/16	-0.89
1/64	-1.47

WEIGHTS is more powerful than ADJUSTMENT (although also more cumbersome): it lets you give a table of explicit weights. This table can be classified by any, or all, of the factors over whose levels the predictions are to be averaged; the levels of remaining factors will be weighted according to the ADJUSTMENT option. Moreover, you can classify the weights by the factors in the CLASSIFY parameter as well, to provide different weightings for different combinations of levels of these factors. However, if you do this, you must use all the factors in the current model to classify the table of weights. Factors that you have included in both the CLASSIFY parameter and in the classification of the table must be in the same order, and must occur after any other factors. If you supply explicit weights in the WEIGHTS option, the setting of the COMBINATIONS option is ignored.

One set of explicit weights you might want to use is known as the population proportions. For example, you might know that these proportions for 'baresoil', 'emerald', and 'emergo' are in the ratios 4:1:1. Then you could specify weights as in the next example (where the levels of BLOCK are weighted equally):

EXAMPLE 8.3.4f

```
  24  TABLE [CLASSIFICATION=Leachate; VALUES=4,1,1] Wtleach
  25  PREDICT [ADJUSTMENT=equal; WEIGHTS=Wtleach] Dilution
```

25...

*** Predictions from regression model ***

The predictions have been standardized by averaging
fitted values over the levels of some factors:

Factor	Weighting policy	Status of weights
Block	Equal weights	Constant over levels of other factors
Leachate	Supplied weights	Constant over levels of classifying factors

Response variate: Logit of percentage hatched

8.3 Linear regression with grouped or qualitative data

```
Dilution
       1       0.02
     1/4      -0.69
    1/16      -1.61
    1/64      -1.49
```

We have been assuming so far that averaging is the appropriate way of combining predicted values over levels of a factor. But sometimes summation is needed, for example in the analysis of counts by log-linear models (8.4.1). You can achieve this by setting the METHOD option to 'total'. The rules about weights and so on still apply, although since there is now no denominator the COMBINATIONS option has no effect.

If a model contains any aliased parameters, then predicted values cannot be formed for some cells of the full table without assuming a value for the aliased parameters. By default, predictions will be formed only if you have set the COMBINATIONS option to 'present'. If you want other types of predictions when there is aliasing, you should set the ALIAS option to 'ignore'. All aliased parameters are then taken to be zero, and fitted values are calculated for all cells of the table from the remaining parameters in the model.

Now here is an example that shows how the above data are augmented by control observations that were included in the experiment: these are observations for which no leachate was used and for which, therefore, no dilution rate was relevant.

EXAMPLE 8.3.4g

```
  1   "
 -2    Summary of analysis of hatching rate of H. Schachtii
 -3    related to leachate and dilution rate: including control.
 -4   "
  5   UNITS [NVALUES=39]
  6   FACTOR [LEVELS=3; VALUES=13(1...3)] Block
  7   FACTOR [LEVELS=!(0,1); LABELS=!T(control,factorial)] Subdivis
  8   &  [LEVELS=!(0...3); LABELS=!T(control,baresoil,emerald,emergo)] Leachate
  9   &  [LEVELS=!(0...4); LABELS=!T(control,'1','1/4','1/16','1/64')] Dilution
 10   OPEN 'HATCH2.DAT'; CHANNEL=2
 11   READ [CHANNEL=2] Leachate,Dilution,Subdivis,Nhatch,Nnohatch

    Identifier    Minimum       Mean    Maximum     Values    Missing
        Nhatch       54.0      610.8     2821.0         39          3    Skew
      Nnohatch       66.0      832.4     2246.0         39          0

 12   CALCULATE %hatch = 100*Nhatch/(Nhatch+Nnohatch)
 13   & [PRINT=summary] Logit = LOG(%hatch/(100-%hatch))

    Identifier    Minimum       Mean    Maximum     Values    Missing
         Logit    -2.5519    -0.5093     3.0939         39          3

 14   MODEL Logit
 15   FIT [PRINT=model,accumulated,estimates] Block+Leachate*Dilution
```

15...

* MESSAGE: Term Dilution can not be fully included in the model
  because 1 parameter is aliased with terms already in the model

* MESSAGE: Term Leachate.Dilution can not be fully included in the model
  because 6 parameters are aliased with terms already in the model

***** Regression Analysis *****

Response variate: Logit
    Fitted terms: Constant + Block + Leachate + Dilution + Leachate.Dilution

*** Accumulated analysis of variance ***

Change	d.f.	s.s.	m.s.	v.r.
+ Block	2	0.4770	0.2385	1.22
+ Leachate	3	40.9233	13.6411	69.92
+ Dilution	3	27.7907	9.2636	47.48
+ Leachate.Dilution	6	11.7468	1.9578	10.03
Residual	21	4.0972	0.1951	
Total	35	85.0350	2.4296	

*** Estimates of regression coefficients ***

	estimate	s.e.	t
Constant	-1.183	0.276	-4.29
Block 2	-0.107	0.186	-0.58
Block 3	-0.290	0.180	-1.61
Leachate baresoil	-0.186	0.361	-0.52
Leachate emerald	0.363	0.361	1.01
Leachate emergo	-0.651	0.361	-1.80
Dilution 1	3.290	0.521	6.32
Dilution 1/4	2.626	0.361	7.28
Dilution 1/16	0.857	0.361	2.38
Dilution 1/64	0	*	*
Leachate baresoil .Dilution 1	-2.799	0.633	-4.42
Leachate baresoil .Dilution 1/4	-2.695	0.510	-5.28
Leachate baresoil .Dilution 1/16	-1.684	0.510	-3.30
Leachate baresoil .Dilution 1/64	0	*	*
Leachate emerald .Dilution 1	0.508	0.633	0.80
Leachate emerald .Dilution 1/4	-0.186	0.544	-0.34
Leachate emerald .Dilution 1/16	0.859	0.510	1.68
Leachate emerald .Dilution 1/64	0	*	*
Leachate emergo .Dilution 1	0	*	*
Leachate emergo .Dilution 1/4	0	*	*
Leachate emergo .Dilution 1/16	0	*	*
Leachate emergo .Dilution 1/64	0	*	*

When the model is fitted, some parameters of the interaction term Leachate.Dilution are aliased, because there are no observations with "control"

Leachate and "non-control" Dilution, and vice versa.

Next we form predictions of the effect of Dilution from this model, based on just those factor combinations that were present in the experiment, and weighting those combinations by their occurrence in the data.

EXAMPLE 8.3.4h

```
 16  PREDICT [COMBINATIONS=present] Dilution

 16.............................................................................

*** Predictions from regression model ***

The predictions have been standardized by averaging
fitted values over the levels of some factors:
        Factor    Weighting policy   Status of weights
      Leachate    Marginal weights   Based on counts of factor levels in data
         Block    Marginal weights   Based on counts of factor levels in data

Response variate: Logit
     Dilution
      control        -1.32
            1         0.95
          1/4         0.03
         1/16        -0.89
         1/64        -1.47
```

The PREDICTION and SE options allow you to save predictions and their standard errors in table structures. If there are several response variates, the setting of these options should be pointers supplying one identifier for each response variate. The VCOVARIANCE option, similarly, allows you to save the variance-covariance matrix of the predictions in a symmetric matrix.

The SAVE option allows you to specify a regression save structure on which you can base predictions. When you do not use this option, the latest regression model is taken as the basis of predictions.

8.4 Generalized linear regression

The regression models described so far in this chapter have all been linear, and the Genstat analysis of them has relied on the Normal error distribution. In this section we describe two generalizations of linear regression: a nonlinear *link function* in the regression equation, and the use of other distributions for the error. These *generalized linear models* have a form that lets Genstat use a simple type of iteration for calculating estimates of the parameters. To fit these models you use mostly the same statements as for linear regression: the only differences concern how you specify other error distributions, and how you control details of the iteration. All that you have to do, in general, is specify the link function and the error distribution in a MODEL statement.

To understand this section, you need to know something about generalized linear models: we describe only how to fit them with Genstat. Subsection 8.4.1 has a summary of many of the essential concepts, but for more information, see McCullagh and Nelder (1983).

This example shows an analysis using a generalized linear model: it fits log-linear models to a contingency table.

EXAMPLE 8.4

```
  1   "
 -2     Unaided distance vision of 7477 women, aged 30 to 39, tested in Royal
 -3     Ordnance factories 1943-46.
 -4     Data from Stuart (1953).
 -5   "
  6   FACTOR [NVALUES=16; LEVELS=4] Right,Left,Symm
  7   & [LEVELS=7] Diagonal
  8   "Vision with right and left eyes was classified 1 (highest) to 4 (lowest)"
  9   GENERATE Right,Left
 10   VARIATE [VALUES= 1520, 266, 124,  66,\
 11                     234,1512, 432,  78,\
 12                     117, 362,1772, 205,\
 13                      36,  82, 179, 492] Count
 14   CALCULATE Symm = 1+ABS(Right-Left)
 15   & Diagonal = 4+Right-Left
 16   MODEL [DISTRIBUTION=poisson] Count
 17   TERMS Right,Left,Symm,Diagonal
 18   "Fit model of independence of left and right eye vision:
-19     if this model fits well, left and right eye vision is unrelated."
 20   FIT [NOMESSAGE=residual] Right,Left
```

20...

***** Regression Analysis *****

  Response variate: Count
      Distribution: Poisson
     Link function: Log
      Fitted terms: Constant, Right, Left

*** Summary of analysis ***
    Dispersion parameter is 1

              d.f.   deviance   mean deviance
Regression   6   2021.   336.8
Residual   9   6672.   741.3
Total   15   8692.   579.5

Change   -6   -2021.   336.8

*** Estimates of regression coefficients ***

                 estimate   s.e.   t
Constant   6.2225   0.0299   207.80

8.4 Generalized linear regression

```
Right 2              0.1325         0.0308          4.30
Right 3              0.2175         0.0302          7.20
Right 4             -0.9181         0.0421        -21.80
Left  2              0.1529         0.0312          4.90
Left  3              0.2736         0.0304          9.00
Left  4             -0.8187         0.0414        -19.78
* MESSAGE: s.e.s are based on dispersion parameter with value 1

  21   "Add symmetric component of interaction between left and right eye vision:
 -22    if this model fits well, there is as much incidence of left eye vision
 -23    being worse than right eye vision as of the reverse."
  24   ADD Symm

24..............................................................................
```

```
***** Regression Analysis *****

  Response variate: Count
      Distribution: Poisson
     Link function: Log
      Fitted terms: Constant, Right, Left, Symm


*** Summary of analysis ***
    Dispersion parameter is 1

               d.f.      deviance      mean deviance
Regression       9        8637.93         959.770
Residual         6          54.40           9.067
Total           15        8692.33         579.489

Change           *       -6617.11              *

   * MESSAGE: The following units have large residuals:
                              1             4.44
                              2            -3.27
                              5            -3.72
                              7             3.98
                             10             2.52
                             11            -3.04
                             14             2.08


*** Estimates of regression coefficients ***

                  estimate        s.e.            t
Constant            7.2833       0.0253       288.29
Right 2            -0.0190       0.0383        -0.49
Right 3             0.0047       0.0422         0.11
Right 4            -0.6924       0.0550       -12.58
Left  2             0.0651       0.0387         1.68
Left  3             0.2166       0.0422         5.13
Left  4            -0.3776       0.0547        -6.91
Symm  2            -1.5895       0.0282       -56.27
Symm  3            -2.5286       0.0521       -48.53
Symm  4            -2.829        0.102        -27.64
* MESSAGE: s.e.s are based on dispersion parameter with value 1

  25   "Add nonsymmetric component of interaction:
 -26    if this model fits well, the interaction between left and right eye
 -27    vision can be explained just by the difference in classification."
  28   SWITCH [PRINT=accumulated] Diagonal,Symm


28..............................................................................
```

```
* MESSAGE: Term Diagonal can not be fully included in the model
  because 1 parameter is aliased with terms already in the model

***** Regression Analysis *****

*** Accumulated analysis of deviance ***

Change                                        mean      mean deviance
                  d.f.      deviance       deviance         ratio
+ Right
+ Left              6       2020.82         336.80          28.68
+ Symm              3       6617.11        2205.70         187.85
- Symm
+ Diagonal          2          7.44           3.72           0.32
Residual            4         46.97          11.74

Total              15       8692.33         579.49

  29   & [PRINT=fittedvalues] Symm,Diagonal


  29........................................................................


***** Regression Analysis *****

*** Fitted values and residuals ***

                                  Standardized
        Unit    Response  Fitted value  residual  Leverage
          1     1520.00      1455.83       4.44      0.93
          2      266.00       317.00      -3.27      0.59
          3      124.00       144.21      -1.84      0.56
          4       66.00        58.96       1.13      0.68
          5      234.00       291.46      -3.72      0.56
          6     1512.00      1524.51      -0.95      0.94
          7      432.00       361.93       3.98      0.60
          8       78.00        78.11      -0.01      0.37
          9      117.00       116.67       0.03      0.51
         10      362.00       318.48       2.52      0.55
         11     1772.00      1816.29      -3.04      0.94
         12      205.00       204.56       0.03      0.54
         13       36.00        43.04      -1.18      0.56
         14       82.00        62.02       2.08      0.32
         15      179.00       184.57      -0.41      0.50
         16      492.00       499.37      -0.59      0.84

Mean             467.31       467.31      -0.05      0.62
```

8.4.1 Introduction to generalized linear models

Generalized linear models are natural generalizations of ordinary linear regression. The expectation part of the linear model can be written as:

$$\text{(1a)} \quad \text{Mean}(y_i) = \alpha + \sum_{j=1}^{k} \beta_j \times x_{ji} \quad i = 1\ldots N$$

Here x_{ji} is the ith observation of the jth explanatory variable and y_i is the ith ob-

8.4 Generalized linear regression

servation of the response variable.

(2a) Variance $(y_i) = \sigma^2$

where σ^2 is constant for all observations.

Furthermore, the variation in each observation y_i is supposed not to be correlated with the variation in the other observations. Usually the model is specialized further so that the observations y_i are assumed to be Normally distributed.

The generalized linear model has the analogous form

(1b) Mean $(y_i) = H(\alpha + \sum_{j=1}^{k} \beta_j \times x_{ji})$ $i = 1 \ldots N$

where $H()$ is a monotonic and differentiable function;

(2b) Variance $(y_i) = \phi \times V(\text{mean}(y_i))$ $i = 1 \ldots N$

where ϕ is a *dispersion parameter*, known or unknown. Usually, a model is specialized further so that the observations y_i have some distribution such as the Normal, Poisson, binomial, or gamma from the exponential family. $V()$ is a differentiable function, called the *variance function*.

Thus linear regression is in the class of generalized linear models, with $H()$ being the identity function, ϕ being σ^2, and $V()$ being constant.

Many other familiar statistical models are in this class too.

(a) The log-linear model for contingency tables is a generalized linear model with $H(y) = \exp(y)$, $\phi = 1$ and $V(y) = y$. The error distribution for the counts is usually stipulated to be Poisson or multinomial.

(b) The model used in the probit analysis of proportions is a generalized linear model with $H(y) = n \times \Phi^{-1}(y)$, where Φ is the cumulative Normal distribution function, $\phi = 1$, and $V(y) = y \times (1 - y/n)$, n being the number of trials of which y respond. The errors are usually supposed to be binomial: they model the number of successes out of a number of trials.

(c) Dilution assays are often analysed by a model that has $H(y) = n \times \exp(-\exp(y))$, $\phi = 1$, and $V()$ as in (b) also being binomial.

You can fit these and other models using the options of the MODEL directive. The DISTRIBUTION option specifies the characteristic form of the variance function $V()$, according to these rules:

Distribution	Variance function, V
Normal	1
Poisson	y
binomial	$y \times (1 - y/n)$
gamma	y^2
inverse Normal	y^3

If you use the binomial distribution, you must put the number of successes into the response variate (or the number of failures), and you must supply the total numbers (that is successes plus failures) in another variate included in the NBINOMIAL parameter of the MODEL directive. For example:

```
VARIATE [VALUES=2,5,6,7] Success
& [VALUES=10,10,8,8] Trial
MODEL [DISTRIBUTION=binomial] Success; NBINOMIAL=Trial
```

When you use the Normal, or the gamma, or the inverse Normal distribution, the dispersion parameter, ϕ, will usually be unknown, and assumed constant over all observations. For the Normal distribution, this is the constant variance, usually written as σ^2. Sometimes, however, you may know a value for the dispersion parameter. For example, you may know that the response variable has a Normal distribution with a variance that you can estimate from previous experiments or surveys. Another example is when you want to use the exponential distribution; this can be specified as a gamma distribution, fixing the dispersion parameter to 1.0.

Genstat also fixes the dispersion parameter at 1.0 for the Poisson and binomial distributions. Another way of thinking about this fixed dispersion is to say that the variance of an observation is a function only of its mean – no estimator of variance is required from the observations as a whole. But you may also sometimes want to include a dispersion parameter even though you are using the binomial or the Poisson. An example is the heterogeneity factor of probit analysis: the distribution of the observations is taken to be "superbinomial", in the sense that the variance is greater than what we would expect for a binomial distribution; specifically, it is taken to be

$$h \times \text{mean} \times (1 - \text{mean}/n)$$

where h is the *heterogeneity factor* (Finney 1971). Data for which a "superbinomial" or "superPoisson" distribution are needed are called *overdispersed*; see McCullagh and Nelder for more details.

You can control whether the dispersion parameter is fixed by using the DISPERSION option of the MODEL directive (8.1.1). Only for the Poisson and binomial distributions it is fixed by default (at 1.0). You can set an explicit value for the other distributions, for example by supplying a known Normal variance, or you can allow it to be estimated for the Poisson or binomial distributions, as in the statement

```
MODEL [DISTRIBUTION=Poisson; DISPERSION=*] Count
```

The LINK option of the MODEL directive specifies the function that relates the explanatory variables to the mean of the response variable. By convention, the function is the inverse of the $H()$ given above. Thus the *link* is a function of the mean which is stipulated to be equal to a linear combination of the explanatory variables. The linear combination itself is known as the *linear predictor*. For example, the model

8.4 Generalized linear regression

the model
$$E(y_i) = \exp(\alpha + \Sigma \beta_j \times x_{ji})$$
is said to have a logarithmic link function. The reason for this convention is that the model is similar to a linear model for the inverse function of the response. For example
$$\log(E(y_i)) = \alpha + \Sigma \beta_j \times x_{ji}$$
is similar to
$$E(\log(y_i)) = \alpha + \Sigma \beta_j \times x_{ji}$$
but is of course not the same.

The link functions that you can use in Genstat are:

	link function, $H^{-1}(f)$	$H(y)$
identity	f	y
logarithm	$\log(f)$	$\exp(y)$
logit	$\log(f/(n-f))$	$n \times \exp(y)/(1+\exp(y))$
reciprocal	$1/f$	$1/y$
power	f^{power}	$y^{(1/\text{power})}$
square root	$f^{1/2}$	y^2
probit	$\Phi(f/n)$	$n \times \Phi^{-1}(y)$
complementary log-log	$\log(-\log(f/n))$	$n \times \exp(-\exp(y))$

By default, the 'power' setting uses the exponent -2; you can specify other values in the EXPONENT option, for example

    MODEL [DISTRIBUTION = gamma; LINK = power; EXPONENT = 1.5] Y

For each of the available distributions, one of the links is known as the *canonical link*. This has special properties. In particular, a model with its canonical link always provides a unique set of parameter estimates, whereas with other models this may not be so. There are also often practical scientific reasons for using the canonical link. Conversely, there may be very good reasons for using a non-canonical link. If you do not set the LINK option, the default is the canonical link of the chosen distribution:

Normal	Identity
Poisson	Log
Binomial	Logit
Gamma	Reciprocal
Inverse Normal	Power, with exponent -2

An offset variable is a variable that appears in the linear predictor without a parameter. It provides for each observation a fixed offset, o_i say, from the estimated constant:

Mean $(y_i) = H(o_i + \alpha + \Sigma \beta_j \times x_{ji})$

You set it by the OFFSET option of the MODEL directive. Offset variables occur naturally in log-linear models for rates where each cell has a different exposure time. They also arise naturally in the standard analysis for dilution assay, involving a complementary log-log link function. The model in the next example is

Npositive$_i$ = Nsamples$_i$ × exp(−Dilution$_i$ × exp(Constant
 + rate × Time$_i$))
 = Nsamples$_i$ × exp(−exp(log(Dilution$_i$) + Constant + rate × Time$_i$))

So the logarithm of the dilution is an offset.

EXAMPLE 8.4.1

```
   1  "
  -2    Results from a dilution experiment on Leptospira autumnalis, showing
  -3    the change of density of the bacteria with time.
  -4    Data from Koch and Tolley (1975).
  -5  "
   6  UNITS [NVALUES=42]
   7  READ [PRINT=data] Time,Dilution,Npositiv

   8     4.08  .01 10     4.08  .001 10     4.08  .0001 0
   9    12     .01 10    12     .001  9    12     .0001 0
  10    23.92  .01 10    23.92  .001  7    23.92  .0001 0
  11    29.83  .01 10    29.83  .001  3    29.83  .0001 1
  12    35.92  .01 10    35.92  .001  3    35.92  .0001 0
  13    42.08  .01 10    42.08  .001  2    42.08  .0001 1
  14    47.92  .01 10    47.92  .001  0    47.92  .0001 0
  15    60     .01  6    60     .001  2    60     .0001 0
  16    66     .01  6    66     .001  0    66     .0001 0
  17    71.92  .01  4    71.92  .001  1    71.92  .0001 0
  18    80.17  .01  4    80.17  .001  1    80.17  .0001 0
  19    88.33  .01  0    88.33  .001  1    88.33  .0001 0
  20    95.92  .01  1    95.92  .001  0    95.92  .0001 0
  21   107.33  .1   9   107.33  .01   1   107.33  .001  0 :

  22  "At each time, ten samples at each of three dilutions of a culture
 -23   were tested for presence or absence of bacteria."
  24  VARIATE [VALUES=42(10)] Nsamples
  25  VARIATE Log; EXTRA=' dilution'
  26  CALCULATE Log = LOG(Dilution)
  27  MODEL [DISTRIBUTION=binomial; LINK=complementary; OFFSET=Log] \
  28     Npositiv; NBINOMIAL=Nsamples
  29  "Fit a model with the effect of dilution only, ignoring time."
  30  FIT [NOMESSAGE=leverage,residuals]
```

30...

***** Regression Analysis *****

 Response variate: Npositiv
  Binomial totals: Nsamples
     Distribution: Binomial
    Link function: Complementary log-log
   Offset variate: Log dilution
      Fitted terms: Constant

```
*** Summary of analysis ***
    Dispersion parameter is 1

                d.f.        deviance        mean deviance
Regression       0              *                *
Residual        41           251.1            6.124
Total           41              *                *


*** Estimates of regression coefficients ***

                    estimate          s.e.             t
Constant              4.8129         0.0975         49.36
* MESSAGE: s.e.s are based on dispersion parameter with value 1

    31  "Fit a linear effect of time on the complementary log-log scale."
    32  FIT Time


32..............................................................................

***** Regression Analysis *****

  Response variate: Npositiv
   Binomial totals: Nsamples
      Distribution: Binomial
     Link function: Complementary log-log
    Offset variate: Log dilution
      Fitted terms: Constant, Time


*** Summary of analysis ***
    Dispersion parameter is 1

                d.f.        deviance        mean deviance
Regression       1              *                *
Residual        40            41.12           1.028
Total           41              *                *

* MESSAGE: The following units have high leverage:
                              2              0.227
                              5              0.191
                             40              0.334


*** Estimates of regression coefficients ***

                    estimate          s.e.             t
Constant              7.928          0.207           38.26
Time                 -0.05157        0.00360        -14.34
* MESSAGE: s.e.s are based on dispersion parameter with value 1
```

8.4.2 The deviance

You can assess how well a linear regression fits by doing an analysis of variance: assuming that the residuals have independent Normal distributions with equal variances, the variance ratio (mean square due to the regression divided by the residual mean square) has an F distribution.

With generalized linear models, there is no similar exact distributional property. However, you can get approximate assessments of the quality of the fit from a statistic called the *scaled deviance*. This is defined as minus twice the log-likelihood ratio between the model you have fitted and a full model that explains all the variation in the data. The scaled deviance has approximately a χ^2_d distribution, d being the number of residual degrees of freedom. The approximation is better for large numbers of observations than for small numbers, and is poor when there are many extreme observations (such as zeroes for the Poisson distribution).

The scaled deviance is a function of the dispersion parameter, and so its distribution depends also on any estimate of that parameter. Usually you make such an estimate from a model that, you judge, explains all systematic variation – a *maximal model*, as in the analysis of variance for linear regression. You can judge the importance of a term in any generalized linear model by considering the difference between the scaled deviances of that model and the model excluding the term. The difference in deviances also has an approximate χ^2_t distribution, where t is the number of degrees of freedom of the term; this approximation is better than that for the scaled deviance itself.

Alternatively, you can consider ratios of mean scaled deviances between competing models, one of which is nested inside the other. (The mean scaled deviance is the scaled deviance divided by the corresponding number of degrees of freedom.) The resulting ratios do not involve the dispersion parameter. Such a ratio has approximately an F distribution – exact for linear regression models with Normal errors.

Genstat reports the *deviance* of the data for each type of model, which is equivalent to the scaled deviance multiplied by the dispersion parameter. The deviance is otherwise known as the log-likelihood ratio statistic.

You can summarize the fit of a sequence of nested models by an *analysis of deviance*, which you interpret in much the same way as an analysis of variance; but do not forget about the approximations we have just noted.

Here are the formulae for the deviance for each distribution; the ith response is represented by y_i, and the corresponding fitted value by f_i:

Normal	$\Sigma(y_i - f_i)^2$
Poisson	$2 \times \Sigma\{y_i \times \log(y_i/f_i) - (y_i - f_i)\}$
Binomial	$2 \times \Sigma\{y_i \times \log(y_i/f_i) + (n_i - y_i) \times \log((n_i - y_i)/(n_i - f_i))\}$
Gamma	$2 \times \Sigma\{(y_i - f_i)/f_i - \log(y_i/f_i)\}$
Inversenormal	$\Sigma\{(y_i - f_i)^2/(y_i \times f_i^2)\}$

Sometimes you cannot get parameter estimates for a model that has been fitted iteratively, unlike the linear model. The most common cause of this with models using the binomial or Poisson distribution is the presence of some observations at

8.4 Generalized linear regression

the extremes (zero for Poisson, zero or *n* for binomial). One or more of the parameters may then need to be infinite to give maximum likelihood: in practice, approximate convergence will usually be achieved with the parameters large but finite (the meaning of "large" being dependent on the link function).

EXAMPLE 8.4.2

```
  1  "
 -2     Comparison of the effect on mice of the analgesic drug Pethidine
 -3     with the effect of a placebo.
 -4     Based on data for Pethidine from Grewal (1952).
 -5  "
  6  UNITS [NVALUES=10]
  7  FACTOR [LABELS=!T(Pethidine,Placebo); VALUES=5(1,2)] Drug
  8  READ [PRINT=data] Logdose,Ntest,Nrespond

  9  0.70 60 13  0.88 85 27  1.00 60 32  1.18 90 55  1.30 60 44
 10  0.70 60  0  0.88 85  0  1.00 60  0  1.18 90  0  1.30 60  0 :

 11  MODEL [DISTRIBUTION=Binomial; LINK=probit] Nrespond; NBINOMIAL=Ntest
 12  "Fit separate probit lines for the two 'drugs'."
 13  FIT [CONSTANT=omit] Drug+Logdose.Drug
```

13...

```
***** Regression Analysis *****

  Response variate: Nrespond
   Binomial totals: Ntest
      Distribution: Binomial
     Link function: Probit
      Fitted terms: Drug + Logdose.Drug

*** Summary of analysis ***
    Dispersion parameter is 1
                d.f.       deviance      mean deviance
 Regression       4           *                *
 Residual         6         1.823          0.3038
 Total           10           *                *

*** Estimates of regression coefficients ***
                            estimate        s.e.           t
 Drug Pethidine              -2.456        0.365        -6.72
 Drug Placebo                 -4.6         26.2         -0.18
 Logdose.Drug Pethidine       2.363        0.351         6.73
 Logdose.Drug Placebo         0.0          25.3          0.00
 * MESSAGE: s.e.s are based on dispersion parameter with value 1
```

Occasionally, the iterative process may converge only very slowly when a parameter needs to be infinite: you can increase the limit on the number of cycles with the RCYCLE directive (8.4.3). Very rarely you might even get divergence, because of a lot of extreme data. This can happen when the initial guesses for the fitted values are very bad, and the deviance appears to increase after the first iteration. But in such cases, the model would be unlikely to fit the data satisfactorily anyway.

Failure to find a solution may occur when estimates from a fit take impossible values. For example, the gamma distribution is defined in the range $(0, \infty)$, but some sets of data may produce an estimated mean that is negative. In such cases, you should consider a different link, or try a new fit omitting those explanatory variables whose parameters were estimated as negative.

8.4.3 The RCYCLE directive

RCYCLE
Controls iterative fitting of generalized linear and nonlinear models and specifies parameters, bounds etc for nonlinear models.

Options

MAXCYCLE	= scalar	Maximum number of iterations; default * gives 10 for generalized linear, 20 for nonlinear models
TOLERANCE	= scalar	Convergence criterion; default 0.0004
FITTEDVALUES	= variate	Initial fitted values for generalized linear model; default *
METHOD	= string	Algorithm for fitting nonlinear model (GaussNewton, NewtonRaphson); default g, but n for scalar minimization

Parameters

PARAMETER	= scalars	Nonlinear parameters in the model
LOWER	= scalars	Lower bound for each parameter
UPPER	= scalars	Upper bound for each parameter
STEPLENGTH	= scalars	Initial step length for each parameter
INITIAL	= scalars	Initial value for each parameter

The parameters of the RCYCLE directive are ignored when generalized linear models are fitted; see 8.5.5 and 8.6.1 for how to use them with nonlinear models.

The MAXCYCLE option allows you to change the limit on the number of cycles in the iteration. Usually, the algorithm converges in four or five cycles, but when there are many extreme observations more cycles may be needed; then the resulting fit is often uninformative.

8.4 Generalized linear regression

The TOLERANCE option controls the criterion for convergence. The iteration stops when the absolute change in deviance in successive cycles is less than the tolerance times D; D is either the number of residual degrees of freedom plus 1.0 (if you have supplied a value for the dispersion parameter) or else is the current value of the deviance (if you have not).

The algorithm has to start by estimating an initial set of fitted values. Genstat usually gets these by a simple transformation of the observed responses. It may be that better estimates are available, for example from a previously fitted model; you can supply them by the FITTEDVALUES option.

The METHOD option is relevant only for nonlinear models, as described in 8.6.1.

8.4.4 Modifications to output and the RKEEP and PREDICT directives

Some aspects of the results of fitting generalized linear models are different from those described for linear regression, because of the iteration. We call any generalized linear model other than linear regression an *iterative* model.

Genstat will analyse only one response variate if the model is iterative; if you specify several variates in the Y parameter of the MODEL directive, Genstat analyses the first one only. (This is because the iterative process depends on successively forming a single modified response variate.)

For most models you can get the total deviance only if you have given a TERMS statement. The exception is linear regression with the Normal distribution: then the total deviance is the total sum of squares, which is always available. With other distributions, you get the total deviance by fitting a null model, with a TERMS statement. Otherwise the summary of the analysis will contain only residual information, as in the example in 8.4.1.

Genstat does not report the percentage variance accounted for by iterative models.

The standard errors of parameter estimates produced by the 'estimates' setting of the PRINT option are only approximate for iterative models; the same applies to the t-statistics, and to the correlations from the 'correlations' setting. The adequacy of the approximation depends on the model and the context, and so you should use these values as a guide only: for example, a "t-statistic" less than 1.0 will usually not be significant and one greater than 3.0 will be significant. You can get a better test of the corresponding parameter by dropping it from the model and then assessing the change in deviance.

Genstat calculates leverages to be displayed by the 'fittedvalues' setting of the PRINT option or stored by the LEVERAGE parameter of the RKEEP directive. The formula for them is:

$$\text{leverage}_i = l_i = u_i \times w_i \times \{X(X'UWX)^{-1}X'\}_{ii} \quad i = 1...N$$

where U is a diagonal matrix consisting of the iterative weights u_i (defined below). These values are also used in the standardization of residuals, according to the formula given in 8.1.1.

By default, the residuals are deviance residuals, as described in 8.1.1: each residual is the signed square root of the contribution to the deviance. See 8.4.2 for the definition of deviance for each distribution. The standardization of the residuals uses the leverages, l_i, as described above, and the weights, w_i, if specified; the weights are 1.0 if the WEIGHT option of MODEL was not set:

$$\text{residual}_i = \text{sign}(y_i - f_i) \times \sqrt{\{w_i \times (\text{deviance contribution})_i / (\text{dispersion} \times (1 - l_i))\}}$$

For example, the deviance residuals for a model with the Poisson distribution are given by:

$$\text{sign}(y_i - f_i) \times \sqrt{\{2 \times (y_i \times \log(y_i/f_i) - (y_i - f_i)) / (1 - l_i)\}}$$

If you set the RMETHOD option of the MODEL directive to 'Pearson' then Genstat forms the residuals by adjusting the ordinary residuals for their estimated variance:

$$\text{residual}_i = (y_i - f_i) / \sqrt{\{w_i / (V(F_i) \times \text{dispersion} \times (1 - l_i))\}}$$

If you chose the binomial distribution, the table produced by the 'fittedvalues' setting includes a column for the binomial totals that you specified in the NBINOMIAL parameter of the MODEL directive.

The 'accumulated' setting of the PRINT option produces an accumulated analysis of deviance for iterative models, just as for linear models except that all contributions from one statement are pooled. The POOL option of the RDISPLAY directive, and of directives like FIT, has no effect for iterative models. Thus you cannot calculate the change in deviance attributable to each individual term unless you add the terms into the model individually. For example, here are statements that provide a full analysis of deviance for two factors A and B and their interaction:

```
TERMS A*B
ADD [PRINT=*] A
& B
& [PRINT=accumulated] A.B
```

The 'monitoring' setting of the PRINT option provides a report on the progress of the iteration:

EXAMPLE 8.4.4

```
  33  FIT [PRINT=monitoring]

33.............................................................

*** Convergence monitoring ***
```

```
Cycle           Deviance        Current parameters
  1             270.8137             5.1934
  2             252.1168             4.7194
  3             251.1110             4.8273
  4             251.0846             4.8101
  5             251.0839             4.8129

Convergence at cycle 5
```

Two of the parameters of the RKEEP directive are relevant only for saving results of iterative models. The LINEARPREDICTOR parameter lets you save the linear predictor; that is

$$p_i = \alpha + o_i + \Sigma \beta_j \times x_i \quad i = 1...N$$

The ITERATIVEWEIGHTS parameter saves the iterative weighting variate that was used in the last cycle of the iteration. It is

$$u_i = V(f_i) \times \{p'_i\}^2$$

where $V()$ is the variance function (8.4.1) and p'_i is the derivative of the linear predictor with respect to the fitted value. The iterative weights do not contain any contribution from the weights that can be specified whether or not the model is iterative by the WEIGHT option of the MODEL directive.

The PREDICT directive forms summaries of the fit of an iterative model as for a linear model. However, note that averaging is done on the scale of the original response variable, not on the scale transformed by the link function. In other words, linear predictors are formed for all the combinations of factor levels and variate values that you have specified by PREDICT, and then transformed by the link function back to the natural scale. This back transformation is useful when you are reporting results, since the tables from PREDICT can then be interpreted as natural averages of means predicted by the fitted model.

Genstat calculates the standard errors of predictions from iterative models by using first-order approximations that allow for the effect of the link function. Thus you should interpret them only as a rough guide to the variability of individual predictions.

8.5 Standard nonlinear curves

The regression models described in this section are neither linear nor generalized linear: they are standard nonlinear curves, relating the response variable to the explanatory variable. The standard curves available in Genstat are those that have proved to be useful in many applications of statistics. They are fitted by a modified Newton method of maximizing the likelihood, using stable forms of parameterization as used in the program MLP (Ross 1987). You can fit other curves as described in 8.6.

The method Genstat uses to fit curves is iterative, using a search procedure to

362   8   *Regression analysis*

find parameter values that maximize the likelihood. The search is much quicker if Genstat knows the shape of the curve: thus, fitting a curve by the methods in this section is more efficient than by those in 8.6. With standard curves you do not usually have to specify starting values for the search, nor to control the course of the search; in contrast, you nearly always have to do these things when you are fitting non-standard curves. For more information about nonlinear curve fitting, see Ratkowski (1983).

The first example shows Genstat fitting the exponential curve

$$y = a + b \times r^x + \varepsilon$$

after first fitting a linear model to test the effect of a blocking factor.

EXAMPLE 8.5

```
  1  "
 -2     Exponential model for the response of sugar cane to nitrogen fertilizer.
 -3  "
  4  UNITS [NVALUES=20]
  5  FACTOR [LEVELS=4] Blocks
  6  & [LEVELS=!(0,50...200)] Fnitrogen
  7  READ [PRINT=data] Blocks,Fnitrogen,Yield

  8  1    0  60    2    0  73    3    0  77    4    0  72
  9  1   50 125    2   50 144    3   50 145    4   50 116
 10  1  100 152    2  100 154    3  100 160    4  100 141
 11  1  150 182    2  150 167    3  150 181    4  150 185
 12  1  200 198    2  200 188    3  200 189    4  200 182   :

 13  "Form a variate, Nitrogen, from the factor, Fnitrogen"
 14  CALCULATE Nitrogen = Fnitrogen
 15  VARIATE [MODIFY=yes] Nitrogen; DECIMALS=0
 16  GRAPH [NROWS=16; NCOLUMNS=61] Yield; Nitrogen
```

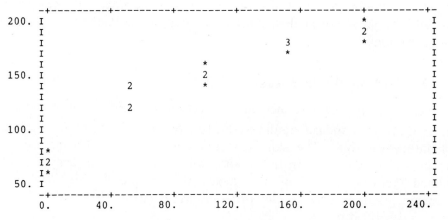

Yield   v. Nitrogen using symbol *

8.5 Standard nonlinear curves

```
 17  MODEL Yield
 18  "Assess the importance of blocks and treatments"
 19  FIT [PRINT=accumulated] Blocks,Fnitrogen
```

19..

***** Regression Analysis *****

*** Accumulated analysis of variance ***

Change	d.f.	s.s.	m.s.	v.r.
+ Blocks	3	322.95	107.65	1.34
+ Fnitroge	4	35392.70	8848.17	109.77
Residual	12	967.30	80.61	
Total	19	36682.95	1930.68	

```
 20  "Fit an exponential response, ignoring blocks"
 21  FITCURVE [PRINT=model,summary,estimates,correlation,fittedvalues] Nitrogen
```

1...

***** Nonlinear regression analysis *****

Response variate: Yield
    Explanatory: Nitrogen
   Fitted Curve: A + B*R**X
    Constraints: R < 1

*** Summary of analysis ***

	d.f.	s.s.	m.s.
Regression	2	35046.	17523.18
Residual	17	1637.	96.27
Total	19	36683.	1930.68

Percentage variance accounted for 95.0

*** Estimates of parameters ***

	estimate	s.e.	Correlations		
R	0.98920	0.00213	1.000		
B	-131.1	10.6	-0.762	1.000	
A	203.0	10.8	0.937	-0.898	1.000

*** Fitted values and residuals ***

Unit	Explanatory	Response	Fitted value	Standardized residual	Leverage
1	0.	60.0	71.9	-1.39	0.24
2	0.	73.0	71.9	0.13	0.24
3	0.	77.0	71.9	0.60	0.24
4	0.	72.0	71.9	0.02	0.24
5	50.	125.0	126.8	-0.20	0.14
6	50.	144.0	126.8	1.89	0.14
7	50.	145.0	126.8	2.00	0.14
8	50.	116.0	126.8	-1.19	0.14
9	100.	152.0	158.7	-0.73	0.10

10	100.	154.0	158.7	-0.51	0.10
11	100.	160.0	158.7	0.14	0.10
12	100.	141.0	158.7	-1.91	0.10
13	150.	182.0	177.3	0.50	0.08
14	150.	167.0	177.3	-1.09	0.08
15	150.	181.0	177.3	0.40	0.08
16	150.	185.0	177.3	0.82	0.08
17	200.	198.0	188.1	1.12	0.18
18	200.	188.0	188.1	-0.01	0.18
19	200.	189.0	188.1	0.11	0.18
20	200.	182.0	188.1	-0.68	0.18
Mean	100.	144.6	144.6	0.00	0.15

8.5.1 The FITCURVE directive

FITCURVE
Fits a standard nonlinear regression model.

Options

PRINT	= strings	What to print (model, summary, accumulated, estimates, correlations, fittedvalues, monitoring); default m,s,e
CURVE	= string	Type of curve (exponential, dexponential, cexponential, lexponential, logistic, glogistic, gompertz, ldl, qdl, qdq); default e
SENSE	= string	Sense of curve (right, left); default r
ORIGIN	= scalar	Constrained origin; default *
NONLINEAR	= string	How to treat nonlinear parameters between groups (common, separate); default c
CONSTANT	= string	How to treat constant (estimate, omit); default e
FACTORIAL	= scalar	Limit for expansion of model terms; default 3
POOL	= string	Whether to pool ss in accumulated summary between all terms fitted in a linear model (no, yes); default n
DENOMINATOR	= string	Whether to base ratios in accumulated summary on rms from model with smallest residual ss or

		smallest residual ms (ss, ms); default s
NOMESSAGE	= strings	Which warning messages to suppress (dispersion, leverage, residual, aliasing, marginality); default *
Parameter		
	formula	Explanatory variate, list of variate and factor, or variate*factor

The parameter provides the identifier of the explanatory variate. In addition, you can include a factor to specify groups of the observations for which you want separate curves to be fitted. If you give just the variate and the factor identifiers, then Genstat constrains the curves for each group to be parallel: that is, they differ only by a constant (the analogy of what in linear regression would be called the intercept). For example,

    FACTOR [LEVELS=2; VALUES=10(1,2)] G
    MODEL Y
    FITCURVE G,X

fits
$$y = a_1 + b \times r^x + \varepsilon$$
for level 1 of G, and
$$y = a_2 + b \times r^x + \varepsilon$$
for level 2.

If you fit the interaction between the variate and the factor, Genstat constrains the curves to have common nonlinear parameters, but estimates all linear parameters without constraint:

    FITCURVE G*X

fits
$$y = a_1 + b_1 \times r^x + \varepsilon$$
for level 1 of G, and
$$y = a_2 + b_2 \times r^x + \varepsilon$$
for level 2.

If you set the NONLINEAR option to 'separate' and you also include in the parameter the variate, the factor and the interaction, then Genstat estimates all parameters independently; only information about variability is pooled:

    FITCURVE [NONLINEAR=separate] X*G

fits

$$y = a_1 + b_1 \times r_1^x + \varepsilon$$
for level 1 of G, and
$$y = a_2 + b_2 \times r_2^x + \varepsilon$$
for level 2.

The PRINT, FACTORIAL, POOL, DENOMINATOR, and NOMESSAGE options are as for FIT.

The CURVE option specifies which of the standard curves is to be fitted. For some of these, the SENSE option lets you choose among several forms. Before we describe them, here is a list for convenient reference:

Exponential
    ordinary exponential $y = a + b \times r^x + \varepsilon$
    double exponential $y = a + b \times r^x + c \times s^x + \varepsilon$
    critical exponential $y = a + (b + c \times x) \times r^x + \varepsilon$
    line plus exponential $y = a + b \times r^x + c \times x + \varepsilon$

Logistic
    ordinary logistic $y = a + c/\{1 + \exp(-b \times (x - m))\} + \varepsilon$
    generalized logistic $y = a + c/\{1 + t \times \exp(-b \times (x - m))\}^{1/t} + \varepsilon$
    Gompertz $y = a + c \times \exp(-\exp(-b \times (x - m))) + \varepsilon$

Rational functions
    linear divided by linear $y = a + b/(1 + d \times x) + \varepsilon$
    quadratic divided by linear $y = a + b/(1 + d \times x) + c \times x + \varepsilon$
    quadratic divided by quadratic $y = a + (b + c \times x)/(1 + d \times x + e \times x^2) + \varepsilon$

The four exponential curves each arise as solutions of linear ordinary differential equations. These represent processes that increase exponentially with time, for example, or that increase with a law of diminishing returns (that is, for which the rate of increase decreases with time).

The default setting of the CURVE option is 'exponential', corresponding to the "asymptotic regression" or Mitscherlich curve:
$$y = a + b \times r^x + \varepsilon$$
An equivalent form of this equation is
$$y = a + b \times \exp(-k \times x) + \varepsilon$$
where $r = \exp(-k)$. The form involving r is used in Genstat to avoid problems with large values of k.

The model has only one nonlinear parameter, r, and Genstat estimates the other parameters by linear regression at each stage of an iterative search for the best estimate of r. The explanatory variate is automatically scaled to avoid problems of its having too small or too large a range of values: such would lead to difficulties

8.5 Standard nonlinear curves

of estimation near the boundary of the allowed values of r.

By default, r is restricted to the range $0 < r < 1$, giving a curve corresponding to the law of diminishing returns. The alternative is $r > 1$, which you can specify by the 'left' setting of the SENSE option. If Genstat finds that the opposite sense to the one you have specified provides a better fit, the sense will be reversed and you will get a warning.

The double exponential curve has the form:

$$y = a + b \times r^x + c \times s^x + \varepsilon$$

You can fit this either with $0 < r < 1$ and $0 < s < 1$ or with $r > 1$ and $s > 1$, specified by the SENSE option as for the exponential curve. The fitting process is unlikely to find a satisfactory solution for this curve unless there are enough data to estimate both components separately: there should be at least four points for which the fast component is larger than the slow component; the fast component corresponds to the smaller of r and s when SENSE = right, and is the larger of r and s when SENSE = left.

Genstat provides two limiting cases of the double exponential as special curves. The critical exponential curve has the form

$$y = a + (b + c \times x) \times r^x + \varepsilon$$

and the line plus exponential curve has the form

$$y = a + b \times r^x + c \times x + \varepsilon$$

Again here, the constraint on the parameter r depends on the setting of the SENSE option as for the exponential curve.

Another type of standard curve is sigmoid and monotonic, and is often used to model the growth of biological subjects. There are three types of such growth curves in Genstat, each a logistic of some sort. The first type is the generalized logistic without any constraints: that is,

$$y = a + c/\{1 + t \times \exp(-b \times (x-m))\}^{1/t} + \varepsilon$$

where a is the lower asymptote, m is the point of inflexion for the explanatory variable, b is a slope parameter, t is a power-law parameter, and $a+c$ is the upper asymptote. To fit this curve you need data both for the steep central part and for both flat parts.

Genstat also provides two special cases of the generalized logistic. The ordinary logistic curve has the form

$$y = a + c/\{1 + \exp(-b \times (x-m))\} + \varepsilon$$

It is sometimes known as the autocatalytic or inverse exponential curve. The same curve can be rewritten in several different forms, so be on the alert for concealed equivalences of apparently different curves: otherwise you might be tempted to use the methods in 8.6, which would be slower.

The other special case is the Gompertz curve:

$$y = a + c \times \exp(-\exp(-b \times (x-m))) + \varepsilon$$

which is non-symmetrical about the inflexion, $x=m$, and has asymptotes at $y=a$ and $y=a+c$.

You can also fit these three growth curves to data in which y decreases as x increases. For the logistic and generalized logistic curves, you are not allowed to constrain the sense of the curve by setting the SENSE option. This is because the sense depends on both the parameters b and c. In fact, the logistic curve with parameters a, b, c, and m is the same as the logistic curve with parameters $(a+c)$, $-b$, $-c$, and m; Genstat will report only one of the two possible versions.

For the Gompertz curve, you can use the 'left' setting of the SENSE option to specify the upside-down Gompertz curve corresponding to $b<0$; otherwise b is constrained to be positive.

FITCURVE provides three rational functions, that are ratios of polynomials. The linear-divided-by-linear curve has the form

$$y = a + b/(1 + d \times x) + \varepsilon$$

which is a rectangular hyperbola; it occurs as the Michaelis-Menten law of chemical kinetics. The quadratic-divided-by-linear curve has the form

$$y = a + b/(1 + d \times x) + c \times x + \varepsilon$$

which is a hyperbola with a non-horizontal asymptote. The quadratic-divided-by-quadratic curve has the form

$$y = a + (b + c \times x)/(1 + d \times x + e \times x^2) + \varepsilon$$

which is a cubic curve having an asymmetric maximum falling to an asymptote. Genstat ignores the SENSE option for all three rational functions.

8.5.2 Distributions and constraints in curve fitting

The curves available with FITCURVE can be fitted in Genstat only with the Normal likelihood. If you set some other distribution in the MODEL statement, you will get a warning message and the distribution will automatically be reset to Normal. However, you can specify a weighted Normal likelihood by providing weights in the WEIGHTS option of the MODEL directive, as for linear regression.

You can set the DISPERSION option if you want Genstat to use a known variance for the distribution of the response variate (8.1.1).

FITCURVE ignores the LINK and EXPONENT options of the MODEL directive, and you are not allowed to set the GROUPS option.

You can constrain the exponential and rational curves to pass through a given point. The ORIGIN option of the FITCURVE directive specifies a value for the response variate corresponding to a zero value of the explanatory variate: if you wanted to specify the response for another value of the explanatory variate you would have to modify the explanatory variate beforehand. For all these standard curves, except the double exponential, the supplied origin corresponds to the expression $(a+b)$; in the double exponential it is $(a+b+c)$. If you constrain the origin in this way, you should probably use some form of weighting, because points near the constraint are likely to vary less than points far away. You can get approximately log-Normal weighting by using a weight variate with values $1/(y-\text{origin})^2$, where y is the response variate.

Another way of constraining the curves is by omitting the constant term – the parameter a in each case. This parameter represents the asymptote: for growth curves with parameter $b>0$ it represents the asymptote as $x \to -\infty$, and for those with $b<0$ it represents the asymptote as $x \to +\infty$. To constrain the asymptote to be other than 0, put the value that you require into every element of the variate in the OFFSET option of the MODEL directive. An example is the exponential curve

$$y = o + b \times r^x$$

where o is the constant value you would give to the offset variate.

8.5.3 Modifications to regression output and the RKEEP directive

The output controlled by the PRINT options of the FITCURVE and RDISPLAY directives for fitted curves is like that for iterative generalized linear models with a Normal distribution (8.4.4). In particular, only one response variable is analysed, standard errors are approximate, and the accumulated summary contains pooled contributions for all terms fitted in one statement.

You cannot get standard errors and correlations for linear parameters in models where you have constrained some parameters of the curve to be equal for all the groups defined by a fitted factor. When you fit separate curves for the groups of a factor, correlations between parameters in different groups are zero and are not shown.

Neither can you get leverages for models in which parameters are constrained to be equal across groups. Genstat therefore does not standardize residuals with respect to the leverages in these models. For other models, the leverages are defined as:

$$\text{leverage}_i = \{DCD'\}_{ii}$$

where D is the matrix of derivatives of the fitted values with respect to the parameters, and C is the variance-covariance matrix of the parameters.

You can display intermediate results of the iteration by the 'monitoring' setting of the PRINT option of the FITCURVE directive. At each cycle, the current parameter values are displayed together with the total number of times the likelihood function has been evaluated (Nfun) and an indication of the state of the search (Move). The possible states are:

Move
0   The current step is acceptable

1 Preconvergence; small adjustments are being made
2 The function is concave in at least one direction
3 Convergence is being approached, but there is distinct curvature
4 A bound has been violated
5 The current step is too large relative to the steplengths
6 Convergence
7 A step has been taken within a boundary plane.

You also get the steplengths used in the search whenever they are changed.

EXAMPLE 8.5.3

```
  22  FITCURVE [PRINT=monitoring] Nitrogen

22.............................................................

Temporary scaling of X by 0.0141

*** Convergence monitoring ***

Cycle Eval Move    Function value   Current parameters
    0    6   0         1658.735         0.50000
                         Steps          0.0100000
    1    9   0         1636.688         0.46167
                         Steps          0.0025000
    2   12   1         1636.598         0.46398
    3   16   6         1636.598         0.46393
```

The search may not converge, particularly if the model being fitted is not suitable for the data. Genstat will give a warning message to indicate why convergence has not been achieved; often you will also get a suggestion for a limiting form of the curve that would be likely to be a more suitable description of the data than the one you have specified. You can find out about the final status of the search by the EXIT parameter of the RKEEP directive. It takes a value according to the following key:

Exit
0 Successful convergence
1 Limit on number of cycles has been reached without convergence
2 Parameter out of bounds
3 Likelihood appears constant
4 Failure to progress towards solution
5 Some standard errors are not available because the information matrix is near singular
6 Curve is close to a limiting form.

With code 6, the limiting form of the curve is described by the warning diagnostic.

Further messages warn you about vertical asymptotes of rational curves. You can use the 'summary' setting of the PRINT option to display the value or values of the explanatory variate for which the fitted curve is infinite. A warning is also printed if an asymptote occurs within the range of the data.

The derivatives of the fitted values with respect to each parameter can be stored in variates using the GRADIENTS parameter of the RKEEP directive. You can use these quantities to assess the relative influence of each observation on a parameter; also you can construct a measure of leverage by summing the gradients for all the parameters.

8.5.4 Modifying models fitted by FITCURVE

You can modify a model fitted by FITCURVE by using the ADD, DROP, or SWITCH directives as for linear models, provided you have given an appropriate TERMS statement before the FITCURVE statement. But the alterations you make must produce a model that would be allowed in the FITCURVE directive: that is, it must contain precisely one variate, either no factor or one factor, and possibly the interaction between the variate and the factor if both are included. The NONLINEAR options of the ADD, DROP, and SWITCH directives correspond to the same option in the FITCURVE directive. Thus you can compare curves between groups of a factor, assessing for example whether they are parallel. You may find convenient here the 'accumulated' setting of the PRINT option of the fitting directives. Here is an example showing such an *analysis of parallelism*.

EXAMPLE 8.5.4

```
  1  "
 -2     Modelling the relationship between dilution and optical density
 -3     for four solutions.
 -4     Data from Bouvier et al. (1985) page 129.
 -5  "
  6  VARIATE [NVALUES=64] Density
  7  READ Density
```

Identifier	Minimum	Mean	Maximum	Values	Missing
Density	0.0350	0.9250	1.9140	64	0

```
 16  FACTOR [LEVELS=4; VALUES=16(1...4)] Solution
 17  VARIATE [VALUES=(30,90,270,810,2430,7290,21870,65610)8] Dilution
 18  VARIATE Log; EXTRA=' dilution'
 19  CALCULATE Log = LOG10(Dilution)
 20  MODEL Density
 21  TERMS Log*Solution
 22  FITCURVE [CURVE=logistic] Log
```

22...

***** Nonlinear regression analysis *****

Response variate: Density

```
      Explanatory: Log dilution
    Fitted Curve: A + C/(1 + EXP(-B*(X - M)))


*** Summary of analysis ***

              d.f.        s.s.          m.s.
Regression      3       34.7306      11.576869
Residual       60        0.2461       0.004102
Total          63       34.9767       0.555186

Change          3      -34.7306     -11.576869

Percentage variance accounted for 99.3

* MESSAGE: The following units have large residuals:
                    52                   -3.22
                    60                   -3.10


*** Estimates of parameters ***

             estimate           s.e.
  B           -2.816           0.139
  M            2.9973          0.0184
  C            1.8633          0.0329
  A            0.0658          0.0184

 23  ADD Solution

23..............................................................................


***** Nonlinear regression analysis *****

Response variate: Density
    Explanatory: Log dilution
 Grouping factor: Solution, constant parameters separate
   Fitted Curve: A + C/(1 + EXP(-B*(X - M)))


*** Summary of analysis ***

              d.f.        s.s.          m.s.
Regression      6       34.8666       5.811096
Residual       57        0.1101       0.001932
Total          63       34.9767       0.555186

Change          3       -0.1360      -0.045324

Percentage variance accounted for 99.7

* MESSAGE: The following units have large residuals:
                    52                   -2.77
                    60                   -2.61


*** Estimates of parameters ***

                       estimate           s.e.
  B                    -2.8157          0.0672
  M                     2.99728         0.00891
  C                     1.86326
  A  Solution 1         0.104972
  A  Solution 2         0.0620968
```

8.5 Standard nonlinear curves

```
  A  Solution 3                         0.0992218
  A  Solution 4                        -0.0108408

     24  & Log.Solution
```

```
24.................................................................
```

***** Nonlinear regression analysis *****

  Response variate: Density
      Explanatory: Log dilution
   Grouping factor: Solution, all linear parameters separate
      Fitted Curve: A + C/(1 + EXP(-B*(X - M)))

*** Summary of analysis ***

	d.f.	s.s.	m.s.
Regression	9	34.8689	3.874328
Residual	54	0.1078	0.001996
Total	63	34.9767	0.555186
Change	3	-0.0024	-0.000791

Percentage variance accounted for 99.6

* MESSAGE: The following units have large residuals:
                       52   -2.80
                       60   -2.64

*** Estimates of parameters ***

```
                            estimate        s.e.
  B                         -2.7764         0.0683
  M                          3.00329        0.00906
  C  Solution 1              1.89126
  A  Solution 1              0.0910467
  C  Solution 2              1.86560
  A  Solution 2              0.0600458
  C  Solution 3              1.87730
  A  Solution 3              0.0917567
  C  Solution 4              1.84565
  A  Solution 4             -0.00366122
```

```
     25  & [PRINT=model,summary,estimates,accumulated; NONLINEAR=separate]
```

```
25.................................................................
```

***** Nonlinear regression analysis *****

  Response variate: Density
      Explanatory: Log dilution
   Grouping factor: Solution, all parameters separate
      Fitted Curve: A + C/(1 + EXP(-B*(X - M)))

*** Summary of analysis ***

	d.f.	s.s.	m.s.
Regression	15	34.95314	2.3302090
Residual	48	0.02358	0.0004913
Total	63	34.97672	0.5551860

```
Change              6      -0.08418    -0.0140307
```

Percentage variance accounted for 99.9

```
* MESSAGE: The following units have large residuals:
                   12                   -2.39
```

*** Accumulated analysis of variance ***

Change	d.f.	s.s.	m.s.	v.r.
+ Log	3	34.7306061	11.5768690	23562.15
+ Solution	3	0.1359714	0.0453238	92.25
+ Log.Solution	3	0.0023730	0.0007910	1.61
+ Separate nonlinear	6	0.0841842	0.0140307	28.56
Residual	48	0.0235840	0.0004913	
Total	63	34.9767189	0.5551860	

*** Estimates of parameters ***

		estimate	s.e.
B	Solution 1	-2.9175	0.0979
M	Solution 1	3.0491	0.0122
C	Solution 1	1.8622	0.0217
A	Solution 1	0.0825	0.0127
B	Solution 2	-2.8285	0.0967
M	Solution 2	2.9924	0.0127
C	Solution 2	1.8572	0.0227
A	Solution 2	0.0693	0.0126
B	Solution 3	-2.8692	0.0968
M	Solution 3	3.0783	0.0125
C	Solution 3	1.8560	0.0220
A	Solution 3	0.0655	0.0131
B	Solution 4	-2.7433	0.0951
M	Solution 4	2.8610	0.0134
C	Solution 4	1.8822	0.0245
A	Solution 4	0.0487	0.0121

8.5.5 Controlling the start of the search with the RCYCLE directive

You can use the RCYCLE directive to supply initial values for the nonlinear parameters: you may want to do that, for example, if you are analysing an experiment on plant growth, and you have good estimates from a non-experimental survey on similar material. Usually, FITCURVE determines a reasonable starting value for each parameter by a short grid search, or some manipulation of the data values: this will not be done if you supply initial values. An example is:

    RCYCLE PARAMETER = Rate; INITIAL = 0.62
    FITCURVE [CURVE = exponential] X

You must give an identifier (here Rate) for each nonlinear parameter in the model to be fitted. Genstat will set it up as a scalar holding the initial value you supplied. It will contain the estimated value of the parameter after the model has been fitted.

If you supply initial values for a logistic curve, you must supply values for all the parameters – nonlinear and linear. In any case, the parameters you supply must be in the order in which Genstat reports them; for example, b, m, c then a for the logistic.

The other parameters of RCYCLE are ignored by FITCURVE: bounds are set up automatically according to the curve chosen and to the way it is parameterized by Genstat (over which you have no control).

You can use the MAXCYCLE option to reset the limit on the number of cycles, but Genstat ignores the METHOD and TOLERANCE options. For all standard curve fitting Genstat uses a modified Newton method (8.6.1).

8.6 General nonlinear regression, and minimizing a function

You can use the methods described in this section to fit any kind of regression. However, check first that the model does not belong to any of the categories we have already looked at, for the appropriate directives are then much more efficient. These categories are linear models, generalized linear models, and the standard curves provided by the FITCURVE directive.

Because the methods described here are very general, they are neither as robust nor as automatic as, for example, the method that is used for fitting linear models.

Optimization is easiest with few parameters, when functions are approximately quadratic, when correlations between parameters are small, and when initial parameter estimates are good.

You can effectively reduce the number of parameters to be optimized by separating linear and nonlinear ones: that is, you can first fit linear parameters, and treat the resulting residual sums of squares as functions of the nonlinear parameters only (8.6.2).

You can often transform parameters to make functions nearly quadratic, especially when the parameters as transformed describe salient features of the data. Problems in using optimization methods are most likely to arise if you neglect to take advantage of the ability to transform parameters. Another source of difficulty is if you try to fit inappropriately many parameters.

You can usually find descriptive statistics based on the data that will provide initial estimates reasonably close to the final parameter estimates. For example, suitably spaced ordinates provide parameters for curve fitting that give much the same likelihood surface whatever curve is being fitted.

For advice on reformulating functions to speed up optimization, see Ross (1970) and Ross (1975). The methods used for nonlinear regression in Genstat are the same as those in MLP, the Maximum Likelihood Program. The MLP Manual (Ross 1987) contains useful advice on alternative ways of specifying models.

8.6.1 Fitting nonlinear models

This subsection describes some preliminary things you must do before fitting a nonlinear model that is not a standard curve. It also describes some aspects of the algorithms that Genstat uses.

Before you fit a nonlinear model, you must use the directives MODEL and RCYCLE to specify the response variate, or a scalar storing the value of a general function; you also use these to specify the nonlinear parameters. After you have model, you can use the directives ADD, DROP, and SWITCH to modify the model, as in the previous sections, and you can use the directives RDISPLAY and RKEEP to display or save the results.

Genstat fits nonlinear regression models by maximizing the likelihood function. The likelihood is usually from a distribution in the exponential family, which you have specified in the DISTRIBUTION option of the MODEL directive. If you want any other form of likelihood, specify how it is to be calculated (8.6.4) and set the FUNCTION option of the MODEL directive to a scalar whose value is assigned by the calculation. You may want to use this device of a non-exponential likelihood to minimize a general function with respect to its parameters; but the number of parameters is limited to six.

Genstat ignores the settings of the LINK and EXPONENT options of the MODEL directive, and you are not allowed to set the GROUPS option; other options and parameters are as for linear regression.

Genstat provides you with two algorithms to optimize the specified function; both work with numerical differences and so do not require you to specify derivatives. The default algorithm is the modified Gauss-Newton method; it takes advantage of the fact that the likelihood function can be expressed as a sum of squares. But you cannot use this method for minimizing a general function as outlined above. The alternative algorithm is the modified Newton method, which you specify by the 'Newton' setting of the METHOD option of the RCYCLE directive. You can use it for any nonlinear model with up to six parameters: thus you cannot use Genstat to minimize general functions with more than six parameters.

You can change the limit on the number of iterations by the MAXCYCLE option of the RCYCLE directive, as for the FITCURVE directive.

You must set the PARAMETER parameter to the identifiers of the nonlinear parameters (scalars) that will be used in calculating the model (8.6.2). There must be at least one nonlinear parameter, and no more than six are allowed for the Newton algorithm. If there are more than six, the Gauss-Newton algorithm will be used without comment.

You can set the LOWER and UPPER parameters to provide fixed bounds for each parameter. By default, the values $\pm 10^9$ are used. You should always set bounds, particularly to avoid such problems as attempting to take the log of a negative number. You can incorporate more general constraints as logical functions within the calculations. For example you can compute an extra term

$$(\text{Constr} > 0) * K * \text{Constr}$$

which imposes a penalty on exceeding the constraint, controlled by setting different values of *K*. Often, the best way to impose a constraint is to reparameterize. For example, if a parameter *p* must be positive, consider replacing *p* by exp(*r*), and allowing *r* to take any value.

The STEPLENGTH parameter can be set to provide initial steplengths for the search. By default the steplength is 0.05 times the initial value of the corresponding parameter, or precisely 1.0 if the initial value is zero. Initial values for each parameter can be specified in the INITIAL parameter. If you set a steplength to zero, Genstat treats the corresponding parameter as being fixed at its initial value. By default, the initial value is taken to be the current value of the scalar, or 1.0 if the value is missing.

If you can calculate a range within which you expect a parameter to lie, then you should choose a steplength that is of the order of one per cent of the length of the range. If steps are too small, numerical differencing may not work; if they are too large, gradients may be unreliable and you will get premature convergence. Genstat tests convergence by the relationship of final adjustments to step lengths.

The more parameters there are to estimate, and the more scattered are the data, the more iterations are required to find the optimum. The maximum number of iterations is by default set to 20, but you can reset this by using the MAXCYCLE option of RCYCLE (8.4.3). However, if convergence fails with a given setting of MAXCYCLE, you should check the data or reparameterize the model before you indiscriminately increase the number of iterations.

When convergence does fail, you get a warning. The only sections of output that are then available are the residual number of d.f., the residual s.s. and m.s., and the estimates of parameters, without standard errors, for the current cycle. You can store a numerical code indicating why convergence failed by using the the EXIT parameter of the RKEEP directive (8.5.3).

For any nonlinear model, you can choose just to evaluate the likelihood for a range of combinations of parameter values, rather than to maximize the likelihood with respect to the parameters. You do this by setting the NGRIDLINES option of FITNONLINEAR (8.6.2). Thus you can study the shape of the likelihood, or indeed of a general function of parameters, perhaps using the CONTOUR directive (7.1.3) or DCONTOUR directive (7.3.3) to produce a picture. The calculated values of the likelihood can be stored in a variate using the GRID parameter of the RKEEP directive (8.1.4).

Now that we have dealt with these preliminaries, we can describe how to fit a nonlinear model by the FITNONLINEAR directive. But first we show how to fit the model that we have already looked at at the beginning of 8.5. Since this is an exponential model, you would of course be much better advised to use the FITCURVE

378   8 *Regression analysis*

directive (as we did earlier). But the comparison helps to elucidate the FITNON-LINEAR directive. A more practical example is given later.

EXAMPLE 8.6.1

```
  1  "
 -2    Exponential model for the response of sugar cane to nitrogen fertilizer.
 -3  "
  4  VARIATE [NVALUES=20] Nitrogen,Yield; DECIMALS=0;*
  5  READ Nitrogen,Yield

     Identifier    Minimum       Mean    Maximum    Values    Missing
       Nitrogen        0.0      100.0      200.0        20          0
          Yield       60.0      144.6      198.0        20          0

 11  "Scale Nitrogen values to avoid numerical problems"
 12  CALCULATE Nitrogen = Nitrogen/100
 13  MODEL Yield
 14  "Fit an exponential response"
 15  RCYCLE Rate; LOWER=0.05; INITIAL=0.5
 16  FITNONLINEAR [PRINT=model,summary,estimates,correlations,fittedvalues; \
 17     CALCULATION=!E(ExpN=Rate**Nitrogen)] ExpN


17............................................................................

***** Nonlinear regression analysis *****

 Response variate: Yield
      Fitted terms: ExpN, Constant


*** Summary of analysis ***

             d.f.          s.s.          m.s.
Regression      2        35046.      17523.18
Residual       17         1637.         96.27
Total          19        36683.       1930.68

Percentage variance accounted for 95.0


*** Estimates of parameters ***

               estimate         s.e.    Correlations
Rate             0.3375       0.0726    1.000
* Linear
ExpN           -131.117
Constant        202.988


*** Fitted values and residuals ***

                                    Standardized
         Unit    Response   Fitted value   residual
            1        60.0          71.9      -0.86
            2        73.0          71.9       0.08
            3        77.0          71.9       0.37
            4        72.0          71.9       0.01
            5       125.0         126.8      -0.13
            6       144.0         126.8       1.24
            7       145.0         126.8       1.31
```

8.6 General nonlinear regression, and minimizing a function

```
         8        116.0        126.8        -0.78
         9        152.0        158.7        -0.49
        10        154.0        158.7        -0.34
        11        160.0        158.7         0.09
        12        141.0        158.7        -1.28
        13        182.0        177.3         0.34
        14        167.0        177.3        -0.74
        15        181.0        177.3         0.27
        16        185.0        177.3         0.56
        17        198.0        188.1         0.72
        18        188.0        188.1         0.00
        19        189.0        188.1         0.07
        20        182.0        188.1        -0.44
Mean              144.6        144.6         0.00
```

8.6.2 Nonlinear regression for models with some linear parameters

FITNONLINEAR
Fits a nonlinear regression model or optimizes a scalar function.

Options

PRINT	= strings	What to print (model, summary, accumulated, estimates, correlations, fittedvalues, monitoring, grid); default m,s,e (default g if NGRIDLINES is set)
CALCULATION	= expression or pointer	Calculation of fitted values or of explanatory variates involving nonlinear parameters; default * (only valid if OWN set)
OWN	= scalar	Option setting for OWN directive if this is to be used rather than CALCULATE; default *
CONSTANT	= string	How to treat constant (estimate, omit); default e
FACTORIAL	= scalar	Limit for expansion of model terms; default 3
POOL	= string	Whether to pool ss in accumulated summary between all terms fitted in a linear model (no, yes); default n
DENOMINATOR	= string	Whether to base ratios in accumulated summary on rms from model with smallest residual ss or smallest residual ms (ss, ms); default s

NOMESSAGE	= strings	Which warning messages to suppress (dispersion, leverage, residual, aliasing, marginality); default *
NGRIDLINES	= scalar	Number of values of each parameter for a grid of function evaluations; default *
SELINEAR	= string	Whether to calculate s.e.s for linear parameters (no, yes); default n
INOWN	= identifiers	Setting to be used for the IN parameter of OWN if used in place of CALCULATE; default *
OUTOWN	= identifiers	Setting to be used for the OUT parameter of OWN if used in place of CALCULATE; default *
Parameters		
	formula	List of explanatory variates and/or one factor to be used in linear regression, within nonlinear optimization

At each stage of the iteration, the FITNONLINEAR directive forms fitted values by carrying out the calculations that you have specified in the CALCULATION option. It then performs a linear regression of the response variate (which you have given in the MODEL statement) on the explanatory variates that you have listed in the parameter. Thus when you are specifying how the calculations are to be done (by CALCULATION), you should use the model parameters that you specified in the latest RCYCLE statement; moreover, you should assign values to at least one of the variates in the parameter of the FITNONLINEAR directive. This separation of linear and nonlinear parameters is available only for the Normal and Poisson likelihoods: in the latter, you are allowed only one variate, and you must omit the constant term; so the only linear parameter will be a scaling parameter.

In the example in 8.6.1, the CALCULATION expression uses the value in the parameter Rate and the values of the variate Nitrogen to define the values of the explanatory variate ExpN. The model that is then fitted is

$$\text{Yield} = A + B * \text{Rate}**\text{Nitrogen}$$

where A and B are determined by linear regression of Yield on ExpN (= $\text{Rate}^{\text{Nitrogen}}$) at each step of the iterative process. You could exclude the constant parameter (A) by setting the option CONSTANT=omit: and you must exclude it for the Poisson likelihood. You are allowed only one explanatory factor for the

8.6 General nonlinear regression, and minimizing a function

Normal likelihood, and no factor at all for the Poisson likelihood. If you do include a factor, a separate constant is fitted for each level. Thus

    FITNONLINEAR ExpN,F

would fit the model

$$\text{Sugar} = A_i + B \times \text{Rate}^{\text{Nitrogen}} \quad i=1 \ldots \text{number of levels of F}$$

The actual constants are fitted, not differences from the first level: you are not allowed to omit the constant term when a factor is included, and if you have given a TERMS statement, the option FULL is ignored.

You can also include variates that are not modified by the calculations. For example

    FITNONLINEAR ExpN,Nitrogen

would fit the model

$$\text{Sugar} = A + B \times \text{Rate}^{\text{Nitrogen}} + C \times \text{Nitrogen}$$

That is, you get the linear plus exponential model which is more readily available with the FITCURVE directive.

If you do not give any variates or factor, then Genstat will not estimate any linear parameters (8.6.3).

The PRINT option is as for the FITCURVE directive except for the 'grid' setting: see below, with the NGRIDLINES option.

You must set one of the CALCULATION and OWN options. The CALCULATION option specifies calculations within an expression, or within a pointer storing a series of expressions. You may have to specify the calculation in several stages – each expression being evaluated at every step of the iteration as if it had been given in the statement

    CALCULATE expression

so you must observe the rules for CALCULATE (5.1.1). An example is

    EXPRESSION Ex[1]; VALUES=!E(Xl,Xr = NORMAL((H + 1, − 1∗X)/SQRT(2∗D∗T))
    & Ex[2]; VALUES=!E(Z = Xl + Xr − 1)
    FITNONLINEAR [CALCULATION=Ex] Z

Here, the CALCULATION option is set to the pointer Ex which points to two expression structures Ex[1] and Ex[2]. The model here is for one-dimensional diffusion.

Alternatively, you can set the OWN option to specify that the calculation is to be done by calling the OWN subroutine. You must provide a version of this subroutine and make it available to Genstat, as specified in 12.2.5. Generally, using OWN is likely to be worthwhile only when calculations are very extensive, or when a particular function is needed often. The setting of the OWN option will be passed to the OWN subroutine in the same way as the setting of the SELECT option of the OWN directive is passed to the OWN subroutine.

The CONSTANT, FACTORIAL, POOL, DENOMINATOR, and NOMESSAGE options are as for the FIT directive.

If you set the NGRIDLINES option to *n*, say (with *n* greater than or equal to 2), then the FITNONLINEAR directive will evaluate the likelihood at a grid of values of the nonlinear parameters, and will not search for an optimum. For each parameter, the distance between the upper and lower bounds (set by the RCYCLE directive) will be divided into $(n-1)$ equal parts, defining a rectangular grid with *n* gridlines in each dimension. The default setting of the PRINT option is 'grid' in this case, and produces a display of the function values. Other settings of the PRINT option are ignored. Using the NGRIDLINES option is illustrated in 8.6.4.

By default, standard errors are calculated only for nonlinear parameters. If you set the SELINEAR option to 'yes', standard errors are also calculated for linear parameters. This involves Genstat increasing the number of dimensions of the search to include the linear parameters – but not till after the solution is found – and estimating the rate of change of the likelihood with respect to all these new dimensions.

The INOWN and OUTOWN options are relevant only when the OWN option is set. Genstat passes their settings to the OWN subroutine in the same way as the settings of the IN and OUT parameters of the OWN directive are passed. The only exception is if a setting is a list, when it is automatically formed into an unnamed pointer. Thus, you can use the OWN calculation to specify only one operation within the OWN subroutine, as opposed to the succession of operations that are carried out when you have given the parameters of the OWN directive several settings (12.2).

8.6.3 Nonlinear regression models with no linear parameters

If a model has no parameters that can be estimated linearly, the fitted values cannot be formed by linear regression on explanatory variates calculated from the nonlinear parameters. Instead, the calculations that you specify in the CALCULATION or OWN option of the FITNONLINEAR directive must form the fitted values explicitly, and must store them in a variate. You must supply the identifier of the variate in advance in the FITTEDVALUES parameter of the MODEL directive; you must not then set the parameter of the FITNONLINEAR directive.

Here is an example.

EXAMPLE 8.6.3a

```
  1  "
 -2     Fitting the Michaelis-Menten equation (the linear-divided-by-linear curve
 -3     of FITCURVE, constrained to go through the origin) with Gamma likelihood.
 -4     The response variable is the velocity of a chemical reaction and the
 -5     explanatory variable is the concentration of catalyst.
 -6  "
  7  VARIATE [VALUES=1,2,3,4,6,8] Catalyst
```

8.6 General nonlinear regression, and minimizing a function

```
   8  & [VALUES=3.5,8.4,9.6,11.2,13.1,18.3] Velocity
   9  "Model is:  Fitted = B/(1-D*X) = P2*X'/(P1+X') where X' = 1/X
 -10  But in these forms the parameters are highly correlated, so reparameterize
 -11  in terms of functions of the fitted values: Q1,Q2 = P2 * 3,6 / (P1 + 3,6)"
  12  EXPRESSION MM; VALUE=!E(Fitted = Q2*Catalyst/(6*(Q2/Q1-1) \
  13                         + Catalyst*(2-Q2/Q1)))
  14  MODEL [DISTRIBUTION=gamma] Velocity; FITTEDVALUES=Fitted
  15  RCYCLE Q1,Q2; INITIAL=11,12
  16  FITNONLINEAR [PRINT=model,summary,estimates,correlation,fittedvalues, \
  17     monitoring; CALCULATION=MM]

17..................................................................

***** Convergence monitoring *****

Cycle Eval Move    Function value    Current parameters
    0    1    0       0.4030414       11.000      12.000
                        Steps          0.55000     0.60000
                        Steps          0.48173     0.68503
    1    4    0       0.04089623       9.8427     14.433
    2    7    0       0.03320438       9.5368     14.797
                        Steps          0.099667    0.20694
    3   10    1       0.03319460       9.5477     14.786
    4   16    6       0.03319414       9.5508     14.793


                        Steps          0.046366    0.094805
    1   22    0       0.03319414       9.5508     14.793


***** Nonlinear regression analysis *****

Response variate: Velocity
     Distribution: Gamma


*** Summary of analysis ***

               d.f.    deviance    mean deviance
Regression       1         *            *
Residual         4      0.06639      0.01660
Total            5         *            *


*** Estimates of parameters ***

          estimate    s.e.    Correlations
Q1          9.551     0.519    1.000
Q2         14.79      1.06     0.543   1.000


*** Fitted values and residuals ***

                                  Standardized
      Unit    Response  Fitted value   residual
         1     3.500      3.951        -0.92
         2     8.400      7.052         1.40
         3     9.600      9.551         0.04
         4    11.200     11.608        -0.28
         5    13.100     14.793        -0.92
         6    18.300     17.146         0.51

Mean          10.683     10.683        -0.03
```

When the fitted values are calculated explicitly, as here, you have available the full range of settings of the DISTRIBUTION option of the MODEL directive, with the exception of the inverse-Normal distribution. Thus, the deviance will be based on the likelihood function of the Normal, Poisson, binomial, gamma, or multinomial distribution, taking account of the settings of the DISPERSION and WEIGHTS options of the MODEL directive. The first four of these distributions were discussed in 8.4.1 and 8.4.2.

The multinomial distribution is used rather differently: it is for fitting distributions (calculating what are often called empirical distribution functions in the statistical literature). The response variate should be a set of counts of observations falling into a series of groups; the fitted values should then be a set of expected counts for the groups, calculated from the distribution being considered. The likelihood is exactly the same as that of the Poisson distribution, given the constraint that $\Sigma f_i = 1$ or $\Sigma f_i = M$, where M is the sum of the counts; but because of the constraint there is one fewer degree of freedom than for the analogous model with the Poisson distribution.

Despite the terminology "multinomial", you should not use this setting of the DISTRIBUTION option to fit models to response variables that take one of a finite set of values for each unit (a generalization of a binomial response variable). You can fit such models using the Poisson likelihood, though you will have to adjust the numbers of degrees of freedom for the constraints.

Here is an example of fitting a distribution to a set of observations produced by the Genstat pseudo-random number generator.

EXAMPLE 8.6.3b

```
  1   "
 -2    Fit a Normal distribution to pseudo-random numbers in the range (0,1)
 -3    transformed with the function NED.
 -4   "
  5   CALCULATE Random = NED(URAND(25384; 50))
  6   "Define bounds to subdivide the observations"
  7   SCALAR  Limit[1...8]; VALUE=-100,-1,-0.6,-0.2,0.2,0.6,1,100
  8   "Form response variate: counts of numbers within specified bounds"
  9   CALCULATE S[1...7] = SUM(Random<=Limit[2...8] .AND. Random>Limit[1...7])
 10   VARIATE [VALUES=S[1...7]] Count
 11   "Set up expression to calculate expected counts for a Normal variable"
 12   & [VALUES=Limit[2...8]] L1
 13   & [VALUES=Limit[1...7]] L2
 14   EXPRESSION [VALUE=P=50*(NORMAL((L1-Mean)/SD)-NORMAL((L2-Mean)/SD))] Normal
 15   MODEL [DISTRIBUTION=multinomial] Count; FITTED=P
 16   RCYCLE Mean,SD; STEPLENGTH=0.02,*; LOWER=*,0.5; INITIAL=0,1
 17   FITNONLINEAR [CALCULATION=Normal]
```

17...

***** Nonlinear regression analysis *****

Response variate: Count

8.6 General nonlinear regression, and minimizing a function

```
    Distribution: Multinomial

*** Summary of analysis ***
    Dispersion parameter is 1
                d.f.       deviance     mean deviance
Regression       1              *             *
Residual         5          1.904         0.3808
Total            6              *             *

*** Estimates of parameters ***
                  estimate          s.e.
Mean               0.068           0.151
SD                 1.024           0.137
```

8.6.4 General nonlinear models

Subsections 8.6.1 to 8.6.3 have dealt with two methods of calculating the likelihood at each step of the iterative search: performing linear regression of the response variate on calculated explanatory variates, and directly comparing the response variate with a calculated fitted-values variate. A third method is to calculate the likelihood explicitly. You can also use this to minimize the value of a function that is not a likelihood at all. Bear in mind that maximizing the likelihood function by the methods described earlier in this chapter is actually achieved by minimizing the deviance, which is minus twice the log likelihood.

You must form the function value explicitly by the calculations in the CALCULATION or OWN option of the FITNONLINEAR directive; and you must store the value in a scalar whose identifier is given in advance in the FUNCTION option of the MODEL directive. Do not set the parameters of the MODEL directive, since otherwise the FUNCTION option will be ignored. Here is an example, in which an awkward test-function is minimized.

EXAMPLE 8.6.4a

```
  1  "
 -2   Finding the minimum of a function of two parameters:
 -3   Rosenbrock's steep-sided valley.
 -4  "
  5  EXPRESSION Rbrock; VALUE=!E(F = 100*(P2-P1*P1)**2+(1-P1)**2)
  6  MODEL [FUNCTION=F]
  7  RCYCLE P1,P2; STEPLENGTH=0.01; INITIAL=-1.2,1
  8  FITNONLINEAR [PRINT=summary,estimates,correlation,monitoring; \
  9     CALCULATION=Rbrock]

9.................................................................

*** Convergence monitoring ***
```

```
Cycle Eval Move    Function value   Current parameters
    0    1    0        24.20000      -1.2000       1.00000
                          Steps     0.0100000     0.0100000
                          Steps     0.0062270     0.016059
    1   10    0         4.730749     -1.1750       1.3800
    2   23    5         4.137648     -1.0187       1.0127
    3   32    0         4.113855     -0.68200      0.35178
    4   41    0         2.596810     -0.61066      0.36781
                          Steps     0.0090182     0.011089
    5   52    5         2.178213     -0.42798      0.14587
    6   61    0         1.666336     -0.25853      0.038123
    7   70    0         1.270496     -0.071433    -0.029901
    8   79    0         0.9110574     0.062527    -0.014035
                          Steps     0.020071      0.0049823
    9   88    0         0.7105070     0.26618      0.029377
   10   97    0         0.4357039     0.34249      0.11148
   11  115    3         0.1248368     0.64669      0.41787
   12  133    3         0.006587805   0.91883      0.84427
                          Steps     0.0073711     0.013566
   13  142    0         0.005520305   0.92570      0.85688
   14  151    0         0.001725509   0.98900      0.97411
                          Steps     0.0017755     0.0035201
   15  160    1         0.000118103   0.98913      0.97838
   16  169    0         0.000001488   0.99938      0.99865
   17  179    6         0.000000377   0.99939      0.99877

                          Steps     0.069988      0.14007
    1  189    0         0.000000377   0.99939      0.99877
```

***** Results of optimization *****

*** Minimum function value: ***

   0.37669E-06

*** Estimates of parameters ***
              sq. root of
    estimate   2nd derivs   Scaled 2nd derivatives
P1   0.999   0.579   1.000
P2   1.00   1.16   0.998   1.000

You will usually want to inspect the shape of the function near the minimum. So next we form a grid of function values using the NGRIDLINES option of FITNON-LINEAR, and display them with the CONTOUR directive (7.1.3). (The same function is illustrated with the DCONTOUR directive in 7.3.3.)

EXAMPLE 8.6.4b

```
  10   "
 -11    Draw a contour map of the function for P1 in (-1.2,1.2), P2 in (0,1.2).
 -12   "
  13    RCYCLE P1,P2; LOWER=-1.2,0; UPPER=1.2,1.2
  14    FITNONLINEAR [NGRIDLINES=6; CALCULATION=Rbrock]
```

14..

8.6 General nonlinear regression, and minimizing a function

```
*** Grid of function values ***

         P1   -1.2000   -0.7200   -0.2400    0.2400    0.7200    1.2000
         P2
      0.0000   212.2000   29.8323    1.8694    0.9094   26.9523  207.4000
      0.2400   148.8400   10.7091    4.8646    3.9046    7.8291  144.0400
      0.4800    97.0000    3.1059   19.3798   18.4198    0.2259   92.2000
      0.7200    56.6800    7.0227   45.4150   44.4550    4.1427   51.8800
      0.9600    27.8800   22.4595   82.9702   82.0102   19.5795   23.0800
      1.2000    10.6000   49.4163  132.0454  131.0854   46.5363    5.8000

 15   RKEEP GRID=Vgrid
 16   MATRIX [ROWS=6; COLUMNS=6] Mgrid; VALUES=Vgrid
 17   CONTOUR [YTITLE='P2'; XTITLE='P1'; INTERVAL=8; \
 18       YLOWER=0; YUPPER=1.2; XLOWER=-1.2; XUPPER=1.2] Mgrid

Contour plot of Mgrid at intervals of     8.000

** Scaled values at grid points **
    1.3250    6.1770   16.5057   16.3857    5.8170    0.7250
    3.4850    2.8074   10.3713   10.2513    2.4474    2.8850
    7.0850    0.8778    5.6769    5.5569    0.5178    6.4850
   12.1250    0.3882    2.4225    2.3025    0.0282   11.5250
   18.6050    1.3386    0.6081    0.4881    0.9786   18.0050
   26.5250    3.7290    0.2337    0.1137    3.3690   25.9250
```

```
            -1.2000   -0.7200   -0.2400    0.2400    0.7200    1.2000
              ,         ,         ,         ,         ,         ,
      1.200-   22   44 6 8 0 2 4 66          66 4 2 0 86   4  22   00-
                222  44 6 8 0 2 4  6666666    4 2 0 8 6 4  22      0
                2222  44 6 8 0 2 44           444  2 0 8 6 4  222
               22222222  4 6  8 0 2     44444    2 0 8 6 4  222
               222222222  4 6 8  0 222     222 0 8 6  4 222           2
                222222222  4 6 8  0     22222    0 8   6 4 222        22
      0.960-   222222222  44 6 88 000        000 8   6 4  22        222-
                2222  2222 4  6 88  000000       8   6 44 22        222
                4  222      22  4  6 888          88  6 44 22        22
                 4  22         22  4  6 888888888    6  4 22        22  4
                 44 22          22  4  6              66  4  2  0   22  4
                  6  4  2         22 44  66666666666    4   22   000  2  4
      0.720-    44 22     0   22 44     666666        4   22    0000   22 4 6-
                 6 4 22    000    22 44              44  22    00000      2  46
                 8 64  2    0000    22   44444444444   22   000000      2 4 68
                  6 4 2    00000    222    44444444    22   0000000  22 46 8
                  08  4 22  000000    222              22    0000000  2 4 68
                   6 4 2  0000000     2222        2222     0000    2 46 80
      0.480-208  4 2 00000000   222222222222     000       00 224 68  -
                  64 2   0000000      222222222      000    00 2 4   802
                 42086 4 2 00000000                 0000      00 2   6
                    64 2 000000000                  0000       0   4  8024
                 64 086  2 0000 00000              000000    00 2 46
                   208  4    000  00000000000000000          00 2   680246
      0.240-86    64 2 000      0000000000000000       000    4           8-
                   42086  2   000     000000000000       00 2  4680246
                     8    4   000      00000000000        000 2   802  0
                   2  6420864 2  000    0000          00      00   46        68
                    08  08  42   0000 0000             0      000 2  68024 0
                     2 64    64 2   00000000           0     000 224     68 4
       0.000-6  0  2086   2   00000000           00 0000   2 468024   2 -
                   ,         ,         ,         ,         ,         ,
```

P2

P1

When you are minimizing a general function, some of the output from FIT-NONLINEAR is irrelevant. Genstat ignores the 'accumulated' and 'fittedvalues' settings, and the 'summary' setting displays only the minimum function value. The 'correlation' setting displays the estimated matrix of second derivatives of the function with respect to the parameters, scaled by the diagonal values. Similarly, in place of the standard errors usually displayed by the 'estimates' setting, you get the square roots of the diagonal values of the second-derivative matrix. These can give a useful indication of the form of the function near the minimum. If the function is a likelihood, you can interpret these as asymptotic standard errors and correlations (not scaled by an estimate of dispersion).

Genstat ignores the CONSTANT option of the FITNONLINEAR directive for general functions, and you must not set the parameter. Similarly, the WEIGHTS and OFFSET options of the MODEL directive are ignored, and the GROUPS option must not be set. The only parameters of the RKEEP directive that you are allowed to use are ESTIMATE, SE, INVERSE, EXIT, GRADIENTS, and GRID. You can get the minimum value of the function in the scalar that you have specified in the FUNCTION option of the MODEL directive.

P.W.L.

9 Analysis of designed experiments

This chapter describes the Genstat directives for estimating effects of treatments and doing an analysis of variance with data from a designed experiment.

In a designed experiment, each treatment is applied to several units, such as plots of land, or animal or human subjects, or samples of material. Usually the treatments are allocated randomly, since the units might not be absolutely identical. This guards against any treatment systematically getting more than its fair share of the best units, which might cause it to appear to be better than the treatments on the less favourable units. It is also one form of justification for the statistical analysis. For a more detailed discussion of why randomization is important, see for example Chapter 5 of Cox (1958).

In the simplest type of investigation, the treatments do not have any particular structure. In a field experiment, for example, they may be several varieties of a crop; in an industrial experiment they could be different types of catalyst. In Genstat you represent treatments like these by a factor. The factor has a level for each treatment; the values of the factor indicate which treatment was applied to each unit.

More complicated are factorial experiments. Here there are several different types of treatment, each represented by a different factor. For example, in an investigation of animal diets, you might wish to vary the amounts both of protein and of carbohydrates; in a fertilizer trial, you might have different levels of both nitrogen and phosphorus. Then the set of treatments is the set of all combinations of the levels of the different factors. Thus if there were a levels of nitrogen and b of phosphorus, there would be $a \times b$ treatments altogether.

The advantage of factorial experiments is that you can look not only at the overall effects of each factor, but also at *interactions* which show how the effects of one factor differ according to other factors (9.1). The overall effects are often called *main effects* (though that does not mean that they have to be the main thing that you are interested in). An interaction would be, for example, nitrogen having a large effect in the absence of phosphorus, but only a small effect in its presence.

You specify which main effects, interactions and other treatment terms are to be included in the model using the TREATMENTSTRUCTURE directive (9.1.1). You can also do more sophisticated modelling of the effects of factors, by partitioning them (and their interactions) into polynomial or other contrasts (9.5): for example, the yield of a crop might increase linearly with the amount of nitrogen.

There can also be structure in the units themselves. In a simple experiment, they are unstructured: that is, they are assumed to come from a single homogeneous population. The treatments can then be allocated to the units at random, without the need to consider any other groupings of the units. This is called a *completely randomized design* (see 9.1). The analysis of experiments where the units do have an underlying structure is described in 9.2. For example, you might expect there to be less variation among animals from the same litter than among different litters. You specify the structure of the units by the BLOCKSTRUCTURE directive; if you omit to do this, Genstat assumes that the units are unstructured.

In an experiment, various measurements will be made to assess how the treatments affect the units. These may be made at the end of the experiment, or while it is still in progress. For example, in a field experiment on potatoes, you might be interested in the yield from each plot, the number of potatoes from each plot, estimates of the percentage areas of potato skin affected by particular diseases, and so on. Analysis of variance allows you to examine only one such measurement at a time. The value measured on each unit (or plot) should be entered into a variate and analysed by the ANOVA directive (9.1.2). After the analysis you can produce further output using the ADISPLAY directive (9.1.3), or you can save some of the quantities that have been calculated during the analysis using the AKEEP directive (9.6).

Sometimes measurements are made before the experiment. For example, the initial blood pressures and other attributes of human subjects might be recorded before the treatments are given. You would want to allow for these baseline readings (or *covariates*) when analysing the effects of the treatments. You specify the variates that are to act as covariates using the COVARIATE directive. By default, the model is assumed to contain no covariates.

The analysis can cope with missing values, either in the variates to be analysed, or in the covariates (9.4). But no factor values should be missing.

In summary, then, the model to be fitted is specified by the BLOCKSTRUCTURE, COVARIATE, and TREATMENTSTRUCTURE directives; the analysis is done by the ANOVA directive; further output can be obtained by the ADISPLAY directive; and information from the analysis can be saved by the AKEEP directive.

There are several other directives that you may find useful. You can use the GET directive (2.3.1) to obtain the current model settings specified by BLOCKSTRUCTURE, COVARIATE, and TREATMENTSTRUCTURE, and you can change them by the SET directive (2.2.1). (But you are not likely to want to do this unless you are writing procedures.) You can use the RESTRICT directive (2.4.1) to restrict the analysis to only a subset of the units. You can specify how many decimal places will be used in the output of tables of means, effects, contrasts and residuals by setting the DECIMALS parameter in the declaration of the variate to be analysed (3.1.2). Moreover, when you are designing an experiment you may wish to use the RANDOMIZE directive (5.3.3) to randomize the allocation of treatments.

The designs that can be analysed by Genstat are said to be *balanced* or, more accurately, to have the property of *first-order balance* defined by Wilkinson (1970) and James and Wilkinson (1971). A brief explanation of the property is given in

9.7, where the method of analysis is explained, but you do not need to understand this in order to use Genstat. Virtually all the standard designs can be analysed, including all the generally-balanced designs of Nelder (1965 a,b). Here are some examples.

(a) All orthogonal designs, whether with a single error term or with several: for example, completely randomized designs, randomized blocks, split plots, Latin and Graeco-Latin squares, split-split plots, and fractional replicates.

(b) All designs with balanced confounding: for example, balanced incomplete blocks, balanced lattices, and Youden squares.

(c) Designs with partial balance, provided the pattern of balance can be specified by pseudo-factors (9.7.3).

Amongst the worked examples available on your computer is a data file showing how to analyse all the worked examples in Cochran and Cox (1957); this should cover most of the designs that you are likely to encounter. Genstat itself detects whether or not your design is balanced, by a process known as the *dummy analysis* (9.7.5). So, if you are unsure about whether or not a particular design can be analysed, try it and see what happens. Unbalanced designs with a single error term can be analysed by the regression directives described in Chapter 8, although the output may be less fully annotated. But you should not try to analyse an unbalanced design without understanding fully what the regression is doing.

9.1 Designs with a single error term

Suppose that you have done an experiment to examine v different treatments, and that the value measured on the jth unit out of r receiving treatment i is y_{ij}. For each treatment i, we suppose that there is an underlying mean value of y that we wish to estimate; we shall write this as m_i. This will not be the value observed because there will be measurement error, there may be uncontrolled differences in the way the different units have been dealt with, and the units themselves may not be uniform. So y_{ij} is assumed to follow the linear model

$$y_{ij} = m_i + \varepsilon_{ij}$$

where ε_{ij}, termed the *residual* for the ijth unit, represents the difference between the true value m_i and the value actually observed. The residuals are assumed to be independently distributed: that is, the size of the residual on one unit is assumed to be unaffected by the residuals on other units. They are also assumed to have a zero mean and a constant variance (so the expected value for the ijth unit is m_i). For some of the properties of analysis of variance, it is necessary to assume also that the residuals each have a Normal distribution.

The process by which values for the parameters m_i are estimated from the observed measurements y_{ij} is known as *least squares*. The estimators $\hat{m}_i$ are chosen to minimize the sum of squares of the estimated residuals:

$$\text{RSS} = \sum_{i=1}^{v} \sum_{j=1}^{r} (y_{ij} - \hat{m}_i)^2$$

You can find details of this process in any standard statistical textbook. For a simple design like this one, the estimate $\hat{m}_i$ of each mean, is simply the average of the values observed on the units with treatment i. However this may not be so in more complicated experiments, for example where there is non-orthogonality (9.7) or where there are covariates (9.3). In such cases Genstat uses the term *mean* to denote the prediction of the mean value for a treatment, rather than its crude average, and we follow the same convention in this chapter.

Analysis of variance also estimates the uncertainty attached to the estimates of the parameters, allowing you to assess whether the treatments genuinely differ in their effects. In simple cases, this involves assessing whether the variation between the units with different treatments is genuinely greater than that between units with the same treatment. To help investigate this, a more common form of the linear model is

$$y_{ij} = \mu + e_i + \varepsilon_{ij}$$

where μ is known as the *grand mean*, and e_i as the *effect* of treatment i. So:

$$m_i = \mu + e_i$$

If the treatments do not differ, the effects (e_i, $i = 1 \ldots v$) will all be zero. To assess this we would fit first a model containing just the grand mean (and residuals), and then a model with the effects as well. The difference between the residual sums of squares of these two models measures whether the treatments differ: this difference is called the sum of squares due to treatments. Conventionally the different sums of squares are presented in a table known as the analysis-of-variance table.

The example below shows the analysis-of-variance table for a rather more complicated experiment, details of which can be found in Snedecor and Cochran (1980, page 305); further output is shown in 9.1.3 and 9.5. The experiment studies the effect of diet on the weight gains of rats. There were six treatments arising from two treatment factors: the source of protein (beef, pork, or cereal), and its amount (high or low). The 60 rats that provided the experimental units were allocated at random into six groups of ten rats, one group for each treatment combination. The model to be fitted in the analysis contains three terms to explain the effects of the treatments: s_i ($i = 1,2,3$) the main effects of the source of protein (beef, pork, or cereal); a_j ($j = 1,2$) the main effects of the amount of protein (high or low); and sa_{ij} the interaction between source and amount of protein.

$$y_{ijk} = \mu + s_i + a_j + sa_{ij} + \varepsilon_{ijk}$$

The parameters a_j make the same contribution to the model irrespective of the source of the protein received by the rat. So they represent the overall effects of the amount of protein. Similarly, the parameters s_i represent the overall effects of the source of protein. If the interaction effects were all zero, we would have a model in which the difference between high and low amounts of protein was the same whatever the source of the protein. Also, the difference between sources of protein would be identical whether at high or low amounts. So the parameters sa_{ij} indicate whether or not these two factors interact: whether we can determine the best source of protein without regard to its amount; likewise whether we can decide the best amount without considering the source. The estimates of the parameters are included in the output under the heading "Tables of effects" (9.1.3).

Genstat prints the analysis-of-variance table in the conventional form, which you can find in statistical textbooks: there is a line for each treatment term, a line for the residual, and a final "Total" line recording the total sum of squares after fitting the grand mean. The first column, "d.f." standing for *degrees of freedom*, records the number of extra independent parameters included when each term is added into the model; thus with the source of protein, there are three parameters (s_1, s_2, s_3) but, since the grand mean μ has already been fitted, they sum to zero and so the degrees of freedom are two. (A full explanation of this too can be found in statistical textbooks.) The second column "s.s." contains the sums of squares. The column "m.s.", standing for *mean square*, has sums of squares divided by numbers of degrees of freedom. You can assess whether a particular treatment term has had an effect by comparing its mean square with the residual mean square: if there has been an effect, then the mean square for the treatment term will be large compared to the residual. The column denoted "v.r." (for *variance ratio*) helps you make these comparisons: it contains the ratio of each treatment mean square to the residual mean square. If the residuals do indeed have independent Normal distributions with zero mean and equal variance, then each such ratio has an F distribution with t and r degrees of freedom, where t is the number of degrees of freedom of the treatment term and r is the number of degrees of freedom of the residual. The corresponding probabilities can be looked up in statistical tables, or you can ask Genstat to calculate them for you, by setting option FPROBABILITY=yes in the ANOVA or ADISPLAY directives. However, you should not interpret these probabilities too rigidly, as the assumptions are rarely more than approximately satisfied; for this reason, Genstat does not print probabilities less than 0.001, but will put "<.001" instead. Also, you should not merely report that a term in an analysis is significant; you should also study its means or its effects to see what their biological (or economic) importance may be, whether their pattern can be explained scientifically, and so on.

EXAMPLE 9.1

```
  1  " 3x2 factorial experiment
 -2      (Snedecor and Cochran 1980, page 305)."
  3  UNITS [NVALUES=60]
```

```
  4   FACTOR [LABELS=!T(beef,cereal,pork); VALUES=(1...3)20] Source
  5   & [LABELS=!T(high,low); VALUES=3(1,2)10] Amount
  6   READ Gain

      Identifier    Minimum      Mean    Maximum     Values    Missing
            Gain      49.00     87.87     120.00         60          0

 17   TREATMENTSTRUCTURE Source*Amount
 18   ANOVA [PRINT=aovtable] Gain

18..........................................................................

***** Analysis of variance *****

Variate: Gain

Source of variation          d.f.        s.s.        m.s.      v.r.
Source                          2       266.5       133.3      0.62
Amount                          1      3168.3      3168.3     14.77
Source.Amount                   2      1178.1       589.1      2.75
Residual                       54     11586.0       214.6
Total                          59     16198.9
```

Before you can do the analysis you must set up factors to define the treatment that was applied to each unit. Here there are two factors, for source and for amount of protein. Also you must form a variate containing the data values y_{ijk} that are to be analysed. The ways in which you can do this (as shown in lines 3 to 6) are described in earlier chapters of this manual.

For the analysis of variance, you must first define the model to be fitted. Here we have a single error term ε_{ijk}: the units have no structure. Consequently you need not give a BLOCKSTRUCTURE statement (9.2.1) but can let it take its default value. If you have already defined some other structure (perhaps for an earlier analysis), you should cancel it by giving either a BLOCKSTRUCTURE statement with a null formula, or else one with a single factor indexing the units (9.2.1). Provided you have no covariates (9.3), the only statement that you need give is TREATMENTSTRUCTURE.

9.1.1 The TREATMENTSTRUCTURE directive

TREATMENTSTRUCTURE
Specifies the treatment terms to be fitted by subsequent ANOVA statements.

No options

Parameter

    formula   Treatment formula, specifies the treatment model terms to be fitted by subsequent ANOVAs

---

9.1 Designs with a single error term

The single unnamed parameter of the TREATMENTSTRUCTURE directive is a formula known as the *treatment formula*. Formulae (1.5.3) are composed of identifier lists and functions, separated by the operators:

    .   +   *   /   //   −   −*   −/

In the formulae for analysis of variance, the identifier lists must only be of factors. Variates and matrices can appear in the functions (to fit polynomials, for example); these are described in 9.5. In this subsection we describe the first four operators, which are those that are used most often. The pseudo-factorial operator //, which occurs only in treatment formulae, is described in 9.7.3. The final three operators are for deletion. Full definitions of all the operators are in 1.5.3.

Genstat expands a formula into a series of model terms, linked by the operator plus (+). Each model term consists of one or more elements, separated from one another by the operator dot (.); in analysis of variance the elements are either factors or functions. You can always specify a formula in this expanded form: the other operators simply provide a more succinct way of writing long formulae. For the formulae defined by TREATMENTSTRUCTURE and by BLOCKSTRUCTURE (9.2.1), this expansion does not take place until the analysis is being done (by ANOVA): TREATMENTSTRUCTURE and BLOCKSTRUCTURE merely store the formulae in their original form. Consequently there are some syntactic errors that will not be found until the ANOVA statement. When Genstat does the expansion, the FACTORIAL option of ANOVA sets a limit on the number of elements in a model term from the treatment formula: any terms with more elements are deleted.

Each model term in the treatment formula corresponds to a treatment term in the linear model. The expanded version of the formula in line 17 of the example is

    Source + Amount + Source.Amount

(So you could have specified this instead of Source*Amount.) Terms with a single factor represent main effects of the factor: for example Source corresponds to the main effects of the source of protein, s_i. Terms with several factors define higher-order effects: for example Source.Amount corresponds to the interaction effects between source and amount of protein, sa_{ij}. However, the meaning of a higher-order term depends on the context: in general, it refers to all those joint effects of the factors in the term that have not been accounted for by preceding terms in the model. So Source.Amount, above, is an interaction because the main effects of source and amount have both been fitted already. But, in the formula

    A + A.B

there are no main effects of B, merely a_i and ab_{ij}, so A.B denotes the fitting of different B effects for each level of A; these are usually called the *B-within-A* effects.

Any redundant terms in a formula are deleted. So, for example,

    A + B + A

becomes

>   A + B

Also, as A.B is defined to include all the joint effects of A and B that are not yet accounted for, the formula

>   A.B + B

becomes just A.B, which already includes the B main effects. Thus the order in which you specify the terms is important.

The operators * and / are termed the crossing and nesting operators respectively. For example,

>   Source * Amount

defines the factors Source and Amount to have a crossed relationship: that is, we wish to examine the effects of each factor individually, and then their interaction. Models containing only crossing are often called factorial models. Another factorial model, but with three factors, is in 9.7.1: the formula is

>   N * K * D

which expands to

>   N + K + D + N.K + N.D + K.D + N.K.D

including not only two-factor interactions, like N.K, but also the three-factor interaction N.K.D. In general, if L and M are two formulae, the definition (1.5.3) is that

>   L*M = L + M + L.M

Nesting (/) occurs most often in block formulae, which are specified by the BLOCKSTRUCTURE directive (9.2). To take an illustration from later in this chapter, for the example analysed in 9.3 the formula is

>   Blocks / Plots

indicating that plots are nested within blocks; so the interest is in block effects and the effects of plots within blocks (see 9.2.1). This is exactly what the operator / provides: the expanded form of the formula is

>   Blocks + Blocks.Plots

The general definition of the slash operator (1.5.3) is that

>   L/M = L + L.M

where L is a model term containing all the factors that occur in L. (The rationale for this is that if M is nested within all the terms in L, it must be nested within all the factors in L.) For example, if you expand the first operator in the formula

>   Blocks/Wplots/Subplots

used to specify a split-plot design (9.2.1), you obtain

>   (Blocks + Blocks.Wplots)/Subplots

9.1 Designs with a single error term

This then expands to

    Blocks + Blocks.Wplots + Blocks.Wplots.Subplots

(which, reassuringly, gives an identical list of terms to those obtained by expanding the second operator before the first operator).

An example of a treatment formula in which there is nesting is the factorial plus added control

    Fumigant/(Dose*Type) = Fumigant + Fumigant.Dose + Fumigant.Type
        + Fumigant.Dose.Type

in which the factorial combinations of dose and type occur within the 'fumigated' level of the factor Fumigant, as explained in 9.3.

The definition of the operator dot (.) with model formulae L and M is that L.M is the sum of all pairwise combinations of a term in L with a term in M. For example

    (A + B.C).(D + E) = A.D + A.E + B.C.D + B.C.E

After expanding the operators dot (.), star (*), and slash (/), Genstat rearranges the list of model terms so that the numbers of factors in the terms are in increasing order. Where several terms contain the same numbers of factors, the terms are put into lexicographical order according to the order in which the factors first appeared in the formula. For example

    (A + C.D + B + A.B) * E

expands to

    A + C.D + B + A.B + E + A.E + C.D.E + B.E + A.B.E

which is reordered to

    A + B + E + A.B + A.E + C.D + B.E + A.B.E + C.D.E

9.1.2 The ANOVA directive

Once you have defined the model, you can analyse the variates containing the data (the *y-variates*) using ANOVA. All the options and parameters are listed here, although some are relevant only to the more complicated designs and analyses described later in this chapter.

ANOVA
Analyses y-variates by analysis of variance according to the model defined by earlier BLOCKSTRUCTURE, COVARIATE, and TREATMENTSTRUCTURE statements.

Options

PRINT	= strings	Output from the analyses of the y-variates, adjusted for any covariates (aovtable, information, covariates,

		effects, residuals, contrasts, means, %cv, missingvalues); default a,i,c,m,mi
UPRINT	= strings	Output from the unadjusted analyses of the y-variates (aovtable, information, effects, residuals, contrasts, means, %cv, missingvalues); default * i.e. no printing
CPRINT	= strings	Output from the analyses of the covariates, if any (aovtable, information, effects, residuals, contrasts, means, %cv, missingvalues); default * i.e. no printing
FACTORIAL	= scalar	Limit on number of factors in a treatment term; default 3
CONTRASTS	= scalar	Limit on the order of a contrast of a treatment term; default 4
DEVIATIONS	= scalar	Limit on the number of factors in a treatment term for the deviations from its fitted contrasts to be retained in the model; default 9
PFACTORIAL	= scalar	Limit on number of factors in printed tables of means or effects; default 9
PCONTRASTS	= scalar	Limit on order of printed contrasts; default 9
PDEVIATIONS	= scalar	Limit on number of factors in a treatment term whose deviations from the fitted contrasts are to be printed; default 9
FPROBABILITY	= string	Printing of probabilities for variance ratios (no, yes); default n
SE	= string	Standard errors to be printed with tables of means, SE = * requests s.e.'s to be omitted (differences, means); default d
TWOLEVEL	= string	Representation of effects in $2**N$ experiments (responses, Yates, effects); default r
DESIGN	= pointer	Stores details of the design for use in

9.1 Designs with a single error term

		subsequent analyses; default *
WEIGHT	= variate	Weights for each unit; default * i.e. all units with weight one
ORTHOGONAL	= string	Whether or not design to be assumed orthogonal (no, yes, compulsory); default n
SEED	= scalar	Seed for random numbers to generate dummy variate for determining the design; default 12345
TOLERANCES	= variate	Tolerances for zero in various contexts; default * i.e. appropriate zero values assumed for the computer concerned
MAXCYCLE	= scalar	maximum number of iterations for estimating missing values; default 20
Parameters		
Y	= variates	Variates to be analysed
RESIDUALS	= variates	Variate to save residuals for each y variate
FITTEDVALUES	= variates	Variate to save fitted values
SAVE	= identifiers	Save details of each analysis for use in subsequent ADISPLAY or AKEEP statements

Before Genstat does any calculations with the y-variates, it does an initial investigation to acquire all the information that it needs for the analysis. Alternatively, you can supply this from an earlier analysis using the DESIGN option.

During this initial investigation Genstat first generates the model, excluding covariates (9.3), by expanding the block and treatment formulae into a list of model terms (9.1.1). For a design with a single error term, you do not have to define the block formula; its use in the definition of more complicated designs is described in 9.2.1. Genstat also finds out whether the treatment formula contains any functions and, if so, forms the contrasts that they define (9.5).

The treatment terms to be included in the model are controlled by the options FACTORIAL, CONTRASTS, and DEVIATIONS. FACTORIAL sets a limit on the number of factors in a treatment term: terms containing more than that number are deleted. CONTRASTS and DEVIATIONS control the inclusion of contrasts, and of deviations from fitted contrasts (9.5). The maximum number of different factors that you can have in the block and treatment formulae is two less than the number of bits that your computer uses to store integers. On most computers this gives you 30 different factors, which should be sufficient for most sensible purposes.

Genstat then checks whether any of the y-variates is restricted (2.4.1). If several variates are restricted, they must all be restricted to the same set of units. Only these units are included in the analysis of each y-variate.

Next Genstat investigates the design: for example, it checks whether each term can be estimated, whether any are non-orthogonal (9.7), which error term is appropriate for each estimated treatment term if the model contains several, and indeed whether the design has the balance required for ANOVA to analyse it. This process, known as the *dummy analysis*, is described in 9.7.5, but you do not have to understand how it works in order to use ANOVA.

Options WEIGHT, ORTHOGONAL, SEED, and TOLERANCES control various aspects of the analysis.

The WEIGHT option allows you to specify a weight for each unit, to define a weighted analysis of variance. You might want to do this if, for example, different parts of the experiment have different variability; each weight would then be proportional to the reciprocal of the expected variance for the corresponding unit. However, unless the weights are fairly systematic, for example to give proportional weighted replication (9.5), the design is unlikely to be balanced.

Genstat has a simplified version of the dummy analysis which you can use to save computing time if all the model terms are orthogonal and if, for every term, all the combinations of its factors were applied to the same number of units (9.7.5). A check is incorporated which will detect non-orthogonality except in particularly complicated designs where terms are aliased. If you set option ORTHOGONAL=yes, Genstat does the simple version unless non-orthogonality is detected, whereupon it gives a warning message and then switches to the full version. The simplified version is done also if ORTHOGONAL=compulsory, but non-orthogonality now causes the analysis to stop altogether, with an error message; this is useful for checking for typing errors in the factor values when you know that the design should otherwise be orthogonal.

Options SEED and TOLERANCES control numerical aspects of the dummy analysis and of the analysis of the y-variates (see 9.7.5).

You can use the DESIGN option to store the details of the model, of the design and of any restrictions of the units, so that Genstat need not recalculate them for future ANOVA statements. The structure in the option is automatically declared as a pointer if you have not declared it already. It points to several other structures which store information about different aspects of the analysis. The only other details that are required for future analyses are the values of the factors in the block and treatment formulae.

If you have not previously declared the design structure, or if it has no values, then the current statement derives and stores the necessary information. If the pointer does already have values, then these are used to do the analysis. In that case, of course, values of the factors in the block and treatment formulae must not have been changed since the design structure was formed. The current settings of options FACTORIAL, CONTRASTS, DEVIATIONS, and WEIGHT are then ignored, as is any change in the restrictions on the y-variates. The DESIGN option is particularly useful with designs where there are many model terms or where there is non-orthogonality, as the dummy analysis may then be time-consuming.

The MAXCYCLE option, which sets a limit on the number of iterations for estimating missing values, is described in 9.4. The other ANOVA options control the printed output, and are described with the ADISPLAY directive (9.1.3). The first parameter of ANOVA, Y, lists the variates whose values are to be analysed. Genstat examines them all and forms a list of units for which any of the y-variates or any covariate (9.3) has a missing value. These units are treated as missing in all the analyses. (This is necessary to avoid having to re-analyse covariates for each y-variate; analysis of covariance is described in 9.3.) However, if your y-variates have different missing units, you may prefer to analyse them with separate ANOVA statements, while saving details of the model and design with the DESIGN option ate for comparing any residual with zero. If the replications or weighted replications were unequal, these would be printed in parallel with the residuals, and the range of standard errors would be printed, the lower value being appropriate for residuals with the maximum replication or weighted replication, and the upper value for those with the minimum replication or weighted replication.

The fitted values from the analysis are defined to be the data values minus the estimated residuals. These too can be saved, using the FITTEDVALUES parameter. In models where there are several error terms, only the final error term is subtracted. If this is not what you want, you can save the other error terms using AKEEP (9.6.1) and subtract them by CALCULATE (5.1).

The last parameter, SAVE, allows you to save the complete details of the analysis in an ANOVA save structure. The ADISPLAY directive lets you use a save structure to produce further output (9.1.3). You can also use it in the AKEEP directive to put quantities calculated from the analysis into data structures which you can then use elsewhere in Genstat (9.6.1). Save structures are pointers, and Genstat declares them automatically. The save structure for the last y-variate analysed is stored automatically, and forms the default for ADISPLAY and AKEEP if you do not provide one explicitly.

Genstat still generates the model and does the dummy analysis even if a y-variate has no values, or if you specify a null entry in the Y list. You then get a *skeleton* analysis-of-variance table, which excludes sums of squares, mean squares and variance ratios; the only other output available is the information summary (9.1.3). You can save a design structure, but no save structure is formed. This is a good way of checking that a design can be analysed, before the experiment is carried out.

9.1.3 The ADISPLAY directive

ADISPLAY
Displays further output from analyses produced by ANOVA.

Options

PRINT	= strings	Output from the analyses of the y-variates, adjusted for any covariates (aovtable, information, covariates, effects, residuals, contrasts, means, %cv, missingvalues); default * i.e. no printing
UPRINT	= strings	Output from the unadjusted analyses of the y-variates (aovtable, information, effects, residuals, contrasts, means, %cv, missingvalues); default * i.e. no printing
CPRINT	= strings	Output from the analyses of the covariates, if any (aovtable, information, effects, residuals, contrasts, means, %cv, missingvalues); default * i.e. no printing
CHANNEL	= scalar	Channel number for output; default * i.e. current output channel
PFACTORIAL	= scalar	Limit on number of factors in printed tables of means or effects; default 9
PCONTRASTS	= scalar	Limit on order of printed contrasts; default 9
PDEVIATIONS	= scalar	Limit on number of factors in a treatment term whose deviations from the fitted contrasts are to be printed; default 9
FPROBABILITY	= string	Printing of probabilities for variance ratios in the aov table (no, yes); default n
SE	= string	Standard errors to be printed with tables of means, SE = * requests s.e.'s to be omitted (differences, means); default d
TWOLEVEL	= string	Representation of effects in 2**N experiments (responses, Yates, effects); default r

Parameters

RESIDUALS	= variates	Variate to save residuals from each analysis
FITTEDVALUES	= variates	Variate to save fitted values
SAVE	= identifiers	Save structure (from ANOVA) storing details of each analysis from which information is to be displayed; if omitted, output is from the most recent ANOVA

The ADISPLAY directive allows you to display further output from one or more analyses of variance, without having to repeat all the calculations. You can store the information from each analysis in a save structure, using ANOVA, and then specify the same structure in the SAVE parameter of ADISPLAY. Several save structures can be listed, corresponding to the analyses of several different variates. They need not all have been produced by the same ANOVA statement nor even be from the same design. Alternatively, if you just want to display output from the last y-variate that was analysed, you need not specify the SAVE parameter in either ANOVA or ADISPLAY: the save structure for the last y-variate analysed is saved automatically, and provides the default for ADISPLAY.

You can obtain variates of residuals and of fitted values, using the RESIDUALS and FITTEDVALUES parameters in the same way as in the ANOVA directive (9.1.2). Note that these two parameters come before SAVE in the parameter list (this is to keep the same ordering of parameters as in ANOVA). So if you wish merely to specify save structures, you must put the equivalent number of null entries into the RESIDUALS list which supplies the list of primary arguments (1.6.2): for example, to print the effects from the save structure Ysave, you must put

    ADISPLAY [PRINT = effects] *; SAVE = Ysave

However, you will usually not need to specify the SAVE parameter, unless you are analysing several variates at a time.

All the options of ADISPLAY also occur with ANOVA, except for CHANNEL which allows you to divert the output to another channel.

The PRINT option selects which components of output are to be displayed. These are all illustrated in this chapter, as indicated in this list.

    aovtable: analysis-of-variance table (9.1, 9.2.1, 9.3, 9.5, 9.7)
    information: information summary, giving details of aliasing and
        non-orthogonality (9.1.3 and 9.7.1)
    covariates: estimates of covariate regression-coefficients (9.3.1)
    effects: tables of estimated treatment parameters (9.1.3)
    residuals: tables of estimated residuals (9.1.3 and 9.2.1)

contrasts: estimated contrasts of treatment effects (9.5)

means: tables of predicted means for treatment terms (9.1.3)

%cv: coefficients of variation and standard errors of individual units (9.1.3 and 9.2)

missingvalues: estimates of missing values (9.4)

The default for PRINT with ADISPLAY is different from that with ANOVA. With ANOVA, the default gives the output that you will require most often from a full analysis: 'aovtable', 'information', 'covariates', 'means', and 'missingvalues'. You are most likely to use ADISPLAY when you are working interactively, to examine one component of output at a time, and it is not obvious that any one component will then be more popular than any other. So the default for ADISPLAY produces no output (that is, PRINT=*). This also means that you do not need to suppress the output explicitly when you are using UPRINT and CPRINT to examine components of output from analysis of covariance (9.3).

The settings 'information', 'covariates', and 'missingvalues' have a slightly different effect with ANOVA than with ADISPLAY. As they are part of the default specified for ANOVA, they will not produce any output unless there is something definite to report. With ADISPLAY you need to request them explicitly, so Genstat will always produce some sort of report. For example, there are no missing values with the variate Gain analysed earlier in this section, there are no covariates, and there is no aliasing or non-orthogonality.

EXAMPLE 9.1.3a

```
  19  ADISPLAY [PRINT=information,covariates,missingvalues]
19.............................................................
***** Information summary *****
All terms orthogonal, none aliased.

***** Covariate regressions *****
No covariates

***** Missing values *****
Variate: Gain
No missing values
```

If you had asked for these three pieces of information by ANOVA, you would not have obtained any output, since there is nothing positive to report.

The other default components produced by ANOVA are the analysis-of-variance table, shown earlier in this section, and the tables of means.

9.1 Designs with a single error term

EXAMPLE 9.1.3b

```
 20  ADISPLAY [PRINT=means]
 20.................................................................
```

***** Tables of means *****

Variate: Gain

Grand mean 87.9

Source	beef	cereal	pork
	89.6	84.9	89.1

Amount	high	low
	95.1	80.6

Source	Amount	high	low
beef		100.0	79.2
cereal		85.9	83.9
pork		99.5	78.7

*** Standard errors of differences of means ***

Table	Source	Amount	Source Amount
rep.	20	30	10
s.e.d.	4.63	3.78	6.55

A table of means is produced for each term in the treatment model. By using the PFACTORIAL option you can exclude tables for terms containing more than a specified number of factors; Genstat does not allow tables to have more than nine factors, so the default value of nine gives all the available tables.

The means are predicted mean values: estimated expected values for each combination of levels in the table, averaged over the levels of other factors. The table for each term is calculated by taking the table of estimated effects for the term and then adding in the estimated effects of all its margins. The grand mean is a margin, as is every term whose factors are a subset of those in the table. For example, the effects of source of protein have only the grand mean as a margin, and so the table of means for Source is calculated by adding the grand mean to each of the Source effects. Source.Amount has three margins; its table of means is formed by adding the grand mean and the main effects of Source and of Amount to the Source.Amount interaction effects. (You can verify this from the tables of effects printed in the next part of the output, below.) An assumption of analysis of variance is that the effects of each error term (or residuals) are independently distributed with zero mean and a common variance (see the initial part of 9.1); so they have predicted values of zero. Consequently, even if a term from the block formula (9.2.1) is a margin of a treatment term, its effects will not be included in the table of means. Similarly, if the deviations from fitted contrasts have been ascribed to error (9.5), these effects are also excluded; the table of means is then said to be *smoothed* (9.5).

In stratified designs (9.2), there may be information on a treatment term in more than one stratum. Genstat uses only the effects from the lowest stratum in which the term is estimated (9.7.1). For some designs you might want to combine all the information about the term. This cannot be done directly by ANOVA, but you can save the necessary information with AKEEP (9.6.1) and then use CALCULATE (5.1). Library procedures are available for some of the more common designs (12.1.1).

Usually this process of prediction produces tables of means that are the same as the averages of the observed values: for example, in the common situation where the design is orthogonal and there are no covariates, the only further requirements for this to happen are that the term for the table must have no block terms as margins nor any of its deviations ascribed to error. In an analysis of covariance, the means are all adjusted to correspond to a common value, namely the grand mean of each covariate (9.3.1). Adjusted means are also produced when there is non-orthogonality: they are adjusted for the effects to which the term or its margins are non-orthogonal (9.7.4).

Below the tables of means, Genstat prints an array of standard errors, usually (as here) one for each table. Unless you request otherwise, each will be an s.e.d. – that is, a standard error for assessing the difference between a pair of means within the table. If you prefer standard errors for the means themselves, set the option SE=means. By putting SE=* you can suppress the standard errors altogether. More than one s.e. or s.e.d. will be given when some of the comparisons between the means in a table have different standard errors, as for example in split-plot designs (9.2.1).

The replication of the means in each table is also printed. In an unweighted analysis of variance, like that above, the replication is the number of units that received each combination of the factors in the table. In a weighted analysis, the weighted replication (wt. rep.) is given: this is the sum of the weights of the units that received each combination. If the replication (or weighted replication) is the same for every combination in the table, it is printed with the standard error; otherwise a table of replications is printed in parallel with the table of means, as illustrated in 9.3.

When the means have different replications, standard errors are presented for three types of comparison: between two means with the minimum replication, between two means with the maximum replication, and between a mean with minimum replication and one with maximum replication. But if, for example, there is only one mean with the minimum replication, the first type of comparison will not arise. If Genstat detects such situations, the appropriate s.e.d. is marked with an X.

EXAMPLE 9.1.3c

```
  21  ADISPLAY [PRINT=effects]
21..................................................................

***** Tables of effects *****

Variate: Gain

Source effects      e.s.e. 3.28    rep. 20

    Source        beef    cereal      pork
                   1.7      -3.0       1.2

Amount response                    -14.5   s.e. 3.78      rep. 30

Source.Amount effects     e.s.e. 4.63    rep. 10

    Source    Amount     high     low
      beef                3.1    -3.1
    cereal               -6.3     6.3
      pork                3.1    -3.1
```

Tables of effects are estimates of treatment parameters in the linear model (9.1). Although effects are less often used than means for summarizing the results of an experiment, they may be useful if you wish to study the model in more detail. The option PFACTORIAL applies to tables of effects in the same way as to tables of means. In this example, there are tables for the Source main effects and the Source.Amount interaction. (The Amount main effects are presented as a *response*, as we explain later.)

Each term is subject to constraints that are generated by the fitting of the terms that come before it in the linear model. The grand mean is fitted first of all. So the sum of the effects, each multiplied by its replication (or weighted replication), is zero within every table. The replication is printed in the header line of the table or, if the replications are unequal, with the table itself. Here the effects within all the tables are equally replicated, and you can check that their sum is zero within each table.

Similarly the table of Source.Amount interaction effects has zero row and column sums because the main effects of Source and Amount have been fitted first.

The header also specifies an e.s.e. or a range of e.s.e.'s for the effects in the table: e.s.e. stands for *effective standard error* – the adjective *effective* denotes that it is appropriate only for comparisons that are unaffected by the constraints within the table. So the e.s.e. for Source is appropriate for obtaining an s.e.d. to assess differences between effects, but not for testing the sum of the effects, nor any individual effect, against zero.

To understand how the e.s.e. arises, we can consider the Source main effects.

(If you do not want to know about this piece of theory, skip this paragraph.) These effects are estimated by

$$s_i = 1/20 \sum_{j=1}^{2} \sum_{k=1}^{10} y_{ijk} - 1/60 \sum_{i=1}^{3} \sum_{j=1}^{2} \sum_{k=1}^{10} y_{ijk}$$

and can be shown to have a variance of $\sigma^2(1/20 - 1/60)$ where 20 is the replication of the Source effects, and 60 is the total number of units. The second term in the formula (which is the estimate of the grand mean) is common to all the estimates, and it is because of this that pairs of effects have a non-zero covariance of $-\sigma^2/60$. The variance of the difference between two effects can be calculated by a familiar formula: it is the sum of the variances of the two effects minus twice their covariance, giving an s.e.d. of $\sqrt{(2\sigma^2/20)}$. However an easier way of deriving this s.e.d. is to notice that, when you subtract one estimate from the other, the second term cancels out to leave the difference between two sums of independent random variables, each with variance $\sigma^2/20$. We can thus refer to each estimated effect as having an effective variance of $\sigma^2/20$ and an effective covariance of zero when calculating the variance of a comparison unaffected by the constraint. The general formula for the e.s.e. is:

e.s.e. = $\sqrt{(\sigma^2/((\text{weighted}) \text{ replication} \times \text{efficiency factor} \times}$

covariance efficiency factor))

The efficiency factor is described in 9.7.1; for an orthogonal term its value is one. Likewise, the covariance efficiency factor is one when there are no covariates (9.3). The variance σ^2 is estimated by the residual mean square of the stratum where the effects are estimated. Strata are explained in 9.2. Here there is only one stratum, so σ^2 is estimated by 214.6 and the e.s.e. is $\sqrt{(214.6/20)}$.

When a factor has only two levels, like Amount above, Genstat prints the difference between the two main effects. This difference is called a *response*. For interaction terms whose factors all have only two levels, there are two forms of response. The choice between them is controlled by the TWOLEVEL option. If you leave the default, TWOLEVEL=response, Genstat calculates the response for an interaction between two factors as the difference between the two main-effect responses, and so on; this is the form described in most textbooks. By putting TWOLEVEL=Yates, you can obtain the form as defined by Yates (1937). Alternatively, put TWOLEVEL=effects if you prefer not to have responses, but to have the effects themselves, as for factors with more than two levels.

EXAMPLE 9.1.3d

```
  22  ADISPLAY [PRINT=residuals]

  22..................................................
```

Tables of residuals *****

Variate: Gain

Units residuals   s.e. 13.90   rep. 1

1	2	3	4	5	6	7
-27.0	12.1	-5.5	10.8	23.1	-29.7	2.0
8	9	10	11	12	13	14
-11.9	-20.5	-3.2	11.1	3.3	18.0	-29.9
15	16	17	18	19	20	21
-3.5	10.8	13.1	-5.7	4.0	25.1	-1.5
22	23	24	25	26	27	28
-15.2	-3.9	7.3	-19.0	9.1	2.5	6.8
29	30	31	32	33	34	35
14.1	2.3	7.0	2.1	2.5	-28.2	-9.9
36	37	38	39	40	41	42
18.3	0.0	-3.9	8.5	-7.2	-9.9	27.3
43	44	45	46	47	48	49
-13.0	-8.9	-8.5	10.8	-16.9	-8.7	17.0
50	51	52	53	54	55	56
0.1	20.5	15.8	5.1	-17.7	11.0	6.1
57	58	59	60			
5.5	-1.2	-25.9	3.3			

Residuals correspond to the error parameters of the linear model (9.1). Here there is a single error term, and thus a single set of residuals. There is no block model (9.2.1) to define factors to index the units of the design, and so each estimated residual is printed with a unit number, under the heading *Units*. The header line shows the replication or weighted replication, and gives a standard error appropriate for comparing any residual with zero. If the replications or weighted replications were unequal, these would be printed in parallel with the residuals, and the range of standard errors would be printed, the lower value being appropriate for residuals with the maximum replication or weighted replication, and the upper value for those with the minimum replication or weighted replication.

EXAMPLE 9.1.3e

```
  23  ADISPLAY [PRINT=%cv]
23..................................................................
```

***** Stratum standard errors and coefficients of variation *****

Variate: Gain

d.f.	s.e.	cv%
54	14.65	16.7

The setting PRINT=%cv displays the residual number of degrees of freedom, the standard error of a single unit of the design, and the *coefficient of variation* (cv%), which is the standard error of a single unit expressed as a percentage of the grand mean. The coefficient of variation is often used as an index of the variability when comparing several experiments on the yields of the same field crop. However, it can be misleading, especially with transformed variables like the logarithm of yield, where the grand mean may even be zero, or with other variables that can take negative values. In designs with several error terms, the same information is presented for each stratum, as shown in 9.2.1. If the units in a stratum have unequal replication or weighted replication, there is no single standard error for a unit; so a missing value is printed instead.

The only component of output that we have not yet illustrated contains the estimates of treatment contrasts, which you can obtain by putting PRINT=contrasts. These are shown in 9.5, together with an explanation of how to control their printing by the options PCONTRASTS and PDEVIATIONS.

With analysis of covariance, you can also print output from the analyses of the covariates and from the analysis of the y-variate ignoring the covariates. This is controlled by options CPRINT and UPRINT respectively, as shown in 9.3.1.

9.2 Designs with several error terms

The units in the designs covered in 9.1 had no structure: they were assumed to be from a single homogeneous population. The randomization was over the design as a whole, without taking account of any groupings of the units, and there was thus a single error term. Often, however, the population of units is not homogeneous. The rats used to study a set of diets might be grouped according to their litter. An agricultural experiment might involve several different fields, or parts of a field, all with different underlying levels of fertility. An industrial experiment might need to be conducted on several different days, with different batches of material. Or you might wish to impose a structure artificially, by trying to form sets of similar units (and perhaps also subsets) with the aim of decreasing the variability of the experiment.

This structure should then be reflected in the way that you do the randomization and apply the treatments. Some examples are described below. Others can be found in textbooks on design of experiments: for example, Cochran and Cox (1957), John and Quenouille (1977), and John (1971).

9.2.1 The BLOCKSTRUCTURE directive

BLOCKSTRUCTURE
Defines the blocking structure of the design and hence the strata and the error terms.

9.2 Designs with several error terms

No options

Parameter

formula	Block model (defines the strata or error terms for subsequent ANOVA statements)

The BLOCKSTRUCTURE directive specifies the underlying (or *blocking*) structure of the design that is to be analysed. Examples of its use are given below and in 9.3 and 9.7. For unstructured designs with a single error term you can omit this directive, as described in 9.1.

In many designs, the units are nested. The simplest is the randomized block design. Here the units are grouped into sets, known as *blocks*, the aim being that units in the same block should be more similar than those in different blocks. The allocation of the treatments is randomized independently within each block. The design thus has two sources of random variation: differences between blocks as a whole, and differences between the units within each block. An example is in 9.3, where the units are plots of land and the blocks are groupings of nearby plots. The block model is

    Blocks/Plots

indicating that the plots are nested within blocks, and thus that there is no special similarity, for example, between the plot numbered 3 in block 1 and plot 3 of the other blocks. The expanded version of the formula is

    Blocks + Blocks.Plots

giving terms for the differences between blocks as a whole, and the differences between the units within each block, as required.

In the simplest form of the randomized block design, there is a single treatment factor, each of whose levels occurs once in every block. More complicated arrangements are possible, but each treatment combination must still occur exactly the same number of times in every block. This means that any differences found between the blocks cannot be caused by differences between treatments. Thus the treatment terms are all estimated between the plots within the blocks. If the blocks have been chosen successfully, the variation within the blocks should be less than that between blocks, and so the treatment estimates will be less variable than if a completely randomized design had been used.

For the example in 9.3, the treatments have the structure

    TREATMENTSTRUCTURE Fumigant/(Dose*Type)

If you look at the first analysis shown in 9.3, which ignores the covariate discussed later in that section, you can see that the analysis of variance is split into two components called *strata*. The Blocks stratum contains the sum of squares between blocks; this all arises from the variability between the blocks. The Blocks.Plots stratum contains the sum of squares for the plots within the blocks; this is partitioned into the sums of squares due to each of the treatment terms, and a residual against which these can be assessed.

Thus, you can deduce the block model from the structure of the units, which should correspond to the way in which the randomization has been done. Genstat expands the block model to form the list of *block* (or *error*) terms, each of which defines a stratum corresponding to one of the sources of variability in the design. Alternatively, if you prefer to deduce the error terms by some other means, as for example if you follow the philosophy of fixed and random effects, you can specify the block model to be the sum of these terms.

In the analysis, Genstat initially partitions the sums of squares according to the block model alone. This gives the total sum of squares for each of the strata. Then it partitions each stratum sum of squares into sums of squares for those treatment terms estimated in that stratum, and a residual which provides an estimate of variability against which these treatment sums of squares should be compared.

In the randomized block design, the treatments are estimated only in the final (bottom) stratum. You would thus get the same sums of squares if you omitted the BLOCKSTRUCTURE statement and put Blocks at the start of the treatment model. In the example, you would put

TREATMENTSTRUCTURE Blocks + Fumigant/(Dose*Type)

The effect would also be the same if you specified this treatment model and retained the block model, because any model term that occurs in both the block and treatment models is deleted from the block model. So Blocks would be deleted and there would then be a single stratum Blocks.Plots. You may prefer this specification as it gives an analysis of variance that looks more conventional. However, the form in the example better reflects the structure of the design, as it correctly identifies Blocks as an error term. It also allows for the possibility of treatments being estimated between blocks, as in the balanced incomplete-block design.

The simplest design in which the treatments are not all estimated in one stratum is the split-plot design. This again is a nested structure. It was originally devised for agricultural experiments where some of the factors can be applied to smaller plots of land than others. But it also occurs in industrial experiments (for example Cox 1958, page 149), in medical experiments (Armitage 1974), and even in the study of cake mixtures (Cochran and Cox 1957, page 299). A well-known example (Yates 1937, page 74; John 1971, page 99) is shown below. There are two treatment factors: three different varieties of oats (line 8), and four levels of nitrogen (line 9). Because of limitations on the machines for sowing seed, different varieties cannot conveniently be applied to plots as small as those that can be used for the different rates of fertilizer. So the design was set up in two stages. First of all, the blocks were each divided into three plots of the size required for the varieties, and the three varieties were randomly allocated to the plots within each block (exactly as in the randomized block design). Then each of these plots, or *whole-plots* as they are usually known, was split into four *sub-plots* (one for each rate of nitrogen), and the allocation of nitrogen was randomized independently within each whole-plot.

To specify the block structure for this design, three factors are required (lines 4 to 6): Blocks to indicate the block (1 to 6) to which each unit belongs, Wplots to indicate the whole-plot (numbered 1 to 3 within each block), and Subplots to identify the sub-plot (numbered 1 to 4 within each whole-plot). You can use the same whole-plot numbers in each block, since the block model (defined below) does not contain any main effect for whole-plots: that is, Genstat will not assume any special similarity between whole-plots with the same numbers. In fact it is best that you do use the same numbering, since otherwise the tables of residuals become very sparse and wasteful of space. In situations like this, it is often convenient to arrange the values of the factors in the block model in a systematic order, for example to reflect positions on the field. This makes patterns in the tables of residuals easier to see. The GENERATE directive (5.5.1) provides a convenient way of specifying their values (line 7).

The design has sub-plots nested within whole-plots, which are themselves nested within the blocks: that is,

    BLOCKSTRUCTURE Blocks/Wplots/Subplots

The block model expands to

    Blocks + Blocks.Wplots + Blocks.Wplots.Subplots

(see 9.1.1 and 1.5.3), giving strata for variation between blocks, between whole-plots within the blocks, and for sub-plots within the whole-plots (within blocks). The treatment model (line 24) specifies terms for the main effects of variety and of nitrogen, and for their interaction (9.1.1).

Just as in the randomized block design, the blocks all contain the same sets of treatments, and so no treatments are estimated in the Blocks stratum. But varieties, which were applied to whole-plots, are estimated in the Blocks.Wplots stratum; in conventional terminology this is called the stratum for whole-plots within blocks. The variance ratio for varieties is calculated by dividing the Variety mean square by the Blocks.Wplots residual mean square. It is easy to see that this is the correct thing to do. When we look to see whether the varieties differ we are really trying to answer the question: "Do the yields from the three sets of whole-plots, on the first of which the variety Victory was grown, on the second Golden rain, and on the third Marvellous, differ by more than the amount that we would expect for any three randomly chosen sets of whole-plots (each set with one whole-plot from every block)?". Technically, variety is said to be *confounded* with whole plots. The terms for Nitrogen, which was applied to sub-plots, and for the Variety-Nitrogen interaction are both estimated in the stratum for sub-plots within whole-plots (Blocks, Wplots, Subplots).

EXAMPLE 9.2.1a

```
  1    " Split-plot design
 -2      (Yates 1937, page 74; also John 1971, page 99)."
  3    UNITS [NVALUES=72]
  4    FACTOR [LEVELS=6] Blocks
  5    & [LEVELS=3] Wplots
  6    & [LEVELS=4] Subplots
  7    GENERATE Blocks,Wplots,Subplots
  8    FACTOR [LABELS=!T(Victory,'Golden rain',Marvellous)] Variety
  9    & [LABELS=!T('0 cwt','0.2 cwt','0.4 cwt','0.6 cwt')] Nitrogen
 10    VARIATE Yield; EXTRA=' of oats'
 11    READ [SERIAL=yes] Nitrogen,Variety,Yield
```

Identifier	Minimum	Mean	Maximum	Values	Missing
Yield	53.0	104.0	174.0	72	0

```
 24    TREATMENTSTRUCTURE Variety*Nitrogen
 25    BLOCKSTRUCTURE Blocks/Wplots/Subplots
 26    ANOVA Yield
```

26..

***** Analysis of variance *****

Variate: Yield of oats

Source of variation	d.f.	s.s.	m.s.	v.r.
Blocks stratum	5	15875.3	3175.1	
Blocks.Wplots stratum				
Variety	2	1786.4	893.2	1.49
Residual	10	6013.3	601.3	
Blocks.Wplots.Subplots stratum				
Nitrogen	3	20020.5	6673.5	37.69
Variety.Nitrogen	6	321.8	53.6	0.30
Residual	45	7968.8	177.1	
Total	71	51985.9		

***** Tables of means *****

Variate: Yield of oats

Grand mean 104.0

Variety	Victory	Golden rain	Marvellous
	97.6	104.5	109.8

Nitrogen	0 cwt	0.2 cwt	0.4 cwt	0.6 cwt
	79.4	98.9	114.2	123.4

9.2 Designs with several error terms

Variety	Nitrogen	0 cwt	0.2 cwt	0.4 cwt	0.6 cwt
Victory		71.5	89.7	110.8	118.5
Golden rain		80.0	98.5	114.7	124.8
Marvellous		86.7	108.5	117.2	126.8

*** Standard errors of differences of means ***

Table	Variety	Nitrogen	Variety Nitrogen
rep.	24	18	6
s.e.d.	7.08	4.44	9.72

Except when comparing means with the same level(s) of
Variety   7.68

The example above shows the default output from ANOVA. Notice that a separate s.e.d. is given for comparisons between means in the variety × nitrogen table when both means are for the same variety. To see why this is necessary, consider how you might calculate the difference between two of the means, using the original data. One way would be to look at each block to find the pairs of sub-plots with these two treatment combinations, and then to calculate the sum of the differences between the values recorded on each pair. If the means are both for the same variety, each pair of sub-plots will be within the same whole-plot; when you take the differences any whole-plot variation then cancels out, to give a smaller s.e.d.

This second section of output shows the tables of residuals and of estimated treatment effects from each stratum, and the coefficients of variation.

EXAMPLE 9.2.1b

```
  27  ADISPLAY [PRINT=effects,residuals,%cv]
```
27..

***** Tables of effects and residuals *****

Variate: Yield of oats

***** Blocks stratum *****

Blocks residuals   s.e. 14.85   rep. 12

Blocks	1	2	3	4	5	6
	31.4	-5.8	3.3	-13.1	-8.1	-7.7

***** Blocks.Wplots stratum *****

Variety effects   e.s.e. 5.01   rep. 24

Variety	Victory	Golden rain	Marvellous
	-6.3	0.5	5.8

Blocks.Wplots residuals   s.e. 9.14   rep. 4

Blocks	Wplots	1	2	3
1		-11.4	14.0	-2.6
2		-9.0	-0.3	9.3

```
          3              5.5       8.2      -13.7
          4            -11.5       4.1        7.4
          5             -9.7      -7.1       16.8
          6             -0.4      -6.5        6.9
```

***** Blocks.Wplots.Subplots stratum *****

Nitrogen effects   e.s.e. 3.14   rep. 18

```
 Nitrogen      0 cwt   0.2 cwt   0.4 cwt   0.6 cwt
               -24.6    -5.1      10.3      19.4
```

Variety.Nitrogen effects   e.s.e. 5.43   rep. 6

```
     Variety  Nitrogen    0 cwt   0.2 cwt   0.4 cwt   0.6 cwt
     Victory              -1.5     -2.9       3.0       1.5
 Golden rain               0.1     -0.9      -0.1       0.9
  Marvellous               1.5      3.8      -2.9      -2.4
```

Blocks.Wplots.Subplots residuals   s.e. 10.52   rep. 1

```
   Blocks   Wplots  Subplots       1        2        3        4
      1       1                   9.2    -19.1     11.5     -1.6
              2                  -5.9     -5.0     10.1      0.8
              3                   8.3    -13.3     17.6    -12.6
      2       1                   1.6     -1.9     -4.7      5.0
              2                   9.6      8.6      5.5    -23.7
              3                   1.0    -19.5     13.8      4.7
      3       1                   0.7      2.6     15.4    -18.7
              2                   5.7      4.0     -7.6     -2.1
              3                  -0.1     -8.1     11.7     -3.5
      4       1                  16.4     -3.3      0.9    -14.0
              2                   9.0    -11.7     10.2     -7.5
              3                   6.0     -5.2      3.1     -3.9
      5       1                 -11.1     -2.3     -7.9     21.2
              2                  16.3    -17.4     11.6    -10.5
              3                   6.1    -11.5     11.8     -6.4
      6       1                  15.3    -10.4      2.6     -7.5
              2                  -6.6    -14.4     23.2     -2.3
              3                  11.1     -8.7      2.6     -5.0
```

***** Stratum standard errors and coefficients of variation *****

Variate: Yield of oats

```
Stratum                    d.f.      s.e.      cv%
Blocks                       5      16.27      15.6
Blocks.Wplots               10      12.26      11.8
Blocks.Wplots.Subplots      45      13.31      12.8
```

There are some designs where the units have a crossed instead of a nested structure. A simple example is the Latin square. This was devised for agricultural experiments to cater for situations where there are fertility trends both along and across the field, but it can be used whenever there are two independent ways of grouping the units: for example time of testing and batch of material, or the litter of the rat and its order by weight within the litter. In field experiments, the plots are arranged in a square, with blocking factors called Rows and Columns. These each have the same number of levels as there are treatments. Values of the single

9.2 Designs with several error terms

treatment factor are arranged so that each level occurs once in each row and once in each column. The block structure has rows crossed with columns: that is,

BLOCKSTRUCTURE Rows*Columns (= Rows + Columns + Rows.Columns)

The treatments are estimated only in the Rows.Columns stratum. Removing variation between rows and between columns should make these estimates less variable. We do not include output from a Latin square, but recommend that you try an example from one of the books listed earlier in this section.

More complicated designs can involve both crossing and nesting, for example:

BLOCKSTRUCTURE Squares/(Rows*Columns)

which is used for replicated Latin squares (John 1971, page 114), quasi-Latin squares (Cochran and Cox 1957, pages 317–324; John and Quenouille 1977, pages 146–152), and lattice squares (Cochran and Cox 1957, pages 483–506; John and Quenouille 1977, page 192). Another example is

BLOCKSTRUCTURE (Rows*Columns)/Subplots

which is for a Latin square with the plots split into sub-plots (Kempthorne 1952, page 378).

If the factors in the block formula do not provide a unique index for every unit of the experiment, the terms in the block model will not account for all the variation. Genstat must then define a final stratum to contain the variation between the sets of units whose levels are the same for each block factor. At the end of the block model, Genstat therefore sets up an extra term containing all the block factors, together with an extra "factor", denoted *units*, which numbers the units within each set. So, for the randomized block design, you could put just

BLOCKSTRUCTURE Blocks

which would then become

BLOCKSTRUCTURE Blocks + Blocks.*units*

likewise, for the split-plot design,

BLOCKSTRUCTURE Blocks/Wplots

would become

BLOCKSTRUCTURE Blocks/Wplots + Blocks.Wplots.*units*

Consequently, if you define no block structure at all, Genstat assumes

BLOCKSTRUCTURE *units*

giving a single source of variation representing random differences between the units; this defines a completely randomized design, as in 9.1. However, you may prefer to define a more meaningful labelling of the units, for example

BLOCKSTRUCTURE Rat

You will also need to do this if you wish to use AKEEP to obtain the residuals as a

9.3 Analysis of covariance

You can do analysis of covariance for any of the designs that can be analysed by ANOVA (9.1). As well as defining the block and treatment models (9.1.1 and 9.2.1), you must also list the covariates, using the COVARIATE directive (9.3.1). Then you can do the analysis by ANOVA (9.1.2), get further output by ADISPLAY (9.1.3), and save information by AKEEP (9.6.1), all exactly as in an ordinary analysis of variance.

The example used in this section illustrates the treatment structure of a factorial arrangement of several types of treatment, as well as a control. This structure of *factorial plus added control* can be useful when you wish to examine several ways of modifying a preparation, and also wish to see what would happen if you applied nothing at all. This experiment was done at Rothamsted in 1935 to study soil fumigants for decreasing the numbers of nematodes (or eelworms as they were then known). Further details are given in Cochran and Cox (1957, pages 45–46), although there the data are analysed untransformed. There were four types of fumigant, each of which was applied in either a single or a double dose. A randomized block design was used, with four blocks of twelve plots. In each block, four plots were untreated (to act as controls), and there was one plot for each dose of each type of fumigant. This first section of output analyses the logarithm of the number of nematode cysts counted in a sample of 400 grammes of soil, taken at the end of the experiment.

EXAMPLE 9.3

```
  1    " Example of a factorial + added control and analysis of covariance
 -2      (Cochran and Cox 1957, page 46).
 -3      A log transformation has been used, and unit 43 has a missing
 -4      value in the y-variate. "
  5    UNITS [NVALUES=48]
  6    FACTOR [LEVELS=4] Blocks
  7    &      [LEVELS=12] Plots
  8    FACTOR [LEVELS=5; LABELS=!T(None,CN,CS,CM,CK)] Type
  9    &      [LEVELS=3; LABELS=!T(None,Single,Double)] Dose
 10    &      [LEVELS=2; LABELS=!T('Not fumigated',Fumigated)] Fumigant
 11    GENERATE Blocks,Plots
 12    READ Dose,Type,Initnem,Finalnem
```

Identifier	Minimum	Mean	Maximum	Values	Missing
Initnem	9.0	128.5	283.0	48	0
Finalnem	80.0	311.7	708.0	48	1

```
 25    CALCULATE Fumigant = NEWLEVELS(Dose; !(1,2,2))
 26    & Initnem,Finalnem = LOG(Initnem,Finalnem)
 27    BLOCKSTRUCTURE Blocks/Plots
 28    TREATMENTSTRUCTURE Fumigant/(Dose*Type)
 29    ANOVA Finalnem
```

9.3 Analysis of covariance

29..

***** Analysis of variance *****

Variate: Finalnem

Source of variation	d.f.(m.v.)	s.s.	m.s.	v.r.
Blocks stratum	3	4.0295	1.3432	
Blocks.Plots stratum				
Fumigant	1	0.6918	0.6918	3.73
Fumigant.Dose	1	0.0650	0.0650	0.35
Fumigant.Type	3	0.6656	0.2219	1.20
Fumigant.Dose.Type	3	0.1212	0.0404	0.22
Residual	35(1)	6.4898	0.1854	
Total	46(1)	11.7582		

***** Tables of means *****

Variate: Finalnem

Grand mean 5.618

Fumigant	Not fumigated	Fumigated
	5.788	5.533
rep.	16	32

Fumigant	Dose	None	Single	Double
Not fumigated		5.788		
Fumigated			5.488	5.578

Fumigant	Type	None	CN	CS	CM	CK
Not fumigated		5.788				
	rep.	16				
Fumigated			5.529	5.370	5.763	5.470
	rep.		8	8	8	8

Fumigant	Dose	Type	None	CN	CS	CM	CK
Not fumigated	None		5.788				
		rep.	16				
Fumigated	Single			5.483	5.280	5.818	5.371
		rep.		4	4	4	4
	Double			5.575	5.461	5.707	5.570
		rep.		4	4	4	4

*** Standard errors of differences of means ***

Table	Fumigant	Fumigant Dose	Fumigant Type	Fumigant Dose Type	
rep.	unequal	16	unequal	unequal	
s.e.d.			0.2153	0.3045	min.rep
	0.1318	0.1522	0.1865	0.2407	max-min
			0.1522X	0.1522X	max.rep

(No comparisons in categories where s.e.d. marked with an X)
(Not adjusted for missing values)

***** Missing values *****

Variate: Finalnem

```
Unit    estimate
 43       5.071
Max. no. iterations 3
```

The block model for this design (line 27) is discussed in 9.2.1. The treatment model requires three factors (lines 8 to 10): Fumigant indicates whether or not the plot has been fumigated with any type of fumigant at all, Type indicates the type of fumigant (if any), and Dose indicates how much was used. If you examine the table of means classified by Fumigant, Dose, and Type, you can see that Dose and Type have a crossed structure within the 'fumigated' level of Fumigant. This suggests a treatment model

    Fumigant/(Dose*Type)

which expands to

    Fumigant + Fumigant.Dose + Fumigant.Type + Fumigant.Dose.Type

As explained in 9.1.1, a term like Fumigant.Dose represents all the joint effects of these two factors, after eliminating any terms that precede it in the model. The main effect Fumigant removes the difference between no fumigant and any positive dose (either single or double). So Fumigant.Dose represents the difference between a single and a double dose. Similarly, Fumigant.Type represents differences between types of fumigant, and Fumigant.Dose.Type represents the interaction between dose and type of fumigant. Notice that one of the units has a missing value; this aspect of the analysis is explained in 9.4.

The numbers of nematodes were also sampled at the start of the experiment, before any treatments were applied. This gives extra information about the plots, which we can incorporate into the analysis by using the original numbers as a covariate. We have transformed the initial numbers to logarithms, in the same way as the final numbers; so the model to be fitted assumes that the final numbers are related to some power of the original numbers.

You can use covariates to incorporate any quantitative information about the units into the model. In field experiments there may often be linear trends in fertility. These can be estimated and removed by fitting a covariate of the position of the plot along the direction of the trend. For a quadratic trend, you would also include a covariate containing the squares of the positions. In experiments on animals, you may wish to use measurements such as the original weight. However, the assumption is always that the y-variate is linearly related to the covariates.

After you have defined variates to contain the measurements that are to act as covariates and done any transformations that may be required, you list them in the COVARIATE directive.

9.3.1 The COVARIATE directive

COVARIATE

Specifies covariates for use in subsequent ANOVA statements.

No options

Parameter

                    variates   Covariates

Covariates are incorporated into the model as terms for a linear regression. Genstat fits the covariates, together with the treatments, in each stratum. This should explain some of the variability of the units in the stratum, and so decrease the stratum residual mean square.

Each treatment combination will have been applied to units whose mean value for each covariate differs from that of other treatment combinations; so even in the absence of any treatment effects, the y-values recorded for the different combinations would not be identical. A further effect of the analysis is to adjust the treatment estimates for the covariates, to correct for this. This adjustment causes some loss of efficiency in the treatment estimation. The remaining efficiency is measured by the *covariance efficiency factor*, shown for each treatment term in the "cov. ef." column of the analysis-of-variance table. The values are in the range zero to one. A value of zero indicates that the treatment contrasts are completely correlated with the covariates: after the covariates have been fitted there is no information left about the treatments. A value of one indicates that the covariates and the treatment term are orthogonal. Usually the values will be around 0.8 to 0.9. A low value should be taken as a warning: either the measurements used as covariates have been affected by the treatments, which can occur when the measurements on covariates are taken after instead of before the experiment (see for example Cochran and Cox 1957, page 90); or the random allocation of treatments has been unfortunate in that some treatments are on units with generally low values of the covariates while others are on generally high ones. The covariance efficiency factor is analogous to the efficiency factor printed for non-orthogonal treatment terms, described in 9.7.1.

For a residual line in the analysis of variance, the value in the "cov. ef." column measures how much the covariates have improved the precision of the experiment. This is calculated by dividing the residual mean square in the adjusted analysis by its value in the unadjusted analysis (which excludes the covariates).

The covariance efficiency factor is used by Genstat in the calculation of standard errors for tables of effects, as shown by the formula in 9.1.3. So, if you want to calculate the net effect of the analysis of covariance on the precision of the estimated effects of a treatment term, you should multiply the covariance efficiency factor of the term by the value printed in the residual line of the stratum where the term is estimated. Where a term has more than one degree of freedom, the adjustment given by the covariance efficiency factor is an average over all the comparisons between the effects of the term. However this adjustment should not differ by much from those required for any particular comparison unless the randomization has been especially unfortunate. For Fumigant in the example, the calculation is 0.99×2.35. So the e.s.e. of the Fumigant effects from the adjusted analysis is less than that from the unadjusted analysis by a factor of $\sqrt{2.3}$.

In the example we have printed tables of means, but no tables of effects. However, since the table of means for Fumigant is calculated merely by adding the grand mean to each entry in its table of effects (9.1.3), the same factor also applies to the s.e.d. of the Fumigant means. For a table of means classified by several factors, Genstat combines the covariance efficiency factors of the effects from which the means are calculated (9.1.3) into a harmonic mean, weighted according to the numbers of degrees of freedom of each term: for example $4/(1/0.99 + 3/0.92)$ for Fumigant.Type.

The adjusted analysis-of-variance table has an extra line in the analysis of each stratum, giving the sum of squares due to the covariates. This is the extra sum of squares that is removed by the covariates after eliminating all that can be ascribed to the treatments. It lets you assess whether there is any evidence that the covariates are required in the model. The line for each treatment term contains the sum of squares eliminating the covariates. It indicates whether there is evidence of any effects of that term, after taking account of the differences in the values of the covariates on the units to which each treatment was applied.

As explained in 9.7.4, when an analysis of variance contains non-orthogonal components, the total sum of squares is given by adding the sum of squares for component 1 ignoring component 2 to that for component 2 eliminating component 1, and so on. Here, however, the sums of squares are for covariates eliminating the treatment terms, and for each treatment term eliminating the covariates. So you will find that the values in the s.s. column of the analysis-of-variance table do not add up to the total.

EXAMPLE 9.3.1

```
 30   COVARIATE Initnem
 31   ANOVA [PRINT=aovtable,covariates,means] Finalnem
```

31...

***** Analysis of variance (adjusted for covariate) *****

Variate: Finalnem
Covariate: Initnem

9.3 Analysis of covariance

```
Source of variation         d.f.(m.v.)      s.s.        m.s.      v.r.   cov.ef.
Blocks stratum
Covariate                       1         3.35292     3.35292    9.91
Residual                        2         0.67657     0.33828            3.97

Blocks.Plots stratum
Fumigant                        1         0.95330     0.95330   12.10    0.99
Fumigant.Dose                   1         0.00020     0.00020    0.00    0.98
Fumigant.Type                   3         1.43634     0.47878    6.07    0.92
Fumigant.Dose.Type              3         0.11913     0.03971    0.50    0.99
Covariate                       1         3.81015     3.81015   48.34
Residual                       34(1)      2.67969     0.07881            2.35

Total                          46(1)     11.75815
```

***** Covariate regressions *****

Variate: Finalnem

```
Covariate                   coefficient       s.e.
Blocks stratum
  Initnem                     0.48            0.153
Blocks.Plots stratum
  Initnem                     0.522           0.0751
```

***** Tables of means (adjusted for covariate) *****

Variate: Finalnem
Covariate: Initnem

Grand mean 5.618

```
  Fumigant  Not fumigated   Fumigated
                5.818         5.518
      rep.       16            32

       Fumigant  Dose    None     Single    Double
  Not fumigated         5.818
      Fumigated                   5.520     5.515

       Fumigant  Type    None      CN        CS        CM        CK
  Not fumigated         5.818
           rep.           16
      Fumigated                   5.783     5.357     5.692     5.239
           rep.                     8         8         8         8

       Fumigant  Dose    Type     None      CN        CS        CM        CK
  Not fumigated  None            5.818
                  rep.             16
      Fumigated  Single                    5.703     5.401     5.768     5.209
                  rep.                       4         4         4         4
                 Double                    5.864     5.313     5.616     5.269
                  rep.                       4         4         4         4
```

*** Standard errors of differences of means ***

```
Table         Fumigant   Fumigant    Fumigant    Fumigant
                         Dose        Type        Dose
                                                 Type
rep.          unequal       16       unequal     unequal
s.e.d.                                0.1449      0.2024     min.rep
               0.0862    0.0999       0.1255      0.1600     max-min
                                      0.1025X     0.1012X    max.rep
```

(No comparisons in categories where s.e.d. marked with an X)
(Not adjusted for missing values)

The method that Genstat uses for analysis of covariance essentially reproduces the method that you would use if you were doing the calculations by hand. First of all, it analyses each covariate according to the block and treatment models. You can print information from these analyses using the CPRINT option of either ANOVA or ADISPLAY. As ADISPLAY (9.1.3) does not constrain you to list save structures that were all produced by the same ANOVA, CPRINT will produce information about the covariate analyses from every save structure that you list; duplicate information will thus be produced if several of the save structures are for analyses involving the same covariates. The output from CPRINT, particularly the analysis-of-variance table, gives you another way of assessing the relationship between treatments and covariates: a large variance ratio for a treatment term in the analysis of one of the covariates would indicate either that the treatment had affected the covariate or that the randomization had been unfortunate (as discussed in the description of cov. ef. above).

Genstat then analyses each y-variate in turn. First of all it does the usual analysis ignoring the covariates. You can control output from this unadjusted analysis by the UPRINT option of ANOVA and ADISPLAY. (So the whole of the output given for the example could have been produced by a single ANOVA statement.) Then the covariates are fitted by linear regression and the full, adjusted, analysis is calculated. Output from the adjusted analysis is controlled by the PRINT option of ANOVA and ADISPLAY. This option has an extra setting, not available for UPRINT and CPRINT: PRINT=covariates prints the regression coefficients of the covariates as estimated in each stratum.

9.4 Missing values

Values from some of the units of an experiment may occasionally fail to be recorded. A laboratory animal may become ill or die during the experiment for reasons unconnected with the treatments. A human subject may withdraw from a clinical trial before it is complete. A plot in a field experiment may become flooded and fail to produce any plants. A value may need to be regarded as missing if a mistake has been made in its recording, or in the way in which the unit was managed during the experiment.

To obtain the exact analysis in such circumstances these units should be excluded, but that would lose the properties such as balance for which the experiment was designed. Consequently techniques have been devised by which missing values are entered for these units, and then estimated during the analysis. The estimates can be printed using the 'missingvalues' setting of the PRINT, CPRINT, or UPRINT options of ANOVA or ADISPLAY:

EXAMPLE 9.4

```
 32  ADISPLAY [PRINT=missingvalues]
 32..........................................................................

***** Missing values (adjusted for covariate) *****

Variate: Finalnem
Covariate: Initnem

 Unit   estimate
   43      5.290

Max. no. iterations 3
```

You can have missing values in the y-variates or the covariates, but not in the block or treatment factors: that is, you should at least know where each missing unit belongs according to the factors of the block model, and what treatments it was scheduled to receive. Genstat regards a unit as missing for all the y-variates listed in an ANOVA statement if it is missing for any one of them, or if it is missing for a covariate. This is because the analysis of covariance requires a missing value in either the y-variate or a covariate to be set missing throughout (Wilkinson 1957); forming the complete list over all the y-variates avoids having to re-analyse the covariates for each y-variate. If you have units where some but not all of the y-variates have missing values, you may prefer to analyse each y-variate separately: for example

```
      FOR Y = Weight,Age,Height
         ANOVA [DESIGN = Dsave] Y
      ENDFOR
```

instead of

```
      ANOVA Weight,Age,Height
```

Use of the DESIGN option (9.1.2) avoids Genstat having to redetermine the structure of the design for each analysis.

Genstat uses the method of Healy and Westmacott (1956). This estimates the missing values by an iterative approach in which they are initially set to the grand mean, then the analysis is repeated with the estimate for each missing unit adjusted each time to set its residual to zero. Genstat also employs the modification discussed by Preece (1971) which over-adjusts each residual to accelerate convergence, but this is discontinued if divergence results instead. Missing cells can occur in higher strata, for example if all the sub-plots in a whole-plot are missing. These missing effects are estimated by a similar iteration of the analysis within the stratum. Likewise missing treatment effects are estimated by minimizing the sum of squares of the treatment term concerned. There is a limit on the number of iterations; by default it is 20, but it can be changed by the MAXCYCLE option of

ANOVA. Genstat decides that the process has converged when the residual sum of squares from the previous iteration exceeds the current residual sum of squares by less than 10^{-5} times the current residual sum of squares. This value of 10^{-5} can be changed using the third value of the variate in the TOLERANCES option of ANOVA. Genstat prints the maximum number of iterations required in any of the strata of the design, along with the estimates of the missing values. Convergence is usually fairly rapid: for the example above, only three iterations were required.

In the analysis of variance, as shown in the example in 9.3, the numbers of degrees of freedom are decreased to take account of the missing units and effects; the number subtracted is shown in brackets. The analysis of variance is only approximate. The residual sums of squares are correct (to within the tolerance of convergence) but the treatment sums of squares will be larger than their correct value. (As a result, the sums of squares in the analysis-of-variance table will no longer sum to the total.) If there are few missing values, this increase is unlikely to be large. The estimated effects and means are correct but the calculation of the standard errors does not take account of the missing units. So some standard errors will be too small. For further details, see for example Cochran and Cox (1957, pages 80–82).

If the model has only one error term, you can obtain the exact analysis using regression (Chapter 8). Alternatively you could use the method of Bartlett (1937), in which a dummy covariate is specified for each missing value with minus one in the missing unit and zero elsewhere. The missing units in the y-variates should be set to zero; the regression coefficients of the covariates then estimate the missing values.

9.5 Contrasts between treatments

Sometimes there may be comparisons between the levels of a treatment factor that you particularly wish to assess. With the three sources of protein in 9.1, you might wish to see whether the animal sources (beef and pork) were uniformly better than the cereal source, or you might suspect that the type of meat made little difference and so wish to compare beef with pork. With factors whose levels represent the application of different amounts of some substance like a fertilizer or a drug, you may wish to model the relationship between the effect and the amount. For example, with the nitrogen fertilizer in 9.2, you might wish to see if the yield of oats increases linearly with the amount of fertilizer; you might also include a quadratic term to check for curvature in the response.

Each of these comparisons can be described by specifying a coefficient for each level of the factor. The estimated value of the contrast is obtained by taking the sum of the coefficients each multiplied by the appropriate effect. For example the contrasts for the source of protein are defined by coefficients:

9.5 Contrasts between treatments

	Source:	beef	cereal	pork
Contrast:	animal versus cereal	0.5	−1.0	0.5
	beef versus pork	1.0	0.0	−1.0

To compare beef with pork you subtract one effect from the other; while for animal versus cereal sources, you subtract the effect of cereal from the mean of the effects of the animal sources.

As illustrated by this example, to represent a comparison between the levels of the factor, the sum of the coefficients must be zero. These two contrasts are also orthogonal: they represent independent comparisons between the effects. This is shown by the fact that the sum of the pairwise products of the coefficients is zero: $0.5 \times 1.0 + (-1.0) \times 0.0 + 0.5 \times (-1.0)$. With polynomial contrasts it is usual to fit orthogonal polynomials, so that the quadratic term represents the effect of adding a quadratic term to a linear polynomial, the cubic represents the effect of adding a cubic term to a quadratic polynomial, and so on (see for example: John 1971, page 50; John and Quenouille 1977, pages 33–36). The coefficients of the orthogonal polynomials to examine the linear and quadratic effects of the Nitrogen factor in 9.2 are

	Nitrogen:	0.0	0.2	0.4	0.6
Contrast:	linear	−0.3	−0.1	0.1	0.3
	quadratic	0.4	−0.4	−0.4	0.4

To examine contrasts like these, you put a function of the factor into the treatment formula, instead of the factor itself. The simplest function, POL, is for fitting polynomial contrasts. It has three arguments: the first specifies the factor, the second is a number or a scalar giving the order of polynomial to be fitted (1 for linear, 2 for quadratic, 3 for cubic, and 4 for quartic), and the third is a variate specifying a numerical value for each level of the factor. Genstat calculates the orthogonal polynomials for you. In the Nitrogen example, the levels are equally spaced and in ascending order of magnitude, but this need not be so. You can omit the third argument if the levels already declared with the factor are suitable. For Nitrogen, the declaration (line 9 in the output shown in 9.2.1) specified only labels, and so the levels are the defaults 1 to 4. The variate Nitlev is defined to supply the correct values (line 28).

EXAMPLE 9.5a

```
  28  VARIATE [VALUES=0,0.2,0.4,0.6] Nitlev
  29  TREATMENTSTRUCTURE POL(Nitrogen; 2; Nitlev) * Variety
  30  ANOVA [PRINT=aov] Yield

30......................................................................

***** Analysis of variance *****

Variate: Yield of oats
```

Source of variation	d.f.	s.s.	m.s.	v.r.
Blocks stratum	5	15875.3	3175.1	
Blocks.Wplots stratum				
Variety	2	1786.4	893.2	1.49
Residual	10	6013.3	601.3	
Blocks.Wplots.Subplots stratum				
Nitrogen	3	20020.5	6673.5	37.69
Lin	1	19536.4	19536.4	110.32
Quad	1	480.5	480.5	2.71
Deviations	1	3.6	3.6	0.02
Nitrogen.Variety	6	321.8	53.6	0.30
Lin.Variety	2	168.4	84.2	0.48
Quad.Variety	2	11.1	5.5	0.03
Deviations	2	142.3	71.2	0.40
Residual	45	7968.8	177.1	
Total	71	51985.9		

In the analysis of variance, the sum of squares for Nitrogen is partitioned into the amount that can be explained by a linear relationship of the yields with nitrogen (the line marked Lin), the extra amount that can be explained if the relationship is quadratic (the line Quad), and the amount represented by deviations from a quadratic polynomial. A cubic term would be labelled as "Cub", and a quartic as "Quart". You are not allowed to fit more than fourth-order polynomials.

The interaction of nitrogen and variety is also partitioned: Lin.Variety lets you assess the effect of fitting three different linear relationships, one for each variety, instead of a single overall linear contrast; Quad.Variety represents three different quadratic contrasts; and Deviations represents deviations from these three quadratic polynomials. You can print the estimated values of the contrasts by putting PRINT=contrasts in either ANOVA or ADISPLAY.

EXAMPLE 9.5b

```
  31  ADISPLAY [PRINT=contrasts]
31..............................................................

***** Tables of contrasts *****

Variate: Yield of oats

***** Blocks.Wplots.Subplots stratum *****

*** Nitrogen contrasts ***

Lin         73.7    s.e.  7.01    ss.div.  3.60

Quad       -65.     s.e. 39.2     ss.div.  0.115

Deviations   e.s.e.  3.14    ss.div. 18.0

   Nitrogen    0 cwt   0.2 cwt   0.4 cwt   0.6 cwt
                 0.1      -0.3       0.3      -0.1
```

9.5 Contrasts between treatments

```
*** Nitrogen.Variety contrasts ***
```

Lin.Variety   e.s.e. 12.1   ss.div. 1.20

Variety	Victory	Golden rain	Marvellous
	7.	2.	-9.

Quad.Variety   e.s.e. 67.9   ss.div. 0.0384

Variety	Victory	Golden rain	Marvellous
	-1.	12.	-11.

Deviations   e.s.e. 5.43   ss.div. 6.00

Nitrogen	Variety	Victory	Golden rain	Marvellous
0 cwt		0.7	0.1	-0.8
0.2 cwt		-2.2	-0.3	2.4
0.4 cwt		2.2	0.2	-2.4
0.6 cwt		-0.7	-0.1	0.8

The table of estimated contrasts for Quad.Variety, for example, gives the differences between the overall contrast of −65 for Quad and the contrasts fitted for the three varieties separately. So the estimated contrast for Golden rain is −65 + 12 = −53. The accompanying "ss. div" value is analagous to the replication in a table of effects: it is the divisor used in calculating the estimated values of the contrasts. This is useful mainly where there is a range of e.s.e.'s for a table of contrasts: the contrasts with the smallest values of the ss. div. are those with the largest e.s.e., and vice versa. The ss. div. of each estimated contrast is the sum of squares of the coefficients of the orthogonal polynomial (or other contrast) used to calculate it, weighted according to the replication (or weighted replication in a weighted analysis of variance). The formula for the e.s.e. is similar to that for tables of effects (9.1.3):

$$\text{e.s.e.} = \sqrt{(\sigma^2 / (\text{ ss. div. } \times \text{ efficiency factor } \times \text{ covariance efficiency factor }))}$$

The variance σ^2 is estimated from the residual mean square of the stratum (9.2) where the contrasts are estimated. The efficiency factor (9.7.1) has the value one for terms that are orthogonal, like those in this design. The covariance efficiency factor (9.3.1) equals one when there are no covariates.

To define your own comparisons, you can use the function REG. The first two arguments are the same as those for POL: the first specifies the factor; the second is a number or scalar giving the number of contrasts to be fitted, which must be in the range 1 to 7. The third argument is a matrix, with a column for each level of the factor. Each row of the matrix specifies the coefficients of one of the contrasts. Genstat orthogonalizes these; so the sum of squares, and the estimate, for the second contrast represent the improvement from fitting the second contrast after the first has already been fitted, and so on. If you use a text to label the rows of the matrix (3.4.1), Genstat will use it to annotate the output. Otherwise the contrasts are labelled Reg1 to Reg7.

The next example shows how to extend the analysis of the example in 9.1, to examine the contrasts between the sources of protein.

EXAMPLE 9.5c

```
 24  MATRIX [ROWS=!T('animal vs cereal','beef vs pork'); COLUMNS=3; \
 25    VALUES=0.5,-1,0.5,1,0,-1] Contrasts
 26  TREATMENTSTRUCTURE REG(Source; 2; Contrasts) * Amount
 27  ANOVA [PRINT=aov,contrasts] Gain

27.........................................................................

***** Analysis of variance *****

Variate: Gain

Source of variation         d.f.      s.s.       m.s.     v.r.
Source                         2     266.5      133.3     0.62
  animal vs cereal             1     264.0      264.0     1.23
  beef vs pork                 1       2.5        2.5     0.01
Amount                         1    3168.3     3168.3    14.77
Source.Amount                  2    1178.1      589.1     2.75
  animal vs cereal.Amount      1    1178.1     1178.1     5.49
  beef vs pork.Amount          1       0.0        0.0     0.00
Residual                      54   11586.0      214.6
Total                         59   16198.9


***** Tables of contrasts *****

Variate: Gain

*** Source contrasts ***

animal vs cereal
      3.0    s.e. 2.67     ss.div. 30.0

beef vs pork
      0.3    s.e. 2.32     ss.div. 40.0

*** Source.Amount contrasts ***

animal vs cereal.Amount   e.s.e. 3.78   ss.div. 15.0

    Amount       high       low
                  6.3       -6.3

beef vs pork.Amount   e.s.e. 3.28   ss.div. 20.0

    Amount       high       low
                  0.0        0.0
```

The main effect and interaction are again partitioned to examine the contrasts of interest. The interaction term "beef vs pork.Amount", for example, allows you to examine whether there is any evidence that the difference between beef and pork varies according to the amount of protein fed to the rat.

Where a term has two or more factors partitioned into contrasts, Genstat will also fit interactions between the contrasts. For example Lin.Lin looks at the linear

9.5 Contrasts between treatments

change in the linear component of each factor with the other. With two REG functions, terms like Reg1.Reg1 or Reg2.Reg1 will appear whose interpretation will depend on exactly what comparisons you have defined. If the partitioning of a factor has a component for deviations, there will also be terms like Dev.Lin, which represents the interaction between the deviations component of the first factor and the linear part of the second factor. You can suppress the fitting of these interactions by using the function POLND instead of POL, or REGND instead of REG. For example, putting POLND(A; 1) instead of POL(A; 1) ensures that no interactions will be fitted between other contrasts and the Dev component of A.

The CONTRASTS option in the ANOVA directive (9.1.2) places a limit on the order of contrast to be fitted. For a term involving a single factor, the orders of successive terms run from one upwards, with the deviations term (if any) numbered highest. So for Nitrogen in the example above, the orders are Lin 1, Quad 2, and Deviations 3; while for Source they are "animal vs cereal" 1, "beef vs pork" 2. In interactions between contrasts, the order is the sum of the orders of the component parts, so Lin.Lin has order 2, Quad.Lin has order 3, Reg1.Quad has order 3, Reg1.Reg3 has order 4, and so on. Where the component is a factor, it contributes one to the sum, so Lin.Variety has order 2. The default value for CONTRASTS is 4.

Option PCONTRASTS sets a limit on the order of the contrasts that are printed by either ANOVA or ADISPLAY (9.1.3); its default value is 9.

If your design has few or no degrees of freedom for the residual, you may wish to regard the deviations from some of the fitted contrasts as error components, and assign them to the residual of the stratum where they occur. You can do this by the DEVIATIONS option of ANOVA (9.1.2); its value sets a limit on the number of factors in the terms whose deviations are to be retained in the model. For example, by putting DEVIATIONS = 1, the deviations from the contrasts fitted to all terms except main effects will be assigned to error. The option PDEVIATIONS in ANOVA or ADISPLAY (9.1.3) similarly controls the printing of deviations: to put PDEVIATIONS = 0, for example, would ensure that no deviations are printed. When deviations have been assigned to error, they will not be included in the calculation of tables of means (9.1.3), which will then be labelled "smoothed". However the associated standard errors of the means are not adjusted for the smoothing.

There are limitations on the models and designs for which Genstat can fit contrasts. In a factorial model, each interaction that is partitioned into contrasts must have equal or proportional replication (or proportional weighted replication in a weighted analysis of variance). Otherwise Genstat gives an error. Here is an example of proportional replication for two factors A and B, giving the numbers of replications for each combination of their levels.

	B:	1	2	3	Total over B
A:					
	1	4	8	12	24
	2	2	4	6	12
Total over A:		6	12	18	36

The fraction of the replication in each cell is the product of the fractions in the marginal total cells: for example the cell for level 1 of A and level 3 of B has 12/36 (= 1/3) of the total replication; the product of the marginal totals for these levels is also 1/3, being 24/36 × 18/36.

An exception to this rule occurs in nested models like the factorial with added control which we discussed in 9.3.1. The table below shows what the replication of the factors Fumigant, Dose, and Type would be if, for illustration, there were also a triple level of dose.

Fumigant:	not fumigated					fumigated				
Type:	none	CN	CS	CM	CK	none	CN	CS	CM	CK
Dose:										
none	16	–	–	–	–	–	–	–	–	–
single	–	–	–	–	–	–	4	4	4	4
double	–	–	–	–	–	–	4	4	4	4
triple	–	–	–	–	–	–	4	4	4	4

The treatment model has

Fumigant/(Dose*Type)

= Fumigant + Fumigant.Dose + Fumigant.Type + Fumigant.Dose.Type

None of the higher-order terms (such as Fumigant.Dose) has either equal or proportional replication. However, within the 'fumigated' level of Fumigant, there is equal replication. So Genstat can fit any contrast of the nested factors (Type and Dose) provided the level 'none' is excluded. For example, you could estimate linear and quadratic contrasts of Dose using only the non-zero doses by:

MATRIX [ROWS=4; COLUMNS=2; VALUES= 0, −1, 0, 1\
 0, 1, −2, 1] Quadcon
TREATMENTSTRUCTURE Fumigant / (REG(Dose; 2; Quadcon) * Type)

But the rows of Quadcon must be specified in orthogonal form. Otherwise the automatic orthogonalization, using the replications of Dose, would produce contrasts involving 'none'.

A further limitation is that contrasts cannot be fitted to terms that involve pseudo-factors (9.7.3). In such situations, the specification of the contrasts is ignored by Genstat.

In nested models, no coherent meaning can be given to contrasts between levels of one of the nested factors if the factor within which it is nested is also partitioned into contrasts. So, for example, the specification

POL(A; 1) / POL(B; 2)

would generate an error.

Finally there is the limitation that no model term that is to be partitioned into contrasts can contain more factors than one third of the number of bits in the integers on your computer. On most computers, integers contain more than 30 bits, so this limit should not be restrictive.

9.6 Saving information from an analysis of variance

Most of the quantities calculated during an analysis of variance can be saved in data structures within Genstat. This allows you to write analyses where the analysis of variance itself is only a component part. One example is the multivariate analysis of variance; further examples are in the procedure library (12.1.1). Alternatively, you may wish to save components of the output (such as tables of means) for plotting, or for printing in the form required for a publication.

You can save variates containing residuals for the final error term of the model, using the RESIDUALS parameter of either ANOVA or ADISPLAY (9.1.2). The FITTEDVALUES parameter of these two directives similarly allows you to save the fitted values. Other components of the output can be saved using AKEEP.

9.6.1 The AKEEP directive

AKEEP
Copies information from an ANOVA analysis into Genstat data structures.

Options

FACTORIAL	= scalar	Limit on number of factors in a model term; default 3
STRATUM	= formula	Model term of the lowest stratum to be searched for effects; default * implies the lowest stratum
SUPPRESSHIGHER	= string	Whether to suppress the searching of higher strata if a term is not found in STRATUM (no, yes); default n
SAVE	= identifier	SAVE structure (from ANOVA) storing details of the analysis; default * gives that from the most recent ANOVA

Parameters

TERMS	= formula	Model terms for which information is required
MEANS	= tables	Table to store means for each term (treatment terms only)
EFFECTS	= tables	Table to store effects (treatment terms only)
PARTIALEFFECTS	= tables	Table to store partial effects (treatment terms only)
REPLICATIONS	= tables	Table to store replications
RESIDUALS	= tables	Table to store residuals (block terms only)
DF	= scalars	Number of degrees of freedom for each term
SS	= scalars	Sum of squares for each term
EFFICIENCY	= scalars	Efficiency factor for each term
VARIANCE	= scalars	Unit variance for the effects of each term
CREGRESSION	= variates	Estimated regression coefficients for the covariates in the specified stratum
CSSP	= symmetric matrices	Covariate sums of squares and products in the specified stratum

AKEEP allows you to copy components of the output from an analysis of variance into standard Genstat data structures. You can save the information from the analysis in a save structure, using the SAVE option of ANOVA (9.1.2) and then specify the same structure in the SAVE option of AKEEP. Alternatively, Genstat automatically stores the save structure from the last y-variate that has been analysed, and this is used as a default by AKEEP if you do not specify a save structure explicitly.

With the TERMS parameter you specify a model formula, which Genstat expands to form the series of model terms about which you wish to save information. As in ANOVA (9.1.2), option FACTORIAL sets a limit on the number of factors in each term. Any term containing more than that limit is deleted.

The subsequent parameters allow you to specify identifiers of data structures to store various components of information for each of the terms that you have specified. If there are components that are not required for some of the terms, insert a missing identifier (*) in the list. For example

AKEEP Source + Amount + Source.Amount; MEANS = *,*,Meangain;\
  SS = Ssource,Samount,Ssbya; VARIANCE = Vsource,*,*

sets up a table Meangain containing the Source by Amount table of means; it

9.6 Saving information from an analysis of variance

forms scalars Ssource, Samount, and Ssbya to hold the sums of squares for Source, Amount, and Source.Amount respectively, and scalar Vsource to store the unit variance for the effects of Source.

The structures to hold the information are defined automatically, so you need not declare them in advance. If you have declared any of the tables already, its classification set will be redefined, if necessary, to match the factors in the table that you wish to store. Thus Meangain here would be redefined to be classified by the factors Source and Amount, if it had previously been declared with some other set of classifying factors. Sizes of variates and symmetric matrices will also be redefined if necessary.

Most of the components are self-explanatory. Tables of means and effects are described in 9.1.2; these are relevant only for treatment terms. Partial effects (which are also available only for treatment terms) differ from the usual effects presented by Genstat, only when there is non-orthogonality. The usual effects of a treatment term are estimated after eliminating the terms that precede it in the model (9.1.1), whereas the partial effects are those that would be estimated after eliminating the subsequent treatment terms as well (9.7.4). Replication tables are described in 9.1.3 and appear in the example in 9.3. The replications will be arranged in a table, even if all the values are identical. Tables of residuals, available for block terms, are illustrated in 9.1.3 and 9.2.1.

Four components can be saved in scalars: sums of squares (9.1), numbers of degrees of freedom (9.1), efficiency factors (9.7.1), and unit variances. The unit variance of a treatment term is the residual mean square of the stratum where the term is estimated, divided by its efficiency factor and covariance efficiency factor. Thus you can calculate the estimated variance of any of the effects of the term by dividing its unit variance by the replication of the effect (9.1.3).

The last two parameters allow you to save information about the covariates (9.3). To save the regression coefficients estimated in a particular stratum, specify the model term of the stratum with the TERMS parameter and a variate with the CREGRESSION parameter. Genstat defines the variate to have a length equal to the number of covariates, and stores the estimated regression coefficients of the covariates in the order in which they were listed in the COVARIATE statement (9.3.1). For the example in 9.3.1, you could put

   AKEEP Blocks.Plots; CREGRESSION = B

to save the regression coefficient estimated for the covariate in the Blocks.Plots stratum; B will be declared implicitly as a variate of length one, as there was only one covariate. The CSSP parameter allows you to obtain sums of squares and products between the covariates for the specified model term. These are arranged in a symmetric matrix. The value in row i on the diagonal is the sum of squares for the term in the analysis of variance that has as its y-variate the ith covariate listed in the COVARIATE statement. The value in row i and column j is the cross-product between the effects estimated for the term in the analysis of variance of covariate i and those estimated for the same term in the analysis of covariate j.

In designs where there is partial confounding, and treatment terms are estimated in more than one stratum (9.7.1), options STRATUM and SUPPRESSHIGHER allow you to specify the strata from which the information is to be taken. This is relevant to tables of effects and partial effects, sums of squares, efficiency factors, unit variances, and sums of squares and products between covariates. By default, Genstat searches all the strata, and takes the information from the lowest of the strata where the term is estimated. If you set the STRATUM option, only strata down to the specified stratum are searched. By setting SUPPRESSHIGHER=yes, you can restrict the search to only that stratum. For the example in 9.7.1,

    AKEEP [STRATUM=Blocks] K.D; EFFECTS=EffKD; EFFICIENCY=EfacKD

would take the effects estimated for K.D in the Blocks stratum, and put them into the table EffKD, and it would put their efficiency factor into the scalar EfacKD.

You cannot save tables of means if you have excluded any stratum from the search. Likewise, tables of residuals and residual sums of squares cannot be saved for any of the excluded strata. If a term is not estimated in any of the strata that are searched, the corresponding data structures are filled with missing values.

If you want to save information about the final stratum in the analysis, you must have specified this stratum explicitly in the block formula, and not have allowed Genstat to specify it for you using the "factor" *units* (9.2.1). Otherwise you will be unable to specify the model term for the stratum. In the split-plot example in 9.2.1 the final stratum is for Blocks.Wplots.Subplots, so its residual sum of squares can be saved in scalar R by

    AKEEP Blocks.Wplots.Subplots; SS=R

However, to be able to save the residual sum of squares for the example in 9.1, the block structure should have been defined explicitly, for example by defining a factor Rat with a different level on each unit and then putting

    BLOCKSTRUCTURE Rat

You could then save the residual sum of squares by

    AKEEP Rat; SS=R

9.7 Non-orthogonality and balance

So far, all the examples in this chapter have all been orthogonal. Each treatment term has been estimated in only one stratum. Any confounding between block and treatment terms has been complete: for example, in the split-plot design in 9.2.1, differences between varieties were completely confounded with whole-plots, and so were estimated only in that stratum.

The ANOVA directive can also analyse designs where there is partial confounding

9.7 Non-orthogonality and balance

or where there is non-orthogonality, provided there is still the necessary property of balance. These concepts are discussed in this section.

9.7.1 Efficiency factors

The example below is of a design where there is partial confounding. Full details are given by Yates (1937, page 21) and by John (1971, page 135). This is an experiment to study the effects of three factors N, K, and D on the yields of King Edward potatoes. The factor levels were as follows.

N: sulphate of ammonia at rates of 0 and 0.45 cwt per acre
K: sulphate of potash at rates of 0 and 1.12 cwt per acre
D: dung at rates of 0 and 8 tons per acre

The treatment formula (line 22) is

N * K * D = N + K + D + N.K + N.D + K.D + N.K.D

There were eight treatment combinations, but the blocks each had only four plots. Consequently some of the treatment terms needed to be confounded between blocks. This was done by confounding N.K.D between blocks 1 and 2, N.K between blocks 3 and 4, N.D between blocks 5 and 6, and K.D between blocks 7 and 8. There was thus only partial confounding: the interaction terms could be estimated within some of the blocks but not others. To illustrate how this was done, we can consider N.K: this represents the difference in the effect of N according to the level of K (and vice versa). Representing the treatment combinations as triplets of letters, giving respectively the level of N (− or n), K (− or k) and D (− or d), this can be written as

{('n − −' + 'n − d') − ('− − −' + '− − d')} − {('nk −' + 'nkd') − ('− k −' + '− kd')}
= ('n − −' + 'n − d' + '− k −' + '− kd') − ('− − −' + '− − d' + 'nk −' + 'nkd')

The combinations in the first pair of brackets all occur in block 3, while those in the second pair all occur in block 4. Thus within blocks 3 and 4 there is no information on N.K; but information is available within the other 6 blocks. Thus N.K is estimated with efficiency 6/8 (= 0.75) in the Blocks.Plots stratum. The difference between the mean of the yields of the plots in block 3 and those in block 4 also provides an estimate of N.K; this represents the remaining 1/4 of the efficiency available for estimating N.K.

If a term is orthogonal, its efficiency factor equals one: the term is estimated with full efficiency in the stratum concerned. The efficiency factors of non-orthogonal terms are listed in the Information Summary obtained by setting option PRINT = information in either ANOVA or ADISPLAY (9.1.3). Terms that are aliased with earlier terms in the model (and so cannot be estimated) are also listed: these have zero efficiency factors.

The efficiency factors are not always so easy to derive and interpret as here: the original definition by Yates (1936) was for the balanced incomplete-block design.

438   9   Analysis of designed experiments

But they always represent the proportion of the information available to estimate a term.

EXAMPLE 9.7.1

```
  1    " A partially confounded factorial experiment
 -2      (Yates 1937, page 21; also John 1971, page 135)."
  3    UNITS [NVALUES=32]
  4    FACTOR [LEVELS=8] Blocks
  5    &      [LEVELS=4] Plots
  6    &      [LEVELS=2; LABELS=!T(_,n)] N
  7    &      [LABELS=!T(_,k)] K
  8    &      [LABELS=!T(_,d)] D
  9    GENERATE Blocks,Plots
 10    READ [PRINT=data,errors] N,K,D; FREPRESENTATION=labels

 11    _ _ _    n k _    n _ d    _ k d    n _ _    _ k _    _ _ d    n k d
 12    n _ _    _ k _    n _ d    _ k d    _ _ _    _ _ d    n k _    n k d
 13    n _ _    _ _ d    n k _    _ k d    _ _ _    _ k _    n _ d    n k d
 14    _ k _    _ _ d    n k _    n _ d    _ _ _    n _ _    _ k d    n k d  :

 15    VARIATE Yield
 16    READ Yield
```

Identifier	Minimum	Mean	Maximum	Values	Missing
Yield	87.0	291.6	471.0	32	0

```
 21    BLOCKSTRUCTURE Blocks/Plots
 22    TREATMENTSTRUCTURE N * K * D
 23    ANOVA Yield
```

23..

***** Analysis of variance *****

Variate: Yield

Source of variation	d.f.	s.s.	m.s.	v.r.
Blocks stratum				
N.K	1	780.1	780.1	3.02
N.D	1	276.1	276.1	1.07
K.D	1	2556.1	2556.1	9.91
N.K.D	1	112.5	112.5	0.44
Residual	3	774.1	258.0	
Blocks.Plots stratum				
N	1	3465.3	3465.3	10.86
K	1	161170.0	161170.0	505.21
D	1	278817.8	278817.8	873.99
N.K	1	28.2	28.2	0.09
N.D	1	1802.7	1802.7	5.65
K.D	1	11528.2	11528.2	36.14
N.K.D	1	45.4	45.4	0.14
Residual	17	5423.3	319.0	
Total	31	466779.7		

***** Information summary *****

Model term   e.f.   non-orthogonal terms

9.7 Non-orthogonality and balance

```
Blocks stratum
  N.K                              0.250
  N.D                              0.250
  K.D                              0.250
  N.K.D                            0.250

Blocks.Plots stratum
  N.K                              0.750  Blocks
  N.D                              0.750  Blocks
  K.D                              0.750  Blocks
  N.K.D                            0.750  Blocks
```

***** Tables of means *****
Variate: Yield

Grand mean 291.6

```
         N              n
         281.2          302.0

         K              k
         220.6          362.6

         D              d
         198.3          384.9

         N         K              k
                   211.3          351.1
         n̄         229.9          374.1

         N         D              d
                   196.5          365.9
         n̄         200.0          404.0

         K         D              d
                   105.4          335.9
         k̄         291.1          434.0

                   K         —              k
         N         D         d              d
                   106.1     316.5     286.9     415.2
         n̄        104.6     355.2     295.3     452.8
```

*** Standard errors of differences of means ***

```
Table                 N         K         D         N
                                                    K
rep.                  16        16        16        8
s.e.d.                6.31      6.31      6.31      8.93
Except when comparing means with the same level(s) of
  N                                                 9.65
  K                                                 9.65

Table                 N         K         N
                      D         D         K
                                          D
rep.                  8         8         4
s.e.d.                8.93      8.93      13.15
Except when comparing means with the same level(s) of
  N                   9.65                13.64
  K                             9.65      13.64
  D                   9.65      9.65      13.64
  N.K                                     14.12
  N.D                                     14.12
  K.D                                     14.12
```

(Notice in the output that the underline symbol has been used instead of minus for the zero level, to avoid having to put quotes around the labels when they are read in lines 11 to 14.)

As we explained in 9.1.3, Genstat calculates a table of means by taking the effects of each term only from the lowest stratum where it is estimated. Thus it would estimate N.K for example only from the Blocks.Plots stratum. The different efficiency factors for the component terms of the two-way and three-way tables of means in the example lead to different standard errors for some comparisons. For example, the s.e.d. for the N.K.D table is 13.15 when comparing means with different levels of all three factors, it is 13.64 if the level of one of the factors is identical for both means, and it is 14.12 if two of the factors are at identical levels.

9.7.2 Balance

The condition of first-order balance required for a design and its specification to be analysable by the ANOVA directive is explained algorithmically by Wilkinson (1970) and mathematically by James and Wilkinson (1971). Essentially it is that the contrasts of each term should all have a single efficiency factor, wherever the term is estimated. In the example in 9.7.1, all the terms have only one degree of freedom, and so represent only one contrast. There is thus no difficulty in verifying that the design is balanced.

Suppose instead that the treatment combinations were represented by a single factor T with eight levels:

  FACTOR [LABELS = !T(' − − −',' − − d',' − k − ',' − kd','n − − ','n − d','nk − ','nkd')] T

The main effect of T would not be balanced: the comparison of levels

  ' − − −' ' − − d' ' − k − ' ' − kd'

with

  'n − −' 'n − d' 'nk − ' 'nkd'

has efficiency factor one in the Blocks.Plots stratum and zero in the Blocks stratum (this contrast is equivalent to the main effect of N in the original specification); but the comparison of levels

  'n − −' 'n − d' ' − k − ' ' − kd'

with

  ' − − −' ' − − d' 'nk − ' 'nkd'

has efficiency 0.25 in the Blocks stratum and 0.75 in the Blocks.Plots stratum (this is equivalent to N.K in the original specification). Thus the main effect of T is not balanced, since in the Block.Plots stratum some of its contrasts have efficiency factor one, while others have efficiency factor 0.75. Genstat can detect this imbalance and will give you an error diagnostic: see later in this section.

For the design to have been balanced for T, a further three pairs of blocks

9.7 Non-orthogonality and balance

would be required. By confounding the comparison corresponding to the main effect of N between the first pair of extra blocks, that for K between the second pair, and that for D between the third pair, all the contrasts of T would be estimated within twelve of the (now) fourteen blocks, and confounded in the other two. The extended design would thus be balanced – as you may wish to verify!

To analyse the original design with a single treatment term T, a more complicated specification is required involving pseudo-factors.

9.7.3 Pseudo-factors

The pseudo-factorial operator // allows you to partition an unbalanced treatment term into pseudo-terms, which are each balanced. In our example, there is a factor T, some of whose contrasts have efficiency one in the Blocks.Plots stratum and zero elsewhere, while others have efficiency 0.25 in the Blocks stratum and 0.75 in the Blocks.Plots stratum. If instead of

 TREATMENTSTRUCTURE T

we specify

 TREATMENTSTRUCTURE T // (N + K + D + N.K + N.D + K.D)

the terms within the brackets that follow the operator // are linked to the term T as pseudo-terms. (Without the brackets, only the term immediately after // would be linked to T.) When the time comes for T to be fitted, the pseudo-terms N, K, D, N.K, N.D, and K.D are fitted first. All the contrasts wholly estimated in the Blocks.Plots stratum are thus removed (by N, K, and D), as well as some of the other contrasts. The remaining contrasts (denoted by T in the information summary) are all estimated with efficiency 0.25 between blocks and 0.75 within blocks. Thus all the pseudo-terms are balanced: those specified explicitly (N, K, D, N.K, N.D, and K.D), and the final pseudo-term which represents the contrasts not accounted for by N, K, D, N.K, N.D, and K.D. So by using the pseudo-factors, the design becomes analysable. In this example all the pseudo-terms represent single degrees of freedom – the final pseudo-term corresponds to the contrast represented earlier by N.K.D – but later we give an example where the pseudo-terms each have several degrees of freedom.

The sums of squares of the pseudo-terms are automatically combined to form the sum of squares for T in the analysis-of-variance table. Similarly the effects are all added together to form the table of means for T.

EXAMPLE 9.7.3a

```
  24   FACTOR [LABELS=!T('   _ _','_  d','  k  _','  k d',\
  25                        'n _ _','n _ d','n k _','n k d')] T
  26   READ [PRINT=data,error] T; FREPRESENTATION=labels
  27   '_ _ _' 'n k _' 'n _ d' '_ k d'   'n _ _' '_ k _' '_ _ d' 'n k d'
  28   'n _ _' '_ k _' 'n _ d' '_ k d'   '_ _ _' '_ _ d' 'n k _' 'n k d'
```

```
29  'n _ ' ' _ _ d' 'n k _ ' ' _ k d'       ' _ _ _ ' ' _ k _ ' 'n _ d' 'n k d'
30  ' _ k _ ' ' _ _ d' 'n k _ ' 'n _ d'     ' _ _ _ ' 'n _ _ ' ' _ k d' 'n k d' :
31  TREATMENTSTRUCTURE T // (N + K + D + N.K + N.D + K.D)
32  ANOVA Yield
```

32..

***** Analysis of variance *****

Variate: Yield

Source of variation	d.f.	s.s.	m.s.	v.r.
Blocks stratum				
T	4	3724.9	931.2	3.61
Residual	3	774.1	258.0	
Blocks.Plots stratum				
T	7	456857.4	65265.3	204.58
Residual	17	5423.3	319.0	
Total	31	466779.7		

***** Information summary *****

Model term	e.f.	non-orthogonal terms
Blocks stratum		
N.K	0.250	
N.D	0.250	
K.D	0.250	
T	0.250	
Blocks.Plots stratum		
N.K	0.750	Blocks
N.D	0.750	Blocks
K.D	0.750	Blocks
T	0.750	Blocks

***** Tables of means *****

Variate: Yield

Grand mean 291.6

T	$-$	d	k	kd	n	$n\ d$	$n\ k$	$n\ k\ d$
N	$\overline{1}$	$\overline{1}$	$\overline{1}$	1	$\overline{2}$	$\overline{2}$	$\overline{2}$	2
K	1	1	2	2	1	1	2	2
D	1	2	1	2	1	2	1	2
	106.1	316.5	286.9	415.2	104.6	355.2	295.3	452.8

*** Standard errors of differences of means ***

```
Table                          T
rep.                           4
s.e.d.                     13.15
Except when comparing means with the same level(s) of
  N                        13.64
  K                        13.64
  D                        13.64
  N.K                      14.12
  N.D                      14.12
  K.D                      14.12
```

9.7 Non-orthogonality and balance

The basic idea, then, is to use each pseudo-term to pick out a set of contrasts whose efficiency factors are all the same, wherever they are estimated. This should be reasonably straightforward, provided you understand how your design has been constructed. A further example is given below. But first we demonstrate that Genstat can indeed detect an unbalanced design. If we do not include the pseudo-factors, the design would be unbalanced. The error message correctly identifies T as the unbalanced term.

EXAMPLE 9.7.3b

```
 33   TREATMENTSTRUCTURE T
 34   ANOVA Yield

******* Fault (Code AN 1). Statement 1 on Line 34
Command: ANOVA Yield

Design unbalanced - cannot be analysed by ANOVA
Model term T (non-orthogonal to term Blocks) is unbalanced, in the Blocks.Plots
stratum.

A fatal fault has occurred - the rest of this job will be ignored
```

The traditional example for pseudo-factors is the partially balanced lattice. This has a single treatment factor, with number of levels equal to the square of some integer, k. To form the design, this factor is arbitrarily represented as the factorial combinations of two pseudo-factors, below called A and B, each with k levels. Further details are given by Yates (1937) or Kempthorne (1952). The example below is a simple lattice, taken from Cochran and Cox (1957, page 406). Here the treatment factor, Variety, has 25 levels. The correspondence between levels of Variety and the two pseudo-factors is:

A:	B:	1	2	3	4	5
1		1	2	3	4	5
2		6	7	8	9	10
3		11	12	13	14	15
4		16	17	18	19	20
5		21	22	23	24	25

The simple lattice has two replicates, each with k blocks of k plots: the block model is

    Rep/Block/Plot = Rep + Rep.Block + Rep.Block.Plot

The main effect of A is confounded with the blocks in the first replicate: block 1 has the five levels of T that correspond to level 1 of A, block 2 has those with level 2, and so on. Similarly, B is confounded with the blocks of the second replicate.

Thus A and B are each confounded with blocks in one out of the two replicates. So they have efficiency 0.5 in the Rep.Block (or blocks-within-replicates) stratum, and 0.5 in the Rep.Block.Plot (or plots-within-blocks) stratum. The treatment model is

Variety//(A + B)

The partially confounded parts of Variety are specified by the two pseudo-terms, A and B, and will be fitted first. The remaining contrasts of Variety correspond to the interaction between A and B, which is all estimated in the Rep.Block.Plot stratum. This final pseudo-term is thus also balanced, so the design can be analysed. This example shows only the analysis-of-variance table. This differs from the analysis given by Cochran and Cox (1957); they do not present the treatment sums of squares between and within blocks, but merely a sum of squares unadjusted for blocks. If the means had been printed, they would also differ from those given by Cochran and Cox, since Genstat does not combine information on effects when terms are estimated in more than one stratum (9.1.3). However, procedures are available in the Library (12.1.1) to do this for many standard designs.

EXAMPLE 9.7.3c

```
  1   " 5x5 Simple lattice
 -2       (Cochran and Cox 1957, page 406)."
  3   UNITS [NVALUES=50]
  4   FACTOR [LEVELS=2] Rep
  5   & [LEVELS=5] Block,Plot,A,B
  6   & [LEVELS=25; VALUES=(1...25),(1,6...21),(2,7...22), \
  7       (3,8...23),(4,9...24),(5,10...25)] Variety
  8   GENERATE Rep,Block,Plot
  9   & [TREATMENTS=Variety; REPLICATES=Rep; BLOCKS=Block] A,B
 10   READ Yield

      Identifier    Minimum      Mean    Maximum    Values    Missing
           Yield       4.00     13.62      30.00        50          0

 13   BLOCKSTRUCTURE Rep/Block/Plot
 14   TREATMENTSTRUCTURE Variety//(A+B)
 15   ANOVA [PRINT=aovtable] Yield
15..........................................................................
***** Analysis of variance *****

Variate: Yield
```

Source of variation	d.f.	s.s.	m.s.	v.r.
Rep stratum	1	212.18	212.18	
Rep.Block stratum				
Variety	8	350.00	43.75	
Rep.Block.Plot stratum				
Variety	24	711.12	29.63	2.17
Residual	16	218.48	13.66	
Total	49	1491.78		

This example also illustrates how to use the GENERATE directive (line 8) to form the values of pseudo-factors, as we promised in 5.5.1. The treatment term to which the pseudo-factors are to be linked is specified by the TREATMENTS option; here this is the main effect of Variety. The factors that identify the replicates are specified by the REPLICATES option, and those that identify the blocks within each replicate are specified by the BLOCKS option. The settings of these two options are model formulae, but Genstat merely scans them to find which factors they contain; so you may find it easiest simply to give the factors as a list. Here the replicates and blocks are identified by the single factors Rep and Block respectively. The single unnamed parameter of GENERATE lists the pseudo-factors. These have as many levels as there are blocks within each replicate. The blocks in the first replicate are used to determine which combinations of the factors in the treatment term correspond to each level of the first pseudo-factor, those in the second replicate are used for the second pseudo-factor, and so on. Here the first pseudo-factor is A, and the five blocks of replicate 1 contain Variety levels 1–5, 6–10, 11–15, 16–20, and 21–25. Thus the plots with varieties 1 to 5 are allocated level 1 of A, and so on. If a treatment combination occurs in more than one block within the same replicate, the level of the corresponding pseudo-factor is not determined uniquely and Genstat will report an error.

9.7.4 Non-orthogonality between treatment terms

The examples earlier in this section illustrate non-orthogonality between treatment and block terms. Balanced designs can also occur where the non-orthogonality is between treatment terms. However the interpretation of the analysis requires more care; indeed there may be information that Genstat is unable to calculate. (Similar difficulties occur in ordinary regression with observational data, see Chapter 8: usually the explanatory variables will not be orthogonal to each other and so their sums of squares, and thus the importance that may be ascribed to them, will depend on the order in which they are fitted.)

Suppose that the treatment model is

A + B + C

that B is non-orthogonal to A, and that C is non-orthogonal to both A and B. Genstat fits the model sequentially. Thus the sum of squares produced for A is for A ignoring B and C: no account is taken of these two factors, which are still to be fitted. With B, A has already been fitted and thus eliminated, whereas C has not. So the sum of squares produced for B is for B eliminating A and ignoring C. The sum of squares for C, which is fitted last, is eliminating both A and B.

Each sum of squares can be expressed as the difference between the residual sums of squares before and after fitting a particular term. So the sums of squares that are presented by Genstat will automatically add to the total sum of squares. Examining these enables you to check whether any of the terms in the model has an effect. However, to be sure that there is an effect of A, for example, that cannot be explained by B and C requires the sum of squares for A eliminating B and C. To obtain this you could redefine the treatment model as either

    B + C + A

or

    C + B + A

but the design would not necessarily be balanced according to these specifications.

Similarly, the effects estimated for each term are eliminating those terms fitted before it, and ignoring those that are still to be fitted. *Partial effects*, defined as the effects of a term eliminating all the other treatment terms, are calculated during the analysis and can be obtained using AKEEP (9.6.1).

A table of means for A.B, if this were in the model, would require the effects for A eliminating B, those for B eliminating A, and those for the interaction A.B. However, with the treatment model A + B + C, the necessary effects for A are not available. Consequently, no means are presented for terms that contain mutually non-orthogonal margins (like A and B for the table A.B).

A maximum of 10 mutually non-orthogonal terms is allowed. For example, term T[10] may be non-orthogonal to T[9], which is non-orthogonal to T[8], and so on down to term T[2], which is non-orthogonal to term T[1]; but to include an extra term T[11] in the sequence would exceed the limit. This limit should be sufficient for any designed experiment. Data with many non-orthogonal terms are, in any case, analysed more efficiently by the regression directives described in Chapter 8.

Note that, if the terms A, B, and C here had been orthogonal, the sum of squares and effects obtained for any one of them would remain the same irrespective of which of the other two terms had been fitted. For example, the sum of squares for A ignoring B and C would be identical to that for A eliminating B and C. Thus each of these three terms could be assessed independently, without regard to the other two. If two terms are far from orthogonal, you may find that the effects of either term ignoring the other are significant, but that neither set of effects is significant when the other term is eliminated. Deciding which of the terms are important may then be very difficult, and you may have to recommend that another experiment be done. This illustrates that orthogonality between treatment terms is not merely a convenience for making the computations more efficient: it also greatly simplifies the interpretation of the results.

9.7.5 The method of analysis

In this subsection we briefly describe the algorithm that is used to do the analysis of variance. However, for most purposes you will not need this information.

The model formulae defined by the BLOCKSTRUCTURE and TREATMENTSTRUCTURE are interpreted by an extension of the algorithm of Rogers (1973); further details are given by Wilkinson and Rogers (1973).

The method used to do the analysis is described in detail by Payne and Wilkinson (1977) and Wilkinson (1970). It operates on a working vector which initially contains the data values, and finally contains the residuals. The terms in the model are fitted by a series of *sweep* operations. Each sweep estimates the effects of a term, and then subtracts them from the current working vector, which then becomes the working vector for the next sweep. The first sweep is for the grand mean. The block terms are fitted next, to give an initial partitioning into strata. Then the treatments are fitted within each stratum.

If a term is orthogonal, its estimated effects are simply the corresponding table of means calculated from the current working vector. If the term is non-orthogonal to any of the terms already fitted, some of the information about the term is unavailable, and its effects are the totals calculated from the current vector, divided by its replication and efficiency factor. For the term to be balanced, the information still available must be the same for all the contrasts between the effects of the term, so that there is a single efficiency factor for all the contrasts. If the term is orthogonal, the efficiency factor is one. A zero efficiency factor indicates that the term is completely aliased with earlier terms in the model, and so cannot be estimated.

A sweep for a non-orthogonal term reintroduces effects for the terms to which it is non-orthogonal. Before sweeping for the next term in the model, these effects are removed by a sequence of *re-analysis* sweeps for the terms concerned. If any term in the re-analysis sequence is itself non-orthogonal, it must itself be followed by its own re-analysis sequence, and so on. Genstat allows for re-analysis sequences to be nested only ten deep, which is why there is the limit of ten mutually non-orthogonal terms (9.7.4).

When there are several strata, the analysis of each one is introduced by a special sweep known as a *pivot*, in which the value in each unit of the working vector is replaced by the corresponding effect calculated for the block term of the stratum. During the analysis of a stratum, the re-analysis sweeps for its own block term take the form of recalculating the effects and repeating the pivot.

The algorithm, unlike multiple regression algorithms, does not distinguish between the individual contrasts of each term (unless you partition it up into pseudo-terms: 9.7.3). This makes the computations more efficient, but it means that only balanced terms can be fitted. The design can be analysed if all the terms in the model are balanced: that is if they each have a single efficiency factor for their effects, in any stratum where they are estimated. The design is then said to have *first-order balance* with respect to the specified model (Wilkinson 1970; James and Wilkinson 1971): for a brief description, see 9.7.2.

A further consequence of the way in which the effects of each term are all fitted together is that, if any part of a term is present in a stratum, Genstat must assume that all its effects can be estimated there. Thus if a term is only partially estimable in a stratum (due to partial aliasing or to partial confounding), the degrees of freedom will be incorrect. Genstat can detect some of the occasions when this occurs, as for example if it results in a negative number of residual degrees of freedom for the stratum, and will then give a warning diagnostic. You can obtain an analysis with the correct numbers of degrees of freedom by using pseudo-factors (9.7.3) to identify the parts of a term that are estimated in the different strata.

Genstat determines the structure of the design by a process known as the *dummy analysis* (9.1.2). This is similar to the analysis of the data, but involves extra sweeps to detect whether each term can be estimated in a particular stratum, and to determine its efficiency factor there. In these sweeps, a near-zero sum of squares is taken to indicate that the term cannot be estimated. However, the test cannot be against an exact value of zero, because computer calculations always involve errors of round-off. Thus Genstat tests against a number slightly larger than zero; this zero limit is calculated as the total sum of squares in the working variate (after removing the grand mean) multiplied by the first element of the variate specified in the TOLERANCE option of ANOVA (9.1.2). By default, this first element contains the value 10^{-7}. A similar limit checks for zero sums of squares in the analysis of the data, but here the multiplier is given in the second element of the TOLERANCE variate; the default value is 10^{-9}.

The working vector for the dummy analysis contains random values from a Cauchy distribution. The starting value for their generation is set by the SEED option of ANOVA (9.1.2). Thus if you have doubts about a particular dummy analysis, for example if you think that a term is incorrectly listed as aliased, you can change the starting value and repeat the analysis with a different working vector.

A simpler and quicker form of the dummy analysis is available for designs that are orthogonal, and for which all the effects of each term have equal replication. (An orthogonal design is one in which each term has efficiency factor either zero or one in each stratum.) This incorporates a check which will detect any non-orthogonality, unless the design is particularly complicated and terms are aliased. The ORTHOGONAL option of ANOVA (9.1.2) allows you to specify whether non-orthogonality should cause Genstat to switch to the full dummy analysis, or to terminate the analysis with an error diagnostic.

R.W.P.

10 Multivariate and cluster analysis

10.1 Introduction

In this chapter we are concerned with statistical methods for simultaneously analysing more than one variable. Very often such methods initially combine information on all the given variables into a measure of association, such as a distance or dissimilarity; so, in a sense, they become univariate. Indeed in some fields of application, notably psychology and the social sciences, a single variable of associations may be observed directly, rather than calculated from more basic information. Multivariate analysis is concerned with two forms of data: (a) information on p variables for each of n samples (this can be called the data matrix); or (b) information, usually presented as a symmetric matrix, giving associations between all pairs of samples or all pairs of variables.

In the simplest cases the data matrix has no further structure, and may be regarded as the multivariate generalization of a simple random sample. Genstat does not have a special data structure for a data matrix; generally you must either list the corresponding variables, or collect them in a pointer (3.3.4). From a data matrix you can calculate the symmetric matrix of sums of squares and products, or alternatively, the correlation matrix of the variables. These are stored in compound data structures, which also contain the means of the variables and other information (3.6.2). However, you can easily extract the basic symmetric matrix from this more general structure (5.7). Genstat has a set of directives for multivariate analyses based on sums of squares and products; these are described in 10.2.

Just as univariate samples may have structure imposed on the units, so may multivariate samples. In canonical variates analysis the units belong to a set of k mutually exclusive groups. For this Genstat lets you calculate the matrix of sums of squares and products, pooled within groups, as well as the means of all the variables in all the groups (5.7.3); these means are held as a set of p variates, each with k values, from which Genstat can calculate a matrix of between-group sums of squares and products. Sums of squares and products arising from more general sample structures are provided by ANOVA (9.6.1).

Correlations and sums of squares and products are elementary examples of how associations can be measured between variables; methods based on such measures are sometimes termed *R-techniques*. Measures of association between units lead to methods known as *Q-techniques*, which are discussed in 10.4 to 10.6. Section 10.3 is concerned with how to calculate symmetric matrices giving similarities or distances between all pairs of samples.

You can think of matrices of distances or dissimilarities as being generated by a cloud of n points in a multidimensional Euclidean space, where the distance between the points representing two samples is or is related to the corresponding distance or dissimilarity in the given matrix. To visualize such a cloud of points is difficult, and much multivariate analysis is concerned with providing approximate graphical representations that are easily interpreted by eye. These representations fall into two main classes: those depending on scatter plots of points in two or, more rarely, three dimensions; and those expressed in the form of networks, especially rooted trees. The plotted distance is usually supposed to approximate to the "true" distance in multidimensional space. Alternatively you may need to examine angle, inner product or area, rather than distance: for example, angles are used to interpret the output from biplots (Gabriel 1971). Apart from the minimum spanning tree given by the HDISPLAY directive, all other standard network-type displays in Genstat are in the form of rooted trees. In particular, you can get dendrograms: these are rooted trees with a scale associated with the nodes. Dendrograms are especially useful for representing hierarchical classifications (10.5).

The directives described in this chapter are for principal components analysis, principal coordinates analysis, canonical variates analysis, orthogonal Procrustes rotation, factor rotation, and various forms of hierarchical and non-hierarchical classification. Other techniques are provided by procedures in the Genstat Procedure Library (12.1.1).

You can also write your own procedures, taking advantage of CALCULATE (5.1) to operate with matrices. You may also find useful the operations of singular value decomposition (5.7.1), spectral (eigenvalue) decomposition (5.7.2), and extracting a sub-matrix from a larger matrix. You can obtain the singular value decomposition for any rectangular real matrix; this is valuable statistically because of its least-squares properties (Eckart and Young 1936). Genstat can calculate the eigenvalues and eigenvectors only of a symmetric matrix, as otherwise the results may be complex numbers. In statistics, non-symmetric matrices often arise in the form $B^{-1}A$, where A and B are both symmetric. Then the eigenvalues of $B^{-1}A$ are real, and are easily found by solving the two-sided algebraic eigenvalue problem $Ax = \lambda Bx$; this case is covered by Genstat (5.7.2). You can extract submatrices by CALCULATE, either with qualified identifiers or by using the ELEMENTS function (5.2.7).

For general reading in applied multivariate analysis see Mardia, Kent, and Bibby (1979) and Gower (1985a) and the further references they cite. Although a little old, Lawley and Maxwell (1971) is still the best short account of factor analysis. For work in classification and cluster analysis, see Gordon (1981).

10.1.1 Rules for printing and saving results from multivariate analyses

Most of the directives for multivariate and cluster analysis allow you both to print and to save results. For some of them, the results can be interpreted as the coordinates of points in a multidimensional space; when this is the case you can usually print and save the results for a subset of the dimensions. This subsection describes the general rules for doing so.

When a directive has the option NROOTS this refers to a set of latent roots each of which is associated with one dimension of the results. You set the NROOTS option to the number of dimensions that you want printed. By default, it is the results corresponding to the largest roots that are printed. But some directives allow you to set option SMALLEST=yes to get the results corresponding to the smallest roots. The setting of the NROOTS option governs all the printed results for which dimensionality is relevant. For example, these statements will read 20 variates of length 100 from input channel 2, and do a principal components analysis:

```
UNIT [100]
POINTER [NVALUES = 20] V
READ V[]
PCP [PRINT = roots,scores,residuals; NROOTS = 5] V
```

The first five latent roots will be printed, as will the first five columns of the matrix of principal component scores. The residuals that will be printed are formed from the principal component scores for dimensions 6 to 20.

The sum of the diagonal elements of a matrix is equal to the sum of its latent roots, and is called the *trace*. Whenever latent roots are printed they are also printed expressed as percentages of the trace, together with the trace itself.

You can save the results of some multivariate directives for a subset of the dimensions. These are always those corresponding to the largest latent roots; to save the results corresponding to the smallest roots, you must first save the full set of results and then extract those corresponding to the dimensions that you want. Before you can save anything from a multivariate analysis, you must already have declared structures to hold the information. All the size attributes of these structures that refer to dimensionality must be consistent: the results to be saved from the analysis must all have the same dimensionality. For example, from a principal components analysis, you cannot directly save the first four latent roots but only the first two columns of scores. If residuals are to be saved, then these will automatically be taken to correspond to the dimensions that have not been saved. But you do not have to save the scores from the analysis in order to save the residuals: you merely have to give some indication of the dimensionality of the saved results; for example, you could specify a structure to save the latent roots and vectors, thus implicitly specifying the number of dimensions. The following statements would save the residuals formed from dimensions 6 to 20, as well as the first five latent roots and vectors, calculated from the data matrix stored in pointer V above; all of the latent roots will be printed.

```
LRV [ROWS = V; COLUMNS = 5] L
MATRIX [ROWS = 100; COLUMNS = 1] R
PCP [PRINT = roots] V; LRV = L; RESIDUALS = R
```

10.2 Analyses based on sums of squares and products

Two of the multivariate methods in Genstat are based on sums of squares and products: principal components analysis and canonical variates analysis. The declaration of SSPM structures is described in 3.6.2, and their formation using the FSSPM directive is described in 5.7.3. Both principal components analysis (10.2.1) and canonical variates analysis (10.2.2) give loadings of a set of variates: the methods of factor rotation (10.2.3) can sometimes help with the interpretation of loadings.

10.2.1 The PCP directive

Principal components analysis finds linear combinations of a set of variates that maximize the variation contained within them, thereby displaying most of the original variability in a smaller number of dimensions. The PCP directive operates on an SSPM, or on a correlation matrix, formed from the variates.

PCP
Performs principal components analysis.

Options

PRINT	= strings	Printed output required (loadings, roots, residuals, scores, tests); default * i.e. no output
NROOTS	= scalar	Number of latent roots for printed output; default * requests them all to be printed
SMALLEST	= string	Whether to print the smallest roots instead of the largest (no, yes); default n
METHOD	= string	Whether to use sums of squares or correlations (ssp, correlation); default s

Parameters

DATA	= pointers or SSPMs	Pointer of variates forming the data matrix or SSPM giving their sums of

10.2 Analyses based on sums of squares and products

		squares and products (or correlations) etc
LRV	= LRVs	To store the principal component loadings, roots and trace from each analysis
SSPM	= SSPMs	To store the computed sum-of-squares-and-products or correlation matrix
SCORES	= matrices	To store the principal component scores
RESIDUALS	= matrices	To store residuals from the dimensions fitted in the analysis (i.e. number of columns of the SCORES matrix, or as defined by the NROOTS option)

You give the input for PCP using the first parameter; this list may have more than one entry, in which case Genstat repeats the analysis for each of the input structures. Rather than supplying an SSPM or correlation matrix explicitly, you can supply a pointer, containing a set of variates representing a data matrix. Genstat will then calculate the sums of squares and products, or the correlations, for the analysis.

For example, these two forms of input are equivalent:

(1)   SSPM [TERMS = Height,Length,Width,Weight] S
      FSSP S
      PCP [1PRINT = roots] S
(2)   PCP [PRINT = roots] !P(Height,Length,Width,Weight)

But the first form does mean that you have the sums of squares and products available for later use, in the SSPM S. Here the pointer is unnamed (1.4.3). But you may wish to use a named POINTER. For example:

POINTER [VALUES = Height,Length,Width,Weight] Dmat
PCP [PRINT = roots] Dmat

By default the PCP directive does not print any results: you use the PRINT option to specify what output you require. The printed output is in five sections, each with a corresponding setting, as illustrated in the examples below.

The columns of the matrices of principal component loadings and scores correspond to the latent roots. Each latent root corresponds to a single dimension, and gives the variability of the scores in that dimension. The loadings give the linear coefficients of the variables that are used to construct the scores in each dimension. This example shows a principal components analysis of four variates of length 12.

EXAMPLE 10.2.1a

```
  1  UNITS [NVALUES=12]
  2  POINTER [VALUES=Height,Length,Width,Weight] Dmat
  3  READ [PRINT=data,errors,summary] Dmat[]

  4  4.1 5.2 1.2 3.1 4.2 1.5 3.2 5.6 2.3 0.2 0.1 0.2
  5  6.2 4.1 4.1 4.1 2.3 6.2 6.3 5.1 0.2 0.9 4.9 7.3
  6  10.1 5.6 3.2 9.4 1.2 9.8 1.0 1.0 6.1 9.7 1.0 3.7
  7  6.1 9.6 9.7 5.5 2.3 5.0 9.4 8.1 4.5 4.9 0.3 1.8 :

     Identifier   Minimum    Mean    Maximum   Values   Missing
     Height        0.200     4.133    10.100     12         0
     Length        0.200     5.225     9.800     12         0
     Width         0.100     3.700     9.700     12         0
     Weight        0.200     4.575     9.400     12         0

  8  PCP [PRINT=roots,scores,loadings,tests] Dmat
8.........................................................,.........................


*****   Principal components analysis   *****

***  Latent Roots  ***

               1            2           3           4
           181.8        130.2        82.5        18.5

***  Percentage variation  ***

               1            2           3           4
            44.01        31.52       19.98        4.49

***  Trace  ***

     413.2

***  Latent Vectors (Loadings)  ***

                    1           2           3           4
     Height    -0.21529     0.37981     0.78747    -0.43506
     Length    -0.25623     0.86524    -0.34389     0.25970
     Width     -0.74104    -0.21726    -0.37937    -0.50964
     Weight    -0.58211    -0.24474     0.34308     0.69537


***  Significance tests for equality of final K roots  ***

     No. (K)      Chi
      Roots     squared      df

        2         4.61        2
        3         7.49        5
        4        10.30        9

***  Principal Component Scores  ***

                    1           2           3           4
          1      2.725       0.870       0.425       0.256
```

10.2 Analyses based on sums of squares and products 455

2	0.714	-3.340	1.875	-0.029
3	6.897	-3.191	0.149	-1.715
4	-0.177	-0.159	1.700	-1.725
5	-2.087	-0.546	-2.585	0.091
6	-0.521	-6.164	-1.130	1.871
7	-3.819	1.518	6.415	1.112
8	3.541	4.306	-4.085	1.354
9	0.940	5.420	0.734	1.074
10	-6.529	3.002	-1.915	-2.134
11	-5.824	-2.992	-2.319	0.285
12	4.139	1.276	0.738	-0.441

The significance tests are for equality of the k smallest roots: l_i ($i = 1, 2, ..., k$). The test statistic is

$$n - \frac{(2p + 11)}{6}\left[\log\left(\frac{1}{k}\sum_{i>k} l_i\right) - \frac{1}{k}\sum_{i>k} \log l_i\right]$$

where n is the number of units and p is the number of variables. Asymptotically, the statistics have a chi-squared distribution with $(k+2)(k-1)/2$ degrees of freedom. If any latent roots are zero, Genstat excludes them from the calculation of the test statistic; the effective value of p is reduced accordingly.

If you omit the NROOTS option, Genstat prints by default the results corresponding to all the latent roots. The number of latent roots is the number of variates involved in the input to PCP. The NROOTS option allows you to print only part of the results, corresponding to the first or last r latent roots. You may then want to print the residuals. The following example prints the results corresponding to the first two latent roots; the residuals are formed from the remaining two columns of scores.

EXAMPLE 10.2.1b

```
  9  PCP [PRINT=scores,residuals; NROOTS=2] Dmat

9.....................................................................

*****  Principal components analysis  *****

***  Principal Component Scores  ***

                     1              2
         1         2.725          0.870
         2         0.714         -3.340
         3         6.897         -3.191
         4        -0.177         -0.159
         5        -2.087         -0.546
         6        -0.521         -6.164
         7        -3.819          1.518
         8         3.541          4.306
         9         0.940          5.420
        10        -6.529          3.002
        11        -5.824         -2.992
        12         4.139          1.276
```

*** Residuals ***

1	0.496
2	1.875
3	1.721
4	2.422
5	2.587
6	2.186
7	6.510
8	4.304
9	1.301
10	2.867
11	2.337
12	0.860

To print results corresponding to the r smallest latent roots, you must set option NROOTS to r and option SMALLEST to 'yes'. Now if residuals are printed they will be formed from the scores corresponding to the largest roots.

The NROOTS and SMALLEST options apply to the latent roots and vectors, the principal component scores and the residuals. So you cannot print directly, for example, the first two columns of scores and the last three columns of loadings. This is rarely required, but can be done by saving the relevant results and printing them separately, as shown in the next example.

The METHOD option of the PCP directive specifies whether the analysis is to be based on the SSPM or on a derived matrix of correlations. The default setting is 'ssp'; if you want to analyse the correlation matrix, you must set METHOD = correlation, as shown in the final example of this subsection. Note that when correlations are analysed the significance-test statistics no longer have an asymptotic chi-squared distribution.

There are four parameters used for saving results. You must already have declared the structures that you supply with these parameters; and these must have correct and consistent dimensions: for further details see 10.1.1.

The SCORES parameter is used to save the principal component scores in a matrix. You must have declared this with number of rows equal to the number of units; if you have declared it with r columns, then the first r columns of the scores will be saved.

EXAMPLE 10.2.1c

```
  10  MATRIX [ROWS=12; COLUMNS=2] PCPscore
  11  PCP [PRINT=roots,residuals; NROOTS=3; SMALLEST=yes] Dmat; SCORES=PCPsc
11..............................................................
```

***** Principal components analysis *****

*** Latent Roots ***

10.2 Analyses based on sums of squares and products

```
                   1          2          3
               130.23      82.54      18.55
*** Percentage variation ***
                   1          2          3
                31.52      19.98       4.49
*** Trace ***
      413.2
*** Residuals ***
       1     2.725
       2     0.714
       3     6.897
       4     0.177
       5     2.087
       6     0.521
       7     3.819
       8     3.541
       9     0.940
      10     6.529
      11     5.824
      12     4.139

  12  PRINT PCPscore

            PCPscore
                   1          2
       1     2.725      0.870
       2     0.714     -3.340
       3     6.897     -3.191
       4    -0.177     -0.159
       5    -2.087     -0.546
       6    -0.521     -6.164
       7    -3.819      1.518
       8     3.541      4.306
       9     0.940      5.420
      10    -6.529      3.002
      11    -5.824     -2.992
      12     4.139      1.276
```

In this example the residuals are printed. They correspond to the first column of scores, and can be compared with the scores printed earlier. You can see that all the residuals are positive: this is because residuals from multivariate analyses generally occupy several dimensions, so they represent distances in multidimensional space and signs cannot be attached to them.

The SSPM parameter saves the SSPM structure used for the analysis. You must have declared it already, and it must be of the correct size. A particularly convenient instance is when you have supplied an SSPM structure as input, but have set METHOD=correlation: the correlation matrix used for the analysis will then be saved in the structure specified by the SSPM parameter.

The LRV parameter is used to save the principal component loadings, the latent roots, and their sum (the trace) in an LRV structure. You must have declared this already and its number of rows must be the same as the number of variates supplied in an input pointer or implied by an input SSPM. If the LRV structure has been declared with fewer columns than rows, the largest roots will be saved, with their corresponding vectors. However, the trace saved as the third component of the LRV structure will contain the sums of all the latent roots, not just those that have been saved. The number of columns of the LRV structure and of the matrix for the scores must be the same, if they are both present.

The RESIDUALS parameter is used to save the principal component residuals in a matrix; you must have declared this with number of rows equal to the number of units, and with one column. If the latent roots and vectors (loadings) are saved from the analysis, the residuals will correspond to the dimensions not saved; the same applies if you save scores. If neither the LRV nor scores are saved, the saved residuals will correspond to the smallest latent roots not printed.

EXAMPLE 10.2.1d

```
  13   LRV [ROWS=Dmat; COLUMNS=2] Latent
  14   MATRIX [ROWS=12; COLUMNS=1] Res
  15   SSPM [TERMS=Dmat[]] Corrmat
  16   PCP [PRINT=roots,scores,tests; METHOD=correlation] Dmat; \
  17       LRV=Latent; SSPM=Corrmat; RESIDUALS=Res

17.................................................................

***** Principal components analysis *****

*** Latent Roots ***

                  1            2            3            4
              1.748        1.209        0.855        0.188

*** Percentage variation ***

                  1            2            3            4
              43.70        30.23        21.37         4.70

*** Trace ***

       4.000


*** Significance tests for equality of final K roots ***

Correlation matrix used -
  test statistics are NOT asymptotically chi-squared

       No. (K)      Chi
        Roots     squared        df

          2         5.78          2
          3         8.55          5
          4        11.87          9
```

10.2 Analyses based on sums of squares and products

```
*** Principal Component Scores ***
              1         2         3         4
    1    0.2477    0.1025    0.0527    0.0317
    2    0.0190   -0.2415    0.2655   -0.0057
    3    0.6772   -0.2301    0.1819   -0.1701
    4   -0.0473    0.0699    0.1737   -0.1709
    5   -0.1305   -0.1479   -0.2612    0.0010
    6   -0.0340   -0.6395    0.0172    0.1726
    7   -0.5549    0.3197    0.5267    0.1211
    8    0.4383    0.2537   -0.4546    0.1423
    9    0.0577    0.5110   -0.0489    0.1215
   10   -0.5569    0.2186   -0.3213   -0.2194
   11   -0.4980   -0.3846   -0.2263    0.0113
   12    0.3816    0.1683    0.0946   -0.0352

 18  PRINT Latent[],Res

         Latent['Vectors']
                   1         2
   Dmat
   Height    -0.3476    0.6121
   Length    -0.1981    0.6896
   Width     -0.6201   -0.3067
   Weight    -0.6749   -0.2359

         Latent['Roots']
                   1         2
               1.748     1.209

 Latent['Trace']       4.000
                        Res
                         1
    1    0.0615
    2    0.2656
    3    0.2490
    4    0.2437
    5    0.2612
    6    0.1734
    7    0.5404
    8    0.4763
    9    0.1310
   10    0.3891
   11    0.2266
   12    0.1009
```

If the variables used to form the SSPM structure are restricted, then the analysis will be subject to that restriction. Similarly, if a pointer to a set of variates is used as input to PCP, then any restriction on the variates will be taken into account by the analysis. If you want principal component scores or residuals to be printed or saved from the analysis, the original data must be available. The matrices to save such results must have been declared with as many rows as the variates have values, ignoring the restriction. You can calculate the analysis from one subset of units, but calculate the scores and residuals for all the units, by using as input to PCP an SSPM structure formed using a weight variate with zeros for the excluded sampling units and unity for those to be included. For example, to exclude a known set of outliers from an analysis, but to print scores for them, these statements could be used:

```
POINTER [NVALUES = 5] V
FACTOR [LABELS = !T(No,Yes)] Outlier
READ [CHANNEL = 2] Outlier,V[]
CALCULATE Wt = Outlier .IN. 'No'
SSPM [V] S
FSSPM [WEIGHT = Wt] S
PCP [PRINT = scores] S
```

10.2.2 The CVA directive

The CVA directive, for canonical variates analysis, operates on a within-group SSPM. This structure contains information on the within-group sums of squares and products, pooled over all the groups; it also contains the group means and group sizes, from which Genstat can derive the between-group sums of squares and products. The directive finds linear combinations of the original variables that maximize the ratio of between-group to within-group variation, thereby giving functions of the original variables that can be used to discriminate between the groups. The squared distances between group means are Mahalanobis D^2 statistics when all the dimensions are used; otherwise they are approximations. You can form exact Mahalanobis distances with the PCO directive (10.4.1).

CVA
Performs canonical variates analysis.

Options

PRINT	= strings	Printed output required (roots, loadings, means, residuals, distances, tests); default * i.e. no output
NROOTS	= scalar	Number of latent roots for printed output; default * requests them all to be printed
SMALLEST	= string	Whether to print the smallest roots instead of the largest (no, yes); default n

Parameters

WSSPM	= SSPMs	Within-group sums of squares and products, means etc (input for the analyses)

10.2 Analyses based on sums of squares and products

LRV	= LRVs	Loadings, roots and trace from each analysis
SCORES	= matrices	Canonical variate means
RESIDUALS	= matrices	Distances of the means from the dimensions fitted in each analysis
DISTANCES	= symmetric matrices	Inter-group-mean Mahalanobis distances

You give the input to CVA using the first parameter; this may contain a list of structures, in which case Genstat repeats the analysis for each of them. The input must be an SSPM structure, declared with the GROUPS option of the SSPM directive (3.6.2) set to a factor giving the grouping of the units. If the variates used to form this SSPM structure are restricted, then the SSPM is restricted in the same way, and so the CVA directive takes account of the restriction. In addition to the first parameter, there are four parameters which can be used to save the results; the structures supplied must have been declared already, and must have correct and consistent sizes (see 10.1.1).

The three options of the CVA directive control the printed output. By default there is no printed output, and so you should set the PRINT option to indicate which sections you want.

Doran and Hodson (1975) give some measurements made on 28 brooches found at the archeaological site of the cemetery at Munsingen. Seven of these variables are used in the next example; they have been transformed by taking logarithms. For a specified grouping of the 28 brooches into four groups, canonical variates analysis is used to determine possible differences among the groups, and which variables contribute to such differences. (These seven variables are also used in the first example of the CLUSTER directive (10.7.1), and the grouping used here is that obtained from CLUSTER).

EXAMPLE 10.2.2a

```
  1  UNITS [NVALUES=28]
  2  POINTER [VALUES=Foot_lth,Bow_ht,Coil_dia,Elem_dia,Bow_wdth, \
  3    Bow_thck,Length] Data
  4  FACTOR [LEVELS=4] Groupno
  5  READ Groupno,Data[]

     Identifier    Minimum      Mean    Maximum    Values   Missing
       Foot_lth      2.398     3.278      4.554        28         0
         Bow_ht      2.079     2.842      3.296        28         0
       Coil_dia      1.792     2.166      2.833        28         0
       Elem_dia      1.099     2.026      2.708        28         0
       Bow_wdth      3.045     4.064      5.176        28         0
       Bow_thck      2.708     3.621      4.357        28         0
         Length      3.296     4.003      4.860        28         0

 35  SSPM [TERMS=Data[]; GROUPS=Groupno] W
 36  FSSP W
```

```
 37  CVA [PRINT=roots,loadings,means,tests] WSSPM=W
37........................................................................
```

***** Canonical variate analysis *****

*** Latent Roots ***

	1	2	3
	4.543	3.777	2.537

*** Percentage variation ***

	1	2	3
	41.85	34.79	23.37

*** Trace ***

   10.86

*** Latent Vectors (Loadings) ***

	1	2	3
1	1.130	-2.656	3.397
2	-0.633	1.631	4.799
3	3.501	-1.708	1.450
4	-2.669	-0.623	-2.802
5	-3.468	-0.758	0.757
6	1.859	-2.028	-2.478
7	-1.279	-0.110	-3.598

*** Significance tests for dimensionality greater than K ***

K	Chi-squared	df
0	97.60	21
1	60.78	12
2	27.16	5

*** Canonical Variate Means ***

	1	2	3
1	2.967	1.998	0.613
2	-0.825	0.122	-1.584
3	1.254	-3.545	0.825
4	-2.835	0.856	2.241

*** Adjustment terms ***

	1	2	3
1	-8.40	-19.90	1.94

The pointer Data is declared on lines 2 and 3 to refer to the seven data variables. The factor Groupno specifies the groups (line 4). The matrix of within-group sums of squares and products is declared and formed on lines 35 and 36. The CVA directive (line 37) specifies that the latent roots, the vectors (loadings), and the means of the canonical variate groups are to be printed, together with values for the significance tests for the latent roots that indicate the number of dimensions required.

10.2 Analyses based on sums of squares and products

If there are g groups, at most $g-1$ independent combinations of the variables can be found to discriminate amongst them. However, if there are fewer than $g-1$ variables, v say, then at most v independent combinations can be calculated. Thus there will be at most $\min(g-1, v)$ non-zero latent roots, with associated loadings and canonical variate scores for the group means. In the example above $\min(g-1, v)$ is 3.

The significance tests that are printed are for a significant dimensionality greater than k, that is for the joint significance of the 1st, 2nd, ..., $(k+1)$th latent roots. This test is printed for $k = 0, 1, \ldots \min(g-1, v)-1$. If the test is "not significant" for $k = r$, then the values of chi-square for $k > r$ should be ignored as the indication is that the remaining dimensions have no interesting structure. The test statistic (Bartlett 1938) is

$$\left[n - g - \tfrac{1}{2}v - g)\right]\left[\sum_i \log(l_i + 1) - \sum_{i>k} \log(l_i + 1)\right]$$

which is asymptotically distributed as chi-squared with $(v-k) \times (g-k-1)$ degrees of freedom. Here n is the number of units, g is the number of groups, v is the number of variables, and l_i is the ith latent root. If the coefficient $[n-g-\tfrac{1}{2}(v-g)]$ is less than zero, there are too few units for the statistics to be calculated and a message is printed to this effect. In any case, the tests should be treated with caution unless $n-g$ is very much larger than v.

The latent vectors, or loadings, are scaled in such a way that the average within-group variability in each canonical variate dimension is 1: thus the within-group variation is equally represented in each dimension. Since the latent roots are the successive maxima of the ratio of between-group to within-group variation, loadings corresponding to roots less than 1 are for dimensions in the canonical variate space that exhibit more within-group variation than between-group variation. In the example, all three roots are greater than 1, suggesting that differences between the four groups exist in all three dimensions; this is in accordance with the significance tests, which indicate a dimensionality greater than 2. It may not be easy to interpret latent vectors but, for example, the second latent vector here contrasts the third variable (the height of the bow of the brooch) with the others. This suggests that the second canonical variate distinguishes brooches with a relatively narrow shape. The directive FACROTATE (10.2.3) may help you to interpret the loadings. However, canonical variates analysis and principal components analysis can still be useful, even if the loadings cannot be interpreted.

The scores for the means are arranged so that their centroid, weighted by group size, is at the origin. This is done by subtracting a constant term, for each canonical variate dimension, from the scores initially formed as a linear combination of the group means of the original variables. For example, the constant term of -19.90 occurs in the second score for the third mean, -3.545, formed as:

$$-2.656\bar{v}_{13} + 1.631\bar{v}_{23} - 1.708\bar{v}_{33} - 0.623\bar{v}_{43} - 0.758\bar{v}_{53} - 2.028\bar{v}_{63} - 0.110\bar{v}_{73} - 19.90$$

where v_{ij} is the mean of the ith variable for the jth group. If you ask for the group mean scores to be printed, then the corresponding constant terms are also printed, as shown in the above example. You can see from the canonical variate means that the second canonical variate separates the third group from the other three.

Results can be printed for a subset of the latent roots by setting the NROOTS and SMALLEST options of CVA. NROOTS specifies the number of roots for which you want the results to be printed. By default these will be the largest roots, unless you set SMALLEST=yes; then the results will be printed for the smallest non-zero roots. When you print a subset of the results, residuals can be formed and printed from the dimensions that are not presented.

If you ask for distances, they are formed from the group mean scores for the canonical variate dimensions that are printed. If results are printed for the full dimensionality, the distances will be Mahalanobis distances between the groups.

The SCORES parameter allows you to save the canonical variate means so that, for example, you can plot them. You may also want to plot canonical variate scores for the units in the data matrix; to do this you must first save the loadings (the latent vectors) using the LRV parameter. The next section of the example shows this for the first two canonical variates.

EXAMPLE 10.2.2b

```
  38  LRV [ROWS=Data; COLUMNS=2] L
  39  MATRIX [ROWS=Groupno; COLUMNS=2] Meanscrs
  40  CVA [PRINT=residuals,distances; NROOTS=2] WSSPM=W; LRV=L; SCORES=Meanscrs

40.......................................................................

***** Canonical variate analysis *****

*** Residuals ***

              1      0.613
              2      1.584
              3      0.825
              4      2.241

*** Inter-group distances ***

        1   0.000
        2   4.231    0.000
        3   5.802    4.215    0.000
        4   5.913    2.140    6.007    0.000
              1        2        3        4

  41  MATRIX [ROWS=28; COLUMNS=Data] Datamat
  42  CALCULATE Datamat$[*;1...7] = Data[]
  43  MATRIX [ROWS=28; COLUMNS=2] Unitscrs
  44  CALCULATE Unitscrs = Datamat*+L[1]
  45  VARIATE [NVALUES=28] Uscores[1,2]
```

10.2 Analyses based on sums of squares and products

```
 46    & [NVALUES=Groupno] Mscores[1,2]
 47  CALCULATE Uscores[1,2]=Unitscrs$[*;1,2]
 48    & Mscores[1,2]=Meanscrs$[*;1,2]
 49    & Uscores[1,2]=Uscores[1,2]-MEAN(Uscores[1,2])
 50  GRAPH [TITLE='Scores for means (*) and units'; \
 51      YTITLE='canonical variate 2'; XTITLE='canonical variate 1'; \
 52      EQUAL=scale; NROWS=37; NCOLUMNS=61] Uscores[2],Mscores[2]; \
 53      X=Uscores[1],Mscores[1]; SYMBOLS=Groupno,*
```

```
                       Scores for means (*) and units

                -+---------+---------+---------+---------+---------+-
              I                                                     I
        4.5 I                                                        I
              I                                                     I
              I                                      1              I
              I                                                     I
              I                                                     I
        3.0 I                                                        I
              I                                                     I
              I         4                         1                 I
    c         I                                     1*              I
    a         I                                     1               I
    n         I                                                     I
    o   1.5 I                                                        I
    n         I                    4 2                              I
    i         I                *  42                                I
    c         I         4         22  2              11             I
    a         I                  2 2                                I
    l         I                                                     I
        0.0 I                       *                                I
    1         I                     :    2                          I
              I              4                                      I
              I                                                     I
    v         I                                                     I
    a         I                     22                              I
    r  -1.5 I                                                        I
    i         I                                                     I
    a         I                                                     I
    t         I                                                     I
    e         I              3         3                            I
       -3.0 I                                                        I
    2         I                                                     I
              I                          *                          I
              I                          3       3                  I
              I                                                     I
       -4.5 I                                                        I
              I                          3                          I
              I                                                     I
              I                                                     I
              I                                                     I
       -6.0 I                                                        I
                -+---------+---------+---------+---------+---------+-
                -6.       -4.       -2.        0.        2.        4.        6.
                                   canonical variate 1
```

The matrix Meanscrs, with its rows indexed by the grouping factor, is declared on line 39 to hold the canonical variate means. The LRV structure L, with rows indexed by the pointer Data, is declared on line 38. Both Meanscrs and L are declared to have two columns corresponding to the first two canonical variates. The CVA statement (line 40) prints only the residuals and distances. The NROOTS option specifies that the results to be printed are for the two largest latent roots; the residuals thus correspond to the remaining roots, here only the third. The printed distances are formed from the first two canonical variate means, and thus are the distances between the group means on the graph.

Lines 41–44 obtain a matrix of canonical variate scores for the units. First, the matrix Datamat is set up to hold the data variates as its columns (lines 41 and 42). The matrix of unit scores is formed by post-multiplying Datamat by the loadings, that is by the first element of the LRV structure; L[1] has as many rows as there are variates (here seven), and as many columns as there are roots to be saved (here two). To plot the scores, you must first copy them from the matrices into variates, as illustrated in lines 45–48. Remember that the canonical variate means are arranged so that their weighted centroid is at the origin. The same adjustment needs to be made to the unit scores; this simply involves subtracting the means from the columns of unit scores (line 49).

For full details of the GRAPH statement (lines 50–53), see 7.1.2. Note that we have set option EQUAL = scale: this specifies that the same range is to be represented on the x and y axes. The NROWS = 37 and NCOLUMNS = 61 option settings specify that the graph is to be six inches square. Thus distance is equally represented in any direction on the graph. This is important for canonical variates analysis, as for other multivariate methods where you want to use graphs to examine the relative distances amongst points. The settings 37 and 61 used here are appropriate for most printers, which print six lines to the inch and ten columns to the inch; however, you should check that this is true for your printer. The settings are inappropriate for displaying graphs on a VDU; suitable settings will vary depending on the type of the VDU, but NROWS = 19 and NCOLUMNS = 57 is reasonable for many.

10.2.3 The FACROTATE directive

Principal components analysis and canonical variates analysis both define a set of dimensions (sometimes called axes) that are linear combinations of the original variables. The individual coefficients of these combinations are called loadings, and can be used to interpret the dimensions. With principal components analysis, the loadings must lie in the range $(-1, 1)$; this is the situation that we discuss in the initial part of this subsection. The situation with canonical variates analysis is slightly different and is described at the end of this subsection.

When several dimensions are considered it is possible to define an equivalent set of new dimensions, whose loadings are linear combinations of the original loadings. If the absolute values of the loadings for a new dimension are either close to 0 or close to 1, you can interpret the dimension as mainly representing only those original variables with large positive (or negative) loadings. You may sometimes want new dimensions determined by loadings like these, because they are easier to interpret. The methods by which these new dimensions can be obtained are gen-

10.2 Analyses based on sums of squares and products

erally known collectively as *factor rotation* because the new dimensions represent a rotation of the axes of the original dimensions. The FACROTATE directive provides two methods of orthogonal factor rotation: varimax rotation and quartimax rotation (Cooley and Lohnes 1971). The default method, varimax rotation, maximizes the variance of the squares of the loadings within each new dimension: the effect of this rotation should be to spread out the squared-loadings to the extremes of their range. Quartimax rotation uses the fourth power of the loadings instead of the second power.

FACROTATE
Rotates factors from a principal components or canonical variates analysis according to either the varimax or quartimax criterion.

Options

PRINT	= strings	Printed output required (communalities, rotation); default * i.e. no output
METHOD	= string	Criterion (varimax, quartimax); default v

Parameters

OLDLOADINGS	= matrices	Original loadings
NEWLOADINGS	= matrices	Rotated loadings for each set of OLDLOADINGS

The first parameter, OLDLOADINGS, specifies a list of matrices which provide the input; the columns of each of these matrices should contain the loadings for the original dimensions. The matrices to save the new loadings are listed with NEWLOADINGS parameter; each of these must have been declared in advance, and must be of the same size as the corresponding OLDLOADINGS matrix. Often it will be convenient to use the same structure for output as was used for input.

One way of supplying the loadings for the original variables is by saving the latent roots and vectors from a principal components analysis using the LRV parameter; the first structure of the LRV is then the matrix of loadings. The example below is similar to the first example of principal components analysis, above; however, here the first two latent roots and vectors are saved and used as input to the FACROTATE directive.

EXAMPLE 10.2.3a

```
  1   UNITS [NVALUES=12]
  2   POINTER [VALUES=Height,Length,Width,Weight] Dmat
```

```
  3  READ [PRINT=errors] Dmat[]

  8  LRV [ROWS=Dmat; COLUMNS=2] Latent
  9  PCP [PRINT=loadings] Dmat; LRV=Latent
```

9..

***** Principal components analysis *****

*** Latent Vectors (Loadings) ***

```
                 1          2          3          4
   Height   -0.21529    0.37981    0.78747   -0.43506
   Length   -0.25623    0.86524   -0.34389    0.25970
    Width   -0.74104   -0.21726   -0.37937   -0.50964
   Weight   -0.58211   -0.24474    0.34308    0.69537

 10  FACROT [PRINT=rotation,communalities] Latent[1]
```

10...

***** Factor rotation *****

*** Communalities ***

```
           1
   1    0.1906
   2    0.8143
   3    0.5963
   4    0.3988
```

*** Rotated factors ***

```
           1          2
   1   -0.0630    0.4320
   2    0.0747    0.8993
   3   -0.7694    0.0660
   4   -0.6312   -0.0172
```

The LRV structure Latent is declared on line 8, and is used on line 9 to save the latent roots and vectors. The full set of latent vectors is printed from the PCP directive to allow you to compare the original loadings with those after rotation. The original loadings seem to tell us that the first new axis is some negative measure of overall size, and that the second is a contrast between the first two variables (Height and Length) and last two (Width and Weight). The new loadings give the first axis as largely consisting of Width and Weight, and the second as largely consisting of Height and Length.

Note that under either method of factor rotation, the total contribution of each of the original variables always remains the same as in the input set of loadings (for mathematical reasons). These contributions are called the *communalities* of the variables, and can be expressed as the sum of the squared loadings: they indicate how much of the variation of each of the original variables is retained in either set of dimensions (whether the original set from the principal components analysis, or the new set from the rotation). For example, the communality for the first variable can be calculated from the set of new dimensions as follows

$$0.1906 = (-0.0630)^2 + (0.4320)^2$$

Equivalently, from the original set, it is

$$0.1906 = (-0.2153)^2 + (0.3798)^2$$

If you keep all the loadings from a principal components analysis, each of the variables will have communality 1. Factor rotation in this case will simply give a set of new loadings, each of which will represent just one of the variables, with loading 1. Thus factor rotation is sensible only if you keep merely the higher-dimensional loadings.

The loadings from canonical variates analysis are not constrained to lie in the range $(-1, 1)$. The factor rotation methods operate in a similar manner as for principal component loadings. Again, the objective is to obtain loading values, such that each is either relatively small or relatively large. Also the communalities of the variables remain the same in the rotated loadings as in the original loadings, and the new loadings are obtained as an orthogonal rotation of the old loadings. However, the complete set of loadings can generally be retained from canonical variates analysis and used for factor rotation, without giving meaningless results. This is because the original dimensions from the canonical variates analysis do not contain all the dimensionality of the original variables, unless the number of variables is less than the number of groups. So a factor rotation of all the dimensions will not merely recover the original variables, as would happen with loadings from principal components analysis. This is illustrated in the next example, where the loadings from the first example of canonical variates analysis (10.2.2) are rotated.

EXAMPLE 10.2.3b

```
  1   UNITS [NVALUES=28]
  2   POINTER [VALUES=Foot_lth,Bow_ht,Coil_dia,Elem_dia,Bow_wdth, \
  3     Bow_thck,Length] Data
  4   FACTOR [LEVELS=4] Groupno
 '5   READ [PRINT=errors] Groupno,Data[]

 35   SSPM [TERMS=Data[]; GROUPS=Groupno] W
 36   FSSP W
 37   LRV [ROWS=Data; COLUMNS=3] L
 38   CVA [PRINT=loadings] WSSPM=W; LRV=L

38..............................................................

***** Canonical variate analysis *****

*** Latent Vectors (Loadings) ***

             L['Vectors']
                  1          2          3
       Data
     Foot_lth   1.130     -2.656      3.397
```

```
       Bow_ht       -0.633      1.631      4.799
       Coil_dia      3.501     -1.708      1.450
       Elem_dia     -2.669     -0.623     -2.802
       Bow_wdth     -3.468     -0.758      0.757
       Bow_thck      1.859     -2.028     -2.478
       Length       -1.279     -0.110     -3.598


39   FACROTATE OLDLOADINGS=L[1]; NEWLOADINGS=L[1]
40   PRINT L[1]

             L['Vectors']
                   1           2          3
       Data
       Foot_lth    -0.135     -4.381      0.810
       Bow_ht      -0.210     -1.670      4.823
       Coil_dia     2.513     -3.254     -0.612
       Elem_dia    -2.560      2.192     -2.001
       Bow_wdth    -3.530      0.111      0.837
       Bow_thck     1.074     -0.471     -3.512
       Length      -1.041      2.606     -2.593
```

Rather than print the rotated loadings directly from the analysis (line 39), the program saves and prints them separately (line 40). This might be appropriate if you intend to calculate canonical variate scores for the units, in the rotated factor space, as was shown in the second example of canonical variates analysis (10.2.2). If you do intend to do this, you will also have to calculate new canonical variate means in the rotated factor space; however, this is easy to do as they are simply the group means of the rotated scores for the units.

10.3 Forming measures of association

As was explained at the beginning of this chapter, many forms of multivariate analysis operate on symmetric matrices that give similarities between all pairs of samples: these are termed *Q-methods*. The FSIMILARITY directive is concerned with forming similarity matrices, essentially using the method described by Gower (1971). The similarity coefficient that is calculated allows variables to be qualitative, quantitative, or dichotomous, or mixtures of these types; values of some of the variables may be missing for some samples. The values of a similarity coefficient vary between zero and unity, though some authors express them as percentages in the range 0–100%; Genstat prints percentages. Two samples have a similarity of unity only when both have identical values for all variables; a value of zero occurs when the values for the two samples differ maximally for all variables. Thus similarity is the complement of dissimilarity, and to convert a similarity s_{ij} into a dissimilarity you can evaluate expressions like $1 - s_{ij}$ or $\sqrt{(1 - s_{ij})}$.

Whether a set of dissimilarities obeys the metric axioms (particularly the triangle inequality), or can be regarded as being generated by distances between pairs of points in a multidimensional Euclidean space, depends on the particular

10.3 Forming measures of association

coefficient and on the data themselves. Genstat can evaluate similarities using many of the standard similarity coefficients for qualitative and quantitative variables; Gower and Legendre (1986) discuss some of the properties of these coefficients. In Genstat the resulting similarity matrices are ordinary symmetric matrices, and you can operate on them using the matrix operations (5.7); their main use in multivariate analysis is for principal coordinates analysis (10.4.1), or other forms of metric scaling or non-metric scaling, or for hierarchical cluster analysis (10.5).

10.3.1 The FSIMILARITY directive

FSIMILARITY
Forms a similarity matrix or a between-group-elements similarity matrix or prints a similarity matrix.

Options

PRINT	= string	Printed output required (similarity); default * i.e. no output
STYLE	= string	Print percentage similarities in full or just the 10% digit (full, abbreviated); default f
METHOD	= string	Form similarity matrix or rectangular between-group-element similarity matrix (similarity, betweengroupsimilarity); default s
SIMILARITY	= matrix	Input or output matrix of similarities; default *
GROUPS	= factor	Grouping of units into two groups for between-group-element similarity matrix; default *
PERMUTATION	= variate	Permutation of units (possibly from HCLUSTER) for order in which units of the similarity matrix are printed; default *
UNITS	= text or variate	Unit names to label the rows of the similarity matrix; default *

Parameters

DATA	= variates	The data values
TEST	= scalars	Test type, defining how each variate is treated in the calculation of the similarity between each unit

| RANGE | = scalars | Range of possible values of each variate; if omitted, the observed range is taken |

FSIMILARITY allows you to form a symmetric matrix of similarities, or a rectangular matrix of similarities between the units in two groups. You can save either form of similarity matrix, using the SIMILARITY option. FSIMILARITY can also be used to print the symmetric matrix of similarities after it has formed it; alternatively, you can input an existing similarity matrix for printing, using the SIMILARITY option.

The three parameters of the FSIMILARITY directive are also used, for the same purposes, in the directives RELATE (10.4.3), HLIST (10.6.1), and HSUMMARIZE (10.6.3). The DATA parameter specifies a list of variates, all of which must be of the same length. If any of the variates is restricted, or if the factor in the GROUPS option is restricted, then that restriction is applied to all the variates. Any restriction on any other variate must be to the same set of units. The dimension of the resulting symmetric matrix of similarities is taken from the number of units that contribute to the similarity matrix. If the DATA parameter is present, then so also must the TEST parameter. The case where the DATA parameter is not present is dealt with below, under discussion of the SIMILARITY option.

The TEST parameter specifies a list of scalars, one for each variate in the DATA parameter list, that define the "type" of each variate. The type of a variate determines how differences in variate values for each unit contribute to the overall similarity between units. For those variates that you want to contribute to the similarity, the type is a number from 1 up to 5: type 1 is appropriate for dichotomous variables, type 2 for qualitative variables, and types 3, 4, and 5 give different ways for handling quantitative variables. If you want to exclude a variate from contributing, then the type of the variate should be given as zero or missing. The form of contribution to the similarity is:

Type	Contribution	Weight
0 or *	0 (exclude)	0
1	if $x_i = x_j = 1$, then 1	1
	if $x_i = x_j = 0$, then 0	0
	if $x_i \neq x_j$, then 0	1
2	if $x_i = x_j$, then 1	1
	if $x_i \neq x_j$, then 0	1
3	City block: $1 - \dfrac{|x_i - x_j|}{\text{range}}$	1

4		Ecological: same as City block, unless $x_i = x_j = 0$	1 0
5		Euclidean: $1 - \left(\dfrac{x_i - x_j}{\text{range}}\right)^2$	1

The measure of similarity is formed by multiplying each contribution by the corresponding weight, summing all these values, and then dividing by the sum of the weights.

The RANGE parameter contains a list of scalars, one for each variate in the DATA list. This allows you to check that the values of each variate lie within the given range. If any variate fails the range check, FSIMILARITY gives an error diagnostic. You can omit the RANGE parameter for all or any of the variates by giving a missing identifier or a scalar with a missing value. If you omit RANGE, Genstat prints the name, the minimum value, and the range for each variate.

The METHOD option controls what type of matrix is produced. METHOD=similarity, the default, gives a symmetric matrix of similarities amongst a single set of units. METHOD=betweengroupsimilarity gives a rectangular matrix of similarities between two sets of units. To form a rectangular matrix of similarities, you must define the grouping of units by setting the GROUPS option (see below).

The PRINT, STYLE, and PERMUTATION options govern the printing of a symmetric matrix of similarities; you cannot use the FSIMILARITY directive to print a rectangular matrix of similarities between group elements. You can either form the similarity matrix within FSIMILARITY, or input it by the SIMILARITY option. To print the similarity matrix you should set option PRINT=similarity. The STYLE option has two settings, 'full' (the default) or 'abbreviated'. The similarity matrix printed in full style has its values displayed as percentages with one decimal place. If you put STYLE=abbreviated, the values of the similarity matrix are printed as single digits with no spaces, the digit being the 10's value of the similarity as a percentage. The PERMUTATION option allows you to specify a variate with values corresponding to the order in which you want the rows of the similarity matrix to be printed. The reordering of the rows is most effective when the permutation arises from a hierarchical clustering and corresponds to the dendrogram order. This is shown in 10.5.1.

EXAMPLE 10.3.1

```
  1   " Data from   Observers Book of Automobiles   1986
 -2
 -3     16 Italian cars and 12 measurements/characteristics
 -4
 -5     1.  engine capacity          c.c.          Engcc
```

```
 -6    2.  number of cylinders              Ncyl
 -7    3.  fuel tank            litres      Tankl
 -8    4.  unladen weight       kg          Weight
 -9    5.  length               cm          Length
-10    6.  width                cm          Width
-11    7.  height               cm          Height
-12    8.  wheelbase            cm          Wbase
-13    9.  top speed            kph         Tspeed
-14   10.  time to 100kph       secs        Stst
-15   11.  carburettor/inj/diesel 1/2/3     Carb
-16   12.  front/rear wheel drive 1/2       Drive
-17        "
 18   UNITS [NVALUES=16]
 19   VARIATE Engcc,Ncyl,Tankl,Weight,Length,Width,Height,Wbase,Tspeed,Stst, \
 20      Carb,Drive,Vct[1...3]
 21   POINTER Cd; VALUES=!P(Engcc,Ncyl,Tankl,Weight,Length, \
 22      Width,Height,Wbase,Tspeed,Stst)
 23   SCALAR Ty2,Ty3,Ty5; VALUE=2,3,5
 24   OPEN 'CARS.DAT'; CHANNEL=2; FILETYPE=input
 25   READ [CHANNEL=2] #Cd,Carb,Drive
```

```
    Identifier    Minimum       Mean    Maximum    Values   Missing
         Engcc        965       2129       5167        16         0
          Ncyl      4.000      5.375     12.000        16         0   Skew
         Tankl      35.00      61.75     120.00        16         0
        Weight        720       1054       1506        16         0
        Length      338.0      410.6      459.0        16         0
         Width      149.0      167.5      200.0        16         0
        Height      107.0      133.7      146.0        16         0
         Wbase      216.0      242.9      266.0        16         0
        Tspeed      134.0      197.8      291.0        16         0
          Stst       4.90      10.29      18.90        16         0
          Carb      1.000      1.438      3.000        16         0   Skew
         Drive      1.000      1.625      2.000        16         0
```

```
 26   TEXT [16] Carname; VALUES=!T('Estate','Arna1.5','Alfa2.5','Mondialqc',\
 27      'Testarossa','Croma','Panda','Regatta','Regattad','Uno',\
 28      'X19','Contach','Delta','Thema','Y10','Spider')
 29   FACTOR [Carname; LEVELS=16] Fcar; VALUES=!(1...16)
 30   SYMMETRICMATRIX [ROWS=Carname] Carsim
 31   " Form similarity matrix between cars."
 32   FSIMILARITY [SIMILARITY=Carsim; PRINT=similarities] #Cd,Carb,Drive; \
 33      TEST=4(Ty3),4(Ty5),2(Ty3),2(Ty2)
```

```
    Variate    Minimum      Range
      Engcc      965.0     4202.0
       Ncyl      4.000      8.000
      Tankl      35.00      85.00
     Weight      720.0      786.0
     Length      338.0      121.0
      Width      149.0       51.0
     Height      107.0       39.0
      Wbase      216.0       50.0
     Tspeed      134.0      157.0
       Stst      4.900     14.000
       Carb      1.000      2.000
      Drive      1.000      1.000
```

Similarity matrix computed, length = 136

** Similarity matrix **

```
Estate      1    ----
Arna1.5     2    95.0   ----
Alfa2.5     3    78.0   76.7   ----
Mondialqc   4    54.7   50.9   73.6   ----
```

10.3 Forming measures of association

```
Testarossa   5    36.2  32.4  55.1  77.0  ----
Croma        6    76.5  75.2  75.4  73.6  51.7  ----
Panda        7    74.7  78.6  55.9  29.4  11.0  55.5  ----
Regatta      8    95.2  94.3  80.3  55.7  37.3  80.9  73.7  ----
Regattad     9    81.2  79.4  65.6  49.4  30.9  74.6  69.2  84.2  ----
Uno         10    82.7  86.6  63.9  37.4  19.0  63.5  91.1  81.7  76.5  ----
X19         11    81.3  80.8  76.2  55.7  40.7  57.8  70.3  76.9  63.3  73.0
Contach     12    42.8  39.0  57.0  67.0  88.3  41.7  17.5  42.2  27.5  25.5
Delta       13    93.1  93.5  80.6  55.0  36.6  79.0  74.4  92.7  77.4  82.4
Thema       14    75.4  74.2  74.3  74.5  51.5  98.0  54.4  79.9  73.5  62.4
Y10         15    82.6  86.1  63.9  37.5  19.0  63.5  91.9  81.7  68.2  88.5
Spider      16    78.6  75.8  81.2  69.1  51.0  71.2  60.0  76.3  69.6  65.3

                    1     2     3     4     5     6     7     8     9    10

X19         11    ----
Contach     12    47.2  ----
Delta       13    77.2  42.5  ----
Thema       14    56.8  41.6  77.9  ----
Y10         15    77.5  25.6  82.0  62.5  ----
Spider      16    86.1  49.2  77.7  70.1  67.6  ----

                   11    12    13    14    15    16
```

In this example, the RANGES parameter has been omitted so the variate names, minima and ranges are printed. Note also that a message is printed giving the number of values of the similarity matrix. The number of missing values that have been formed is also printed, if there are any.

You use the GROUPS option to specify a partition of the units into two groups, by giving a factor with two levels. The units with level 1 of the factor correspond to the rows of the matrix, while the units with level 2 correspond to the columns. As already explained, you cannot print this matrix using the FSIMILARITY directive; you must save the matrix and then use PRINT (4.2).

The UNITS option allows you to label the rows of the output similarity matrix if the variates of the DATA parameter do not have any unit labels, or if you want to use different labels from those labelling the units of the variates. This labelling also applies to the rows and columns of a matrix of similarities between group elements.

10.3.2 The REDUCE directive

Sometimes you may want to regard an $n \times n$ similarity matrix S as being partitioned into $b \times b$ rectangular blocks. For example, the cars in 10.3.1 could be classified by their manufacturer. You might then want to form a reduced matrix of similarities, between the different manufacturers instead of between the individual members of the full set of cars. A further example is when there are b soil samples for each of which information is recorded on several soil horizons, possibly different in the different samples. The n sampling units are the full set of horizons that have been observed for the soil samples. The similarity matrix S can be computed for these in the usual way (10.3.1). However you may be more interested in obtaining a reduced similarity matrix between the b soil samples. To do this you have to arrange for each of the b^2 blocks of the full matrix to be replaced by a single value. Each diagonal block must be replaced by unity. Several possibilities exist for replacing the off-diagonal blocks: for example, the maximum, minimum, or mean similarity within the block. Alternatively you can take the view that at least the first horizons of each of two soil samples should agree; you would then replace the block by its first value. Rayner (1966) suggested a more complex method which recognized that certain horizons might be absent from some soil samples; this leads to finding successive optimal matches, conditional on the constraint that one horizon cannot match a horizon that has already been assigned to a higher level; after finding these optima, an average is taken for each horizon. This is termed the *zigzag* method. Again Genstat produces a symmetric similarity matrix, which you may operate on subsequently by the matrix operations or by the appropriate multivariate directives.

REDUCE
Forms a reduced similarity matrix (referring to the GROUPS instead of the original units).

Option

PRINT	= string	Printed output required (similarities); default * i.e. no output
METHOD	= string	Method used to form the reduced similarity matrix (first, last, mean, minimum, maximum, zigzag); default f

Parameters

SIMILARITY	= symmetric matrices	Input similarity matrix
REDUCEDSIMILARITY	= symmetric matrices	Output (reduced) similarity matrix
GROUPS	= factors	Factor defining the groups
PERMUTATION	= variates	Permutation order of units (for METHOD = f, l, or z)

The SIMILARITY parameter specifies the similarity matrix for the full set of n observations; this must be present and have values. The REDUCEDSIMILARITY parameter specifies an identifier for the reduced similarity matrix, of order b; this will be declared implicitly if you have not declared it already. The factor that defines the

10.3 Forming measures of association

classification of the units into groups must be specified by the GROUPS parameter. The units can be in any order, so that for example the units of the first group need not be all together nor given first. The labels of the factor label the reduced similarity matrix.

The PERMUTATION parameter, if present, must specify a variate. It defines the ordering of samples within each group, and so must be specified for methods 'first', 'last', and 'zigzag'. Within each group, the unit with the lowest value of the permutation variate is taken to be the first sample, and so on. Genstat will, if necessary, use a default permutation of one up to the number of rows of the similarity matrix.

If you set option PRINT=similarities, the values of the reduced symmetric matrix are printed as percentages.

The METHOD option specifies how the reduced similarity matrix is to be formed. In this example, the similarity matrix for each car is reduced to a similarity matrix for each manufacturer as represented by the factor Maker. The METHOD option is set to 'mean'. The resulting matrix is printed, and finally stored in the symmetric matrix Makersim.

EXAMPLE 10.3.2

```
34    " Form reduced similarity matrix for makers."
35    FACTOR [LEVELS=6; LABELS=!t(Fiat,'Alfa Romeo',Lancia,Ferrari,Lamborghini,\
36        Pinninfarina)] Maker; VALUES=!(2,2,2,4,4,1,1,1,1,1,1,5,3,3,3,6)
37    SYMMETRICMATRIX [ROWS=Maker] Makersim
38    REDUCE [PRINT=similarities; METHOD=mean] Carsim; \
39        REDUCEDSIMILARITY=Makersim; GROUPS=Maker

Reduced   mode 3    mean

** Reduced similarity matrix **

Fiat           1    ----
Alfa Romeo     2    78.0   ----
Lancia         3    76.6   80.4   ----
Ferrari        4    41.0   50.5   45.7   ----
Lamborghini    5    33.6   46.3   36.5   77.6   ----
Pinninfarina   6    71.4   78.5   71.8   60.0   49.2   ----
                     1      2      3      4      5      6
```

10.3.3 Forming associations using CALCULATE

An appropriate similarity coefficient can be calculated by FSIMILARITY (10.3.1) for most sets of data. However, many different coefficients of similarity, or distance, have been suggested (see, for example, Gower and Legendre 1986). FSIMILARITY does not cover all of these, but you will generally be able to form them by using CALCULATE (5.1). Sometimes you may need to convert similarities to dissimilarities (distances), or vice versa. This can be done in many ways; the most common are $D = 1 - S$ and $D = \sqrt{(1-S)}$, but $D = -\log(S)$ can also be useful. So there are also situations where you may need to transform such matrices using CALCULATE. For example, by putting

    FSIMILARITY [SIMILARITY = Smat] V[1...9]; TEST = 5

the symmetric matrix Smat will contain similarities constructed from Euclidean squared distances standardized by the ranges of the variates. If you do not want standardization by range, Euclidean distances can be obtained from the PCO directive (10.4.1); but these may then have to be transformed to similarities, for example if you want to use hierarchical cluster analysis (10.5). If Smat has been obtained from the PCO directive, its values should be squared first, to get Euclidean squared distances, and then transformed to similarities:

    CALCULATE Smat = Smat*Smat
    & Smat = 1 − Smat/MAX(Smat)

The FSIMILARITY directive allows variates of different types; for example, dichotomous variates (with values 0 or 1) can have the TEST parameter set as 1 or 2 to give respectively the Jaccard or the simple matching coefficients of similarity. Other variates with values on a continuous scale can have the TEST parameter set as 3 or 5 giving respectively the city-block or standardized Euclidean coefficients. When both types of variates are present, the resulting similarities will be a weighted average of the component similarities. For example, with five dichotomous variates, Binary[1...5], and three continuous variates, Cont[1...3]

    FSIMILARITY [SIMILARITY = Mixed] Binary[1...5],Cont[1...3]; TEST = (1)5,(3)3

will give the similarity matrix Mixed as a weighted average of the Jaccard similarity matrix constructed from Binary[1...5] and the city-block similarity matrix constructed from Cont[1...3]. If, instead of the city-block coefficient, you want to use the unstandardized Euclidean coefficient, you must construct this yourself, as shown above, and then do the averaging:

    SYMMETRIC [ROWS = N] Jaccard,Euclid,Mixed
    FSIMILARITY [SIMILARITY = Jaccard] Binary[1...5]; TEST = 1
    PCO Cont; DISTANCES = Euclid
    CALCULATE Euclid = Euclid*Euclid
    & Euclid = 1 − Euclid/MAX(Euclid)
    & Mixed = (5*Jaccard + 3*Euclid)/8

Gower (1985b) lists 15 different similarity coefficients that have been used for dichotomous variables. Of these, only the simple-matching and Jaccard coefficients can be formed directly with FSIMILARITY; these are the most commonly used. However, a further seven similarity coefficients can be formed using either, or both, of these two. For example, for the five variates Binary[1...5] the Czekanowski coefficient can be calculated from the Jaccard coefficient, using these statements:

```
FSIMILARITY [SIMILARITY=Jaccard] Binary[1...5]; 1
CALCULATE Czekanow = 2 * Jaccard / (1 + Jaccard)
```

Gower (1985b) gives details of the other relationships.

The city-block and Euclidean measures of distance are special cases of the Minkowski distance, which for some positive value of t is:

$$d_{ij} = \left[\sum \left(\frac{|x_{ik} - x_{jk}|}{r_k}\right)^t\right]^{1/t}$$

where r_k is usually the range of the kth variable. Although similarities derived from this distance cannot be formed with FSIMILARITY directly, the symmetric matrix Minkwski giving such similarities can be formed from the variates X[1...p] using these statements:

```
CALCULATE Minkwski = 0
FOR Thisx = X[1...p]
FSIMILARITY [SIMILARITY=Temp] Thisx; 3
CALCULATE Minkwski = Minkwski + Temp**t
ENDFOR
CALCULATE Minkwski = EXP(LOG(Minkwski)/t)
```

10.4 Ordination from associations

The term *ordination* is used mainly in biometrics, particularly in ecology, where it usually refers to attempts to order a set of objects along some environmental gradient. Archeologists use the term *seriation* to refer to the same set of techniques, whilst the phrase *multidimensional scaling* is used in some other areas. There is no fixed statistical terminology for these methods; however, they have in common an attempt to "order" a set of objects in one dimension with a generalization to give some useful distribution of the objects in multidimensional space. Several of the well-known ordination methods are available in Genstat, for example the principal components analysis of 10.2.1, and correspondence analysis for which there is a procedure in the Genstat Procedure Library (12.1.1). These methods operate with data in the form of a data matrix or a two-way table. A more general method is principal coordinates analysis, or metric scaling, which operates with data in the form of a symmetric matrix of associations; you can produce the associations by the methods of 10.3. Principal coordinates analysis is provided in Genstat by the PCO directive.

Given a symmetric matrix, A say, with values representing the associations amongst a set of n units, principal coordinates analysis (Gower 1966) attempts to find a set of points for the n units in a multidimensional space so that the squared distance between the ith and jth points is given by:

$$d_{ij} = a_{ii} + a_{jj} - 2a_{ij}$$

If A is a similarity matrix then a_{ii} and a_{jj} are both equal to 1, and so this is equivalent to:

$$d_{ij} = 2 \times (1 - a_{ij})$$

Thus similar units are placed close together and dissimilar units are further apart.

Often the data consist of distances rather than similarities (10.3.3). If B is such a matrix, so that b_{ij} gives the observed distance between the ith and jth units, then the preliminary transformation

$$A = - B \times B / 2$$

will give points with inter-point squared distance

$$\begin{aligned} d_{ij} &= a_{ii} + a_{jj} - 2a_{ij} \\ &= 0 + 0 - 2 \times (b_{ij} \times b_{ij} / 2) \\ &= b_{ij}^2 \end{aligned}$$

Therefore the analysis will give points that generate the supplied distances.

The coordinates of the points are arranged so that their centroid, or mean position, is at the origin. Furthermore they are arranged relative to their principal axes, so that the first dimension of the solution gives the best one-dimensional fit to the full set of points, the first two dimensions give the best two-dimensional fit, and so on. The analysis also gives the distances of the points from their centroid, the origin. Associated with each dimension of the set of coordinates is a latent root which is the sum of squares of the coordinates of all the points in that dimension.

For n units, if there is an exact solution it will be in at most n-1 dimensions. However, such a solution may not always be available, because the matrix of distances derived from the associations may not be Euclidean: that is, the distances may not be reproducible by points in a Euclidean space of any number of dimensions. If an incomplete solution results, either because the Euclidean property does not hold or because not all the dimensions are to be used, then a residual can be calculated for each unit; this residual is the difference between (a) the distance from the point for that unit in the incomplete solution to the centroid, and (b) the equivalent distance derived from the original data. When the Euclidean property does not hold, some of the residuals may be complex numbers; Genstat represents these as missing values.

If you regard a set of p variables of length n as giving the coordinates of a set of n points in p dimensions, then you can construct the symmetric matrix with values that give the Euclidean distance between the n points (for example B above). If this matrix is then transformed to an association matrix as

$$A = -B \times B/2$$

the principal coordinates analysis of the association matrix will give identical

10.4 Ordination from associations

results to a principal components analysis of the original set of variables.

Another special case of principal coordinates analysis occurs when a within-group SSPM structure is to be analysed. Now you can calculate Mahalanobis squared distances amongst the group means as

$$d_{ij} = (x_i - x_j)' W^{-1} (x_i - x_j)$$

where x_i is the row vector of means for the ith group, and W is the pooled within-group covariance matrix. These squared distances can be transformed to associations, and used as input to principal coordinates analysis to obtain an ordination of the groups. In general, results from this will be different from those of canonical variates analysis, since the ordination operates on a Mahalanobis distance matrix unweighted by group size, whereas the CVA directive operates on a matrix of between-group sums of squares and products, weighted by group size.

Having obtained an ordination, you may sometimes want to add points to the ordination for additional units. For example, with canonical variates analysis, Genstat gives the scores for the group means; you may want to add points to the group-mean ordination for each of the units. As shown in the example of 10.2.2, it is easy to take the data for the new units, apply the centring of the analysis, and use the loadings matrix to get coordinates for the new units.

When you use principal coordinates analysis to analyse an association matrix, there is no loadings matrix; so the method illustrated in 10.2.2 cannot be used to calculate coordinates. However, if you know the squared distances of the new units from the old, the technique of Gower (1968) can be used to add points to the ordination for the new units. You can do this in Genstat by using the ADDPOINTS directive, together with results saved from the preceding PCO directive.

10.4.1 The PCO directive

PCO
Performs principal coordinates analysis, also principal components and canonical variates analysis (but with different weighting from that used in CVA) as special cases.

Options

PRINT	= strings	Printed output required (roots, scores, loadings, residuals, centroid, distances); default $*$ i.e. no output
NROOTS	= scalar	Number of latent roots for printed output; default $*$ requests them all to be printed
SMALLEST	= string	Whether to print the smallest roots instead of the largest (no, yes);

		default n
Parameters		
DATA	= identifiers	These can be specified either as a symmetric matrix of similarities or transformed distances or, for the canonical variate analysis, as an SSPM containing within-group sums of squares and products etc. or, for principal components analysis, as a pointer containing the variates of the data matrix
LRV	= LRVs	Latent vectors (i.e. coordinates or scores), roots and trace from each analysis
CENTROID	= diagonal matrices	Squared distances of the units from their centroid
RESIDUALS	= matrices	Distances of the units from the fitted space
LOADINGS	= matrices	Principal component loadings, or canonical variate loadings
DISTANCES	= symmetric matrices	Computed inter-unit distances calculated from the variates of a data matrix, or inter-group Mahalanobis distances calculated from a within-group SSPM

The PCO directive is used for principal coordinates analysis. This method encompasses principal components analysis and a form of canonical variates analysis as special cases as explained above.

There are six sections of output from PCO:

    roots   prints the latent roots and trace;
    scores   prints the principal coordinate scores;
    loadings   when the directive is being used for principal components analysis or canonical variates analysis, this specifies that the loadings from the analysis are to be printed;
    residuals   prints the residuals, this is relevant only if results are to be printed corresponding to only some of the latent roots;
    centroid   prints the distances (not squared distances) of each unit from their overall centroid;

10.4 Ordination from associations

distances   prints the matrix of inter-unit distances (not squared distances).

The settings of the NROOTS and SMALLEST options control the printed output of roots, scores, loadings, and residuals as described in 10.1.1. When an association matrix is being analysed, the maximum (and default) setting for NROOTS is n, the number of rows and columns of the input matrix. The printed inter-unit distances are unaffected by the setting of the NROOTS option.

In its simplest form, the PCO directive needs to be supplied with a symmetric matrix, with values giving the associations amongst a set of objects; you must also tell it what results to print. The DATA parameter supplies the symmetric matrix of associations. Option PRINT specifies what is to be printed.

Nathanson (1971) gives squared distances amongst ten types of galaxy: those of an elliptical shape, eight different types of spiral galaxy, and irregularly-shaped galaxies. The spiral types vary from those that are mainly made up of a central core (coded as types SO and SBO) to those that are extremely tenuous (Sc and SBc). The example below uses these data to form an ordination of the ten galaxy types.

EXAMPLE 10.4.1a

```
  1  TEXT [VALUES=E,SO,SBO,Sa,SBa,Sb,SBb,Sc,SBc,I] Galaxies
  2  SYMMETRICMATRIX [ROWS=Galaxies] Galaxy
  3  READ [PRINT=data,errors] Galaxy

  4  0
  5  1.87 0
  6  2.24 0.91 0
  7  4.03 2.05 1.51 0
  8  4.09 1.74 1.59 0.68 0
  9  5.38 3.41 3.15 1.86 1.27 0
 10  7.03 3.85 3.24 2.25 1.89 2.02 0
 11  6.02 4.85 4.11 3.00 2.13 1.71 1.45 0
 12  6.88 5.70 5.12 3.72 3.01 2.97 1.75 1.13 0
 13  4.12 3.77 3.86 3.93 3.27 3.77 3.52 2.79 3.29 0 :

 14  CALCULATE Galaxy = -Galaxy/2
 15  PCO [PRINT=roots,scores,centroid] Galaxy

 15.....................................................................

***** Principal coordinates analysis *****

***   Latent Roots   ***
```

	1	2	3	4	5
	6.662	3.058	1.267	1.171	0.737
	6	7	8	9	10
	0.516	0.381	0.291	0.109	-0.000

***   Percentage variation   ***

	1	2	3	4	5

46.94	21.55	8.93	8.25	5.19
6	7	8	9	10
3.64	2.69	2.05	0.77	-0.00

*** Trace ***

14.19

*** Latent vectors (coordinates) ***

	1	2	3	4	5
1	-1.3965	0.6742	-0.4808	-0.2564	-0.0072
2	-1.0082	-0.1916	0.2521	-0.0488	-0.2665
3	-0.8176	-0.3197	0.2581	-0.2306	-0.1209
4	-0.1744	-0.6571	0.0324	0.0699	0.5732
5	-0.0114	-0.5111	-0.0315	0.1844	0.2450
6	0.4237	-0.4417	-0.5654	0.5320	-0.2897
7	0.8244	-0.3341	0.5082	-0.2136	-0.3104
8	0.9375	0.2451	-0.3141	-0.0592	-0.1534
9	1.1167	0.4324	-0.1205	-0.5713	0.2104
10	0.1057	1.1036	0.4615	0.5937	0.1195

	6	7	8	9
1	-0.0422	0.1080	0.1334	-0.1166
2	0.3960	-0.1314	0.0950	0.1501
3	-0.3759	-0.0046	-0.3260	0.0324
4	-0.1177	0.1796	0.1790	0.0944
5	0.1582	-0.3563	-0.0828	-0.1802
6	0.0839	0.2260	-0.1229	0.0145
7	-0.0376	0.1792	0.1915	-0.1381
8	-0.3087	-0.3187	0.1693	0.0968
9	0.2703	0.0838	-0.1915	0.0410
10	-0.0263	0.0344	-0.0450	0.0058

* Vectors corresponding to zero or negative roots are not printed *

*** Centroid distances ***

1	2	3	4	5
1.657	1.181	1.074	0.940	0.740
6	7	8	9	10
1.065	1.132	1.140	1.392	1.346

Line 2 declares a symmetric matrix to hold the galaxy data; the rows (and columns) are labelled by the codes from Nathanson (1971). Line 14 transforms the data to associations; the data are already in the form of squared distances, and so there is no need to square them. Line 15 specifies that the PCO directive is to print the latent roots, the scores for the 10 galaxy types, and their distances from their centroid. The first two latent roots are much larger than the others, and so we can infer that a good ordination of the galaxy types can be found from the first two columns of scores (i.e. dimensions).

Ignoring for the moment the score for the irregular galaxies (0.1057), the first column of scores follows a trend from the elliptical galaxies, through the densely packed spiral types, to the tenuous spiral types. The irregularly shaped galaxies are placed somewhere near the middle of the others on this first principal axis.

The second axis places the irregular galaxies at the top of the ordination; the

10.4 Ordination from associations

other types again roughly follow a trend, but now it is curved. Remember that at most nine dimensions are needed to obtain an exact solution for 10 points; so here the last latent root is zero, and only nine columns of scores are printed.

Instead of a symmetric matrix of associations, the input to PCO can be a pointer whose values are the identifiers of a set of variates. Now the PCO directive will construct the matrix of inter-unit squared distances, and will base the analysis on associations derived from this. As described above, this is equivalent to a principal components analysis; however, the results are derived by analysing the distance matrix rather than an SSPM. When there are more units than variates, using PCO for principal components analysis is less efficient than using the PCP directive; however, if there are more variates than units the PCO directive is more efficient.

When PCO is used for principal components analysis, all the variates must be of the same length and none of their values may be missing; any restrictions on the variates are ignored.

Suppose that we have data, as parts per million, for 12 chemical elements measured on eight insects. Analysing the 12 variates with the PCP directive will form the matrix of sums of squares and products for the 12 variates, and use that for the analysis. In the next example the more efficient approach is adopted, analysing the 8-by-8 inter-insect distance matrix instead.

EXAMPLE 10.4.1b

```
  1   UNITS [NVALUES=8]
  2   POINTER Elements; VALUES=!P(Na,Mg,P,S,Cl,K,Ca,Zn,Fe,Si,Al,Cu)
  3   READ Elements[]

   Identifier    Minimum       Mean    Maximum    Values    Missing
           Na      137.0      266.6      408.0         8          0
           Mg      481.0      627.2      889.0         8          0
            P       1227       1437       1740         8          0
            S      412.0      590.6      786.0         8          0
           Cl      115.0      201.8      432.0         8          0
            K       1344       1690       2352         8          0
           Ca      28.00      71.63     127.00         8          0
           Zn      0.000      7.625     15.000         8          0
           Fe       9.00      26.12      47.00         8          0
           Si       8.00      22.00      38.00         8          0
           Al       1.00      14.50      30.00         8          0
           Cu       0.00      13.13      30.00         8          0

 12   CALC Elements[] = LOG(Elements[]+1)
 13   PCO [PRINT=roots,scores,distances] Elements

13.............................................................

***** Principal coordinates analysis *****

***   Latent Roots   ***

                    1           2           3           4           5
                25.960      11.437       3.795       1.549       0.790
```

```
                          6              7
                      0.617          0.056
```

*** Percentage variation ***

```
                  1              2              3              4              5
              58.73          25.87           8.58           3.50           1.79
                  6              7
               1.39           0.13
```

*** Trace ***

```
       44.20
```

*** Latent vectors (coordinates) ***

```
                   1              2              3              4              5
      1      -1.0057         1.9782        -0.8397         0.4943         0.0315
      2       2.6013        -0.2070        -0.4511         0.4229        -0.3377
      3       2.2071         1.7375         0.7367        -0.6271         0.0455
      4      -1.7203        -0.6858        -0.8343        -0.7711        -0.3456
      5      -1.6349         0.0188        -0.0597        -0.0440         0.6029
      6      -1.3564        -0.8063         0.7473         0.3134        -0.1868
      7      -1.1926        -0.2210         0.9985         0.1934        -0.1663
      8       2.1015        -1.8145        -0.2976         0.0183         0.3565
                   6              7
      1      -0.1066         0.0856
      2       0.1168        -0.1242
      3       0.0908         0.0243
      4      -0.0282        -0.0018
      5       0.1580        -0.1229
      6       0.4868         0.0779
      7      -0.5364        -0.0393
      8      -0.1813         0.1003
```

*** Distance matrix ***

```
   1    0.000
   2    4.263    0.000
   3    3.764    2.573    0.000
   4    3.060    4.529    4.894    0.000
   5    2.361    4.388    4.362    1.607    0.000
   6    3.291    4.204    4.502    2.030    1.520    0.000
   7    2.929    4.124    4.072    2.253    1.584    1.228
   8    4.967    1.909    3.780    4.161    4.196    3.859
          1        2        3        4        5        6
   7    0.000
   8    3.939    0.000
          7        8
```

The data are defined on lines 1 and 2, and input on line 3. You can see from the report from READ that the amounts of the 12 elements differ considerably from each other. Often with such data, logarithms are taken before any analysis; this has been done on line 12. The PRINT option in the PCO statement (line 13) requests printing of the latent roots, the scores for the eight insects, and the matrix of inter-insect distances. These are shown above. You should note that the printed distances are not squared distances, even though the analysis has been calculated from squared distances.

The third type of input to PCO is an SSPM structure. This must be a within-

10.4 Ordination from associations

group SSPM: that is, you must have set the GROUP option of the SSPM directive (3.6.2) when the SSPM was declared. Now the PCO directive will calculate the Mahalanobis distances amongst the group means, and base the analysis on them. As described above, this will give results similar to a canonical variates analysis. The representation of distances will be better than that of CVA, but CVA will be better if you are interested in loadings for discriminatory purposes.

As an example of this, we analyse the same data as in the examples of CVA (10.2.2). These consist of seven variables measured on 28 brooches; the brooches are classified into four groups.

EXAMPLE 10.4.1c

```
  1   UNITS [NVALUES=28]
  2   POINTER [VALUES=Foot_lth,Bow_ht,Coil_dia,Elem_dia,Bow_wdth, \
  3     Bow_thck,Length] Data
  4   FACTOR [LEVELS=4] Groupno
  5   READ Groupno,Data[]

    Identifier    Minimum       Mean    Maximum     Values    Missing
      Foot_lth      2.398      3.278      4.554         28          0
        Bow_ht      2.079      2.842      3.296         28          0
      Coil_dia      1.792      2.166      2.833         28          0
      Elem_dia      1.099      2.026      2.708         28          0
      Bow_wdth      3.045      4.064      5.176         28          0
      Bow_thck      2.708      3.621      4.357         28          0
        Length      3.296      4.003      4.860         28          0

 35   SSPM [TERMS=Data[]; GROUPS=Groupno] W
 36   FSSP W
 37   PCO [PRINT=roots,scores,distances] W

37..............................................................

***** Principal coordinates analysis *****

***   Latent Roots   ***

              1         2         3
         0.5437    0.1417    0.0365

***   Percentage variation   ***

              1         2         3
          75.31     19.63      5.05

***   Trace   ***

         0.7219

***   Latent vectors (coordinates)   ***

                    1         2         3
         1     0.4010   -0.0024    0.1287
         2     0.1524   -0.2415   -0.1038
         3    -0.5980   -0.0418    0.0540
         4     0.0446    0.2857   -0.0788
```

```
*** Distance matrix ***
    1    0.0000
    2    0.4160    0.0000
    3    1.0026    0.7924    0.0000
    4    0.5031    0.5388    0.7334    0.0000
             1         2         3         4
```

The first part of the example, in lines 1–36, is the same as in the first example of CVA (10.2.2). The PCO statement (line 37) prints the latent roots, the scores (that is canonical variate means for the four groups), and the matrix of inter-group Mahalanobis distances. These are shown above. The printed distances are Mahalanobis distances, not Mahalanobis squared distances.

The second and subsequent parameters of PCO allow you to save the results. The output structures must have been declared before the PCO directive, and their sizes must be consistent (see 10.1.1). The number of units that determine the attributes of the output structures differs according to the input to PCO. For a symmetric matrix the number of units is the number of rows of the matrix, and for a pointer it is the number of values of each variate in the pointer; while for an SSPM the number of units is the number of groups.

The latent roots, scores, and trace can be saved in an LRV structure using the LRV parameter. The number of rows of the LRV must equal the number of units. The number of columns of the LRV, if present, indicates the required dimensionality (see 10.1.1).

The distances of the units from their centroid can be saved in a diagonal matrix using the CENTROID parameter. The diagonal matrix must have the same number of values as the number of units, defined above.

The residuals (10.1.1) can be saved in a matrix using the RESIDUALS parameter. The matrix must have the same number of rows as the number of units, defined above, and must have one column. The residuals are formed from the dimensions that have not been saved.

If the input to PCO is a pointer or an SSPM, the principal component or canonical variate loadings can be saved in a matrix using the LOADINGS parameter. The number of rows of the matrix must equal the number of variates, either those specified by an input pointer, or those specified in the SSPM directive for an input SSPM structure. The number of columns of the matrix must agree with the dimensionality of the corresponding structure specified in the LRV parameter list.

The inter-unit distances can be saved in a symmetric matrix using the DISTANCES parameter. The number of rows of the symmetric matrix must be the same as the number of units, defined above.

To illustrate how results can be saved from PCO, the next example repeats the analysis of the squared distances amongst the 10 galaxy types, and saves the latent roots and vectors in order to draw a graph of the ordination. The resulting graph

10.4 Ordination from associations

is shown in Figure 10.4.1.

EXAMPLE 10.4.1d

```
16  LRV [ROWS=Galaxies; COLUMNS=2] L2
17  PCO Galaxy; LRV=L2
18  PRINT L2[1]
```

```
              L2['Vectors']
                    1           2
     Galaxies
          E     -1.3965      0.6742
         S0     -1.0082     -0.1916
        SB0     -0.8176     -0.3197
         Sa     -0.1744     -0.6571
        SBa     -0.0114     -0.5111
         Sb      0.4237     -0.4417
        SBb      0.8244     -0.3341
         Sc      0.9375      0.2451
        SBc      1.1167      0.4324
          I      0.1057      1.1036
```

```
19  VARIATE [NVALUES=Galaxies] PCOscore[1,2]
20  CALCULATE PCOscore[1,2] = L2[1]$[*;1,2]
21  FRAME 1; YLOWER=0; YUPPER=1; XLOWER=0; XUPPER=1
22  AXES [EQUAL=scale] 1
23  PEN 1; COLOUR=1; SYMBOLS=Galaxies
24  DEVICE 0
25  OPEN 'galaxies.grd'; 1; GRAPHICS
26  DGRAPH [TITLE='Principal coordinate analysis'; WINDOW=1; KEYWINDOW=0]  \
27      PCOscore[2]; X=PCOscore[1]; PEN=1
```

Line 16 declares the LRV structure (3.6.1) L2; the rows of this are indexed by the labels stored in the text Galaxies; only two latent roots, and their corresponding vectors, will be stored. No results are printed by the PCO statement (line 17) as the PRINT option has not been set; the LRV parameter stores in L2 the first two latent roots, the corresponding columns of scores for the galaxy types and the trace. The first structure of the LRV, the matrix of latent vectors, is printed by line 18. Variates PCOscore[1,2] are set up, in lines 19 and 20, to hold the two columns of scores. The statements on lines 21–25 initialise the high-quality graphics (7.2); the DGRAPH statement (lines 26 and 27) draws the graph (Chapter 7). As shown in Figure 10.4.1 the points are marked by the labels stored in the text Galaxies.

FIGURE 10.4.1

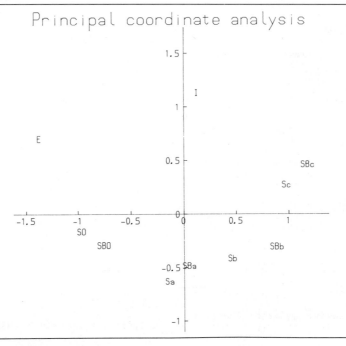
Principal coordinate analysis

10.4.2 The ADDPOINTS directive

ADDPOINTS
Adds points for new objects to a principal coordinates analysis.

Option

PRINT	= strings	Printed output required (coordinates, residuals); default * i.e. no output

Parameters

NEWDISTANCES	= matrices	Squared distances of the new objects from the original points
LRV	= LRVs	Latent roots and vectors from the PCO analysis
CENTROID	= diagonal matrices	Centroid distances from the PCO analysis
COORDINATES	= matrices	Saves the coordinates of the additional points in the space of the

10.4 Ordination from associations

RESIDUALS	= matrices	original points Saves the residuals of the new objects from that space

The input to ADDPOINTS is specified by the first three parameters. The NEWDISTANCES parameter specifies an $s \times n$ matrix containing squared distances of the s new units from the n old units. The LRV and CENTROID parameters specify structures defining the configuration of old units; these have usually been produced by a PCO directive (10.4.1).

The PRINT option controls the printed output; by default nothing is printed. The option has two settings:

    coordinates   prints the coordinates of the new points;
    residuals   prints the residual distances of the new units from the coordinates in the space of the old units.

For example, suppose that three original objects are equidistant, with a squared distance of four units amongst them. An ordination of these squared distances will place the points at the corners of an equilateral triangle of side two units. The coordinates of the three points will be $(-0.5774, 1.0000)$, $(-0.5774, -1.0000)$, and $(1.1547, 0.0000)$. Now suppose that a new object is known to be equidistant from the original objects, at some squared distance d from them. If d is $4/3$ the new object can be located precisely at the centroid of the three original points (that is at the origin), and all the distances in the system will be satisfied exactly. However if $d > 4/3$, it would be possible to satisfy all the distances in three dimensions by placing the new object at a squared distance of $d - 4/3$ above, or below, the plane in which the original points lie. The fitted coordinates in the space of the original objects will be the projection of the new point onto the plane (that is, at the centroid of the original points); the residual for the new object will be the square root of $d - 4/3$. If $d < 4/3$ the new distances can be satisfied only by introducing an imaginary third dimension in which squared distance is negative: the fitted coordinates will be the same as above, but the residual will be a complex number, which the ADDPOINTS directive will print and store as a missing value.

The other parameters can be used to save the results (see also 10.1.1). The COORDINATES parameter allows you to specify an $s \times k$ matrix to save the coordinates for the new units; the residuals can be saved in an $s \times 1$ matrix using the RESIDUALS parameter. The value k is determined by the dimensionality of the input coordinates from the preceding PCO directive.

As an example of the ADDPOINTS directive, we can use the data from 10.4.1 on the different galaxy types, and construct an ordination of the eight spiral forms. Then points for the irregular and elliptical types are added to this ordination. First we need to extract from the data the symmetric matrix of distances for the spiral types and also a matrix giving the distances of the two other types from the spiral types (lines 30 and 32). Remember that the input distances were transformed ready for the PCO in 10.4.1; this transformation is also appropriate for the distances amongst the spiral types used as input to PCO in line 36. However, the ADDPOINTS directive requires squared distances so the reverse transformation is required for the distances of the irregular and elliptical galaxy types from the spiral types (line 33).

EXAMPLE 10.4.2

```
 28   TEXT Gname2,Gname8; VALUES=!T(E,I),!T(SO,SBO,Sa,SBa,Sb,SBb,Sc,SBc)
 29   SYMMETRICMATRIX [ROWS=Gname8] G8
 30   CALCULATE G8 = Galaxy$[!(2...9)]
 31   MATRIX [ROWS=Gname2; COLUMNS=Gname8] G2
 32   CALCULATE G2 = Galaxy$[!(1,10);!(2...9)]
 33   & G2 = -2 * G2
 34   LRV [ROWS=Gname8; COLUMNS=2] L8
 35   DIAGONALMATRIX [ROWS=Gname8] C8
 36   PCO [PRINT=roots,scores; NROOTS=2] G8; LRV=L8; CENTROID=C8
```

36..

***** Principal coordinates analysis *****

*** Latent Roots ***

```
         L8['Roots']
             1           2
          5.006       1.359
```

*** Percentage variation ***

```
         L8['Roots']
             1           2
         55.57       15.08
```

*** Trace ***

```
 L8['Trace']
      9.009
```

*** Latent vectors (coordinates) ***

```
         L8['Vectors']
              1            2
 Gname8
     SO    1.1349       0.3400
    SBO    0.9757       0.3429
     Sa    0.4472      -0.3033
    SBa    0.2690      -0.2879
     Sb   -0.2233      -0.8094
    SBb   -0.5441       0.2237
     Sc   -0.9135      -0.0016
    SBc   -1.1460       0.4955
```

```
 37   ADDPOINTS [PRINT=coordinates,residuals] G2; LRV=L8; CENTROID=C8
```

37..

***** Adding points to a principal coordinates analysis *****

*** Coordinates of added points ***

```
              1         2
     1    1.1082    0.3191
     2   -0.1708    0.2411
```

*** Residuals ***

```
     1    1.524
     2    1.549
```

10.4.3 The RELATE directive

One way of interpreting the principal coordinates obtained from a similarity matrix is by relating them to the original variates of the data matrix. For each coordinate and each data variate, an F statistic can be computed as if the variate and the coordinate vector were independent. This is not the case but, although the exact distribution of these pseudo F values is not known, they do serve to rank the variates in order of importance of their contribution to the coordinate vector.

Qualitative variates are treated as grouping factors, and the mean coordinate for each group is calculated. Only 10 groups are catered for; group levels above 10 are combined. The pseudo F statistic gives the between-group to within-group variance ratio. Missing values are excluded.

Quantitative variates are grouped on a scale of 0–10 (where zero signifies a value up to 0.05 of the range), and mean coordinates for each group are calculated. The printed pseudo F statistic is for a linear regression of the principal coordinate on the ungrouped data variate, after standardizing the data variate to have unit range; the regression coefficient is also printed.

RELATE
Relates the observed values on a set of variates to the results of a principal coordinates analysis.

Options

COORDINATES	= matrices	Points in reduced space; no default i.e. this option must be specified
NROOTS	= scalar	Number of latent roots for printed output; default * requests them all to be printed

494   *10   Multivariate and cluster analysis*

Parameters		
DATA	= variates	The data values
TEST	= scalars	Test type, defining how each variate is treated for calculation of the similarity
RANGE	= scalars	Range of possible values of each variate; if omitted, the observed range is taken

The parameters of the RELATE directive are the same as those of the FSIMILARITY directive, and are described in 10.3.1. However, you do not need to supply the complete list of data variates (with their corresponding types and ranges), only those that you wish to relate to the PCO results. In the example below, to save space, we examine only two of the original variates.

The COORDINATES option must be present and must be a matrix. It represents the units in reduced space. Usually the coordinates will be from a principal coordinates analysis (10.4.1). The number of rows of the matrix must match the number of units present in the variates, taking account of any restriction.

The output from RELATE can be extensive. You may not be interested in relating the variates to the higher dimensions of the principal coordinates analysis, for example the fourth and fifth, even though you may have saved these in the coordinate matrix. The NROOTS option allows you to print the results for, say, the first three dimensions, as in the example below. If you do not use the NROOTS option, the RELATE directive prints information for all the saved dimensions: that is, for the number of columns of the coordinates matrix.

EXAMPLE 10.4.3

```
 40   " Produce output from ordination of Carsim and RELATE
 -41      matrix of coordinates to the original variates "
 42   LRV [ROWS=Carname; COLUMNS=6] Carpco; VECTORS=Carvec
 43   PCO [PRINT=roots] Carsim; LRV=Carpco
```

43..

***** Principal coordinates analysis *****

*** Latent Roots ***

```
           1          2          3          4          5
       2.3107     0.8529     0.5118     0.3865     0.2988
           6          7          8          9         10
       0.1916     0.1371     0.1047     0.0882     0.0731
          11         12         13         14         15
       0.0603     0.0487     0.0410     0.0360     0.0195
          16
       0.0000
```

10.5 Hierarchical cluster analysis

```
*** Percentage variation ***
                    1           2           3           4           5
                44.77       16.53        9.92        7.49        5.79
                    6           7           8           9          10
                 3.71        2.66        2.03        1.71        1.42
                   11          12          13          14          15
                 1.17        0.94        0.79        0.70        0.38
                   16
                 0.00

*** Trace ***
        5.161

* Some roots are negative - non-Euclidean distance matrix *


44   RELATE [COORDINATES=Carvec; NROOTS=3] Weight,Carb; TEST=Ty3,Ty2

   Variate       Minimum       Range
       Weight      720.0       786.0
         Carb      1.000       2.000


**** Relate principal coordinates to original data ****

   Variate    1        Weight
Data scaled down by factor of   0.0127

                   F       -        0        1        2        3        4        5        6
Counts                     0        1        2        2        3        2        1        2
Vector 1   291.9    0.0000  -0.4122  -0.4363  -0.1801  -0.2074  -0.0677   0.0995   0.1142
Vector 2     0.4    0.0000   0.1405   0.2085   0.1676  -0.1048   0.0043  -0.4562  -0.2378
Vector 3     0.0    0.0000  -0.1960  -0.1919   0.1732   0.0096   0.2098  -0.0943   0.0863

                   7        8        9       10
Counts             0        0        2        1
Vector 1     0.0000   0.0000   0.6337   0.8072
Vector 2     0.0000   0.0000   0.1011   0.1428
Vector 3     0.0000   0.0000  -0.0522  -0.1889

** Regression coeffts.      0.0016   -0.0002    0.0000

   Variate    2        Carb
                   F       -        0        1        2
Counts                     0       10        5        1
Vector 1     3.2    0.0000  -0.1405   0.3226  -0.2084
Vector 2     3.6    0.0000   0.1065  -0.1721  -0.2044
Vector 3     0.1    0.0000   0.0058   0.0082  -0.0985
```

In this example, the coordinates for the cars in a reduced space of six dimensions are saved in the matrix, Carvec. The first three coordinates account for 71.2% of the trace.

10.5 Hierarchical cluster analysis

One of the main uses of similarity matrices is for hierarchical cluster analysis, as provided by the HCLUSTER directive. The aim of cluster analysis is to arrange the

n sampling units into more or less homogeneous groups. How this is done can vary; the directive offers a selection of possibilities. The general strategy is best appreciated in geometrical terms, with the n sampling units represented by points in a multidimensional space. In *agglomerative* methods, these points initially represent n separate clusters, each containing one member. At each of $n-1$ stages, two clusters are fused into one bigger cluster, until at the final stage all units are fused into a single cluster: this process can be represented by a hierarchical tree whose nodes indicate what fusions have occurred. The methods fuse the two closest clusters and vary in how *closest* is defined. In *single-linkage* cluster analysis, *closest* is defined as the smallest distance between any two samples from different clusters; in *centroid* clustering it is the smallest distance between cluster centroids; and so on (see Gordon 1981 for a full discussion).

Genstat will display the tree fitted to a given similarity matrix, and provides a scale showing at what level of similarity the fusions have occurred; such a scaled tree is termed a *dendrogram*. The endpoints of the dendrogram correspond to the units in some permuted order; you can save this order, for example to use in the FSIMILARITY directive (10.3.1). Of course, a hierarchical tree does not itself provide a classification. This can be derived by cutting the dendrogram at some arbitrary level of similarity; each cluster then consists of those samples occurring on the same detached branch of the dendrogram. A factor can be formed to indicate cluster membership.

10.5.1 The HCLUSTER directive

HCLUSTER
Performs hierarchical cluster analysis.

Options

PRINT	= strings	Printed output required (dendrogram, amalgamations); default * i.e. no output
METHOD	= string	Criterion for forming clusters (singlelink, nearestneighbour, completelink, furthestneighbour, averagelink, mediansort, groupaverage); default s
CTHRESHOLD	= scalar	Clustering threshold at which to print formation of clusters; default * i.e. determined automatically

Parameters

SIMILARITY	= symmetric matrices	Input similarity matrix for each cluster analysis
GTHRESHOLD	= scalars	Grouping threshold where groups are formed from the dendogram
GROUPS	= factors	Stores the groups formed
PERMUTATION	= variates	Permutation order of the units on the dendrogram
AMALGAMATIONS	= matrices	To store linked list of amalgamations

The SIMILARITY parameter provides the input to HCLUSTER, and must specify a list of symmetric matrices. These matrices can be formed by FSIMILARITY (10.3.1), by REDUCE (10.3.2), or by CALCULATE (5.1 and 10.3.3). Missing values are allowed in the similarity matrix only for the single linkage method.

The GTHRESHOLD and GROUPS parameters must be either both present or both absent. When you are deriving a classification, the level of similarity at which the dendrogram is to be cut is specified by the scalar value in the GTHRESHOLD parameter. The level is given as a percentage similarity. The resulting cluster membership is saved in a factor, whose identifier is specified by the GROUPS parameter. The factor will be declared implicitly, if necessary, and it will have its number of levels set to the number of clusters formed and its number of values taken from the number of rows of the corresponding symmetric matrix.

The PERMUTATION parameter is an output structure that allows you to save the order in which the units appear on the printed dendrogram. Genstat sets it up to be a variate implicitly, and its number of values is taken from the number of rows of the corresponding structure in the SIMILARITY parameter. Conventionally, the first unit on the dendrogram is unit 1 and so the first value of the variate of permutations will be 1.

The AMALGAMATIONS parameter is also an output structure. You can use it to store information about the order in which the units form groups, and at what level of similarity. At any stage in the process of agglomeration, each group is represented by the unit with the smallest unit number: for example, a group containing units 2, 5, 17, and 22 is represented by unit 2. This means that the final merge is always between a group indexed by unit 1 and a group indexed by another unit. Since there are $n-1$ stages of agglomeration, the output structure is a matrix with number of rows one less than the number of rows of the input similarity matrix. Each row represents a joining of two groups and consists of three values. The first two values are the numbers indexing the two groups that are joining, and the third value is the level of similarity. So the matrix has three columns. This structure will be declared implicitly, if necessary.

HCLUSTER prints two pieces of information. The first gives details of each amalgamation, followed by a list of clusters that are formed at decreasing levels of similarity. The second is the dendrogram. You can select which of these are printed, using the PRINT option. If METHOD = singlelink and the PRINT setting includes 'amalgamations', the minimum spanning tree (10.6.2) will be printed instead of the stages at which the clusters merge. This is because information from forming the minimum spanning tree is used to form the single linkage clustering.

The METHOD option has seven possible settings; these determine how the similarities amongst clusters are redefined after each merge. The default 'singlelink', which has synonym 'nearestneighbour', gives single linkage. The setting 'completelink' (synonym 'furthestneighbour') defines the distance between two clusters as the maximum distance between any two units in those clusters. The setting 'averagelink' defines the similarity between a cluster and two merged clusters as the average of the similarities of the cluster with each of the two. For 'groupaverage', an average is taken over all the units in the two merged clusters. Median sorting (Gower 1967) is best thought of in terms of clusters being represented by points in a multidimensional space; when two clusters join, the new cluster is represented by the midpoint of the original cluster points.

The CTHRESHOLD option is a scalar which allows you to define the levels of decreasing similarity at which the lists of clusters are printed with their membership. The decreasing levels of similarity are formed by repeatedly subtracting the CTHRESHOLD value from the maximum similarity of 100%. For example, setting CTHRESHOLD = 10 will list the clusters formed at 90% similarity, 80%, and so on. At each level, those units that have not joined any group are also listed. If you do not set this option, the default value will be calculated from the range of similarities at which merges occur, to give between 10 and 20 separate levels.

EXAMPLE 10.5.1

```
  45   HCLUSTER [PRINT=dendrogram; METHOD=averagelink] Carsim; \
  46      GTHRESHOLD=70; GROUPS=Cargrp; PERMUTATION=Carperm; \
  47      AMALGAMATIONS=Caramalg

**** Average linkage cluster analysis ****

**** Dendrogram ****
 ** Levels   100.0   90.0   80.0   70.0   60.0   50.0   40.0

Estate           1    ..
Regatta          8    ..)..
Arna1.5          2    .....)
Delta           13    .....).....
Panda            7    .....        )
Y10             15    .....)..     )
Uno             10    ........)..)..
Regattad         9    ...............).....
Alfa2.5          3    ...............     )
X19             11    ........            )    )
Spider          16    ........).....)......)..
Mondialqc        4    .................        )
Croma            6    ..                )       )
Thema           14    ..)..............)......)........
```

```
Testarossa       5   ........                      )
Contach         12   ........)....................)............
   48  FSIMILARITY [PRINT=similarity; SIMILARITY=Carsim; \
   49      PERMUTATION=Carperm; STYLE=abbreviated]

** Abbreviated similarity matrix **

Estate           1   -
Regatta          8   9-
Arna1.5          2   99-
Delta           13   999-
Panda            7   7777-
Y10             15   88889-
Uno             10   888898-
Regattad         9   8877667-
Alfa2.5          3   78785666-
X19             11   878777767-
Spider          16   7777566688-
Mondialqc        4   55552334756-
Croma            6   787756677577-
Thema           14   7777566775779-
Testarossa       5   33331113545755-
Contach         12   443412225446448-
```

10.6 Directives associated with hierarchical clustering

This section describes the directives that help you interpret hierarchical cluster analyses.

10.6.1 The HLIST directive

HLIST lists the values of the data matrix in a condensed form, either in their original order or, more usefully, in the order determined by a cluster analysis (10.5.1). This representation can be very helpful for revealing patterns in the data, associated with clusters, or for an initial scan of the data to pick out interesting features of the variates.

HLIST
Lists the data matrix in abbreviated form.

Options

GROUPS	= factor	Defines groupings of the units; used to split the printed table at appropriate places and to label the groups; default *
UNITS	= text or variate	Names for the rows (i.e. units) of the table; default *

Parameters

DATA	= variates	The data values
TEST	= scalars	Test type, defining how each variate is treated for calculation of the similarity between each unit
RANGE	= scalars	Range of possible values of each variate; if omitted, the observed range is taken

The parameters of the HLIST directive are the same as those of the FSIMILARITY directive (10.3.1), and are described there. The DATA and RANGE parameters are treated in the same way for HLIST as for FSIMILARITY, but TEST acts slightly differently.

For the TEST parameter, which governs the type of the variate, HLIST distinguishes only between qualitative variates (types 1 and 2) and quantitative variates (types 3, 4, and 5). The values of qualitative variates are printed directly. If the range of a quantitative variate is greater than 10, the printed values are scaled to lie in the range 0 to 10. This scaling is done by subtracting the minimum value from the variate, dividing by the range and then multiplying by 10. If the range is less than 10, the values are printed unscaled; so variates with values that are all less than 1 will appear as 0 in the abbreviated table. The values are printed with no decimal places, and in a fieldwidth of 3. In this example, you can see the effect of scaling the quantitative variates, and not scaling the qualitative variates.

EXAMPLE 10.6.1a

```
  50   HLIST [UNITS=Carname] #Cd,Carb,Drive; \
  51      TEST=4(Ty3),4(Ty5),2(Ty3),2(Ty2)

       Variate       Minimum        Range
         Engcc         965.0       4202.0
          Ncyl         4.000        8.000
         Tankl         35.00        85.00
        Weight         720.0        786.0
        Length         338.0        121.0
         Width         149.0         51.0
        Height         107.0         39.0
         Wbase         216.0         50.0
        Tspeed         134.0        157.0
          Stst         4.900       14.000
          Carb         1.000        2.000
         Drive         1.000        1.000

  **** Variates listed by current grouping ****
             Variate    1    2    3    4    5    6    7    8    9   10   11   12
                Type    3    3    3    3    5    5    5    5    3    3    2    2
               Range   10    8   10   10   10   10   10   10   10   10    2    1
```

10.6 *Directives associated with hierarchical clustering*   501

```
Estate       1      1   0   1   3   6   2   6   5   2   4   0   1
Arna1.5      2      1   0   1   1   5   2   8   5   2   3   0   1
Alfa2.5      3      3   2   1   5   7   2   8   7   4   2   0   0
Mondialqc    4      5   4   6   9   9   5   4   9   7   1   1   0
Testarossa   5      9   8  10  10   9   9   1   7  10   0   1   0
Croma        6      2   0   4   5   9   5   9  10   4   2   1   1
Panda        7      0   0   0   0   0   0  10   0   0   8   0   1
Regatta      8      1   0   2   3   7   3   8   5   2   3   0   1
Regattad     9      1   0   2   3   7   3   8   5   1  10   2   1
Uno         10      0   0   0   0   2   1   9   4   0   8   0   1
X19         11      1   0   1   2   4   1   2   0   2   4   0   0
Contach     12     10   8  10   9   6  10   0   5   9   0   0   0
Delta       13      1   0   1   3   4   2   7   6   3   2   0   1
Thema       14      2   0   4   5  10   5   9  10   5   1   1   1
Y10         15      0   0   1   0   0   0   9   0   2   4   0   1
Spider      16      2   0   1   4   6   2   4   2   3   2   1   0
```

The UNITS option allows you to change the labelling of the units in the table, as shown above. You can specify a text or a pointer or a variate.

You can use the GROUPS option to specify a factor that will split the units into groups. The table from HLIST is then divided into sections corresponding to the groups. If the factor has labels, these are used to annotate the sections; otherwise a group number is used. For example:

EXAMPLE 10.6.1b

```
  52   HLIST [GROUPS=Maker; UNITS=Carname] #Cd,Carb,Drive; \
  53       TEST=4(Ty3),4(Ty5),2(Ty3),2(Ty2)

Variate          Minimum         Range
        Engcc      965.0        4202.0
         Ncyl      4.000         8.000
        Tankl      35.00         85.00
       Weight      720.0         786.0
       Length      338.0         121.0
        Width      149.0          51.0
       Height      107.0          39.0
        Wbase      216.0          50.0
       Tspeed      134.0         157.0
         Stst      4.900        14.000
         Carb      1.000         2.000
        Drive      1.000         1.000


**** Variates listed by current grouping ****

          Variate    1   2   3   4   5   6   7   8   9  10  11  12
          Type       3   3   3   3   5   5   5   5   3   3   2   2
          Range     10   8  10  10  10  10  10  10  10  10   2   1

Fiat
Croma        6      2   0   4   5   9   5   9  10   4   2   1   1
Panda        7      0   0   0   0   0   0  10   0   0   8   0   1
Regatta      8      1   0   2   3   7   3   8   5   2   3   0   1
Regattad     9      1   0   2   3   7   3   8   5   1  10   2   1
Uno         10      0   0   0   0   2   1   9   4   0   8   0   1
X19         11      1   0   1   2   4   1   2   0   2   4   0   0
```

```
Alfa Romeo
Estate          1    1   0   1   3   6   2   6   5   2   4   0   1
Arna1.5         2    1   0   1   1   5   2   8   5   2   3   0   1
Alfa2.5         3    3   2   1   5   7   2   8   7   4   2   0   0

Lancia
Delta          13    1   0   1   3   4   2   7   6   3   2   0   1
Thema          14    2   0   4   5  10   5   9  10   5   1   1   1
Y10            15    0   0   1   0   0   0   9   0   2   4   0   1

Ferrari
Mondialqc       4    5   4   6   9   9   5   4   9   7   1   1   0
Testarossa      5    9   8  10  10   9   9   1   7  10   0   1   0

Lamborghini
Contach        12   10   8  10   9   6  10   0   5   9   0   0   0

Pinninfarina
Spider         16    2   0   1   4   6   2   4   2   3   2   1   0
```

10.6.2 The HDISPLAY directive

You can use the HDISPLAY directive to print ancillary information that is useful for interpreting cluster analyses, and to save information for use elsewhere in Genstat, for example for plotting.

HDISPLAY
Displays results ancillary to hierarchical cluster analyses: matrix of mean similarities between and within groups, a set of nearest neighbours for each unit, a minimum spanning tree, and the most typical elements from each group.

Options

| PRINT | = strings | Printed output required (neighbours, tree, typicalelements, gsimilarities); default t |

Parameters

SIMILARITY	= symmetric matrices	Input similarity matrix for each cluster analysis
NNEIGHBOURS	= scalars	Number of nearest neighbours to be printed
NEIGHBOURS	= matrices	Matrix to store nearest neighbours of each unit
GROUPS	= factors	Indicates the groupings of the units (for calculating typical elements and mean similarities between groups)

10.6 Directives associated with hierarchical clustering

TREE	= matrices	To store the minimum spanning tree (as a series of links and corresponding lengths)
GSIMILARITY	= symmetric matrices	To store similarities between groups

The SIMILARITIES parameter specifies a list of symmetric similarity matrices. These are operated on, in turn, to produce the output requested by the PRINT option and to save the information specified by other parameters. Since the interpretations of the remaining parameters are closely linked to the different settings of the PRINT option, each setting is discussed below with the relevant parameters.

The NNEIGHBOURS parameter gives a list of scalars indicating how many neighbours will appear in the printed table of nearest neighbours.

The NEIGHBOURS parameter is a list of identifiers in which you can store a table of nearest neighbours. If they are already declared, they must be matrices. Otherwise they will be declared implicitly as matrices. The rows of the matrices correspond to the units; there should be an even number of columns. The values in the odd-numbered columns represent the neighbouring units in order of their similarity, while the values in the even-numbered columns are the corresponding similarities. If you have previously declared the matrix and it does not have enough columns, then NEIGHBOURS stores as many neighbours as possible. If there is an odd number of columns in the matrix, the last column is not filled. If the output matrix is being declared implicitly, the number of columns is twice the value of the scalar given in the NNEIGHBOURS parameter.

If the PRINT option includes the setting 'neighbours', Genstat prints a table of nearest neighbours for every sample, together with their values of similarity. The number of neighbours printed is determined by the value of the scalar in the NNEIGHBOURS parameter. This information is also useful for interpreting clusters and ordinations. If you set the PRINT option to 'neighbours' and the scalar in the NEIGHBOURS parameter does not have a value, then the table is not printed. In this example, the table is printed for three nearest neighbours, and the matrix Carneig is given values corresponding to the first two nearest neighbours.

EXAMPLE 10.6.2a

```
 54   MATRIX [ROWS=Carname; COLUMNS=4] Carneig
 55   HDISPLAY [PRINT=neighbours] Carsim; NNEIGHBOURS=3; NEIGHBOURS=Carneig

**** Neighbours table ****
Carsim
Estate          1          8   95.2        2   95.0       13   93.1
Arna1.5         2          1   95.0        8   94.3       13   93.5
Alfa2.5         3         16   81.2       13   80.6        8   80.3
Mondialqc       4          5   77.0       14   74.5        6   73.6
Testarossa      5         12   88.3        4   77.0        3   55.1
```

```
      Croma            6        14    98.0        8    80.9       13    79.0
      Panda            7        15    91.9       10    91.1        2    78.6
      Regatta          8         1    95.2        2    94.3       13    92.7
      Regattad         9         8    84.2        1    81.2        2    79.4
      Uno             10         7    91.1       15    88.5        2    86.6
      X19             11        16    86.1        1    81.3        2    80.8
      Contach         12         5    88.3        4    67.0        3    57.0
      Delta           13         2    93.5        1    93.1        8    92.7
      Thema           14         6    98.0        8    79.9       13    77.9
      Y10             15         7    91.9       10    88.5        2    86.1
      Spider          16        11    86.1        3    81.2        1    78.6

   56  PRINT Carneig

                         Carneig
                               1            2             3            4
           Carname
             Estate        8.000        0.952         2.000        0.950
             Arna1.5       1.000        0.950         8.000        0.943
             Alfa2.5      16.000        0.812        13.000        0.806
           Mondialqc       5.000        0.770        14.000        0.745
          Testarossa      12.000        0.883         4.000        0.770
              Croma       14.000        0.980         8.000        0.809
              Panda       15.000        0.919        10.000        0.911
            Regatta        1.000        0.952         2.000        0.943
           Regattad        8.000        0.842         1.000        0.812
                Uno        7.000        0.911        15.000        0.885
                X19       16.000        0.861         1.000        0.813
            Contach        5.000        0.883         4.000        0.670
              Delta        2.000        0.935         1.000        0.931
              Thema        6.000        0.980         8.000        0.799
                Y10        7.000        0.919        10.000        0.885
             Spider       11.000        0.861         3.000        0.812
```

The GROUPS parameter specifies a factor to divide the units of each similarity matrix into clusters. You may have formed the factor from a previous hierarchical cluster analysis (10.5.1). This parameter must be present if you set the PRINT option to 'typicalelement' or 'gsimilarities'.

If the PRINT option includes the setting 'typicalelement', Genstat prints the average similarity of each group member with the other group members. This is to help you identify typical members of each group: typical members will have relatively large average similarities compared to those of the other members. Within each group, members are printed in decreasing order of average similarity. In this example, the cars are listed in the order of their mean similarity with the other members of the group to which they belong.

EXAMPLE 10.6.2b

```
   57  HDISPLAY [PRINT=typical] Carsim; GROUPS=Maker

**** Most typical members ****
Carsim

Fiat
  Regatta          8      79.5
  Uno             10      77.2
```

10.6 Directives associated with hierarchical clustering

```
Regattad        9    73.6
Panda           7    72.0
X19            11    68.3
Croma           6    66.5

Alfa Romeo
Estate          1    86.5
Arna1.5         2    85.8
Alfa2.5         3    77.3

Lancia
Delta          13    80.0
Y10            15    72.3
Thema          14    70.2

Ferrari
Testarossa      5    77.0
Mondialqc       4    77.0

Lamborghini
Contach        12   100.0

Pinninfarina
Spider         16   100.0
```

The GSIMILARITY parameter specifies a list of symmetric matrices in which you can save the mean between-group and within-group similarities. Any structure that you have not declared already will be declared implicitly to be a symmetric matrix with number of rows equal to the number of levels of the factor in the GROUPS parameter.

If the PRINT option includes the setting 'gsimilarities', Genstat prints the mean similarities between-groups and within-groups. Self-similarities are excluded. This example forms the group similarity matrix based on the groups in the factor Maker, prints the matrix and saves the values in the symmetric matrix Cargsim.

EXAMPLE 10.6.2c

```
  58  HDISPLAY[PRINT=gsimilarity] Carsim; GROUPS=Maker; GSIMILARITY=Cargsim

**** Mean similarities between and within groups ****
Carsim

** Between and within groups similarity matrix **

Fiat           1    72.8
Alfa Romeo     2    78.0  83.2
Lancia         3    76.6  80.4  74.1
Ferrari        4    41.0  50.5  45.7  77.0
Lamborghini    5    33.6  46.3  36.5  77.6  ----
Pinninfarina   6    71.4  78.5  71.8  60.0  49.2  ----

                     1     2     3     4     5     6
  59  PRINT Cargsim

                   Cargsim
         Fiat       0.7282
```

```
Alfa Romeo      0.7799      0.8321
    Lancia      0.7664      0.8040      0.7414
   Ferrari      0.4098      0.5049      0.4569      0.7696
Lamborghini     0.3361      0.4627      0.3653      0.7764      1.0000
Pinninfarina    0.7141      0.7850      0.7183      0.6003      0.4923

                 Fiat    Alfa Romeo     Lancia     Ferrari  Lamborghini

Pinninfarina    1.0000

                Pinninfarina
```

The TREE parameter specifies a matrix that allows you to save the minimum spanning tree. The matrix is set up with two columns and number of rows equal to the number of units. For each unit, the value in the first column is the unit to which that unit is linked on its left; the second column is the corresponding similarity. The first unit is not linked to any unit on its left, as it is always the first unit on the tree; so the first row of the matrix contains missing values.

Setting the PRINT option to 'tree' prints the minimum spanning tree associated with the similarity matrix specified by the first parameter. The minimum spanning tree (MST) is not a Genstat structure, but it can be kept in the form described above: that is, in a matrix with two columns. An MST is a tree connecting the n points of a multidimensional representation of the sampling units. In a tree every unit is linked to a connected network and there are no closed loops; the special feature of the MST is that, of all trees, it is the one whose links have minimum total length. The links include all those that join nearest neighbours; the MST is closely related to single linkage hierarchical trees (10.5.1). Minimum spanning trees are also useful if you superimpose them on ordinations (10.4) to reveal regions in which distance is badly distorted; if neighbouring points, as given by the MST, are distant in the ordination then something is badly wrong (see Gower and Ross 1969). In this example, the MST is printed and then saved in the structure Cartree which has been declared implicitly as a matrix.

EXAMPLE 10.6.2d

```
 60   HDISPLAY [PRINT=tree] Carsim; TREE=Cartree

**** Minimum spanning tree ****
Carsim

       Estate   Arna1.5        Uno      Panda        Y10
          1......    2......   10......    7......  15
              (    95.0  (   86.6  (   91.1       91.9
              (         (
              (         (      Delta
              (         (...... 13
              (              93.5
              (
              (  Regatta     Croma      Thema  Mondialq  Testaros   Contach
```

10.6 *Directives associated with hierarchical clustering*   507

```
    (......  8......  6...... 14......  4......  5...... 12
    (   95.2 (   80.9     98.0     74.5     77.0     88.3
    (        (
    (        ( Regattad
    (        (......  9
    (              84.2
    (
    (        X19   Spider  Alfa2.5
    (...... 11...... 16......  3
          81.3      86.1     81.2
 ** Total length      1304.8
 61 PRINT Cartree

                  Cartree
                     1         2
        Carname
         Estate      *         *
         Arna1.5     1.000     0.950
         Alfa2.5    16.000     0.812
      Mondialqc     14.000     0.745
      Testarossa     4.000     0.770
          Croma      8.000     0.809
          Panda     10.000     0.911
         Regatta     1.000     0.952
        Regattad     8.000     0.842
            Uno      2.000     0.866
            X19      1.000     0.813
        Contach      5.000     0.883
          Delta      2.000     0.935
          Thema      6.000     0.980
            Y10      7.000     0.919
         Spider     11.000     0.861
```

10.6.3 The HSUMMARIZE directive

The directive HSUMMARIZE helps you to see which clusters, if any, are distinguished by each variate. It requires a factor to define the clusters, as well as the original data variates, together with their types and, optionally, their ranges. From this it prints a frequency table for each variate. Each table is classified by the grouping factor and the different values of the variate. For qualitative variates (types 1 and 2) the values are integral, and for each group Genstat calculates an interaction statistic labelled chi-squared. This statistic does not have a significance level attached to it, but it does draw attention to groups for which the distribution is markedly different from the overall distribution.

For quantitative variates (types 3–5) values are rounded to the nearest point on an 11-point scale (0–10). The interaction statistic is analogous to Students t, and it draws attention to the groups for which the mean variate value is markedly different from the overall means (again with no significance level attached). Missing values are ignored in the computation of these statistics.

HSUMMARIZE

Forms and prints a group by levels table for each test together with appropriate summary statistics for each group.

Option		
GROUPS	= factor	Factor defining the groups; no default i.e. this option must be specified
Parameters		
DATA	= variates	The data values
TEST	= scalars	Test type, defining how each variate is treated for calculation of the similarity between each unit
RANGE	= scalars	Range of possible values of each variate; if omitted, the observed range is taken

The parameters of the HSUMMARIZE directive are the same as those of the FSIMILARITY directive (10.3.1). As with HLIST (10.6.1), the HSUMMARIZE directive distinguishes only between qualitative variates (types 1 and 2) and quantitative variates (types 3, 4, and 5). The GROUPS option specifies a factor that splits the units into clusters.

As the output from this directive can be very long, only two tables are shown in the next example; these illustrate the difference between tables for qualitative and quantitative variates. The grouping factor is taken from the HCLUSTER example in 10.5.1. Each entry in the table gives the number of units from a particular group that have a particular value of the variate.

EXAMPLE 10.6.3

```
 62  HSUMMARIZE [GROUPS=Cargrp] Weight,Carb; TEST=Ty3,Ty2

Variate         Minimum     Range
     Weight      720.0      786.0
       Carb      1.000      2.000

**** Grouped data frequency tables for each variate ****

     Variate  1   Weight
Data scaled down by factor of 0.0127

              *   0   1   2   3   4   5   6   7   8   9  10  Total   Mean       T
Group    1    0   3   1   0   4   0   0   0   0   0   0          8   1.63   -1.99
Group    2    0   0   0   1   0   1   1   0   0   0   0          3   3.67   -0.11
Group    3    0   0   0   0   0   0   2   0   0   1   0          3   6.33    1.33
Group    4    0   0   0   0   0   0   0   0   0   1   1          2   9.50    2.48

    Total     0   3   1   1   4   1   3   0   0   2   1         16   3.88
```

```
Variate   2   Carb

              *   0   1   2   Total   Chi-sq
Group   1     0   7   0   1      8     3.80
Group   2     0   2   1   0      3     0.20
Group   3     0   0   3   0      3     6.60
Group   4     0   1   1   0      2     0.40

Total         0  10   5   1     16
```

10.7 Non-hierarchical classification

A common statistical problem is to divide the units of a data set into some number of mutually exclusive groups, or classes. Usually you would hope that the groups will be reasonably homogeneous, and distinct from each other. When you do not know the most natural number of classes in advance, you might be interested in several classifications into different numbers of groups: you can then inspect these, and make a decision about the most acceptable number of groups. One way of achieving such groupings is to take the results of a hierarchical classification (10.5), and cut the dendrogram at appropriate levels to obtain groupings into several numbers of classes. However, the statistical properties of the resulting groups are not at all clear, and the hierarchical nature of the groupings into various numbers of classes can impose undue constraints. An alternative approach is to optimize some suitably chosen criterion directly from the data matrix, to obtain one or more non-hierarchical classifications.

Non-hierarchical classification methods differ according to the criterion that they optimize and in the algorithm used to search for an optimum value of the chosen criterion. In Genstat one of four different criteria may be optimized, and the optimization algorithm uses one of two different strategies.

Which criterion to choose depends on the type of data. Suppose first that they can be considered as being a mixture of k multi-normal distributions, with the same variance-covariance matrix. Then the maximum likelihood estimate of this matrix is given when the grouping into k classes minimizes the determinant of the within-class variance-covariance matrix, pooled over the k groups (Friedman and Rubin 1967); in other words, the optimization criterion is to minimize this determinant.

When only two groups are to be formed, the criterion above is equivalent to maximizing the Mahalanobis distance between the two classes. However, when the number of groups to be formed is greater than two, maximizing the total Mahalanobis distance between the classes will generally give different results to minimizing the determinant of the pooled within-class dispersion matrix. To maximize the total Mahalanobis distance is the second available criterion.

The third criterion maximizes the total Euclidean distance between the classes; this is equivalent to minimizing the total within-class sum of squares: that is, the trace of the pooled within-class dispersion matrix. This third criterion can be thought of as a simpler variant of the first, that does not rely on the assumptions of multi-normality or equal within-class dispersion.

The fourth criterion gives maximal predictive classification (Gower 1974). It is relevant when all the data are binary: that is, when they take only two values, usually designated by zero and one. Within each class, the *class predictor* is defined to be a list with one entry for each variate: the ith entry is whichever value (zero or one) is more frequent in the class for the ith variate. The criterion, W, to be maximized is the sum over the classes of the number of agreements between units of each class and their class predictor. When several different classifications give the same maximum value for W, a subsidiary criterion B is minimized. Whereas W measures within-class homogeneity, B measures between-class heterogeneity: it is the sum of the number of correct predictions for each unit when predicted by any of the class predictors of the classes other than the one to which the unit is assigned.

The algorithm used in Genstat to search for optimal values of the chosen criterion proceeds as follows. Starting from some initial classification of the units into the required number of groups, the algorithm repeatedly transfers units from one group to another so long as such transfers improve the value of the criterion. When no further transfers can be found to improve the criterion, the algorithm switches to a second stage which examines the effect of swopping two units of different classes. The algorithm alternates between the two types of search until neither gives any improvement. Searching for swops is computationally more expensive than searching for transfers, so only one swop is performed each time before the algorithm switches to search for transfers. However, using only swops has the advantage that the group sizes remain constant: if this is what you want, you can direct Genstat to search only for swops.

There is no guarantee that the classification resulting from the above algorithm will be globally optimal: to be sure of that, you would need to try all possible classifications of the units into the required number of groups. All that is known is that no improvement can be made to the criterion by either of the types of transfer strategy. The chance that the algorithm will produce a near-optimal classification can be much improved by providing a good initial classification. You could obtain this from a hierarchical classification method, or by examining a set of principal component scores from the data. The effect of trying different initial classifications can be interesting, and provides some information on the closeness to optimality.

10.7.1 The CLUSTER directive

CLUSTER
Forms a non-hierarchical classification.

Options

PRINT	= strings	Printed output required (criterion, optimum, units, typical, initial); default * i.e. no output
DATA	= matrix or pointer	Data from which the classification is formed, supplied as a units-by-variates matrix or as a pointer containing the variates of the data matrix
CRITERION	= string	Criterion for clustering (sums, predictive, within, Mahalanobis); default s
INTERCHANGE	= string	Permitted moves between groups (transfer, swop); default t (implies s also)
START	= factor	Initial classification; default * i.e. splits the units, in order, into NGROUPS classes of nearly equal size

Parameters

NGROUPS	= scalars	Numbers of classes into which the units are to be classified: note, the values of the scalars must be in descending order
GROUPS	= factors	Saves the classification formed for each number of classes

By default the CLUSTER directive will not print any results; you must set the PRINT option to indicate which sections of output you want, and whether these should also be printed for the initial classification. The possible settings are as follows.

    criterion   prints the optimal criterion value.
    optimum   prints the optimal classification.
    units   prints the data with the units ordered into the optimal classes.
    typical   prints a typical value for each class: for maximal predictive classification this is the class predictor;

initial for the other methods it is the class mean.
    if this is set the requested sections of output are also printed for the initial classification.

The DATA option supplies the data to be classified: the single structure must be either a matrix, with rows corresponding to the units and columns to the variables, or a pointer whose values are the identifiers of the variates in the data matrix. Note that CLUSTER always operates on a matrix, and so will copy the variate values into a matrix if you supply a pointer as input; thus for large data sets it is better to supply a matrix.

The CRITERION option specifies which criterion CLUSTER is to optimise, the default being 'sums'. The four settings are:

sums   maximize the between-group sum of squares;
predictive maximal predictive classification;
within   minimize the determinant of the pooled within-class dispersion matrix;
mahalanobis maximize the total Mahalanobis squared distance between the groups.

The INTERCHANGE option specifies which types of interchange (transfers or swops) are to be used. The default is 'transfer', which implies that both transfers and swops are used, since a swop is simply two transfers. If you set INTERCHANGE=swop, only swops are used. If INTERCHANGE=* the algorithm does not attempt to improve the classification from the initial classification; you might want this, in conjunction with the PRINT=initial setting, to display the results for an existing classification which you do not wish to improve.

The START option should be used to supply a factor to define the initial classification. If this option is not specified, CLUSTER will divide the units, in order, into roughly equal-sized groups. For example, with 97 units to be classified into 10 groups, the first 10 units will be put into the first group, the 11th to 20th into the second group, and so on; the last three groups will contain only nine units each.

The first parameter, NGROUPS, is used to specify the number of classes to be formed. Any single-valued structure can be supplied here. Often you would want several classifications from a single data set, into different numbers of groups. In this case the NGROUPS parameter should be a list of the numbers of groups in descending order. For the initial classification of the second classification, CLUSTER takes the optimal classification from the first number of groups, and does some reallocation of units to make a smaller number of groups. This is repeated, as often as required, to provide initial classifications for all the later analyses; hence the need to specify the numbers in descending order.

The second parameter, GROUPS, is used to specify a list of identifiers of factors to save the optimal classifications. Each of these must have been declared already,

10.7 Non-hierarchical classification

and have number of levels equal to the corresponding value from the NGROUPS list.

Doran and Hodson (1975) give some measurements made on 28 brooches found at the archeological site of the cemetery at Munsingen. Seven of these variables, transformed to logarithms, are used in this example.

EXAMPLE 10.7.1a

```
  1  UNITS [NVALUES=28]
  2  POINTER [VALUES=Foot_lth,Bow_ht,Coil_dia,Elem_dia,Bow_wdth, \
  3     Bow_thck,Length] Data
  4  READ Data[]

     Identifier    Minimum      Mean   Maximum    Values   Missing
       Foot_lth      2.398     3.278     4.554        28         0
         Bow_ht      2.079     2.842     3.296        28         0
       Coil_dia      1.792     2.166     2.833        28         0
       Elem_dia      1.099     2.026     2.708        28         0
       Bow_wdth      3.045     4.064     5.176        28         0
       Bow_thck      2.708     3.621     4.357        28         0
         Length      3.296     4.003     4.860        28         0

 33  CLUSTER [PRINT=criterion,optimum,initial; DATA=Data] 5,4,3

33..........................................................................

*****  Non-hierarchical Clustering   *****

***  Sums of Squares criterion   ***

***  Initial classification   ***

***  Number of classes = 5

***  Class contributions to criterion   ***

          1          2          3          4          5
       7.623      5.335      1.434      6.251      7.286

***  Criterion value = 27.93006

***  Classification of units   ***

     1    1    1    1    1    1    2    2    2    2    2    2    2    3
     3    3    3    3    3    4    4    4    4    4    5    5    5    5
     5    5

***  Optimum classification   ***

***  Number of classes = 5

***  Class contributions to criterion   ***

          1          2          3          4          5
       2.205      1.715      1.965      2.361      2.633
```

*** Criterion value = 10.87892

*** Classification of units ***

```
4  5  3  1  1  5  5  2  1  1  4  2  3
3  3  3  3  2  3  4  2  2  3  1  1  5
5  4
```

*** Initial classification ***

*** Number of classes = 4

*** Class contributions to criterion ***

```
    1       2       3       4
  2.205   3.839   6.580   2.361
```

*** Criterion value = 14.98485

*** Classification of units ***

```
4  2  3  1  1  3  3  2  1  1  4  2  3
3  3  3  3  2  3  4  2  2  3  1  1  3
3  4
```

*** Optimum classification ***

*** Number of classes = 4

*** Class contributions to criterion ***

```
    1       2       3       4
  4.394   1.715   3.670   3.119
```

*** Criterion value = 12.89720

*** Classification of units ***

```
4  3  1  1  1  3  3  2  1  4  4  2  1
1  1  1  1  2  3  4  2  2  1  1  1  3
3  4
```


*** Initial classification ***

*** Number of classes = 3

*** Class contributions to criterion ***

```
    1       2       3
 11.931   4.174   3.670
```

*** Criterion value = 19.77412

*** Classification of units ***

```
1  3  1  1  1  3  3  2  1  1  1  2  1
1  1  1  1  2  3  1  2  2  1  1  1  3
3  2
```

```
***  Optimum classification  ***
***  Number of classes = 3
***  Class contributions to criterion  ***
           1            2            3
       15.279        1.714        2.633
***  Criterion value = 19.62666
***  Classification of units  ***
     1    3    1    1    1    3    3    2    1    1    1    2    1
     1    1    1    1    2    1    1    2    2    1    1    1    3
     3    1
```

The seven variables, represented by the pointer Data, are defined on lines 1 and 2 and are given values in line 3. The PRINT option of the CLUSTER statement (line 33) specifies that the criterion value and optimal classification are to be printed, and that the criterion value and initial classification are to be printed before the transfer and swop algorithm is used. The criterion to be optimized is the default, namely the minimum sum of squares within groups. The DATA option supplies the seven variables, via their pointer. The first parameter specifies that classifications are to be formed into five, then four, then three, groups.

No initial classification has been supplied, so the CLUSTER directive assigns the units to five classes, as described above. Thus the first six units are in class 1, and so on. This classification is printed near the beginning of the output from CLUSTER. It is preceded by the value of the minimum within-class sum of squares criterion for this classification, and a break-down of this value into the contributions from each class; each such contribution is the sum of squares within a class. At the optimal classification, Genstat prints the criterion value obtained, and its contributions from each class. You can see that the optimal classification obtained is quite different from the initial classification: in fact only 12 of the 28 units are in the same class that they started in.

To obtain an initial classification into four groups the CLUSTER directive reassigns each unit in group 5 to the nearest group: there are five such units, and four of them are closest to group 3. If you examine the initial and optimal classifications into four groups, and the optimal classification into five groups, you will see that many of the units of group 3 have transferred to group 1. This suggests that the optimal fifth group has become the third group; and that the old third and first groups have merged. The initial classification into three groups is similarly formed by reassigning the units in the fourth optimal group: of the five units involved, four are reassigned to group 1. This suggests that group 1 is becoming dominant. In fact little improvement is made to the criterion by forming the optimal classification for three groups; only two units move, both to the first group.

The second example of non-hierarchical classification illustrates the maximal predictive criterion. Remember that this method has a subsidiary criterion, B, as well as the main criterion W. The criterion W measures within-class consistency, and has separate contributions from each class; the criterion B measures between-class distinctness and has a contribution from all possible pairs of groups.

EXAMPLE 10.7.1b

```
  1  POINTER [NVALUES=4] Y
  2  VARIATE [NVALUES=30] Y[ ]
  3  READ [SERIAL=yes] Y[ ]

     Identifier    Minimum       Mean    Maximum     Values    Missing
           Y[1]     0.0000     0.5333     1.0000         30          0
           Y[2]     0.0000     0.4667     1.0000         30          0
           Y[3]     0.0000     0.5000     1.0000         30          0
           Y[4]     0.0000     0.5000     1.0000         30          0

  8  FACTOR [LEVELS=2; NVALUES=30] Optimum[2]
  9  &      [LEVELS=5] Optimum[5]
 10  CLUSTER [PRINT=criterion,optimum,typical,units; DATA=Y; \
 11      CRITERION=predictive] NGROUPS=5,2; GROUPS=Optimum[5,2]


11................................................................

***** Non-hierarchical Clustering *****

***   Maximal Predictive criterion   ***

***   Equally optimum classifications   ***

*** Criterion value = 104.00000
*** Criterion B     =  49.00000


         3   4   2   1   1   5   3   4   3   5   1   1   2
         4   3   5   5   1   3   4   2   5   2   5   3   5
         3   1   5   1

         3   4   2   1   1   5   3   4   3   5   1   1   2
         3   3   5   5   4   3   4   2   5   2   5   3   5
         3   1   5   1


***   Optimum classification   ***

***   Number of classes = 5

***   Class contributions to criterion   ***

             1           2           3           4           5
         25.00       14.00       28.00       12.00       25.00

*** Criterion value = 104.00000

***   Class contributions to criterion B   ***
```

10.7 Non-hierarchical classification

```
              1          2          3          4          5
     1      0.000     10.000      3.000     13.000     18.000
     2      6.000      0.000     10.000     10.000      2.000
     3      4.000     20.000      0.000     14.000     12.000
     4      6.000      9.000      6.000      0.000      3.000
     5     17.000      7.000     15.000     11.000      0.000
```

*** Criterion B = 49.00000
*** Classification of units   ***

```
     3    4    2    1    1    5    3    4    3    5    1    1    2
     3    3    5    5    1    3    4    2    5    2    5    3    5
     3    1    5    1
```

*** Class predictors   ***

```
          1    2    3    4
     1    0    1    0    0
     2    0    0    1    1
     3    1    0    1    1
     4    0    0    0    1
     5    1    1    0    0
```

*** Units rearranged into class order   ***

Group 1
```
        1          2          3          4
     0.0000     1.0000     0.0000     0.0000
     0.0000     1.0000     1.0000     0.0000
     0.0000     1.0000     1.0000     0.0000
     0.0000     1.0000     0.0000     0.0000
     0.0000     0.0000     0.0000     0.0000
     0.0000     1.0000     0.0000     0.0000
     0.0000     1.0000     0.0000     0.0000
```

Group 2
```
        1          2          3          4
     0.0000     0.0000     1.0000     1.0000
     0.0000     0.0000     1.0000     1.0000
     0.0000     1.0000     1.0000     1.0000
     0.0000     0.0000     1.0000     0.0000
```

Group 3
```
        1          2          3          4
     1.0000     0.0000     1.0000     1.0000
     1.0000     0.0000     1.0000     1.0000
     1.0000     1.0000     1.0000     1.0000
     1.0000     0.0000     0.0000     1.0000
     1.0000     0.0000     1.0000     1.0000
     1.0000     1.0000     1.0000     1.0000
     1.0000     1.0000     1.0000     1.0000
     1.0000     0.0000     1.0000     1.0000
```

Group 4
```
        1          2          3          4
     0.0000     0.0000     0.0000     1.0000
     0.0000     0.0000     0.0000     1.0000
     0.0000     0.0000     0.0000     1.0000
```

Group 5
```
        1          2          3          4
     1.0000     0.0000     0.0000     0.0000
     1.0000     1.0000     0.0000     0.0000
     1.0000     0.0000     0.0000     0.0000
     1.0000     1.0000     1.0000     0.0000
```

```
            1.0000       0.0000       0.0000       0.0000
            1.0000       0.0000       0.0000       0.0000
            1.0000       1.0000       0.0000       1.0000
            1.0000       1.0000       1.0000       0.0000
```

*** Optimum classification ***

*** Number of classes = 2

*** Class contributions to criterion ***

```
              1            2
            43.00        44.00
```

*** Criterion value = 87.00000

*** Class contributions to criterion B ***

```
                      1            2
            1       0.000       18.000
            2      17.000        0.000
```

*** Criterion B = 35.00000

*** Classification of units ***

```
    2   2   2   1   1   1   2   2   2   1   1   1   2
    2   2   1   1   1   2   2   2   1   2   1   2   1
    2   1   1   1
```

*** Class predictors ***

```
                     1     2     3     4
               1     1     1     0     0
               2     1     0     1     1
```

*** Units rearranged into class order ***

Group 1
```
              1            2            3            4
            0.0000       1.0000       0.0000       0.0000
            0.0000       1.0000       1.0000       0.0000
            1.0000       0.0000       0.0000       0.0000
            1.0000       1.0000       0.0000       0.0000
            0.0000       1.0000       1.0000       0.0000
            0.0000       1.0000       0.0000       0.0000
            1.0000       0.0000       0.0000       0.0000
            1.0000       1.0000       1.0000       0.0000
            0.0000       0.0000       0.0000       0.0000
            1.0000       0.0000       0.0000       0.0000
            1.0000       0.0000       0.0000       0.0000
            1.0000       1.0000       0.0000       1.0000
            0.0000       1.0000       0.0000       0.0000
            1.0000       1.0000       1.0000       0.0000
            0.0000       1.0000       0.0000       0.0000
```

Group 2
```
              1            2            3            4
            1.0000       0.0000       1.0000       1.0000
            0.0000       0.0000       0.0000       1.0000
            0.0000       0.0000       1.0000       1.0000
            1.0000       0.0000       1.0000       1.0000
            0.0000       0.0000       0.0000       1.0000
            1.0000       1.0000       1.0000       1.0000
```

10.7 Non-hierarchical classification

```
      0.0000        0.0000        1.0000        1.0000
      1.0000        0.0000        0.0000        1.0000
      1.0000        0.0000        1.0000        1.0000
      1.0000        1.0000        1.0000        1.0000
      0.0000        0.0000        0.0000        1.0000
      0.0000        1.0000        1.0000        1.0000
      0.0000        0.0000        1.0000        0.0000
      1.0000        1.0000        1.0000        1.0000
      1.0000        0.0000        1.0000        1.0000

 12   TABULATE [PRINT=counts; CLASSIFICATION=Optimum[5,2]; MARGIN=yes]
```

	Count		
Optimum[2]	1	2	Count
Optimum[5]			
1	7	0	7
2	0	4	4
3	0	8	8
4	0	3	3
5	8	0	8
Count	15	15	30

Lines 1–3 define and read the data, using the pointer Y to specify four variates each of 30 values. On lines 8 and 9 two factors are declared to hold the optimal classifications into two and five groups, respectively. The required non-hierarchical classifications are specified on lines 10 and 11. For each classification the criterion values are printed, together with the optimal classification, the typical units for each group, and the full data set with the units classified into their groups. The GROUPS parameter has been used to specify factors to hold the optimal classifications.

When the CLUSTER directive has found an optimal classification, it will report all the classifications that it can find with the same optimum (provided that you have asked for the optimal classification to be printed). Several equivalent optimal classifications may often occur with maximal predictive classification, and may occur occasionally with the other criteria. When equally optimal classifications are reported, they are preceded by the criterion value together with the value of the subsidiary criterion (if relevant). If you compare the various optimal classifications printed in the example above, you can see that there is some ambiguity over the allocation of the 14th and 18th units.

After the details of the equally optimal classifications, Genstat prints the breakdown of the W and B criteria for the optimal classification that was found first. The (i,j)th cell of the table of class contributions to criterion B shows the number of correct predictions for units in group i when predicted by the class predictor of class j. For example, amongst the four units in the second group, six dichotomous values (out of 16) are correctly predicted by the first class predictor. You can check this quite easily by comparing the first class predictor (0,1,0,0) with the printed units of group 2.

The results for maximal predictive classification into two groups show a loss of within-class consistency, but improved between-class distinctness. Gower (1974) gives suggestions on how such difficulties may be resolved; for example, maximizing $W-B$ would lead to choosing the five-group classification. One preliminary to comparing two classifications is to tabulate them. This has been done on line 12, using as input the factors saved from the CLUSTER statement (for details of the TABULATE directive see 5.8.1). The table printed at the end of the output shows that the first group of the classification into two groups is formed from groups 1 and 5 of the five-group classification; group 2 is formed from groups 2, 3 and 4.

As mentioned already, the results of non-hierarchical classification can vary considerably according to the initial classification. The next example illustrates this, using the same data as the previous example.

EXAMPLE 10.7.1c

```
 13    CLUSTER [PRINT=criterion; DATA=Y; CRITERION=predictive] NGROUPS=6,5

13..........................................................................

***** Non-hierarchical Clustering *****

***  Maximal Predictive criterion  ***

***  Optimum classification  ***

***  Number of classes = 6

***  Class contributions to criterion  ***

              1           2           3           4           5           6
           18.00       24.00       19.00       22.00        4.00       22.00

***  Criterion value = 109.00000

***  Class contributions to criterion B  ***

                  1           2           3           4           5
         1     0..000       7.000      12.000       8.000      12.000
         2     7..000       0.000      17.000       9.000       9.000
         3    11.000       14.000      0.000      11.000       9.000
         4    12.000        6.000     10.000       0.000      12.000
         5     2..000       1.000      2.000       2.000       0.000
         6    10.000        8.000     2.000      10.000      14.000
                  6
         1     8.000
         2    11.000
         3     1.000
         4    14.000
         5     2.000
         6     0.000

***  Criterion B = 50.60000

***  Optimum classification  ***
```

```
***  Number of classes = 5

***  Class contributions to criterion  ***
          1           2           3           4           5
        18.00       24.00       19.00       22.00       24.00

***  Criterion value = 107.00000

***  Class contributions to criterion B  ***
                1           2           3           4           5
        1     0.000       7.000      12.000       8.000       8.000
        2     7.000       0.000      17.000       9.000      11.000
        3    11.000      14.000       0.000      11.000       1.000
        4    12.000       6.000      10.000       0.000      14.000
        5    12.000       9.000       4.000      12.000       0.000

***  Criterion B = 48.75000
```

The CLUSTER statement (line 13) specifies that only the criterion value is to be printed, and not the detailed classifications. The number of groups to be formed is first six, then five; thus the initial classification is different from that of the previous example. The criterion values are both only slightly better than previously ($W = 107.0$ and $B = 48.75$ compared with $W = 104.0$ and $B = 49.0$); however the contributions from the individual classes are quite different. This example illustrates the difference that the choice of initial classification can make, even with a relatively small number of units. In the previous example the initial classification was the default partition into five groups, whereas in this example it is the classification into six groups, with the sixth group being dispersed.

10.8 Procrustes rotation

Multivariate analyses often give the coordinates of a set of points in some multidimensional space. Typically these are obtained so that certain features of the underlying data are represented by the distances between the points in the multidimensional space. One example is principal components analysis, where the distance amongst the principal component scores represents the Pythagorean distances between the values in the data matrix. Another example is canonical variates analysis, where the distance between the canonical variate scores for the means is the Mahalanobis distance between the groups. The distances amongst a set of points do not change if the origin of the coordinate system is shifted, nor do they change if the axes of the coordinate system are rotated.

Suppose that two sets of points are obtained for the same set of objects but with respect to different coordinate systems. For example, two sets of data concerning the same set of objects may be analysed using principal components analysis to give two sets of principal component scores. Alternatively, one set of data may be analysed using two different methods, again giving two sets of points for the same set of objects. The question that now arises is: can the two sets of points be related to each other without disturbing the relationships contained inside the sets? Since the properties of distance are unchanged by a shift of origin or a rotation of the axes, this question is equivalent to asking whether the coordinate system for one set of points can be shifted and rotated so that they match, as well as possible, the coordinates of the other set of points.

FIGURE 10.8a

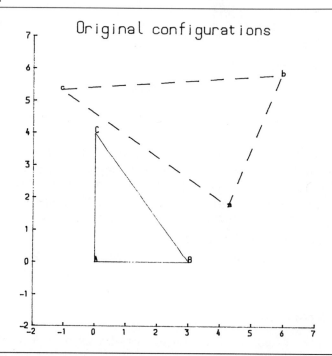

Procrustes rotation, of which there are several variants (Gower 1984), addresses this problem; orthogonal Procrustes rotation is the method most commonly used, and is provided by the ROTATE directive. Figures 10.8a-d show orthogonal Procrustes rotation for a small example; the Genstat program that was used to produce them is given in an example below. Suppose that there are two sets of coordinates for n points in r dimensions contained in the $n \times r$ matrices X and Y. The X-set is arbitrarily supposed to be a fixed configuration, and the Y-configuration is to be shifted and rotated so that it best matches the X-set. Here *best* means minimizing the sum of the squared distances between the points in the X-set and the matching shifted and rotated points in the Y-set. Figure 10.8a shows two sets of three points in two dimensions. The fixed configuration is labelled with upper-case letters, joined by full lines: the points are at the corners of the right-angled triangle with sides (3, 4, 5). The lower-case letters, joined by broken lines, show the other configuration of three points. The best translation (shift of origin) makes the centroids for the two sets of points coincide; this is easily done by translating both sets of points so that their centroids are at the origin. After translation, to find the best rotation involves doing a singular value decomposition (see, for example, Digby and Kempton 1987). Figure 10.8b shows the result: the configurations now have their centroids at the origin; the triangle shown by broken lines has been rotated clockwise through about 60 degrees.

FIGURE 10.8b

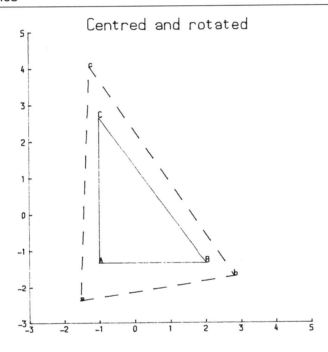

After translation and rotation the goodness of fit can be assessed by the residual sum of squares, which is the sum of squared distances between each X-point and the corresponding Y-point, after translation and rotation. Sometimes the relationships contained inside X and inside Y are similar but are expressed on different scales. For example, you can see in Figure 10.8b that the fixed configuration is smaller. So you might now want the coordinates in the Y-set to be stretched or contracted by a scaling factor; this can be estimated by least squares. But least-squares scaling should not be used if X and Y are known to be on comparable scales: for example, they may both have come from canonical variates analysis and thus express Mahalanobis distance. Figure 10.8c shows the result of an orthogonal Procrustes rotation, allowing for a least-squares scaling factor.

FIGURE 10.8c

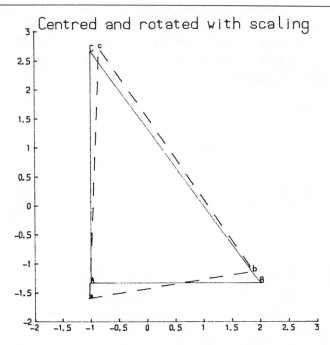

When you cannot say which configuration of points is the fixed set, you might want to know about the results of both Procrustes rotations. The best translation remains the same: both configurations of points are translated so that their centroids coincide, typically at the origin. If the best rotation of Y to X is given by the orthogonal matrix H, then the best rotation of X to Y is the transpose of H. If least-squares scaling is not used, the two residual sums of squares will be the same. However, if scaling is used, then in general these residuals will differ; you can overcome this by arranging that the two configurations of points, after translation, have the same sum of squares: a convenient value is unity. This initial scaling is particularly desirable when several configurations are to be compared pair by pair. In Figure 10.8d you can see the result of centring both configurations at the origin, then scaling them so that their sums of squares are one, and then applying the best rotation; no least-squares scaling is used.

FIGURE 10.8d

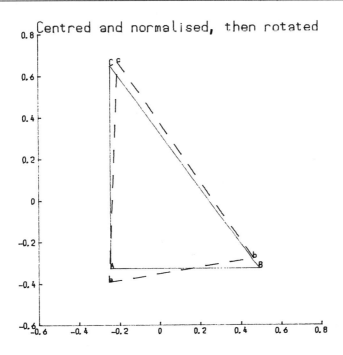

In general, the best rotation of Y to X may contain a reflection. Usually this is acceptable; however, you may sometimes want to stipulate that the rotation should be a pure rotation and not contain any reflection (Gower 1975).

Above we have assumed that the two matrices of coordinates have the same number of columns: that is, that the dimensionalities of the two multidimensional spaces are the same. If they differ, Genstat pads out the smaller matrix with columns of zero values, so that it matches the larger.

10.8.1 The ROTATE directive

ROTATE
Does a Procrustes rotation of one configuration of points to fit another.

Options

PRINT	= strings	Printed output required (rotations, coordinates, residuals, sums); default * i.e. no output

SCALING	= string	Whether or not isotropic scaling is allowed (no, yes); default n
STANDARDIZE	= strings	Whether to centre the configurations (at the origin), and/or to normalize them (to unit sum of squares) prior to rotation (centre, normalize); default c,n
SUPPRESS	= string	Whether to suppress reflection (no, yes); default n

Parameters

XINPUT	= matrices	Inputs the fixed configuration
YINPUT	= matrices	Inputs the configuration to be fitted
XOUTPUT	= matrices	To store the (standardized) fixed configuration
YOUTPUT	= matrices	To store the fitted configuration
ROTATION	= matrices	To store the rotation matrix
RESIDUALS	= matrices	To store distances between the (standardized) fixed and fitted configurations
RSS	= scalars	To store the residual sum of squares

The ROTATE directive provides orthogonal Procrustes rotation. You must set the parameters XINPUT and YINPUT, which specify respectively the fixed configuration and the configuration that you want to be translated and rotated; these are called X and Y above. The other parameters, described below, are used for saving results from the analysis. For X and Y to refer to the same set of objects they must have the same number of rows, and each object must be represented by the same row in both X and Y. If the XINPUT matrix is $n \times p$ and the YINPUT matrix is $n \times q$, Genstat does the analysis using matrices that are $n \times r$, where r is max(p, q). The smaller matrix is expanded with columns of zeros, as explained above.

The PRINT option specifies which results you want to print; the settings are as follows.

    coordinates   specifies that the fixed and fitted configurations are to be printed; note that the fixed configuration is printed after any standardization (see below), and the fitted configuration is printed after standardization and rotation.

    residuals   prints the residual distances of the points in the fixed

	configuration from the fitted points; this is after any standardization and rotation.
rotations	prints the orthogonal rotation matrix.
sums	prints an analysis of variance giving the sums of squares of each configuration, and the residual sum of squares; if scaling is used, the least-squares scaling factor is also printed.

The other three options of the ROTATE directive control the form of analysis. The SCALING option specifies whether you want least-squares scaling to be applied to the standardized YINPUT matrix when finding the best fit to the fixed configuration. You should set SCALING=yes if you want scaling; Genstat will then print the least-squares scaling factor with the analysis of variance. By default there is no scaling.

The STANDARDIZE option specifies what preliminary standardization is to be applied to the XINPUT and YINPUT matrices. It has settings:

centre	centre the matrices to have zero column means;
normalize	normalize the matrices to unit sums of squares.

The default is STANDARDIZE=centre,normalize. The initial centring ensures that the configurations are translated to have a common centroid, and thus automatically provides the best translation of Y to match X. The normalization arranges that the residual sum of squares from rotating X to Y is the same as that for rotating Y to X. Switching off both centring and standardization is rarely advisable, but can be requested by putting STANDARDIZE=*.

The SUPPRESS option is used to prevent any reflection. The default setting is 'no', which allows reflection to take place.

The final five parameters are for saving results from the analysis. The structures used to save results must have been declared before the analysis, and must have the correct type and size.

The structures used in the XOUTPUT and YOUTPUT parameters can be the same structures as used for input. But you must take account of the rules about consistent sizes (10.1.1). These rules cause no difficulty when the input matrices have the same number of columns; but if the numbers differ, then the smaller matrix cannot be used to save results.

Saving the residual sum of squares is useful when more than two configurations of points are to be compared. If there are k configurations, and if all the pairwise rotations are done without scaling, then you can assemble the residual sums of squares into a symmetric matrix of order k. You can then treat this as a matrix of squared distances amongst the k configurations and analyse it, for example, using principal coordinates analysis (10.4.1).

The next example shows the program that produced Figures 10.8a-d. For details of the statements used to produce the high-quality graphs, see Sections 7.2 and 7.3. Each DGRAPH statement is preceded by four EQUATE statements (5.3.1), which are used to obtain input variates for DGRAPH. The two configurations of three points are represented by the matrices Fixed and Tofit, declared in lines 1 and 2. The first of these gives the coordinates of the vertices of a right-angled triangle with sides (3, 4, 5); the second gives the coordinates of the configuration that is to be rotated. They are shown in Figure 10.8a, with vertices labelled (A, B, C) and (a, b, c). Two matrices, Xout and Yout, are declared on line 19 for saving the results of each of the Procrustes rotations.

EXAMPLE 10.8.1

```
 1   MATRIX [ROWS=3; COLUMNS=2] Fixed,Tofit; !(0,0,3,0,0,4), \
 2     !(4.250,1.750,5.917,5.814,-1.113,5.333)
 3   FRAME 1...4; YLOWER=2(0.55,0.0); YUPPER=2(1.0,0.45); \
 4     XLOWER=(0.0,0.55)2; XUPPER=(0.45,1.0)2
 5   AXES 1...4; YLOWER=-2,-3,-2,-0.6; YUPPER=7,5,3,0.8; \
 6     XLOWER=-2,-3,-2,-0.6; XUPPER=7,5,3,0.8; \
 7     YORIGIN=-2,-3,-2,-0.6; XORIGIN=-2,-3,-2,-0.6
 8   PEN 1,2; COLOUR=1; LINESTYLE=1,2; METHOD=line; \
 9     SYMBOLS=!T(A,B,C),!T(a,b,c); JOIN=given
10   OPEN 'ROTATE.GRD'; CHANNEL=1; FILETYPE=graphics
11   DEVICE 0
12   VARIATE [NVALUES=4] xg[1,2],yg[1,2]
13   EQUATE [OLDFORMAT=!(1,-1)] Fixed; NEWSTRUCTURES=xg[1]
14   & [OLDFORMAT=!(-1,1)] Fixed; NEWSTRUCTURES=xg[2]
15   & [OLDFORMAT=!(1,-1)] Tofit; NEWSTRUCTURES=yg[1]
16   & [OLDFORMAT=!(-1,1)] Tofit; NEWSTRUCTURES=yg[2]
17   DGRAPH [TITLE='Original configurations'; WINDOW=1; K=0; SCREEN=clear] \
18     xg[2],yg[2]; X=xg[1],yg[1]; PEN=1,2
19   MATRIX [ROWS=3; COLUMNS=2] Xout,Yout
20   ROTATE [PRINT=coordinates,residuals,rotations; STANDARDISE=centre] \
21     Fixed; YINPUT=Tofit; XOUTPUT=Xout; YOUTPUT=Yout

21.................................................................

***** Procrustes rotation *****

*** Orthogonal Rotation ***

                1          2
      1    0.52020    -0.85405
      2    0.85405     0.52020

*** Fixed Configuration ***

             Xout
                1          2
      1   -1.000     -1.333
      2    2.000     -1.333
      3   -1.000      2.667

*** Fitted Configuration ***

             Yout
                1          2
      1   -1.536     -2.378
      2    2.802     -1.688
```

10.8 Procrustes rotation

```
            3    -1.266      4.066
```

*** Residuals ***

```
                   1
          1     1.174
          2     0.877
          3     1.424
```

```
 22   EQUATE [OLDFORMAT=!(1,-1)] Xout; NEWSTRUCTURES=xg[1]
 23   & [OLDFORMAT=!(-1,1)] Xout; NEWSTRUCTURES=xg[2]
 24   & [OLDFORMAT=!(1,-1)] Yout; NEWSTRUCTURES=yg[1]
 25   & [OLDFORMAT=!(-1,1)] Yout; NEWSTRUCTURES=yg[2]
 26   DGRAPH [TITLE='Centred and rotated'; WINDOW=2; KEYWINDOW=0; SCREEN=keep] \
 27       xg[2],yg[2]; X=xg[1],yg[1]; PEN=1,2
 28   ROTATE [PRINT=coordinates,residuals,rotations,sums; SCALING=yes; \
 29       STANDARDIZE=centre] Fixed; YINPUT=Tofit; XOUTPUT=Xout; YOUTPUT=Yout

29.............................................................................
```

***** Procrustes rotation *****

*** Orthogonal Rotation ***

```
                   1           2
          1     0.52020    -0.85405
          2     0.85405     0.52020
```

*** Fixed Configuration ***

```
                 Xout
                   1           2
          1    -1.000      -1.333
          2     2.000      -1.333
          3    -1.000       2.667
```

*** Fitted Configuration ***

```
                 Yout
                   1           2
          1    -1.028      -1.592
          2     1.876      -1.130
          3    -0.847       2.722
```

*** Residuals ***

```
                   1
          1     0.2603
          2     0.2383
          3     0.1623
```

*** Sums of Squares ***

```
Fitted Configuration           16.5158
Residual                        0.1509
------------------------------------
Fixed Configuration            16.6667
```

*** Least-squares Scaling factor =   0.6695

```
 30   EQUATE [OLDFORMAT=!(1,-1)] Xout; NEWSTRUCTURES=xg[1]
 31   & [OLDFORMAT=!(-1,1)] Xout; NEWSTRUCTURES=xg[2]
 32   & [OLDFORMAT=!(1,-1)] Yout; NEWSTRUCTURES=yg[1]
```

```
 33     & [OLDFORMAT=!(-1,1)] Yout; NEWSTRUCTURES=yg[2]
 34   DGRAPH [TITLE='Centred and rotated with scaling'; WINDOW=3; KEYWINDOW=0; \
 35     SCREEN=keep] xg[2],yg[2]; X=xg[1],yg[1]; PEN=1,2
 36   ROTATE [PRINT=coordinates,residuals,rotations,sums] \
 37     Fixed; YINPUT=Tofit; XOUTPUT=Xout; YOUTPUT=Yout
```

37..

```
***** Procrustes rotation *****

*** Orthogonal Rotation ***

              1         2
    1      0.52020   -0.85405
    2      0.85405    0.52020

*** Fixed Configuration ***

            Xout
              1         2
    1     -0.2449   -0.3266
    2      0.4899   -0.3266
    3     -0.2449    0.6532

*** Fitted Configuration ***

            Yout
              1         2
    1     -0.2530   -0.3918
    2      0.4616   -0.2780
    3     -0.2085    0.6698

*** Residuals ***

              1
    1      0.06567
    2      0.05622
    3      0.04003

*** Sums of Squares ***

Fitted Configuration       1.0000
Residual                   0.0091
------------------------------------
Fixed Configuration        1.0000

* Sums of squares will total if scaling is used
```

```
 38   EQUATE [OLDFORMAT=!(1,-1)] Xout; NEWSTRUCTURES=xg[1]
 39     & [OLDFORMAT=!(-1,1)] Xout; NEWSTRUCTURES=xg[2]
 40     & [OLDFORMAT=!(1,-1)] Yout; NEWSTRUCTURES=yg[1]
 41     & [OLDFORMAT=!(-1,1)] Yout; NEWSTRUCTURES=yg[2]
 42   DGRAPH [TITLE='Centred and normalised, then rotated'; WINDOW=4; \
 43     KEYWINDOW=0; SCREEN=keep] xg[2],yg[2]; X=xg[1],yg[1]; PEN=1,2
```

The first Procrustes rotation is specified on lines 20 and 21. The standardization option requests centring, but no normalization. The output shows the fixed configuration (after centring), the fitted configuration (after centring and rotation),

10.8 Procrustes rotation

and the residuals. The orthogonal rotation, applied after the centring, is of the form

$$\begin{bmatrix} \cos(\text{theta}) & \sin(\text{theta}) \\ -\sin(\text{theta}) & \cos(\text{theta}) \end{bmatrix}$$

and represents a clockwise rotation through theta = 58.65 degrees: this is fairly obvious if you compare Figures 10.8a and 10.8b. The residuals are the distances of the fitted points from the fixed points: that is, the distances of the points labelled a, b, and c from those labelled A, B, and C in Figure 10.8b.

The second rotation is specified on lines 28 and 29. Here the setting SCALING = yes requests least-squares scaling. The PRINT option includes an extra setting, 'sums', to print the sums of squares from the analysis. The rotation matrix and the fixed configuration are as before. However, the fitted configuration is now closer to the fixed configuration, as you can see from the residuals and from Figure 10.8c.

The sum of squares for each configuration is the sum of the squared distances of all the points in the configuration from the origin; remember that the origin is the centroid of the configuration. The residual sum of squares is the sum of squared distances of the points a, b, and c from their counterparts A, B, and C in Figure 10.8c. Since least-squares scaling has been used, the scaling factor is printed as part of the output. Here the fitted configuration has been reduced to about 2/3 of its original size. You can see that the sum of squares of the fixed configuration equals the total of the sum of squares of the fitted configuration and the sum of squares of the residuals. This always happens if least-squares scaling is used (unless a reflection is suppressed).

The third rotation (lines 36 and 37) has the default settings of the SCALING and STANDARDIZE options, so least-squares scaling is not used. But both configurations have been centered, and have then been normalized to have unit sum of squares. The rotation is the same as before. The only change in the fixed configuration is that it has been normalized. However, although Figures 10.8c and 10.8d appear very similar (allowing for the change of scale caused by the normalization), the fitted configuration is nevertheless slightly different. The scaling has caused the printed residuals to be much smaller than before. You can see that each configuration has unit sums of squares. The fitted sum of squares plus the residual sum of squares is thus no longer the same as the sum of squares for the fixed points; a message is printed to that effect.

P.G.N.D.
J.C.G.
R.P.W.

11 Analysis of time series

A *time series* in Genstat is a sequence of observations at equally spaced points in time. Each time series is stored in a variate for which the unit number indexes the time points. Genstat cannot deal explicitly with unequal spacing in time. So if you have such a sequence, you will have to do some form of adjustment or interpolation before using the methods described here.

Usually you will want to describe or model the structure of a series. You can do this without reference to any other variable than the series itself, by examining the relationship between successive measurements. But you can also treat a time series as a response variable, which is related to present and past values of explanatory variables that are also time series. *Forecasts* of time series can be derived from these relationships. You can also use *filters* to adjust time series so as to change their structure.

Most of this chapter describes how to analyse time series by the methods advocated by Box and Jenkins (1970). They recommend a modelling procedure involving three stages: model selection (a term used here in preference to that used by Box and Jenkins, which is "identification"), model estimation, and model checking (again, used here in preference to "verification"). This chapter also provides the basic techniques for spectral analysis, as described by Bloomfield (1976).

Section 11.1 describes how to derive sample statistics from time series, such as *autocorrelations*: these help you select time-series models. Section 11.2 shows how to calculate the *Fourier transform*, which can be useful for displaying cyclical behaviour; it also describes how to construct the *periodogram*, often called the sample spectrum. Section 11.3 describes *autoregressive integrated moving-average* (ARIMA) models, using the notation of Box and Jenkins. It also describes how these are used as *univariate models*: that is, models to describe the behaviour of a single series. There are directives to let you have access to the results of estimation, so that you can check models. Once a model has been fitted, you can make forecasts of the future behaviour of the series.

Section 11.4 shows how to fit regression models between time series, using an ARIMA model to represent correlated errors. Section 11.5 shows how to extend this to general *transfer functions* between series: again you can estimate, check, and forecast. Section 11.6 covers the *filtering* of time series by transfer-function models, as used for example in exponential smoothing or seasonal adjustment. Filtering can also be done by ARIMA models, as used in *pre-whitening*. Section 11.7 presents some ways of displaying the properties of the fitted models, such as the theoretical autocorrelations of ARIMA models.

The index for a time-series variate goes from 1 to N, N being the number of observations. However for defining Fourier transformations, the conventional index is $t = 0 \ldots (N-1)$; we adhere to this.

11.1 Correlation

CORRELATE
Forms correlations between variates, autocorrelations of variates, and lagged cross-correlations between variates.

Options

PRINT	= strings	What to print (correlations, autocorrelations, partialcorrelations, crosscorrelations); default *
GRAPH	= strings	What to display with graphs (autocorrelations, partialcorrelations, crosscorrelations); default *
MAXLAG	= scalar	Maximum lag for results; default * i.e. value inferred from variates to save results

Parameters

SERIES	= variates	Variates from which to form correlations
LAGGEDSERIES	= variates	Series to be lagged to form cross-correlations with first series
AUTOCORRELATIONS	= variates	To save autocorrelations, or to provide them to form partial autocorrelations if SERIES = *
PARTIALCORRELATIONS	= variates	To save partial autocorrelations
CROSSCORRELATIONS	= variates	To save cross-correlations
TEST	= scalars	To save test statistics
VARIANCES	= variates	To save prediction error variances
COEFFICIENTS	= variates	To save prediction coefficients

The most straightforward use of the CORRELATE directive is to calculate correlation coefficients between a set of variates. Here is an example in which the correlations are displayed by the PRINT option in a lower-triangular matrix:

11.1 Correlation

EXAMPLE 11.1

```
  1   "
 -2    Display the correlations of the integers up to 20, squared, and cubed.
 -3   "
  4   VARIATE [VALUES=1...20] Integer
  5   CALCULATE Square,Cube = Integer,Square * Integer
  6   & Random = URAND(8573; 20)
  7   CORRELATE [PRINT=correlations] Integer,Square,Cube,Random

*** Correlation matrix ***

        Integer   1.000
         Square   0.971   1.000
           Cube   0.922   0.986   1.000
         Random  -0.158  -0.151  -0.168   1.000

                Integer  Square    Cube  Random
```

CORRELATE can only display the correlations. If you want to store them, use the FSSPM directive (5.7.3) and the CORRMAT function of the CALCULATE directive (5.2.4).

You may find the correlation coefficients useful when you are doing linear regression (8.1); but they provide little information to help select a model for a time series. More useful are the autocorrelations of the series with itself lagged by particular time intervals. The set of autocorrelations for all possible lags is the *autocorrelation function*. You can derive the *partial autocorrelation function* from it. To look at the relationship between series, you should use the *cross-correlation function* between the one series and the other lagged by the various intervals.

How to interpret the correlation functions is described by many standard books about time series. The books by Anderson (1976) and Nelson (1973) are introductory texts, but do not cover the whole range of models covered in this chapter. The book by Box and Jenkins (1970) gives a full description.

11.1.1 Autocorrelation

You can use the CORRELATE directive to display the sample autocorrelation function of a series, either as a table of numbers, or as a graph – called a *correlogram*. In either case, you must specify the maximum lag for which the autocorrelation is to be calculated, m say. You can do this either by setting the MAXLAG option to m, or by pre-defining the length of a variate to $m+1$ and including it in the AUTOCORRELATIONS parameter to store the calculated values. If you do not specify the maximum lag, CORRELATE will give a fault diagnostic. Here is an example, printing and graphing the correlations up to lag 50 of a series of observations of numbers of airline passengers:

EXAMPLE 11.1.1

```
  8  "
 -9    Show ACF of differenced and logged numbers of airline passengers 1949-59.
-10    Data from Box and Jenkins (1970) page 304.
-11  "
 12  VARIATE [NVALUES=132] Apt
 13  OPEN 'airline.dat'; CHANNEL=2
 14  READ [CHANNEL=2] Apt

     Identifier    Minimum      Mean    Maximum     Values    Missing
            Apt      104.0     262.5      559.0        132          0

 15  CALCULATE Dlapt = DIFFERENCE(LOG(Apt))
 16  CORRELATE [MAXLAG=50; GRAPH=autocorrelations] Dlapt
```

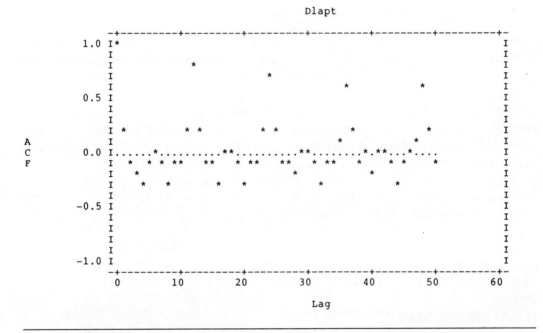

Genstat includes the autocorrelation at lag 0 in the autocorrelation function: it is always unity. The formula used for the sample correlation at lag k is

$$r_k = (1 - k/n) \times C_k / C_0$$

where

$$C_k = \sum_{t=1}^{n-k} \{(y_t - \bar{y}) \times (y_{t+k} - \bar{y})\} n_k$$

The number n_k is the number of terms included in the sum. The series can contain missing values, but this sum excludes any product that involves any missing values at all. The value of $\bar{y}$ is the ordinary sample mean of the whole series, and n is the number of values of the series excluding missing values. You can restrict a series, but the restricted set must consist of a contiguous set of units. Thus, you can look at the autocorrelation function derived from just the first section of a series, or from just the last section, or from a section in the middle; but you cannot use restriction to exclude a section from the middle of the series, or to exclude just individual observations.

The AUTOCORRELATIONS parameter allows you to save the calculated autocorrelations. If you want to display a correlogram in a different form to the standard one produced by the GRAPH option, then you must save the autocorrelations and give a GRAPH statement (7.1.2) explicitly.

The TEST parameter allows you to save a statistic that can be used to test the hypothesis that the true autocorrelation is zero for positive lags. It is defined as

$$S = n \times \sum_{k=1}^{m} r_k^2$$

Provided n is large and m is much smaller than n, then under the null hypothesis, S has a chi-squared distribution with m degrees of freedom. Thus, a large value of S is evidence of statistically significant autocorrelation in a time series.

You can calculate autocorrelation functions for several series in one statement by giving a list of series in the SERIES parameter.

11.1.2 Partial Autocorrelation

Genstat forms partial autocorrelations from an autocorrelation function. The value at lag k is defined as

$$\text{corr}(y_t, y_{t-k} \mid y_{t-1}, y_{t-2} \ldots y_{t-k+1})$$

and is denoted by $\phi_{k,k}$ because you can think of it as being synonymous with the last in the set of coefficients in the autoregressive prediction equation:

$$y_t = c + \phi_{k,1} \times y_{t-1} + \ldots + \phi_{k,k} \times y_{t-k} + e_{k,t}$$

Genstat calculates these coefficients recursively for $k = 1 \ldots m$ by

$$\phi_{k,k} = (r_k - \phi_{k-1,1} \times r_{k-1} - \ldots - \phi_{k-1,k-1} \times r_1) / v_{k-1}$$

$$\phi_{k,j} = \phi_{k-1,j} - \phi_{k,k} \times \phi_{k-1,k-j}, \quad j = 1 \ldots k-1$$

$$v_k = v_{k-1} / (1 - \phi_{k,k}^2)$$

It starts with $v_0 = 1$, the quantity v_k being

$$\text{variance}(e_{k,t}) / \text{variance}(y_t)$$

Partial correlations give you a valuable alternative way of displaying the autocorrelation structure of a series. You can display the partial autocorrelation function either as a table of numbers, or as a graph – just as for the autocorrelation function. Two methods are available for doing this. You can give the series in the SERIES parameter, in which case the autocorrelations are formed first, automatically, and then the partial autocorrelations are derived from them. Or you can set the SERIES parameter to *, and provide autocorrelations in the AUTOCORRELATIONS parameter. You might have previously formed these autocorrelations by CORRELATE, or by TSUMMARIZE (11.7.1) which lets you form the theoretical autocorrelations of a time-series model. You must specify the maximum lag, either by setting the MAXLAG option, or by pre-defining the length of a variate included in the AUTOCORRELATIONS or PARTIALCORRELATIONS parameter. The next example uses the theoretical autocorrelations of a simple ARIMA model; see 11.3.1 and 11.7.3 for explanation of the TSM and TSUMMARIZE statements respectively.

EXAMPLE 11.1.2

```
  17  "
 -18    Calculate partial autocorrelations from an AR[2] model.
 -19  "
  20  TSM AR[2]; ORDERS=!(2,0,0); PARAMETERS=!(1,15,2.5,0.5,-0.5)
  21  TSUMMARIZE [MAXLAG=12] AR[2]; AUTO=Acf_AR[2]
  22  PRINT Acf_AR[2]

   Acf_AR[2]
       1.0000
       0.3333
      -0.3333
      -0.3333
       0.0000
       0.1667
       0.0833
      -0.0417
      -0.0625
      -0.0104
       0.0260
       0.0182
      -0.0039

  23  CORRELATE [PRINT=partialcorrelations] *; AUTOCORRELATIONS=Acf_AR[2]
```

*** Correlations ***

Unit	PACF
1	1.000
2	0.333
3	-0.500
4	0.000
5	0.000
6	0.000
7	0.000
8	0.000
9	0.000
10	0.000
11	0.000
12	0.000
13	0.000

11.1 Correlation

You can save the partial autocorrelation function with the PARTIALCORRELATIONS parameter. Moreover, you can set the VARIANCES and COEFFICIENTS parameters to save the *prediction-error variances* $v_0 \ldots v_m$, and the *prediction coefficients* 1, $\phi_{m,1}$ ... $\phi_{m,m}$ for the maximum lag m. Genstat sets the first coefficient to 1, and also the first element of the partial autocorrelation sequence to 1: you will find this to be a useful convention for the lag 0 values.

CORRELATE will print a warning if you include missing values in an autocorrelation function that you have supplied, or if for some other reason the autocorrelations are invalid. In particular, if a partial autocorrelation value is obtained outside the range $(-1, 1)$, then Genstat truncates the sequence at the previous lag.

11.1.3 Cross-correlation

You can calculate cross-correlations between two series by including one series in the SERIES parameter and the other in the LAGGEDSERIES parameter. You must define the maximum lag, as for autocorrelation. You can graph or tabulate the resulting function. The next example shows the correlation between one series and the later values of a second series, along with the correlation of the second series with later values of the first. Thus these statements display correlation with both future and past values:

EXAMPLE 11.1.3

```
  24   "
 -25      Display cross-correlations between CO2 production and input rate of
 -26      a gas furnace.
 -27   "
  28   OPEN 'furnace.dat'; CHANNEL=3
  29   VARIATE [NVALUES=296] CO2,Inrate
  30   READ [CHANNEL=3; SERIAL=yes; SETNVALUES=yes] CO2,Inrate

       Identifier    Minimum       Mean    Maximum    Values   Missing
              CO2      45.60      53.51      60.50       296         0
            Inrate   -2.71600   -0.05674    2.83400       296         0

  31   CORRELATE [MAXLAG=40; GRAPH=crosscorrelations] Inrate,CO2; CO2,Inrate
```

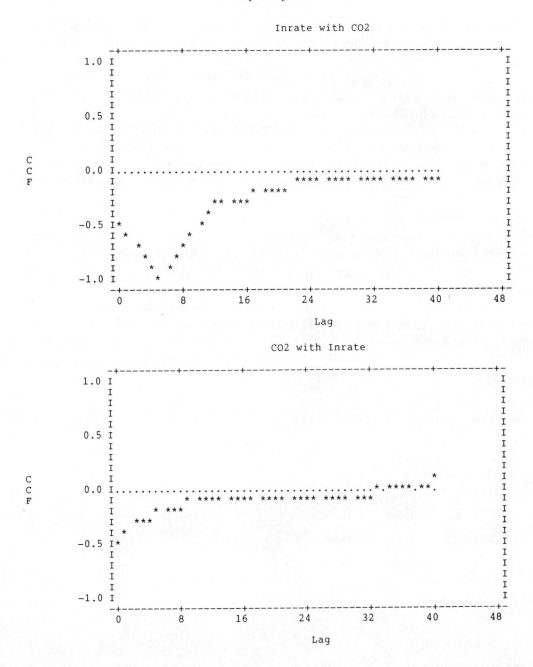

Missing values are allowed, as for autocorrelation. Here is the formula that Genstat uses for the sample cross-correlation between the first series x_t and the lagged series y_t at lag k:

$$r_k = (1 - k/n) \times C_k / (s_x \times s_y)$$

where

$$C_k = \sum_{t=1}^{n-k} \{(x_t - \bar{x}) \times (y_{t+k} - \bar{y})\}/n_k$$

The series x, and y, may be of different lengths. The sum includes all possible terms, but excludes any product containing missing values; the number n_k is the number of terms included in the sum. The values of $\bar{x}$ and $\bar{y}$ and are the sample means, and s_x, s_y are the sample standard deviations. The number n is the minimum of the number of values of x and of y, excluding missing values. You can restrict either series to a set of contiguous units: if both are restricted, their restrictions must match.

You can save the cross-correlation function using the CROSSCORRELATIONS parameter. You can also save a test statistic using the TEST parameter: you can use it in the same way as the statistic described in 11.1.1 to test for lack of lagged cross-correlation in one direction of the relationship between two series. However the test is valid only if each of the series has a zero autocorrelation function. Cross-correlations take precedence in the storage. Thus if you request both autocorrelations and cross-correlations in a single CORRELATE statement, then the stored test statistic will relate to the cross-correlations: that for the autocorrelations will not be stored.

11.2 Fourier transformation

This section describes various types of Fourier transformation. You can use these to do most types of spectral analysis by a few Genstat statements. You may want to put these into procedures (6.3) to use them repeatedly. General procedures for standard types of analysis will be incorporated in the standard Library (12.1.1).

The Fourier or spectral analysis of time series is described comprehensively by Bloomfield (1976) and Jenkins and Watts (1968). The Fourier transformation of a series calculates the coefficients of the sinusoidal components into which the series can be analysed. There are four types of transformation described below, which are appropriate for different types of symmetry in the series. You may often want the length of the variate holding the supplied series to determine implicitly a natural grid of frequencies at which values of the transform are calculated. Genstat will do this if you have not previously declared the identifier supplied for the transform. But you may alternatively want to determine the transform at a finer grid of frequencies, and you can achieve this by declaring a transform variate that is as long as you require. You can do this only for the two types of Fourier transform that apply to real series.

You can recover the series corresponding to a particular transform; that is, you can invert a transformation.

The conventional index for the series that is being transformed is 0 ... (N-1) in the defining formulae, so that the first element corresponds to the origin for the sinusoidal components in the analysis.

FOURIER
Calculates cosine or Fourier transforms of real or complex series.

Option

PRINT	= strings	What to print (transforms); default *

Parameters

SERIES	= variates	Real part of each input series
ISERIES	= variates	Imaginary part of each input series
TRANSFORM	= variates	To save real part of each output series
ITRANSFORM	= variates	To save imaginary part of each output series
PERIODOGRAM	= variates	To save periodogram of each transform

Series of real numbers are stored in single variates, and series of complex numbers in pairs of variates. You can use the FOURIER directive to calculate the cosine transform of the real series $\{a_t, t = 0 \ldots N\text{-}1\}$ as stored in a variate A:

   FOURIER [PRINT=transform] A

You calculate the Fourier transform of the complex series $\{a_t + ib_t, t = 0 \ldots N-1\}$ by storing the values a_t in one variate, A say, the corresponding values b_t in another, B say, and giving the statement:

   FOURIER [PRINT=transform] A; ISERIES=B

You can restrict the series in either the SERIES or ISERIES parameter to a contiguous set of units – as for the CORRELATE directive (11.1). Genstat applies the transformation only to the restricted series of values. Similarly, you may supply restricted variates in the TRANSFORM and ITRANSFORM parameters to save the transform: Genstat will then carry out the transformation so as to supply the required number of values (if that is possible according to the rules at the end of 11.2.2). There must be no missing values in the variates in the SERIES or ISERIES parameters, unless you exclude them by a restriction.

Genstat carries out the Fourier transformation by using a fast algorithm which relies on the order of the transformation being highly composite (de Boor 1980). In practice, an appropriate order is a round number such as 300 or 6000, consisting of a digit followed by zeroes. If, however, the order has a large prime factor, the transformation may take much longer. For example, a transformation of order 499 is about 25 times slower than one of order 500. In the description below,

11.2.1 Cosine transformation of a real series

You will usually want to apply this to calculate the spectrum from a set of autocorrelations. Suppose the variate R contains the values $r_0 \ldots r_n$, and the variate F is to hold the calculated values $f_0 \ldots f_m$ of the spectrum. These values correspond to angular frequencies of $\pi \times j/m$; that is, periods of $2m/j$, for $j = 0 \ldots m$.

You apply the transformation by putting

    FOURIER R; TRANSFORM=F

If F is not previously declared, this statement sets it up automatically as a variate with $n+1$ values (so $m = n$). If F has been declared, then m must be greater than or equal to n; otherwise Genstat will re-declare F to have $n+1$ values.

The transform is defined when $m > n$ by

$$f_j = r_0 + \sum_{k=1}^{n} \{2r_k \times \cos(\pi \times j \times k/m)\}$$

When $m = n$ the final term in this sum is

$$r_n \times \cos(\pi \times j) = r_n \times (-1)^j$$

and it appears without the multiplier 2. The order of the transformation is $2m$.

If R contains sample autocorrelations, you must multiply it by a variate holding a lag window in order to obtain a smooth spectrum estimate (see Bloomfield 1976, page 166; or Jenkins and Watts 1968, page 243).

Here is an example, in which a set of autocorrelations are smoothed with the "Parzen" window before forming the spectrum:

EXAMPLE 11.2.1

```
  1   "
 -2   Spectral analysis of a series of observations of day-length, 1821-1970.
 -3   Data from Shi-fang et al. (1977).
 -4   "
  5   VARIATE [VALUE=1821 ... 1970] Year
  6   UNIT Year
  7   OPEN 'daylength.dat'; CHANNEL=2
  8   READ [CHANNEL=2] Lday

   Identifier    Minimum       Mean    Maximum    Values   Missing
         Lday    -347.00      63.88     421.00       150         0

  9   CALCULATE Diff1day = DIFFERENCE(Lday)
 10   CORRELATE [MAXLAG=25; PRINT=autocorrelations] Diff1day; \
 11      AUTOCORRELATIONS=Auto
```

*** Correlations ***

```
      Unit        Auto
       1         1.000
       2         0.755
       3         0.515
       4         0.204
       5         0.094
       6        -0.010
       7        -0.038
       8        -0.059
       9        -0.013
      10         0.011
      11        -0.032
      12        -0.117
      13        -0.192
      14        -0.213
      15        -0.211
      16        -0.130
      17        -0.055
      18         0.055
      19         0.095
      20         0.117
      21         0.090
      22         0.082
      23         0.047
      24        -0.006
      25        -0.069
      26        -0.103

 12   VARIATE [VALUE=0, 0.01 ... 0.5] Frequency
 13   & [VALUE=0 ... 25] Weight
 14   CALCULATE Weight = Weight / 26
 15   & Weight = 2 * (1-Weight)**3 * (Weight.GE.0.5) +\
 16          ( 1 - 6 * Weight*Weight*(1-Weight) ) * (Weight.LT.0.5)
 17   & Auto = Auto * Weight
 18   VARIATE [NVALUES=Frequency] Spectrum
 19   FOURIER Auto; TRANSFORM=Spectrum
 20   GRAPH [NROWS=21; NCOLUMNS=61] Spectrum; Frequency
```

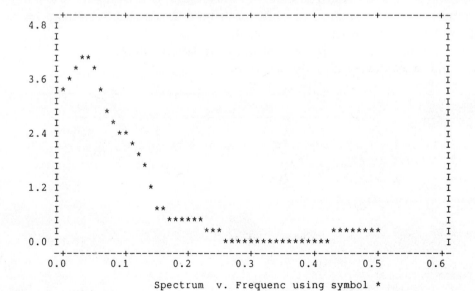

11.2.2 Fourier transformation of a real series

You would usually apply this to calculating the *periodogram* of a time series. Suppose the variate X of length N contains the supplied series values $x_0 \ldots x_{N-1}$. The result of the transformation is a set of coefficients $a_0 \ldots a_m$ of the cosine components and $b_0 \ldots b_m$ of the sine components of the series, held in variates A and B, say. Normally the number of such components is related to the length of the series by taking $m = N/2$ if N is even or $m = (N-1)/2$ if N is odd. Then the coefficients correspond to angular frequencies of $2\pi \times j/N$, which is the same as saying that they correspond to periods N/j for $j = 0 \ldots m$. Since by definition $b_0 = 0$, and $b_m = 0$ if N is even, there are N "free" coefficients in A and B (which you can think of as the real and imaginary parts of a complex transform with values $a_j + ib_j$). You can save the periodogram values $p_0 \ldots p_m$ in a variate P, say: these are the squared amplitudes of the sinusoidal components, and are calculated by Genstat as $p_j = a_j^2 + b_j^2$.

You obtain the transform by putting

FOURIER X; TRANSFORM = A; ITRANSFORM = B; PERIODOGRAM = P

If you want only the periodogram, put

FOURIER X; PERIODOGRAM = P

If you have not previously declared A, Genstat sets it up here as a variate of length $m+1$ where m has the default value defined above. If you have previously declared A, it should have length greater than or equal to $m+1$; otherwise Genstat declares it here to have this length. In any case, B and P should have the same length as A, and will be declared (or re-declared) if necessary.

In the usual case when A, B, or P has the default length $m+1$, the transform is defined by:

$$a_j = \sum_{t=0}^{N-1} \{x_t \times \cos(2\pi \times j \times t/N)\}, \qquad j = 0 \ldots m$$

$$b_j = \sum_{t=0}^{N-1} \{x_t \sin(2\pi \times j \times t/N)\}, \qquad j = 0 \ldots m$$

In this case, the order of the transformation is N. If A, B, and P have length $m'+1$ with $m' > m$, Genstat computes the results at a finer grid of frequencies $2\pi \times j/N'$, $j = 0 \ldots m'$ where $N' = 2m'$. These replace $2\pi \times j/N$ in the above defining sums. The upper limit on the sums remains as $N-1$, although internally Genstat treats it as $N'-1$ with the extra values of $x_N \ldots x_{N'-1}$ being taken as zero. The order of the transformation is then N'.

There are various conventions used for scaling the periodogram with factors $2/m$, $1/m$, or $1/\pi m$. You can apply these by using a CALCULATE statement (5.1) after the transformation. You may also want to apply mean correction to the series before calculating the periodogram.

11.2.3 Fourier transformation of a complex series

This is the most general form of the Fourier transformation, the other three types being essentially special cases in which some coefficients are zero or have a symmetric structure. Suppose variates X and Y contain values $x_0 \ldots x_{N-1}$ and $y_0 \ldots y_{N-1}$, which may be viewed as the real and imaginary parts of the series $\{x_t + iy_t, t = 0 \ldots N-1\}$. The results of the transformation are coefficients $a_0 \ldots a_{N-1}$ and $b_0 \ldots b_{N-1}$ which can be held in variates A and B: these may similarly be considered as parts of complex coefficients $a_t + ib_t$, $t = 0 \ldots N-1$.

You can do the transformation by putting

FOURIER SERIES=X; ISERIES=Y; TRANSFORM=A; ITRANSFORM=B

Both X and Y must be variates with the same length N. Similarly A and B must have length N, and if they do not Genstat will declare (or re-declare) them as variates of length N. The order of the transformation is N.

The results are defined by

$$a_j = \sum_{t=0}^{N-1} \{x_t \times \cos(2\pi \times j \times t/N) - y_t \times \sin(2\pi \times j \times t/N)\}$$

$$b_j = \sum_{t=0}^{N-1} \{x_t \times \sin(2\pi \times j \times t/N) + y_t \times \cos(2\pi \times j \times t/N)\}$$

or equivalently in complex form by

$$(a_j + ib_j) = \sum_{t=0}^{N-1} \{(x_t + iy_t) \times \exp(i2\pi \times j \times t/N)\}$$

The complex transform can be used in cross-spectral analysis.

You can view a Fourier transformation as an orthogonal matrix transformation. Hence its inverse is another Fourier transformation (apart from some simple scaling). You can use this to calculate convolutions. In particular, the correlations of a time series can be obtained by applying the inverse cosine transformation to the periodogram. The next example shows that a repeated Fourier transformation returns the original series – with appropriate scaling.

EXAMPLE 11.2.3

```
  21  "
 -22   Repeat a Fourier transformation on random numbers.
 -23  "
```

```
24  SCALAR Nvalues; VALUE=25
25  CALCULATE Rstart,Istart = URAND(6672,0; Nvalues)
26  FOURIER Rstart; ISERIES=Istart; TRANSFORM=Rmiddle; ITRANSFORM=Imiddle
27  CALCULATE Rmiddle,Imiddle = Rmiddle,Imiddle * 1,-1 / SQRT(Nvalues)
28  FOURIER Rmiddle; ISERIES=Imiddle; TRANSFORM=Rfinish; ITRANSFORM=Ifinish
29  CALCULATE Rfinish,Ifinish = Rfinish,Ifinish * 1,-1 / SQRT(Nvalues)
30  PRINT Rstart,Istart,Rmiddle,Imiddle,Rfinish,Ifinish; DECIMALS=4
```

Rstart	Istart	Rmiddle	Imiddle	Rfinish	Ifinish
0.4236	0.6865	2.5847	-2.6468	0.4236	0.6865
0.4458	0.7316	0.0363	-0.5219	0.4458	0.7316
0.3443	0.5548	-0.1036	-0.2434	0.3443	0.5548
0.0174	0.7045	0.4952	0.2670	0.0174	0.7045
0.0388	0.7507	-0.0092	-0.3748	0.0388	0.7507
0.7562	0.9707	0.0938	-0.2235	0.7562	0.9707
0.3171	0.7538	-0.0380	0.0790	0.3171	0.7538
0.5931	0.6838	-0.0152	0.4113	0.5931	0.6838
0.9229	0.0015	-0.0863	-0.4419	0.9229	0.0015
0.9485	0.5462	0.1806	-0.2726	0.9485	0.5462
0.3938	0.1294	-0.2906	0.0565	0.3938	0.1294
0.6251	0.4935	0.1896	-0.0188	0.6251	0.4935
0.4973	0.7353	0.1773	0.2825	0.4973	0.7353
0.1379	0.2087	-0.1829	-0.3463	0.1379	0.2087
0.2643	0.6310	-0.4662	-0.2856	0.2643	0.6310
0.9029	0.1571	0.2154	0.3256	0.9029	0.1571
0.3597	0.1690	-0.2685	-0.0011	0.3597	0.1690
0.6736	0.4674	-0.1093	0.3781	0.6736	0.4674
0.7469	0.2263	-0.1546	-0.3584	0.7469	0.2263
0.9657	0.8123	-0.2118	-0.0051	0.9657	0.8123
0.0724	0.4666	-0.0544	0.0780	0.0724	0.4666
0.4650	0.6966	0.1744	-0.0849	0.4650	0.6966
0.5000	0.5380	0.5275	0.4453	0.5000	0.5380
0.6257	0.7017	-0.2402	0.3814	0.6257	0.7017
0.8854	0.4171	-0.3260	-0.3117	0.8854	0.4171

11.2.4 Fourier transformation of a conjugate sequence

The easiest way of understanding this is as the reverse of the transformation of a real series (11.2.2), with the roles of the series and the transform interchanged. For the true inverse transformation some simple scaling is also required.

Thus if variates A and B of length $m+1$ are supplied containing values $a_0 \ldots a_m$ and $b_0 \ldots b_m$, which may be viewed as parts of complex coefficients $a_j + ib_j$, the result of the transformation is a single real series $x_0 \ldots x_{N-1}$ held in a variate X of length N.

You may declare X previously with a length $N = 2m$ or $N = 2m+1$ (corresponding to the case N even or odd in 11.2.2). The value of b_0 must then be zero; also if $N = 2m$, the value of b_m must be zero. If either of these conditions is not satisfied, Genstat sets the values of these elements to zero and gives a warning. If X has not been previously declared (or has been declared with a length equal to neither $2m$ nor $2m+1$), then it is declared (or re-declared) with a length governed by whether b_m is 0: $N = 2m$ if $b_m = 0$, or $N = 2m+1$ if $b_m \neq 0$. The value of b_0 is checked to be zero as before.

You can obtain the transform using the statement

FOURIER SERIES = A; ISERIES = B; TRANSFORM = X

The definition of the transform is, in the case $N = 2m+1$,

$$x_t = a_0 + \sum_{j=1}^{m} 2\{a_j \times \cos(2\pi \times t \times j/N) + b_j \times \sin(2\pi \times t \times j/N)\}$$

In the case $N = 2m$, the final term in the sum is simply

$$a_m \times \cos(\pi \times t) = a_m \times (-1)^t$$

and it appears without the multiplier 2. The order of this transformation is N.

11.3 ARIMA modelling

An ARIMA model is an equation relating the present value y_t of an observed time series to past values. The equation includes lagged values not only of the series itself, but also of an unobserved series of *innovations*, a_t; you can interpret the latter as the error in predicting y_t from past values $y_{t-1}, y_{t-2}...$. The usual statistical model is that the innovations are a series of independent Normal deviates with mean zero and constant variance. You can estimate the innovations by the residuals after fitting the model. A time-series model is specified by three things: the orders, which are the numbers of such lagged values that appear in the equation; the parameters, which are the associated coefficients; and, optionally, the actual lags of the values, if these differ from the progression 1 ... m, where m is the number of lags.

For example, consider the model

$$\nabla y_t - c = \phi_1 \times (\nabla y_{t-1} - c) + a_t - \theta_1 \times a_{t-1} - \theta_2 \times a_{t-2}$$

This equation is for the first differences, ∇y_t, of the data, and so has *differencing order* $d = 1$. The *constant term* c represents the mean of ∇y_t. The model has *autoregressive order* $p = 1$ with one parameter ϕ_1, and *moving-average order* $q = 2$ with parameters θ_1 and θ_2.

Here is a Genstat program to fit this model to a series of length 150, and produce forecasts of the next 10 points:

EXAMPLE 11.3

```
  1   "
 -2   Fit an ARIMA(1,1,2) model to the series of daylengths, 1821-1970.
 -3   Display the correlations, check the residuals, and forecast till 1980.
 -4   Data from Shi-fang et al. (1977).
 -5   "
  6   OPEN 'daylength.dat'; CHANNEL=2
  7   READ [CHANNEL=2; SETNVALUES=yes] Daylength

       Identifier    Minimum      Mean    Maximum    Values   Missing
         Daylengt    -347.00     63.88     421.00       150         0

  8   TSM Erp; ORDERS=!(1,1,2)
  9   ESTIMATE Daylength; TSM=Erp
```

```
  9....................................................................
```

***** Time-series analysis *****

  Output series: Daylengt
  Noise model: Erp
                              autoregressive   differencing   moving-average
            Non-seasonal   1   1   2

                    d.f.   deviance
  Residual   145   36959.

*** Autoregressive moving-average model ***

Innovation variance 251.9

                        ref.   estimate   s.e.
  Transformation   0   1.00000   FIXED
  Constant   1   3.98   4.52

* Non-seasonal; differencing order 1

                   lag ref.   estimate   s.e.
  Autoregressive   1   2   0.380   0.104
  Moving-average   1   3   -0.5565   0.0897
                    2   4   -0.6194   0.0794

```
   10  TDISPLAY [PRINT=correlations]


  10....................................................................
```

***** Time-series analysis *****

*** Correlations ***

```
    1   1.000
    2   0.007   1.000
    3   0.004   0.661   1.000
    4  -0.008   0.495   0.557   1.000
            1       2       3       4

   11  TKEEP RESIDUALS=Erpres
   12  CORRELATE [MAXLAG=50; GRAPH=autocorrelations] Erpres
```

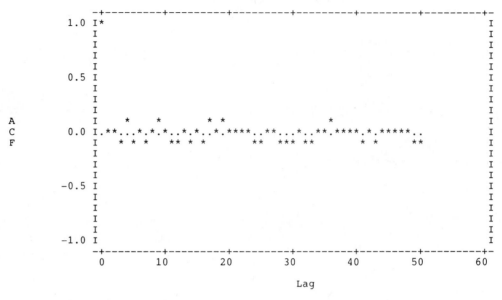

```
   13  FORECAST [MAXLEAD=10]


13..............................................................................


*** Forecasts ***

Maximum lead time: 10

      Lead time    forecast    lower limit    upper limit
           1         297.0        270.9          323.1
           2         305.8        248.9          362.7
           3         311.6        216.6          406.5
           4         316.2        188.3          444.2
           5         320.5        164.4          476.5
           6         325.         144.           505.
           7         329.         126.           531.
           8         333.         111.           555.
           9         337.          96.           577.
          10         341.          83.           598.
```

The TSM statement specifies the orders (p,d,q) of the model as $(1,1,2)$, and names the model Erp (for Earth rotation period). The parameters of the model could also have been specified here; but they have been omitted because they have yet to be estimated. The values for c, ϕ_1, θ_1, and θ_2 are therefore set by Genstat to zero (the default).

The ESTIMATE statement fits the model to the series by an iterative process, the maximum number of iterations and the convergence criterion also taking default values in this case. The results display the estimated *innovation variance* (or residual variance) and estimates of the other model parameters together with their standard errors. Note that the model also allows for a transformation parameter, which by default is not estimated and has the fixed value of 1.0 indicating no transformation.

The TDISPLAY statement prints the correlations between the parameter estimates. Note that each estimate has previously appeared with a reference number that indicates the row or column of the correlation matrix with which it is associated.

The TKEEP statement accesses the variate of residuals a_t; these can also be thought of as the estimated innovations. CORRELATE is used to graph their autocorrelations as a way of checking the model.

Finally the FORECAST statement prints the forecasts of the next 10 values of the series together with their 90% probability limits.

You can use the RESTRICT directive (2.4.1) to fit models to unbroken sub-series of the data. Genstat automatically estimates missing values in a time series together with the model parameters: all these estimates are allowed for in the number of degrees of freedom.

Further examples of all the directives are in 11.3.6.

11.3.1 ARIMA models for time series

The TSM directive is described in 3.6.3. In this section we describe how to use it for ARIMA models, which correspond to the default setting of its MODEL option (MODEL = arima). The definition of transfer-function models is described in 11.5.1.

In many applications you will need only a simple form of the directive, such as:

TSM Erp; ORDERS = !(1,1,2)

Notice that TSM simply sets up a Genstat structure which you can refer to by its identifier with directives such as ESTIMATE. It can for example be saved in a backing-store file (4.5) for further use. In that sense it is analogous to a TERMS statement (8.2.2), which sets up a maximal model for regression analysis, or a TREATMENTSTRUCTURE statement (9.1.1), which sets up a treatment model for analysis of variance.

If a TSM identifier, say Erp, has been declared, then you can print the whole model in a descriptive format with the statement:

PRINT Erp

You can refer to the variates corresponding to the ORDERS, PARAMETERS, and LAGS parameters of the TSM directive by Erp[1], Erp[2], and Erp[3], or for example by Erp['Orders']. Thus the autoregressive order can be assigned to a scalar P by:

CALCULATE P = Erp[1]$[1]

You can change the values of a TSM at any time, for example by CALCULATE statements. Genstat checks that the TSM values specify a valid model whenever they are used in a time-series directive such as ESTIMATE. Be careful, however: you could get strange results if you changed the values of the model between the ESTIMATE and FORECAST statements in the above example.

Using the notation of Box and Jenkins (1970), the simple non-seasonal ARIMA model for the time series y_t is

$$\phi(B)\{\nabla^d y_t^{(\lambda)} - c\} = \theta(B) a_t$$

where B is the *backward shift operator*, ∇ is the *differencing operator*, and

$$\phi(B) = 1 - \phi_1 \times B - \ldots - \phi_p \times B^p$$
$$\theta(B) = 1 - \theta_1 \times B - \ldots - \theta_q \times B^q$$

The parameter λ specifies a Box-Cox power transformation defined by

$$y_t^{(\lambda)} = (y_t^\lambda - 1) / \lambda, \quad \lambda \neq 0$$
$$y_t^{(0)} = \log(y_t)$$

But in the default case when λ is fixed and not estimated, the value $\lambda = 1$ implies no transformation: then $y_t^{(1)} = y_t$ rather than $y_t - 1$. If $\lambda \neq 1$ or if λ is to be estimated, then Genstat will not let you have values of $y_t \leq 0$. The usual case however is that $\lambda = 1$ and is not to be estimated, so that y_t may take any values.

The ORDERS parameter is a list of variates, one for each of the models. For each simple ARIMA model, the variate contains the three values p, d, and q.

The PARAMETERS parameter is a list of variates, one for each of the models. For each simple ARIMA model, the variate contains $(3+p+q)$ values: λ, c, σ_a^2, $\phi_1 \ldots \phi_p$, $\theta_1 \ldots \theta_q$. You must always include the first three parameters. The parameter σ_a^2 is the innovation variance.

Whenever a TSM is used, Genstat checks its values. The orders must all be non-negative. The parameters λ and c can take any values, but σ_a^2 must be non-negative. The next $p+q$ values specify the autoregressive and moving-average parameters: they must satisfy the stationarity and invertibility conditions for ARIMA models (see Box and Jenkins 1970). An exception is that before estimation the model parameters may be unset, in which case Genstat sets them to default values. You can omit the PARAMETERS parameter, in which case an unnamed structure is defined to contain the default values. However, you should usually specify the variate of parameters, and if possible assign good preliminary values before estimation (see 11.7.1).

For convenience of setting the values of parameters, you may wish first to declare scalars or variates containing the separate components:

```
SCALAR Lam,C,Ivar; VALUES=1,4,200
VARIATE [VALUES=0.4] Phi
& [VALUES=-0.5,-0.6] Theta
```

Then to pack these into the parameter variate, you can put

```
VARIATE [VALUES=Lam,C,Ivar,#Phi,#Theta] Erpar
```

11.3 ARIMA modelling

Similarly, in order to extract the components after estimation, you can use the EQUATE directive (5.3.1):

EQUATE Erpar; NEWSTRUCTURES=!P(Lam,C,Ivar,Phi,Theta)

The LAGS parameter is a list of variates, one for each of the models. For each simple ARIMA model, this variate contains $p+q$ values, one corresponding to each of the autoregressive and moving-average parameters. Genstat then modifies the ARIMA model by defining

$$\phi(B) = 1 - \phi_1 \times B^{l_1} - \ldots - \phi_p \times B^{l_p}$$

$$\theta(B) = 1 - \theta_1 \times B^{m_1} - \ldots - \theta_q \times B^{m_q}$$

The LAGS parameter for this model contains $l_1 \ldots l_p, m_1 \ldots m_q$. The sequences of lags $l_1 \ldots l_p$ must be positive integers that are strictly increasing; the default values are $1 \ldots p$ if LAGS is not set. The same rule applies to $m_1 \ldots m_q$.

The seasonal ARIMA model for the time series y_t is an extension of the simple model, to the form

$$\phi(B)\Phi(B^s)\{\nabla^d \nabla_s^D y_t^{(\lambda)} - c\} = \theta(B)\Theta(B^s)a_t$$

where the extra, seasonal, operators associated with seasonal period s are of three types:

$$\Phi(B^s) = 1 - \Phi_1 \times B^s - \ldots - \Phi_P \times B^{Ps}$$

which is seasonal autoregression of order P;

$$\nabla_s^D$$

which is seasonal differencing of order D; and

$$\Theta(B^s) = 1 - \Theta_1 \times B^s - \ldots - \Theta_Q \times B^{Qs}$$

which is seasonal moving average of order Q.

You must extend the ORDERS parameter to contain p, d, q, P, D, Q, and s. Even when $p = d = q = 0$, they must be present. The seasonal orders must satisfy $P \geq 0$, $D \geq 0$, $Q \geq 0$ and $s \geq 1$.

You must also extend the PARAMETERS parameter to contain:

$$\lambda, c, \sigma_a^2, \phi_1 \ldots \phi_p, \theta_1 \ldots \theta_q, \Phi_1 \ldots \Phi_P, \Theta_1 \ldots \Theta_Q$$

You can modify the seasonal model to allow other lags:

$$\Phi(B^s) = 1 - \Phi_1 \times B^{L_1} - \ldots - \Phi_Q \times B^{L_P}$$

$$\Theta(B^s) = 1 - \Theta_1 \times B^{M_1} - \ldots - \Theta_Q \times B^{M_Q}$$

The sequence of lags $L_1 \ldots L_P$ must be strictly increasing and must be positive-integer multiples of the period s; the default values are $s, 2s \ldots Ps$. The same rules apply to $M_1 \ldots M_Q$.

For any seasonal model, you must extend the LAGS parameter, if supplied, to contain:

$$l_1 \ldots l_p, m_1 \ldots m_q, L_1 \ldots L_P, M_1 \ldots M_Q$$

You can use multiple seasonal periods, by extending the variate of ORDERS with further seasonal orders P', D', Q', and s'. You must correspondingly extend the variates of PARAMETERS and LAGS. You can set the seasonal periods to 1, and so you can estimate non-seasonal models with factored operators.

You can declare an ORDERS variate to have more values than is necessary, provided that the extra values are filled with zeroes, and that the number of values is $3+4k$, k being the number of seasonal periods. The same applies to PARAMETERS and LAGS variates, except that Genstat ignores the extra values whatever they may be. Thus you can extend a simple model to a seasonal model, simply by resetting the extra values.

Finally note that you can use the same ORDERS, PARAMETERS, and LAGS variates in more than one TSM.

11.3.2 The ESTIMATE directive

ESTIMATE
Estimates parameters in Box-Jenkins models for time series.

Options

PRINT	= strings	What to print (model, summary, estimates, correlations, monitoring); default m,s,e
LIKELIHOOD	= string	Method of likelihood calculation (exact, leastsquares, marginal); default e
CONSTANT	= string	How to treat constant (estimate, fix); default e
RECYCLE	= string	Whether to continue from previous estimation (no, yes); default n
WEIGHTS	= variate	Weights; default *
MVREPLACE	= string	Whether to replace missing values by their estimates (no, yes); default n
FIX	= variate	Defines constraints on parameters (ordered as in each model, tf models first): zeros fix parameters, parameters with equal numbers are constrained to be equal; default *

MAXCYCLE	= scalar	Maximum number of iterations; default 15
TOLERANCE	= scalar	Criterion for convergence; default 0.0004
SAVE	= identifier	To name save structure, or supply save structure with transfer-functions; default * i.e. transfer-functions taken from latest model

Parameters

SERIES	= variate	Time series to be modelled (output series)
TSM	= TSM	Model for output series
BOXCOXMETHOD	= string	How to treat transformation parameter in output series (fix, estimate); if omitted, transformation parameter is fixed
RESIDUALS	= variate	To save residual series

The main use of ESTIMATE is to fit parameters to time-series models. But you can use it also to initialize for the FORECAST directive, even when the model parameters are already known. In many applications of estimating a univariate ARIMA model, you will need only a simple form of the directive, such as:

ESTIMATE Daylength; TSM = Erp

Examples of ESTIMATE are at the beginning of 11.3 and in 11.3.6.

The SERIES parameter specifies the variate holding the time series to which the model is to be fitted.

The TSM parameter specifies the ARIMA model that is to be fitted. This TSM must already have been declared and its orders must have been set; similarly its lags must have been set if they have previously been declared. But the parameters of the model need not previously have been declared or set. The default values of the parameters are all zero, except that the default of the transformation is 1.0 if it is not to be estimated (see BOXCOXMETHOD and FIX below). If you do specify parameters, Genstat replaces any missing values by the default values. All the parameters that are to be estimated are reset by Genstat to the default if any group of autoregressive or moving-average parameters do not satisfy the required conditions for stationarity or invertibility. After Genstat has finished estimating, the parameters of the TSM contain the estimated values.

The BOXCOXMETHOD parameter lets you estimate the transformation parameter λ.

The RESIDUALS parameter lets you save the estimated innovations (or residuals). As explained in the description of the LIKELIHOOD option, the residuals are calculated for $t = t_0 \ldots N$, where $t_0 = 1+p+d-q$ for a simple ARIMA model. If $t_0 > 1$, missing values will be inserted for $t = 1 \ldots t_0-1$.

The PRINT option controls printed output. If you specify 'monitoring', then at each cycle of the iterative process of estimation, Genstat prints the *deviance* (11.3.3) for the current model, together with the model parameters. The format is simple with the minimum of description, to let you judge easily how quickly the process is converging; an example is given in 11.4. The other settings of PRINT control output at the end of the iterative process. If you specify 'model', then the model is briefly described, giving the identifier of the series and the time-series model, together with the orders of the model. If you specify 'summary', the *deviance* of the final model is printed, along with the residual number of degrees of freedom. If you specify 'estimates', the estimates of the model parameters are printed in a descriptive format, together with their estimated standard errors and reference numbers. If you specify 'correlations', the correlations between estimates of parameters are printed, with reference numbers to the parameters; see 11.3.4 for an example.

You use the LIKELIHOOD option to specify the criterion that Genstat should minimize to obtain the estimates of the parameters: see below.

11.3.3 Technical introduction to how Genstat fits ARIMA models

You may want to skip to the CONSTANT option (p. 558) if you are doing fairly routine work.

The first step in deriving the likelihood for a simple model is to calculate

$$w_t = \nabla^d y_t - c, \quad t = 1+d \ldots N$$

This has a multivariate Normal distribution with dispersion matrix $V\sigma_a^2$, where V depends only on the autoregressive and moving-average parameters. The likelihood is then proportional to

$$\{\sigma_a^{2m} \times |V|\}^{-\frac{1}{2}} \exp\{-w'V^{-1}w/2\sigma_a^2\}$$

where $m = N-d$. In practice Genstat evaluates this by using the formula

$$w'V^{-1}w = W + \sum_{t=t_0}^{N} a_t^2 = S$$

where $t_0 = 1+d+p-q$. The term W is a quadratic form in the p values $w_{1+d-q} \ldots w_{p+d-q}$: it takes account of the starting-value problem for regenerating the innovations a_t, and avoids losing information as would happen if the process used only a conditional sum-of-squares function. If $q > 0$, then Genstat introduces unobserved values of $w_{1+d-q} \ldots w_d$ in order to calculate the sum S. Genstat uses linear least-squares to calculate these q starting values for w, thus minimizing S. We shall call them *back-forecasts*, though if $p > 0$ they are in fact computationally convenient linear functions of the proper back-forecasts. We shall call S the *sum-of-squares function*: it is the sum of the quadratic form and the sum-of-squares term, and is identical to the value expressed by Box and Jenkins as

$$\sum_{t=-\infty}^{N} a_t^2$$

using infinite back-forecasting; that is, using:

$$W = \sum_{t=-\infty}^{t_0-1} a_t^2$$

The values a_t for $t = t_0 \ldots N$ agree precisely with those of Box and Jenkins.

To clarify all this, consider examples with no differencing; that is, $d = 0$. If $p = 0$ and $q = 1$ then $W = 0$ and $t_0 = 0$, and one back-forecast w_0 is introduced. If $p = 1$ and $q = 0$ then $W = (1-\phi_1^2) \times w_1^2$ and $t_0 = 2$, and no back-forecasts are needed. If $p = q = 1$ then $W = (1-\phi_1^2) \times w_0^2$ and $t_0 = 1$, and so one back-forecast w_0 is needed. In this case the proper back-forecast is in fact $w_0/(1 - \theta_1 \times \phi_1)$.

The value of $|V|$ is a by-product of calculating W and the back-forecast. For example, if $p = 0$ and $q = 1$, then

$$|V| = (1 + \theta_1^2 + \ldots + \theta_1^{2N})$$

If $p = 1$ and $q = 0$,

$$|V| = 1/(1 - \phi_1^2)$$

and if $p = q = 1$,

$$|V| = 1 + (\phi_1 - \theta_1)^2 \times (1 + \theta_1^2 + \ldots + \theta_1^{2N-2})/(1 - \phi_1^2)$$

Concentrating the likelihood over σ_a^2 by setting $\sigma_a^2 = S/m$ yields a value proportional to $\{|V|^{1/m} \times S\}^{-m/2}$.

The default setting of the LIKELIHOOD option is 'exact'. In this case the concentrated likelihood is maximized, by minimizing the quantity

$$D = |V|^{1/m} \times S$$

which is called the *deviance*.

The setting 'leastsquares' specifies that Genstat is to minimize only the sum-of-squares term S. This criterion corresponds to the back-forecasting sum-of-squares used by Box and Jenkins, and will in many cases give the same estimates as the exact likelihood. However, some discrepancy arises if the series is short or the model is close to the invertibility boundary. This is because of limitations on the back-forecasting procedure, as described in the algorithms of Box and Jenkins. The deviance value D that Genstat prints is, with this setting, simply S. The setting 'marginal' is described in 11.4.

When you use exact likelihood, the factor $|V|^{1/m}$ reduces bias in the estimates of the parameter: you would get bias if you used 'leastsquares' instead. However, $|V|^{1/m}$ is generally close to one, unless the series is short or the model is either seasonal or close to the boundaries of invertibility or stationarity. The 'leastsquares' setting is therefore adequate for most long, non-seasonal sets of data; Genstat can also handle it much more quickly.

When you specify that Genstat is to estimate the parameter λ of the Box-Cox transformation, Genstat also includes the Jacobian of the transformation in the likelihood function. The result is an extra factor $G^{-2(\lambda-1)}$ in the definition of the deviance, G being the geometric mean of the data, $\left\{\prod_{t=1}^{N} y_t\right\}^{1/N}$. Note that this is not included unless λ is being estimated, even if $\lambda \neq 1$.

You can treat differences in $m \times \log(D)$ as a chi-squared variable in order to test nested models: doing so is supported by asymptotic theory, and by experience with models that have moderately large sample sizes. Similarly, you can select between different models by using $m \times \log(D) + 2k$ as an information criterion, k being the number of estimated parameters. But both of these test procedures are questionable if the estimated models are close to the boundaries of invertibility or stationarity.

You can use the CONSTANT option to specify whether Genstat is to estimate the constant term c in the model. If the setting 'fix' is used, then the constant is held at the value given in the supplied model; this need not be zero.

The RECYCLE option allows a previous ESTIMATE statement to continue: it can save a lot of computing time. If the setting is 'yes', then the most recent ESTIMATE statement is continued, unless you specify in the SAVE option a save structure from some other ESTIMATE statement. The SERIES and TSM settings are taken from the previous ESTIMATE statement: Genstat ignores any that you specify in the present statement. Most of the settings of other parameters and options are carried over from the previous statement, and new values are ignored. But there are some exceptions. You can change the RESIDUALS variate; you can reset MAX-CYCLE to the number of further iterations you require; and you can change the settings of TOLERANCE and PRINT. You can also change the values of the variate in the WEIGHTS option: thus you can get reweighted estimation. You can change the values of the SERIES itself, although you can not change missing values; if the MVREPLACE option was previously set to 'yes', then you must restore the missing values in the SERIES before you use the new ESTIMATE statement.

The WEIGHTS option includes in the likelihood a weighted sum-of-squares term

$$\sum_{t=t_0}^{N} w_t \times a_t^2$$

where w_t, $t = 1 \ldots N$ is held in the WEIGHTS variate. The values of w_t must be strictly positive. If $t_0 < 1$ then w_t is taken as 1 for $t < 1$.

The MVREPLACE option allows you to specify that any missing values in the

11.3 ARIMA modelling

time-series are replaced by their estimates after estimation. But Genstat always in fact estimates the missing values, irrespective of the setting of MVREPLACE: thus you can always obtain the estimates by using TKEEP (11.3.5).

The FIX option allows you to place simple constraints on parameter values throughout estimation. The variate specifies a numbering for the parameters, and must contain integers corresponding to each parameter in the order in which they appeared in the parameters variate of the TSM; but the innovation variance is excluded. If a number is set to 0, the corresponding parameter is constrained to remain at its initial setting. If a number is not 0, and is unique in the variate, the parameter is estimated without special constraint. If two or more numbers are equal, the corresponding parameters are constrained to be equal throughout estimation. The number you set by FIX appears as the reference number of the parameter in the printed model and correlation matrix. This option overrides any setting of CONSTANT and BOXCOXMETHOD. Here is an example of the FIX option, used to constrain some of the parameters in the model fitted at the beginning of this section.

EXAMPLE 11.3.3

```
  14   "
 -15   Fix parametersin ARIMA(1,1,2) model for daylength:
 -16   transformation fixed at 1, Constant unconstrained, AR parameter fixed
 -17   at previous estimate, MA parameters constrained to be equal.
 -18   "
  19   ESTIMATE [FIX=!(0,1,0,2,2)] Daylength; TSM=Erp
```

19...

***** Time-series analysis *****

Output series: Daylengt
Noise model: Erp

	autoregressive	differencing	moving-average
Non-seasonal	1	1	2

	d.f.	deviance
Residual	147	37102.

*** Autoregressive moving-average model ***

Innovation variance 249.5

	ref.	estimate	s.e.
Transformation	0	1.00000	FIXED
Constant	1	3.97	4.51

* Non-seasonal; differencing order 1

	lag	ref.	estimate	s.e.
Autoregressive	1	0	0.380134	FIXED
Moving-average	1	2	-0.5906	0.0596
	2	2	-0.5906	0.0596

The MAXCYCLE option specifies the maximum number of iterations to be performed. You may want to set this to 0 when you use ESTIMATE to initialize for forecasting and you do not want to change the parameters of the TSM (11.3.6). The same setting may also be useful in calculating the deviance for a grid of parameter values: you may want to plot the likelihood function, rather than merely calculate its maximum.

The TOLERANCE option specifies the convergence criterion. Genstat decides that convergence has occurred if the fractional reduction in the deviance in successive iterations is less than the specified value, provided also that the search is not encountering numerical difficulties that force the step length in the parameter space to be severely limited. You can use monitoring to judge whether, for all practical purposes, the iterations have converged. Genstat gives warnings if the specified number of iterations is completed without convergence, or if the search procedure fails to find a reduced value of the deviance despite a very short step length. Such an outcome may be due to complexities in the likelihood function that make the search difficult, but can be due to your specifying too small a value for TOLERANCE.

The SAVE option allows you to save the *time-series save structure*, as produced by ESTIMATE: you can use it in a further ESTIMATE statement with RECYCLE = yes, or in a FORECAST statement. For example, by default Genstat uses this structure to communicate information from ESTIMATE to FORECAST. But it is over-written when a further ESTIMATE statement is executed, unless you have used this option to save it. You can access parts of the save structure by the directives TDISPLAY and TKEEP.

You can access the current time-series save structure by the SPECIAL option of the GET directive (2.3.1), and you can re-set it by the TSAVE option of the SET directive (2.2.1).

11.3.4 The TDISPLAY directive

TDISPLAY
Allows further display after an analysis by ESTIMATE.

Options

PRINT	= strings	What to print (model, summary, estimates, correlations); default m,s,e
CHANNEL	= scalar	Channel number for output; default * i.e. current output channel
SAVE	= identifier	Save structure to supply fitted model; default * i.e. that from last model fitted

11.3 ARIMA modelling

> **No parameters**

You can use this directive to print output from an ESTIMATE statement: you may want to do so if you did not keep this output, or if you find that you did not request enough information in it. But you will not be able to get all of the results if you used the ESTIMATE statement only to initialize for forecasting (11.3.6).

The PRINT option has the same interpretation as in ESTIMATE, except that information is not available to monitor convergence. Here is an example: it is a copy of part of the example given at the beginning of 11.3.

EXAMPLE 11.3.4

```
  10  TDISPLAY [PRINT=correlations]
```

```
10..............................................................

***** Time-series analysis *****

*** Correlations ***

    1   1.000
    2   0.007   1.000
    3   0.004   0.661   1.000
    4  -0.008   0.495   0.557   1.000
            1       2       3       4
```

You can use the CHANNEL option to print the results via a specified output channel (Chapter 4).

You can use the SAVE option to specify a time-series save structure that was set up by any previous ESTIMATE statement. TDISPLAY will then let you display the results of that statement.

11.3.5 The TKEEP directive

> **TKEEP**
> Saves results after an analysis by ESTIMATE.
>
> **Option**
>
> SAVE = identifier   Save structure to supply fitted model; default * i.e. that from last model fitted

Parameters

OUTPUTSERIES	= variate	Output series to which model was fitted
RESIDUALS	= variate	Residual series
ESTIMATES	= variate	Estimates of parameters
SE	= variate	Standard errors of estimates
INVERSE	= symmetric matrix	Inverse matrix
VCOVARIANCE	= symmetric matrix	Variance-covariance matrix of parameters
DEVIANCE	= scalar	Residual deviance
DF	= scalar	Residual degrees of freedom
MVESTIMATES	= variate	Estimates of missing values in series
SEMV	= variate	S.e.s of estimates of missing values
COMPONENTS	= pointer	Variates to save components of output series

An ESTIMATE statement produces many quantities that you may want to use to assess, interpret, and apply the fitted model. The TKEEP directive allows you to get access to these quantities, and to retain them in named structures. But you will not be able to get all of the results if you used the ESTIMATE statement only to initialize for forecasting (11.3.6).

The OUTPUTSERIES parameter specifies the variate that you gave originally in the SERIES parameter of the ESTIMATE statement. You can omit it here.

You use the RESIDUALS parameter to save the residuals in a variate, exactly as in ESTIMATE.

The ESTIMATES parameter specifies a variate to hold the estimated parameters of the TSM. Each estimated parameter is represented once; but the innovation variance is omitted entirely. Genstat includes only the first of a set of parameters that you have constrained to be equal by the FIX option of ESTIMATE. The order of the parameters otherwise corresponds to their order in the variate of parameters in TSM, and is unaffected by any numbering used in the FIX option.

The SE parameter specifies a variate to hold the standard errors of the estimated parameters of the TSM. The values correspond exactly to those in the ESTIMATES variate.

The INVERSE parameter specifies a symmetric matrix to hold the product $\{X'X\}^{-1}$, where X is the most recent design matrix derived from the linearized least-squares regressions that were used to minimize the deviance. The ordering of the rows and columns corresponds exactly to that used for the ESTIMATES variate.

11.3 ARIMA modelling

The VCOVARIANCE parameter specifies a symmetric matrix to hold the estimated variance-covariance matrix, $\hat{\sigma}_a^2 \times \{X'X\}^{-1}$, of the TSM parameters. The ordering of the rows and columns corresponds exactly to that used for the ESTIMATES variate.

The DEVIANCE parameter specifies a scalar that holds the final value of the deviance criterion defined by the LIKELIHOOD option of ESTIMATE.

The DF parameter specifies the residual number of degrees of freedom, defined for a simple ARIMA model by $N-d-$(number of estimated parameters). If a seasonal model is used, this number is further reduced by Ds.

The MVESTIMATES parameter specifies a variate to hold estimates of the missing values of the series, in the order they appear in the series. You can thereby obtain forecasts of the series, by extending the SERIES in ESTIMATE with a set of missing values. This is less efficient than using the FORECAST directive, but it does have the advantage that the standard errors of the estimates take into account the finite extent of the data, and also the fact that the model parameters are estimated.

The SEMV parameter specifies a variate to hold the estimated standard errors of the missing values of the series, in the order they appear in the series.

The COMPONENTS parameter is used when there are explanatory variables, and is described in 11.5.4.

You can use the SAVE option to refer back to the ESTIMATE statement that you want to re-access: you specify a time-series save structure as set up by that ESTIMATE statement.

11.3.6 The FORECAST directive

FORECAST
Forecasts future values of a time series.

Options

PRINT	= strings	What to print (forecasts, sfe, limits); default f,l
CHANNEL	= scalar	Channel number for output; default * i.e. current output channel
ORIGIN	= scalar	Number of known values in FORECAST variate; default 0
MAXLEAD	= scalar	Maximum lead time beyond the origin, i.e. number of forecasts; default * i.e. value taken from length of FORECAST variate
PROBABILITY	= scalar	Probability level for confidence limits; default 0.9

UPDATE	= string	Whether existing forecasts are to be updated (no, yes); default n
FORECAST	= variate	To save forecasts of output series; default *
SETRANSFORM	= variate	To save standard errors of forecasts (on transformed scale, if defined); default *
LOWER	= variate	To save lower confidence limits; default *
UPPER	= variate	To save upper confidence limits; default *
SFE	= variate	To save standardized forecast errors; default *
COMPONENTS	= pointer	Variates to save components of forecast
SAVE	= identifier	Save structure to supply fitted model; default * i.e. that from last model fitted
Parameters		
FUTURE	= variates	Future values of input series
METHOD	= strings	How to treat future values of input series (observations, forecasts); if omitted, they are treated as observations

In many applications of forecasting for univariate ARIMA models, you will need only a simple form of the directive, such as

FORECAST [MAXLEAD=10]

This will cause Genstat to print the required forecasts. However, you must have used a previous ESTIMATE statement to specify the time series to be forecast, and the model to be constructed. Once you have used ESTIMATE, you may give successive FORECAST statements to incorporate new observations of the time series, and to produce forecasts from the end of the new data.

You can use a limited mode of ESTIMATE to initialize for FORECAST. This limited mode saves space. Such savings would be important if you wanted regularly to forecast many time series: you can use the SAVE option to preserve the time-series save structures. You must either supply the ARIMA model here, or else the model must previously have been estimated and you must give here a further ESTIMATE statement with the option settings MAXCYCLE=0 and PRINT=* or

11.3 ARIMA modelling

PRINT = model. Genstat will not change any of the parameter values of the model in this limited mode of ESTIMATE – not even the innovation variance. Furthermore, many of the internal structures will no longer be available via TDISPLAY or TKEEP.

The 'summary', 'estimates', and 'correlations' settings of the PRINT option are valid when you use ESTIMATE to initialize for FORECAST, but Genstat will not use the limited mode if you use these settings. Genstat then estimates the innovation variance, even if MAXCYCLE = 0; also all structures are available for printing and saving with TDISPLAY and TKEEP.

The formal parameters of FORECAST are relevant only when the time-series model incorporates explanatory variables; they are described in 11.4.3.

The best way to understand the options of FORECAST is by example. The next example illustrates how to use ESTIMATE to initialize for FORECAST, applied to a series of 132 points and using a previously estimated model:

EXAMPLE 11.3.6a

```
   1   "
  -2   Forecast number of airline passengers in 1960, using a seasonal
  -3   ARIMA model whose parameters have already been estimated,
  -4   and based on numbers observed 1949-59.
  -5   Data from Box and Jenkins (1970) page 304.
  -6   "
   7   OPEN 'airline.dat'; CHANNEL=2
   8   UNITS [NVALUES=132]
   9   READ [CHANNEL=2] Apt

       Identifier    Minimum      Mean   Maximum    Values   Missing
             Apt      104.0      262.5     559.0       132         0

  10   VARIATE [VALUES=0,1,1, 0,1,1,12] Ord
  11   & [VALUES=0,0,0.00143, 0.34, 0.54] Par
  12   TSM Airpass; ORDERS=Ord; PARAMETERS=Par
  13   ESTIMATE [MAXCYCLE=0; PRINT=model] Apt; TSM=Airpass

13.............................................................................

***** Time-series analysis *****

  Output series: Apt
    Noise model: Airpass
                         autoregressive    differencing    moving-average
             Non-seasonal              0               1                 1
             Period   12               0               1                 1

  14   FORECAST [MAXLEAD=12; FORECAST=Fcst12]

14.............................................................................


*** Forecasts ***

Maximum lead time: 12
```

Lead time	forecast	lower limit	upper limit
1	419.6	394.3	446.5
2	398.9	370.2	429.7
3	466.7	428.6	508.1
4	454.4	413.5	499.5
5	473.9	427.5	525.3
6	547.6	490.1	611.8
7	623.3	553.8	701.5
8	631.7	557.4	716.0
9	527.2	462.1	601.4
10	462.8	403.1	531.2
11	407.1	352.6	470.2
12	452.7	389.7	525.8

The FORECAST option specifies that the forecast values are to be held in the variate Fcst12: you could then, for example, display them graphically.

Now suppose that a further set of observations of the time series has become available, for example a variate New6 containing the next six values of the series. In order to revise the forecasts, you can incorporate this new information as follows:

EXAMPLE 11.3.6b

```
 15   "
-16    Read observed numbers for January to June 1960, and give revised
-17    forecasts for these months with standardized forecast errors.
-18   "
 19   OPEN 'airline2.dat'; CHANNEL=3
 20   READ [CHANNEL=3; SETNVALUES=yes] New6
```

Identifier	Minimum	Mean	Maximum	Values	Missing
New6	391.0	449.2	535.0	6	0

```
 21   FORECAST [PRINT=sfe; ORIGIN=6; MAXLEAD=0; FORECAST=New6]
```

21..

*** Forecasts ***

Forecast origin: 6

Maximum lead time: 0

Lead time	forecast	s.f.e.
-5	417.0	-0.16
-4	391.0	-0.42
-3	419.0	-2.46
-2	461.0	2.39
-1	472.0	0.33
0	535.0	-0.40

The PRINT option now causes Genstat to print the standardized errors of the forecast. These are the innovation values that are generated as each successive new observation is incorporated, divided by the square root of the TSM innovation variance. They provide a useful check on the continuing adequacy of the model. For example, excessively large values (compared to the standard Normal distribution) may indicate that you should revise the model.

The ORIGIN option specifies the number of new values to be incorporated, the last of these becoming the origin from which any further forecasts might be made. The MAXLEAD option is set to 0, thus preventing new forecasts being produced. The FORECAST option is used to specify the variate containing the new observations of the time series.

Revised forecasts of the next six values of the series can then be produced by a further statement:

EXAMPLE 11.3.6c

```
 22    "
-23       Forecast for July to December 1960.
-24    "
 25    FORECAST [MAXLEAD=6; UPDATE=yes; FORECAST=Fcst6]
```

25..

*** Forecasts ***

Maximum lead time: 6

Lead time	forecast	lower limit	upper limit
1	612.1	575.2	651.4
2	620.4	575.8	668.4
3	517.7	475.5	563.7
4	454.5	413.5	499.5
5	399.8	360.7	443.2
6	444.6	397.9	496.7

You use the UPDATE option to incorporate new observations – advancing the origin for future forecasts to the end of whatever new observations have been supplied in previous FORECAST statements since the last ESTIMATE statement. Thus, the FORECAST statement above incorporates the six values supplied in the variate New6 in the previous FORECAST statement. The UPDATE option allows you to alternate between incorporating new observations and producing new forecasts, as observations become available.

You can give a single FORECAST statement both to incorporate new observations and to produce new forecasts:

EXAMPLE 11.3.6d

```
  12   "
 -13   Incorporate new observations and forecast ahead in one statement.
 -14   "
  15   FORECAST [ORIGIN=6; MAXLEAD=6; FORECAST=New6fcst6]
```

 15...

*** Forecasts ***

  Forecast origin: 6

Maximum lead time: 6

Lead time	forecast	lower limit	upper limit
-5	417.0	*	*
-4	391.0	*	*
-3	419.0	*	*
-2	461.0	*	*
-1	472.0	*	*
0	535.0	*	*
1	612.1	575.2	651.4
2	620.4	575.8	668.4
3	517.7	475.5	563.7
4	454.5	413.5	499.5
5	399.8	360.7	443.2
6	444.6	397.9	496.7

The ORIGIN option here specifies that the first six values of the FORECAST variate are to be incorporated as new observations. The MAXLEAD option specifies that the next six values are to hold the forecasts produced from the last of these new values.

You may use this form of statement repeatedly, but the ORIGIN value must increase (or stay the same) unless the UPDATE option is used. Without the setting UPDATE=yes, successive statements will incorporate only those new values of the series that occur beyond the previous ORIGIN and up to the new ORIGIN: the forecasts are revised from the new ORIGIN. Setting UPDATE=yes lets you act as though the previous ORIGIN value were zero.

The PROBABILITY option determines the width of the error limits on the forecast. The specified value is the probability that the actual value will be contained within the limits at any particular lead time. Note that the limits do not apply simultaneously over all lead times.

The SETRANSFORM option specifies a variate for holding the standard errors that Genstat used in calculating the error limits of the forecasts, starting at lead time 1. These are the standard errors of the transformed series, according to the value of the Box-Cox transformation parameter; they are functions of the model only, not of the data.

The LOWER option specifies a variate for holding the lower limits of the forecasts. This must be the same length as the FORECAST variate. The FORECAST directive puts values of the lower limit in the variate, matching the forecasts in the FORECAST variate. The UPPER option is analogous for the upper limits on the forecast. Note that the limits are constructed as symmetric percentiles, assuming Normality of the transformed time series. Similarly, the forecast is a median value – not necessarily the mode or the mean, unless the transformation parameter is 1.0.

The SFE option specifies a variate for holding the standardized errors of the forecasts: see above. The variate must be the same length as the FORECAST variate. The FORECAST directive places values of the errors in the variate, matching the new observations in the FORECAST variate.

The COMPONENTS option is relevant only when the time-series model incorporates explanatory variables, and is described in 11.5.5.

The SAVE option specifies the identifier for the time-series save structure used by the FORECAST directive. You must earlier have named this structure by the SAVE option in the ESTIMATE statement that was used to initialize for FORECAST. The option lets you preserve the structure, for example in backing store: you can use it in later FORECAST statements to revise the forecasts. You do not have to use this option if the initializing ESTIMATE statement and subsequent FORECAST statements all concern the same series and model, because Genstat automatically uses the same structure throughout.

11.4 Regression with autocorrelated (ARIMA) errors

At the beginning of Chapter 8, we stated that regression analysis is not valid if the residuals cannot be assumed to be independent. When modelling observations of a variable that are taken at successive points in time, it is likely that there will be some dependence. A simple check for this is to fit a regression model as in Chapter 8, and then calculate the sample autocorrelation function (11.1.2) of the residuals from the regression. If you think that there might be appreciable autocorrelation, then you should try fitting the regression model using an ARIMA model for the errors, as described in this section. We shall use as an example a time series y_t of daily gas demand (corrected for the effects of days of the week), and a corresponding indicator x_t of the coldness of the days, compiled from temperature, windspeed, and so on. The next example fits a regression between the variates Demand and Coldness which hold 104 consecutive values of the two series. A first-order autoregressive model, AR(1), is specified for the errors: that is, the model is

$$y_t = c + b \times x_t + e_t$$
$$e_t = \phi_1 \times e_{t-1} + a_t$$

where a_t is the series of independent innovations of the errors e_t. We have used the 'monitoring' setting of the PRINT option in the ESTIMATE statement to show the course of the convergence.

EXAMPLE 11.4a

```
  1  "
 -2     Regress daily gas demand on coldness, using an AR(1) model for errors.
 -3  "
  4  OPEN 'demand.dat','cold.dat'; CHANNEL=2,3
  5  READ [CHANNEL=2; SETNVALUES=yes] Demand
```

Identifier	Minimum	Mean	Maximum	Values	Missing
Demand	239.3	348.7	471.8	104	0

```
  6  & [CHANNEL=3] Coldness
```

Identifier	Minimum	Mean	Maximum	Values	Missing
Coldness	-117.30	-49.87	42.60	104	0

```
  7  TSM Erm; ORDERS=!(1,0,0)
  8  TRANSFERFUNCTION Coldness
  9  " Monitor convergence."
 10  ESTIMATE [PRINT=monitoring,estimates] Demand; TSM=Erm; BOXCOX=estimate
```

10..

*** Convergence monitoring ***

Cycle	Deviance	Current parameters			
1	12803380.	0.	1.00000	0.	0.
2	8684909.	1.7447	1.2880	499.04	-0.45365
3	209142.5	3.0618	1.2890	748.64	0.75300
4	38869.89	5.3578	1.3048	1517.3	0.83100
5	28399.29	5.3906	1.3060	1925.6	0.79741
6	27741.35	5.6954	1.3124	1921.7	0.73305
7	27618.37	5.6692	1.3059	1883.2	0.72642
8	27601.49	5.6040	1.3032	1858.0	0.71415
9	27571.62	5.2138	1.2902	1734.7	0.71193
10	27538.84	4.7622	1.2747	1599.7	0.71021
11	27506.65	4.4128	1.2613	1493.9	0.70996
12	27475.54	4.0239	1.2457	1376.4	0.70890
13	27445.39	3.7474	1.2332	1291.5	0.70889
14	27416.51	3.4102	1.2173	1188.4	0.70777
15	27388.23	3.1955	1.2057	1121.4	0.70800

```
******* Warning (Code TS 21). Statement 1 on Line 10
Command: ESTIMATE [PRINT=monitoring,estimates] Demand; TSM=Erm; BOXCOX=estimate


TS21 The iterative estimation process has not converged
The maximum number of cycles is 15
```

***** Time-series analysis *****

*** Transfer-function model 1 ***

Delay time 0

	ref.	estimate	s.e.
Transformation	0	1.00000	FIXED

```
Constant                     0          0.          FIXED

* Non-seasonal; no differencing

                  lag ref.     estimate        s.e.
Moving-average     0   1         2.898        0.896


*** Autoregressive moving-average model ***

Innovation variance 2488.

                      ref.     estimate        s.e.
Transformation         2         1.1894       0.0514
Constant               3         1029.         270.

* Non-seasonal; no differencing

                  lag ref.     estimate        s.e.
Autoregressive     1   4         0.7067       0.0714
```

The TSM statement specifies the AR(1) model for the errors. The TRANSFERFUNCTION statement here merely specifies the explanatory variate. You could use this directive to specify a response model that includes lagged effects of the explanatory variate (11.5.2), but in the example, the response model is a simple linear regression: this is the default.

The warning shows that the convergence criterion has not been reached within 15 iterations. To satisfy the criterion, we could either increase the limit on the number of iterations by setting the option MAXCYCLE=25, say, or initialize the parameters to rough estimates of the parameters in the model, perhaps using the FTSM directive (11.7.1 and 11.7.2). The statements that follow ESTIMATE in this program use the best parameter values found by ESTIMATE, without further comment.

The ESTIMATE statement simultaneously estimates the regression coefficients c and b and the AR parameter ϕ_1. Also in this case, a Box-Cox transformation is estimated for the response variate, Demand. Note in the printed results that the estimate of b appears under "Transfer-function model 1", as a moving-average parameter at lag 0. By default, Genstat fixes the transformation and constant parameters associated with the explanatory variables to be 1 and 0. But you could alternatively estimate these parameters, as described in 11.5.

The constant term c in the regression is included in the results for the autogressive moving-average model; likewise for the transformation parameter of the demand variable, and the estimate of ϕ_1.

You can get forecasts of the demand series, by specifying future values of the explanatory variable. For example, let the variate Newcold contain the next seven values of coldness. Then forecasts of the next seven demand values are obtained as follows.

EXAMPLE 11.4b

```
 11   "
-12    Forecast gas demand for the next week, given values for coldness.
-13   "
 14   READ [CHANNEL=3; SETNVALUES=yes] Newcold

      Identifier    Minimum       Mean    Maximum     Values    Missing
        Newcold      -138.3     -102.3      -75.6          7          0

 15   FORECAST [MAXLEAD=7] Newcold

15.................................................................


*** Forecasts ***

Maximum lead time: 7

      Lead time        forecast   lower limit   upper limit
              1           318.6         290.9         346.0
              2           294.3         259.6         328.1
              3           313.9         277.0         350.0
              4           324.5         286.5         361.7
              5           278.4         238.5         317.2
              6           261.7         221.0         301.2
              7           299.2         259.5         338.0
```

Genstat constructs the forecasts by calculating the predicted linear response at the Newcold values, and adding to it the forecast values of the autocorrelated errors. The forecast limits take this into account.

In practice you would be unlikely to know the future values of explanatory variables. Exceptions include the variable having a fixed deterministic form such as in a trend, or a cycle, or an intervention variable; or when the variable is under the control of the experimenter, as when sales are related to prices; or when the analysis is retrospective, as in this example. You can predict the explanatory variables in various ways. For example, ordinary weather forecasts are in practice used to forecast gas demand. You cannot usually include in the error limits of the forecast the uncertainties in predicting the explanatory variables. These uncertainties would usually be assessed by trying out different future values of the explanatory variables. Thus the FORECAST statement in the example could be repeated with a variety of future values. But there is one case where you can allow for the uncertainty of predicting the explanatory variables. This is when the future values of the explanatory variables are predictions obtained using univariate ARIMA models. Then you can allow for the errors by setting the ARIMA parameter of the TRANSFERFUNCTION directive, and the METHOD parameter of the FORECAST directive.

11.4.1 The TRANSFERFUNCTION directive

TRANSFERFUNCTION
Specifies input series and transfer function models for subsequent estimation of a model for an output series.

Option

SAVE	= identifier	To name time-series save structure; default *

Parameters

SERIES	= variates	Input time series
TRANSFERFUNCTION	= TSMs	Transfer-function models; if omitted, model with 1 moving-average parameter, lag 0
BOXCOXMETHOD	= strings	How to treat transformation parameters (fix, estimate); if omitted, they are treated as fixed
PRIORMETHOD	= strings	How to treat prior values (fix, estimate); if omitted, they are treated as fixed
ARIMA	= TSMs	ARIMA models for input series

For regression with autocorrelated errors, you would use the directive to specify the variates that are to be the explanatory variables in a subsequent ESTIMATE statement. Thus in many applications you will need only a simple form of the directive, such as

    TRANSFERFUNCTION Coldness

The SERIES parameter specifies a list of variates holding the time series of explanatory variables.

The BOXCOXMETHOD parameter lets you estimate separate power transformations for the explanatory variables. Thus the variable x_t is transformed to

$$x_t^{(\lambda)} = (x_t^{\lambda} - 1) / \lambda, \quad \lambda \neq 0$$
$$x_t^{(0)} = \log(x_t)$$

The default is no transformation, corresponding to $x_t^{(\lambda)} = x_t$. You can choose whether the transformations are to be fixed or estimated, by specifying one string for each explanatory variable.

The ARIMA parameter allows you to associate with each explanatory variable a univariate ARIMA model for the time-series structure of that variable. If you think such a model is inappropriate, then you should give a missing value in place of the TSM identifier, or not set this parameter. You can use these models in any subsequent FORECAST statement to incorporate in the error limits of the forecasts an allowance for uncertainties in the predicted explanatory variables; the allowance assumes that the future values of the explanatory variables are forecasts obtained using these ARIMA models (11.4.3).

The TRANSFERFUNCTION and PRIORMETHOD parameters are not relevant in this context, and are described in 11.5.2.

The SAVE option allows you to name the time-series save structure that a TRANSFERFUNCTION statement creates. You can use this identifier in a later ESTIMATE statement, and eventually in a FORECAST statement. Genstat uses this save structure by default to communicate information from TRANSFERFUNCTION to ESTIMATE. But if you do not name it, Genstat overwrites it when a new TRANSFERFUNCTION statement is given: that is why you might want to use this option to save it.

11.4.2 Extensions to the ESTIMATE directive for regression with ARIMA errors

The SERIES parameter now specifies the response variate, and the TSM parameter specifies the ARIMA model for the errors. Note however, that the transformation parameter of this ARIMA model is used to define a transformation for the response variable, not the errors, and the BOXCOXMETHOD parameter controls its estimation.

The constant term in the ARIMA model corresponds to the usual regression constant term only if there is no differencing specified by the ARIMA model; otherwise it is equivalent to a constant term in a regression between the differenced series.

The PRINT option is the same as described in 11.3.2. But note that the regression estimates for the explanatory variables are printed in a sequence of simple transfer-function models, followed by the ARIMA error model: see the example above.

The LIKELIHOOD option settings 'exact' and 'leastsquares' are essentially the same as for univariate ARIMA modelling in 11.3. The likelihood for the model is defined as that of the univariate error series e_t which is defined in general by

$$e_t = y_t - b_1 \times x_{1,t} - \ldots - b_m \times x_{m,t}$$

(the x_i being m explanatory variables). The constant term therefore appears in the model after any differencing of e_i; for example

$$\nabla e_i = c + (1 - \theta_1 B) \times a_t$$

You can get bias in the estimates of the parameters of an ARIMA model because the regression is estimated at the same time as the variances and correlations. Guard against this by specifying LIKELIHOOD = marginal. Doing so can be particularly important if the series are short or if you use many explanatory variables (Tunnicliffe Wilson 1987). The deviance is now defined to be

$$D = S \times \{|X'V^{-1}X| \times |V|\}^{1/m}$$

where m is reduced by the number of regressors (including the constant term) and the columns of X are the differenced explanatory series: the other terms are as in the exact likelihood described in 11.3.3.

You can use this setting also for univariate ARIMA modelling, when the constant term is the only explanatory term. Furthermore, Genstat deals with missing values in the response variate by doing a regression on indicator variates; these too are included in the X matrix. However, you cannot use marginal likelihood and estimate a transformation parameter in either the transfer-function model or an ARIMA model. Neither can you use it if you set the FIX option in ESTIMATE. In these cases Genstat automatically resets the LIKELIHOOD option to 'exact'.

At every iteration, the regression coefficients are the maximum-likelihood estimates conditional upon the estimated values of the parameters of the ARIMA model: these are also the generalized least-squares estimates, conditioned in the same way. This is so even if MAXCYCLE = 0; that is, the coefficients of the regression are re-estimated even at iteration 0. Therefore you must not use the 'marginal' setting in the limited mode (11.3.5) to initialize for FORECAST.

You can use the CONSTANT = fix setting with marginal likelihood. You can compare deviance values that were obtained using marginal likelihood only for models with the same explanatory variables and the same differencing structure in the error model.

You can use the FIX option to impose constraints across any or all of the parameters of the regression and the ARIMA model. In order to do this, you will probably first have to note the printed results from a statement without the FIX option set; then you should give a second statement with the option set. The variate specified in the FIX option must have one element for each parameter that is printed with a reference number. These are, in order, three parameters for each explanatory variate, followed by the ARIMA model parameters. Genstat then uses the variate to provide a parameter numbering as described for the FIX option in 11.4.2. Note that this numbering overrides the BOXCOXMETHOD parameter and the CONSTANT option. Thus you can constrain the transformation parameters to be equal for all or some of the variables. You can also estimate a constant term for an input series. For details of this see 11.5.3.

The results of ESTIMATE that you can access with TDISPLAY and TKEEP are essentially the same as for univariate models. Thus the variate of parameter estimates and associated structures now refer to the whole set of parameters in the order they are printed. The variate of missing-value estimates holds first the values from the response variate, and then those from the explanatory variate, in the order in which they appear in the SERIES parameter of TRANSFERFUNCTION.

11.4.3 Extensions to the FORECAST directive for regression with ARIMA errors

A FORECAST statement for regression with ARIMA errors must be preceded by a TRANSFERFUNCTION statement and an ESTIMATE statement: these initialize the save structure of the time series that is to be used by FORECAST. You can specify the limited mode of initialization by using ESTIMATE as described in 11.3.6.

You use the FUTURE parameter to specify a list of variates, corresponding to the list of variates in the series parameter of TRANSFERFUNCTION. These variates must have the same length. They hold future values of the explanatory variables to be used either for constructing forecasts of the response variable, or for incorporating new observations so as to revise the forecasts. How to use these future values is similar to how to use the FORECAST variate as described in 11.3.6. For example, let Fcdem be a variate of length seven in the example from the beginning of 11.4. The statement

    FORECAST [MAXLEAD = 7; FORECAST = Fcdem] FUTURE = Newcold

causes forecasts of the next week's demand figures to be placed in Fcdem. Suppose that in a week's time, the actual demand had been recorded and was held in the variate Newdem. Then in order to revise the forecasts, you must first incorporate this new information by:

    FORECAST [ORIGIN = 7; MAXLEAD = 0; FORECAST = Newdem] \
        FUTURE = Newcold

Note that if Newcold had previously contained forecasts from an ARIMA model, say, you would have to alter it to contain the recorded values before this statement. You can get revised forecasts of the next week's demand by once more amending Newcold, to hold the values for the coming week, and then using

    FORECAST [UPDATE = yes; MAXLEAD = 7; FORECAST = Fcdem] \
        FUTURE = Newcold

An alternative to the previous two statements would be to use variates of length 14, with Newcold holding the seven values just recorded followed by the seven values for the coming week. Similarly let Newdem hold the last seven days' demand, with its final seven values being missing. Then the statement

    FORECAST[ORIGIN = 7;MAXLEAD = 7;FORECAST = Newdem]FUTURE = Newcold

will incorporate the first seven values (up to the ORIGIN setting) of each variate, and use the last seven values (specified by MAXLEAD) of Newcold to place revised forecasts in the last seven values of Newdem.

You can use the METHOD parameter when some or all of the future values of the explanatory variables are forecasts obtained using univariate ARIMA models. You can amend the error limits of the forecasts for the response variable to allow for the uncertainty in these future values, but you need to assume that there is no cross-correlation between the errors in these predictions. The list of strings specified by the METHOD parameter indicates for each explanatory variable wheth-

11.5 Multi-input transfer-function models

er such an allowance should be made. The future values of a series are by default treated as known values if no corresponding ARIMA model is present, or if the transformation parameter of the ARIMA model is not equal to the value used in the regression model for that series. You can change the settings of the METHOD parameter in successive FORECAST statements.

11.5 Multi-input transfer-function models

A transfer-function model allows for lagged effects of an explanatory variable on the response variable, as well as for autocorrelated errors. Using the notation of Box and Jenkins (1970), including a transfer-function model with an ARIMA model for a response variable gives the equation

$$y_t = v(B)x_t + \psi(B)a_t$$

where we shall now call y_t the *output series* and x_t the *input series*. You can have several input series, so we shall call the full model for y_t a *multi-input* model, corresponding to the term "multiple regression" used in Chapter 8. Writing $y_t = z_t + n_t$ where $z_t = v(B)x_t$ and $n_t = \psi(B)a_t$, we shall call z_t the *component* due to input x_t, and n_t the *noise component*. An ARIMA TSM is used to represent the structure of n_t, and a transfer-function TSM to represent the structure of z_t as a function of x_t.

For example, consider the lagged response, with $|\delta| < 1$:

$$y_t = \omega \times (x_{t-1} + \delta \times x_{t-2} + \delta^2 \times x_{t-3} + \ldots) + n_t$$

Then

$$v(B) = \omega \times B / (1 - \delta \times B).$$

Here is a Genstat program to fit this model to a series of length 40, and produce forecasts of the next eight points:

EXAMPLE 11.5

```
  1    "
 -2    One-input transfer-function model relating level of gilts to profits.
 -3    "
  4    VARIATE [VALUES=1...40] Time
  5    UNITS Time
  6    "Read data on gilts and profits from separate files."
  7    OPEN 'GILTS.DAT','PROFITS.DAT'; CHANNEL=2,3
  8    READ [CHANNEL=2] Gilts
```

Identifier	Minimum	Mean	Maximum	Values	Missing
Gilts	-26.253	1.037	27.971	40	0

```
  9    & [CHANNEL=3] Profits
```

Identifier	Minimum	Mean	Maximum	Values	Missing
Profits	-1.80720	0.02747	1.48670	40	0

```
 10    "Set up transfer-function model with delay time 1 and one AR-type
-11    parameter."
```

```
 12  TSM [MODEL=transfer]  Tf; ORDERS=!(1,1,0,0); PARAMETERS=!(1,0,0,0.1)
 13  TRANSFERFUNCTION Profits; TRANSFER=Tf
 14  "Set up ARIMA model for the noise, with one AR parameter."
 15  TSM Ar; ORDERS=!(1,0,0); PARAMETERS=!(1,0,0,0)
 16  ESTIMATE Gilts; TSM=Ar
```

16..

***** Time-series analysis *****

```
  Output series: Gilts
    Noise model: Ar
                          autoregressive    differencing    moving-average
          Non-seasonal          1                0                0
  Input series:
Profits          ; transfer function: Tf
          Non-seasonal          1                0                0


                 d.f.    deviance
    Residual      36       900.6
```

*** Transfer-function model 1 ***

Delay time 1

```
                      ref.     estimate       s.e.
Transformation         0        1.00000      FIXED
Constant               0        0.           FIXED
```

* Non-seasonal; no differencing

```
                  lag ref.     estimate       s.e.
Autoregressive     1   1        0.6273      0.0805
Moving-average     0   2        8.74        1.16
```

*** Autoregressive moving-average model ***

Innovation variance 24.52

```
                      ref.     estimate       s.e.
Transformation         0        1.00000      FIXED
Constant               3       -1.06         2.87
```

* Non-seasonal; no differencing

```
                  lag ref.     estimate       s.e.
Autoregressive     1   4        0.740       0.118
```

```
 17  "Save the components of the series in variates."
 18  TKEEP COMPONENTS=!P(Fprofits,Noise)
 19  GRAPH [NROWS=21; NCOLUMNS=61] Fprofits,Gilts; Time; METHOD=line,point
```

11.5 Multi-input transfer-function models

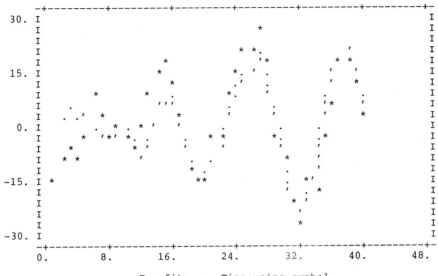

Fprofits v. Time using symbol .
Gilts   v. Time using symbol *

```
 20   "Read future values of profits, and forecast corresponding gilts."
 21   READ [CHANNEL=3; SETNVALUES=yes] Nprofits

      Identifier   Minimum      Mean   Maximum   Values   Missing
       Nprofits    -1.1645   -0.1374    0.4904        8         0

 22   FORECAST Nprofits

22............................................................

*** Forecasts ***

Maximum lead time: 8

     Lead time      forecast   lower limit   upper limit
            1          -6.50        -14.64          1.65
            2         -10.37        -20.51         -0.24
            3         -17.20        -28.27         -6.12
            4         -16.11        -27.67         -4.55
            5         -12.25        -24.06         -0.44
            6          -4.39        -16.34          7.56
            7           1.10        -10.93         13.13
            8           4.14         -7.93         16.21
```

In this program, the first TSM statement defines the orders of the transfer-function model, the initial values of parameters δ and ω being given as 0.0 and 0.1 respectively. The second TSM statement defines the autoregressive error structure. The TRANSFERFUNCTION statement then specifies the input series to be Profits, and gives the associated transfer-function model. The ESTIMATE statement specifies the output series and the noise model.

After the model has been estimated, the TKEEP statement accesses the two components of Gilts. The first of these is graphed together with Gilts, to reveal how well the output series has been modelled by the input series.

Finally, new values of the input series are used to construct forecasts of the output series, using the FORECAST directive.

11.5.1 Declaring transfer-function models

The basic structure of the TSM directive, and of the models that it defines, is given in 11.3.1. Here we describe the ORDERS, PARAMETERS, and LAGS variates for the option setting MODEL = transferfunction.

The simple non-seasonal transfer-function model relates a component z_t of the output series to the corresponding input series x_t, by the equation

$$\delta(B) \nabla^d z_t = \omega(B) B^b \{x_t^{(\lambda)} - c\}$$

where

$$\delta(B) = 1 - \delta_1 \times B - \ldots - \delta_p \times B^p$$
$$\omega(B) = \omega_0 - \omega_1 \times B - \ldots - \omega_q \times B^q$$

The integer $b \geq 0$ defines a pure *delay*, and the integer $d \geq 0$ defines the order of differencing in the transfer function.

The parameter λ specifies a Box-Cox power transformation for the input series, and the parameter c specifies a reference level for the transformed input. There is no mean correction of the input series when transfer-function models are estimated, and you should use a value of c close to the series mean so as to improve the numerical conditioning of the estimation procedure. However, if the input series x_t is trend-like rather than stationary, then you could alternatively use a value for c close to the early series values, because this reduces transient errors that arise when the transfer function is applied. The PRIORMETHOD parameter of TRANSFERFUNCTION, described below, provides further means of handling these transients.

The parameters λ and c are not estimated unless you specify otherwise by the BOXCOXMETHOD parameter of TRANSFERFUNCTION or the FIX option of ESTIMATE. Often c in the transfer-function model is aliased with the constant term in the ARIMA errors, and so they should not both be estimated. In some circumstances, however, they both could be estimated, for example in a differenced transfer-function model with stationary noise.

The ORDERS parameter for the simple transfer-function model described above specifies a variate containing the four values b, p, d, and q.

The PARAMETERS parameter specifies a variate containing $3+p+q$ values: λ, c, $\delta_1, \ldots \delta_p$, $\omega_0, \omega_1 \ldots \omega_q$. You must always include the parameters λ, c, and ω_0. When you use a transfer-function model, Genstat will check its parameter values. In particular the operator $\delta(B)$ must satisfy the stability or stationarity condition.

The LAGS parameter is optional, and may be used to change the lags associated with the parameters, from the default values of $1 \ldots p$, $1 \ldots q$. The variate of lags

11.5 Multi-input transfer-function models

contains values corresponding to the parameters $\delta_1 \ldots \delta_p, \omega_1 \ldots \omega_q$. They have the same interpretation as the lags in ARIMA models, and must satisfy the same conditions as specified in 11.3.1. Note that there is no lag associated with ω_0, because the delay b provides the necessary flexibility for this.

You can also have seasonal extensions of transfer-function models:

$$\delta(B)\Delta(B^s)\nabla^d \nabla_s^D z_t = \omega(B)\Omega(B^s)B^b\{x_t^{(\lambda)} - c\}$$
$$\Delta(B^s) = 1 - \Delta_1 \times B^s - \ldots - \Delta_P \times B^{Ps}$$
$$\Omega(B^s) = 1 - \Omega_1 \times B^s - \ldots - \Omega_Q \times B^{Qs}$$

Note that there is no Ω_0 coefficient, because ω_0 is always present in the model and provides sufficient flexibility.

The ORDERS parameter here contains b, p, d, q, P, D, Q, and s, and the PARAMETERS parameter contains λ, c, $\delta_1 \ldots \delta_p$, $\omega_0 \ldots \omega_q$, $\Delta_1 \ldots \Delta_P$, $\Omega_1 \ldots \Omega_Q$. You can analogously extend the LAGS parameter. You can have extensions to multiple seasonal periods, as for ARIMA models.

11.5.2 Extensions to the TRANSFERFUNCTION directive for multi-input models

This directive specifies several input series and the associated transfer-function models: these are to be used in a subsequent ESTIMATE statement that fits a multi-input model to an output series.

The SERIES and BOXCOXMETHOD parameters are as described in 11.4.1.

The TRANSFERFUNCTION parameter specifies the transfer-function TSMs that are to be associated with the input series. A missing value in place of a TSM identifier causes Genstat to treat the corresponding input series as a simple explanatory variable, equivalent to a transfer-function model with orders (0,0,0,0).

The PRIORMETHOD parameter specifies, for each input series, how Genstat is to treat the transients associated with the early values of the transfer-function response. In calculating the input component z_t from the input x_t, Genstat has to make assumptions about the unknown values of x_t which came before the observation period. The default is that x_t (or generally $x_t^{(\lambda)}$) is assumed to be equal to the reference constant c of the transfer-function model. The pattern of the transient can be controlled by introducing a number $\max(p+d, b+q)$ of nuisance parameters to represent the combined effects of all earlier input values on the observed output. Setting the value of PRIORMETHOD to 'estimate' specifies that these nuisance parameters are estimated so as to minimize the transients. But you should be careful in using this. Often all you will have to do is make a sensible choice of the reference constant c. Estimating the transients is best done as a final stage in refining the model: earlier, you could get poor numerical conditioning; moreover, the transients could be contaminated by dominant features such as trends or cycles in the output series.

11.5.3 Extensions to the ESTIMATE directive for multi-input models

This fits a multi-input model to output series that have a specified model for the output noise. The input series and transfer-function models must have been specified in an earlier TRANSFERFUNCTION statement.

The PRINT option is the same as before, but note that the transfer-function models are printed in a descriptive format similar to the ARIMA model, with parameter reference numbers used throughout.

The LIKELIHOOD option settings 'exact' and 'leastsquares' are similar to the settings described in 11.4.2 for regression with ARIMA errors. For example, with a single input, the likelihood is defined as that for the univariate noise series n_t, calculated as $n_t = y_t - z_t$.

The marginal likelihood is permitted only when all the transfer-function models are equivalent to simple regression.

You can use the FIX option as described in 11.3.2 and 11.4.2, to impose constraints among the parameters while the model is being estimated. These constraints operate here across the whole set (in order) of the parameters of the transfer-function models and of the ARIMA model, excluding the innovation variance. Thus you can use this option to estimate the constant term in a transfer-function model (but bear in mind the remarks in 11.5.1 about possible aliasing).

11.5.4 Extensions to the TKEEP directive for multi-input models

After a multi-input model has been fitted using ESTIMATE, you can use the COMPONENTS parameter to access the components of the output series that are due to the various input series: you can also access the output noise. In simple regression, the input components are proportional to the input series. But the component resulting from a transfer-function model may be quite different from this. You can examine these components separately, or sum them to show the total fit to the output series that is explained by the input series. Note that the fitted values may appear to be offset from that output series, because the constant term is part of the noise component, and so is not included. You may want to examine the output noise component. For example, if you thought that the ARIMA model for the output noise was inadequate, you could investigate the noise component with the univariate ARIMA modelling procedure described earlier in this chapter.

11.5.5 Extensions to the FORECAST directive for multi-input models

FORECAST for multi-input models is the same as for regression models with ARIMA errors (11.4.3). But it does have one further useful option.

The COMPONENTS option specifies a pointer to variates in which you can save components of future values of the output series. There is a variate for each input component and for the output noise component. These variates correspond exactly to the variates that you specified by the FUTURE parameter for the input series, and by the FORECAST variate for the output series; corresponding lengths must match. The values that the variates hold can therefore be components of the forecasts of the output series, or can be new observations. You can use them to investigate the structure of forecasts.

If the input series ARIMA model and the transfer-function model have differing transformation parameters, then the METHOD option reverts to its default action of treating the values of any future input series as known quantities rather than forecasts.

11.6 Filtering time series

Filtering is a means of processing one time series so as to produce a new series. The purpose is usually to reveal some features and remove other features of the original series. Filters in Genstat are one-sided: that is, each value in the new series depends only on present and past values of the original series. However, you can do two-sided filtering by using the SHIFT and REVERSE functions of CALCULATE (5.2.1).

A *filter* is defined by a time-series model. For example, consider the exponentially weighted moving average (EWMA) filter

$$y_t = (1 - \lambda) \times y_{t-1} + \lambda \times x_t$$

which smoothes x_t to produce y_t. Then

$$y_t = \{(1 - \lambda) / (1 - \lambda \times B)\} x_t$$

You can represent this by a transfer function applied to x_t. Here is a Genstat program that applies this filter to smooth a time series of annual temperatures in Central England, taking $\lambda = 0.9$:

EXAMPLE 11.6

```
  1   "
 -2       Filter a series of Central-England temperatures.
 -3   "
  4   VARIATE [VALUES=1...100] Time
  5   UNITS Time
  6   OPEN 'cet100.dat'; CHANNEL=2
  7   READ [CHANNEL=2] Cetemp

      Identifier    Minimum      Mean    Maximum     Values    Missing
         Cetemp       7.400     9.308     10.600        100          0

  8   TSM [MODEL=transfer] Ewma; ORDERS=!(0,1,0,0); PARAMETERS=!(1,0,0.9,0.1)
  9   FILTER Cetemp; NEWSERIES=Smtemp; FILTER=Ewma
 10   PEN 1,2; METHOD=point,line; COLOUR=1
 11   DGRAPH [TITLE='Smoothing temperature'] Cetemp,Smtemp; Time; PEN=1,2
```

The graph from the DGRAPH statement is in Figure 11.6.

FIGURE 11.6

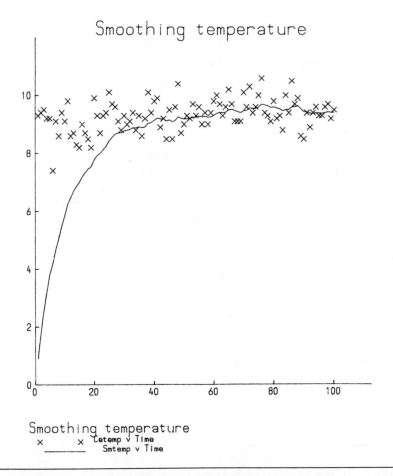

In this example the filter is defined by a transfer-function model. You can alternatively use an ARIMA model to define a filter, in which case the model *pre-whitens* the series. Suppose, for example, an AR(1) model is specified, with parameter ϕ_1; then the result of applying it to a series x_t is to generate a series a_t :

$$a_t = x_t - \phi_1 \times x_{t-1}$$

Such an operation is usefully applied to whiten a series before calculating its spectrum, or to whiten a pair of series before calculating their cross-correlation.

11.6.1 The FILTER directive

FILTER
Filters time series by time-series models.

11.6 Filtering time series

Option		
PRINT	= strings	What to print (series); default *
Parameters		
OLDSERIES	= variates	Time series to be filtered
NEWSERIES	= variates	To save filtered series
FILTER	= TSMs	Models to filter with respect to
ARIMA	= TSMs	ARIMA models for time series

The OLDSERIES and NEWSERIES parameters specify respectively the time series to be filtered, and the series that result from filtering. A new series must not have the same identifier as the series from which it was calculated. Genstat interprets any missing values in the old series as zero. But if you use the ARIMA parameter (see below), then Genstat replaces them by interpolated values when it calculates the filtered series; the missing values remain in the old series.

The FILTER parameter specifies the TSMs to be used for filtering. If the TSM is a transfer-function model (11.5.1), then the new series y_t is calculated from the old series x_t by

$$y_t = \{\omega(B)B^b / \delta(B)\nabla^d\}x_t$$

The filter does not use the power transformation nor the reference constant. This lets you apply a single filter conveniently to a set of time series, for which different transformations and different constants might be appropriate. You can always use the CALCULATE directive to apply a transformation to a series before using FILTER.

If the TSM is an ARIMA model (11.3.1), then the new series a_t is calculated from the old series y_t by

$$a_t = \{\phi(B)\nabla^d / \theta(B)\}y_t$$

Note that the TSM does not have to be the model appropriate for y_t. Again, Genstat ignores the parameters λ, c, and σ_a^2; you can set them to 1,0,0, for example.

The ARIMA parameter specifies a time-series model for the old series. The purpose is to reduce transient errors that arise in the early part of the new series: these arise because Genstat does not know the values of the old series that came before those that have been supplied. If you do not use this parameter, then Genstat takes these earlier values to be zero. This can cause unacceptable transients which can only be partially removed by procedures such as mean-correcting the old series. If you do use the ARIMA parameter, then Genstat uses the specified model to estimate (or back-forecast) the values of the old series earlier than those that have been supplied.

You do not have to have a good ARIMA model for the old series in order to achieve worthwhile reductions in the transients. Thus a model with orders (0,1,1) and parameters (1,0,0,0.7) would estimate the prior values to be constant, at a level that is a backward EWMA of the early values of the series. Here is a continuation of the previous example in which the ARIMA parameter is used:

EXAMPLE 11.6.1

```
  12  "
 -13    Filter the temperatures, using an ARIMA model to reduce transients.
 -14  "
  15  TSM Back; ORDERS=!(0,1,1); PARAMETERS=!(1,0,0,0.7)
  16  FILTER Cetemp; NEWSERIES=Smtemp; FILTER=Ewma; ARIMA=Back
  17  DGRAPH [TITLE='Smoothing with reduced transients'] \
  18     Cetemp,Smtemp; Time; PEN=1,2
  19  STOP
```

The new graph produced by the DGRAPH statement is in Figure 11.6.1.

FIGURE 11.6.1

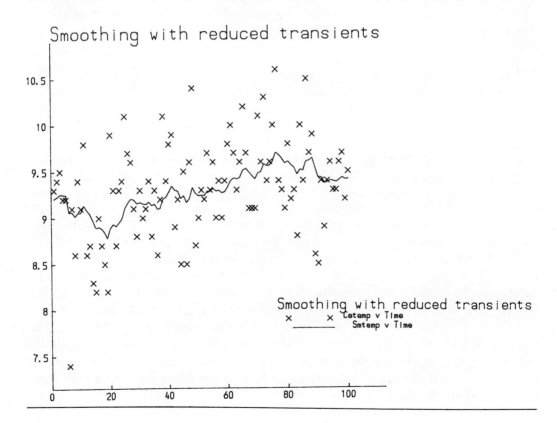

For a seasonal monthly time series, an appropriate ARIMA model could have orders (0,1,1,0,1,1,12) and parameters (1,0,0,0.7,0.7). However you must give the supplied model a transformation parameter $\lambda = 1$. Any other value for λ breaks the assumption of linearity that underlies the calculations for correcting the transients. The constant term in the ARIMA model can be non-zero, and should be if that is appropriate for the old series. Note that the ARIMA model does not define the filter.

If you specify the ARIMA parameter, then Genstat uses this model to interpolate any missing values in the old series before it calculates the new series. Suppose for example that the filter is the identity, defined by a transfer-function model with orders (0,0,0,0) and parameters (1,0,0); then the new series will be the old series with any missing values replaced.

11.7 Forming preliminary estimates and displaying models

The ESTIMATE directive (11.3.2) carries out a lot of computation to find the best estimates of the parameters of a time-series model. The amount of computation can be reduced if you find rough initial values for the parameters, especially when there are many of them. You can get Genstat to do this by using the FTSM directive. FTSM obtains moment estimators of a simple kind, by solving equations between the unknown parameters of the ARIMA or transfer-function model and the autocorrelations or cross-correlations calculated from the observed time series. Sometimes these equations have no solution, or their solution provides values inconsistent with the constraints demanded of the parameters. If so, Genstat sets the corresponding parameters to missing values. The form of the directive is the same for ARIMA and transfer-function models, but the interpretation is slightly different. So we describe the two cases separately.

The TSUMMARIZE directive helps you investigate time-series models by displaying various characteristics. These are the theoretical autocorrelation function of an ARIMA model, and the pi-weights and psi-weights; also the impulse-response function of a transfer-function model. TSUMMARIZE can derive the expanded form of a model, in which all seasonal terms are combined with the non-seasonal term.

11.7.1 Preliminary estimation of ARIMA model parameters

FTSM
Forms preliminary estimates of parameters in time-series models.

Option

PRINT = strings   What to print (models); default *

Parameters

TSM	= TSMs	Models whose parameters are to be estimated
CORRELATIONS	= variates	Auto- or cross-correlations on which to base estimates for each model
BOXCOXTRANSFORM	= scalars	Box-Cox transformation parameter
CONSTANT	= scalars	Constant term
VARIANCE	= scalars	Variance of ARIMA model, or ratio of input to output variance for transfer model

A typical FTSM statement might be

    FTSM [PRINT=model] Yatsm; CORRELATIONS=Yacf; BOXCOX=Ytran;\
        CONSTANT=Ymean; VARIANCE=Yvar

You must previously have declared the time-series model Yatsm to be of type ARIMA with appropriate orders, and lags if you need to specify them. Genstat takes this model to be associated with observations of a time series y_t. The aim of the directive is to set the values of the variate of model parameters equal to preliminary estimates derived from the variate Yacf and scalars Ytran, Ymean, and Yvar.

The variate Yacf should contain sample autocorrelations $r_0 \ldots r_m$. You should obtain these from the original time series, stored in variate Y say, by first using the CALCULATE directive to transform Y according to the Box-Cox equations with transformation parameter Ytran (if indeed you do want a transformation). You should then form the differences of the transformed series, according to the degrees of differencing already set in the model: use the function DIFFERENCE with the CALCULATE directive (5.2.1). Finally, you should store in Yacf the autocorrelations of the resulting series, using the AUTOCORRELATIONS parameter of the CORRELATE directive (11.1.2). Often you will already have done these operations in order to produce Yacf for selecting a model.

At the same time, you can supply the scalars Ytran, Ymean, and Yvar to set the first three elements of the parameters variate of Yatsm; these cannot be set using Yacf alone. The scalar Ytran should be the parameter used to transform Y, and Genstat will copy it into the first element of the variate of parameters. Genstat will copy the scalar Ymean into the second element, which is the constant term of the model; the recommended value for this is the sample mean of the series from which Yacf is calculated, but you may prefer the value 0. The scalar Yvar is used to set the innovation variance, which is the third element of the variate of parameters. The recommended value is the sample variance of the series from which Yacf is calculated. If you set Yvar to 1.0, then Genstat will set the innovation variance to the variance ratio Variance(e)/Variance(y), as estimated from

11.7 Forming preliminary estimates and displaying models

Yacf according to the model.

If any of the BOXCOX, CONSTANT, or VARIANCE parameters are not set, then Genstat will leave unchanged the corresponding value in the variate of parameters of the model. The only exception to this rule is if a parameter is missing. Then Genstat initially sets the transformation parameter to 1.0 (corresponding to no transformation), and the constant to 0.0; the innovation variance is left missing.

11.7.2 Preliminary estimation of transfer-function model parameters

A typical FTSM statement for a transfer-function model might be

FTSM [PRINT=model] Xytsm; CORRELATIONS=Xyccf; BOXCOX=Xtran;\
CONSTANT=Xmean; VARIANCE=Xyvratio

You must previously have declared the time-series model Xytsm to be of type 'transferfunction' with appropriate orders, and lags if you need to specify them. Genstat assumes that this model represents the dependence of an output series y_t on an input series x_t in a multi-input model. The directive sets the values of the parameters of the model equal to preliminary estimates derived from Xyccf, Xtran, Xmean, and Xyvratio.

You should put into the variate Xyccf an estimate of the impulse-response function of the model, from which Genstat will derive the parameters. This estimate is usually a sample cross-correlation sequence $r_0 \ldots r_m$ obtained from variates Y and X containing observations of y_t and x_t according to one of the following four rules:

(a) In the simple case, the differencing orders of Xytsm are all zero, and you do not want to use any Box-Cox transformation of either y_t or x_t. Then the cross-correlations should be those between variates Alpha and Beta, say, derived from X and Y by filtering (or pre-whitening), as described in 11.6.2. The ARIMA model that you used for the filter should be the same for X and Y, and you should choose it so that the values of Alpha represent white noise.

(b) If the differencing orders of Xytsm are not zero, then before you calculate the cross-correlations you should further difference the series Beta as specified by these orders.

(c) If a Box-Cox transformation is associated with y_t, you should apply it to Y before the filtering. However this transformation parameter must not be associated with Xytsm: you should assign it to the univariate ARIMA model that you have specified for the error term (11.3.2).

(d) If a Box-Cox transformation is associated with x_t, it must be the same as the one you used in the ARIMA model for x_t from which the series Alpha was derived. The scalar Xtran must contain this transformation parameter. Genstat copies it into the first element of the parameter variate of Xytsm. If the Box-Cox parameter is unset, Genstat leaves unchanged the transformation parameter of

Xytsm; it is set to 1.0 if it was originally missing.

Genstat copies the scalar Xmean into the second element of the variate of parameters. The recommended value is the sample mean of X after any transformation has been applied. If you do not set the CONSTANT parameter, Genstat leaves unchanged the constant parameter of Xytsm; it is set to 0.0 if it was originally missing.

You use the scalar Xyvratio to obtain the correct scaling of non-seasonal moving-average parameters in Xytsm. All the other autoregressive parameters and moving-average parameters are invariant under scale changes in y_t and x_t. You should set the scalar to the ratio of the sample variances of the variates from which the cross-correlations were calculated; that is, Variance(Beta)/ Variance(Alpha). If you do not set it, Genstat uses the value 1.0.

You can use FTSM to go backwards from autocorrelations to the original time-series model. Apply it to the autocorrelations that were constructed from a time-series model by means of TSUMMARIZE (11.7.3): then it will exactly recover the parameters of the model, provided the model is non-seasonal. If the model contains seasonal parameters, with seasonal period s, then the parameters will not be recovered exactly. But they will be in one special circumstance: that is, when the non-seasonal part of the model, considered in isolation from the seasonal part, has a theoretical autocorrelation function that is zero beyond lag $s/2$. Otherwise, the non-seasonal and seasonal parts of the model "interact", and so Genstat loses accuracy in the recovered parameters. When you use sample autocorrelations, this loss of accuracy tends to be small in comparison with the sampling fluctuations of the estimates. But if s is small, say $s = 4$ for quarterly data, the loss could be serious. Exactly the same considerations apply to transfer-function models.

11.7.3 The TSUMMARIZE directive

TSUMMARIZE
Displays characteristics of time series models.

Options

PRINT	= strings	What to print (autocorrelations, expansion, impulse, piweights, psiweights); default *
GRAPH	= strings	What to display with graphs (autocorrelations, impulse, piweights, psiweights); default *
MAXLAG	= scalar	Maximum lag for results; default 0

Parameters

TSM	= TSMs	Models to be displayed
AUTOCORRELATIONS	= variates	To save theoretical autocorrelations
IMPULSERESPONSE	= variates	To save impulse-response function
STEPFUNCTION	= variates	To save step function from impulse
PIWEIGHTS	= variates	To save pi-weights
PSIWEIGHTS	= variates	To save psi-weights
EXPANSION	= TSMs	To save expanded models
VARIANCE	= scalars	To save variance of each TSM

For an ARIMA model in the TSM parameter, you can set only the AUTOCORRELATIONS, PSIWEIGHTS, and PIWEIGHTS parameters. Also, you can set the IMPULSERESPONSE parameter only for a transfer-function model. You can set the EXPANSION parameter for either type of model. The TSMs in any TSUMMARIZE statement must be completely defined; that is, you must have set the orders and parameters, and the lags if you are using them. The only exceptions are that Genstat takes the transformation parameter to be 1.0 if it is missing, and that the innovation variance of an ARIMA model need not be set.

The MAXLAG option specifies the maximum lag to which Genstat is to do calculations: this applies to autocorrelations, psi-weights, pi-weights, and impulse responses. If it is unset, the maximum lag is taken to be implicitly defined as the length of the first variate in the parameters. Genstat reports a fault if the maximum lag is not defined.

You can set the PRINT and GRAPH options independently of the parameters: these store results, and display the various characteristics of models.

The AUTOCORRELATIONS parameter allows you to store the theoretical autocorrelation function of an ARIMA model. Such a model uniquely defines an autocorrelation function whose values $r_0 \ldots r_m$ are assigned by Genstat to the variate R, where m is the maximum lag. If the model has differencing parameters $d = D = 0$, then the autocorrelation function is that of a series y_t that follows this model.

If either $d > 0$ or $D > 0$, then the theoretical autocorrelations are calculated as if $d = D = 0$, and so they correspond to those of the differenced y_t series. This is because the autocorrelations of y_t are undefined for differenced non-stationary models.

EXAMPLE 11.7.3

```
  1  "
 -2   Display the autocorrelations of an AR[2] model.
 -3   (The partial autocorrelations of this model were shown in Section 11.1.2)
 -4  "
  5  TSM AR[2]; ORDERS=!(2,0,0); PARAMETERS=!(1,15,2.5,0.5,-0.5)
  6  TSUMMARIZE [MAXLAG=12; PRINT=autocorrelations] AR[2]

  6.........................................................................

*** Summary of model AR[2]
    Lag           ACF
      0         1.000
      1         0.333
      2        -0.333
      3        -0.333
      4         0.000
      5         0.167
      6         0.083
      7        -0.042
      8        -0.063
      9        -0.010
     10         0.026
     11         0.018
     12        -0.004
```

The PSIWEIGHTS parameter allows you to store the theoretical psi-weights $\psi_0 \ldots \psi_m$ of an ARIMA model. You will find these useful in constructing error limits for forecasts. These weights are defined by

$$1 + \psi_1 \times B + \psi_2 \times B^2 + \ldots = \theta(B) / \{\phi(B)\nabla^d\}$$

An example where they might be used is in a non-seasonal ARIMA model (11.3.1).

The PIWEIGHTS parameter allows you to store the theoretical pi-weights $\pi_0 \ldots \pi_m$ of an ARIMA model: these show explicitly how past values contribute to a forecast. The weights are defined by

$$1 - \pi_1 \times B - \pi_2 \times B^2 - \ldots = \{\phi(B)\nabla^d\}/\theta(B)$$

The IMPULSERESPONSE parameter allows you to store the theoretical impulse-response function, $v_0 \ldots v_m$, of a transfer-function model. This function can help you interpret the model. The sequence is defined for a non-seasonal transfer-function model by

$$v_0 + v_1 \times B + v_2 \times B^2 + \ldots = \omega(B)B^b / \{\delta(B)\nabla^d\}$$

11.7.4 Deriving the generalized form of a time-series model

For an ARIMA model you can combine into one generalized autoregressive operator all the differencing operators, the non-seasonal autoregressive operators,

11.7 Forming preliminary estimates and displaying models

and the seasonal autoregressive operators. The non-seasonal and seasonal moving-average operators may similarly be combined.

Normally you would want this expanded model to help you understand a series. But you may want to re-estimate the parameters in the expanded model, to test whether the differencing operators or seasonal factors unnecessarily constrain the structure of the original model. Here is an example:

EXAMPLE 11.7.4

```
   7  "
  -8     Expand the seasonal ARIMA model used for modelling the number of airline
  -9     passengers in Section 11.3.6.
 -10  "
  11  VARIATE [VALUES=0,1,1, 0,1,1,12] Ord
  12  & [VALUES=0,0,0.00143, 0.34, 0.54] Par
  13  TSM Airpass; ORDERS=Ord; PARAMETERS=Par
  14  PRINT Airpass
```

Airpass

Innovation variance 0.001430

```
                        parameter
Transformation              0.
Constant                    0.
```

* Non-seasonal; differencing order 1

```
                 lag    parameter
Moving-average    1     0.340000
```

* Seasonal; period 12; differencing order 1

```
                 lag    parameter
Moving-average   12     0.540000
```

```
  15  TSUMMARIZE [PRINT=expansion] Airpass
```

15...

*** Expansion of model Airpass

*** Autoregressive moving-average model ***

Innovation variance 0.001430

```
                        parameter
Transformation              0.
Constant                    0.
```

* Non-seasonal; no differencing

```
                 lag    parameter
Autoregressive    1      1.00000
                 12      1.00000
                 13     -1.00000
```

```
Moving-average     1      0.340000
                  12      0.540000
                  13     -0.183600
```

If you have not previously defined one of the identifiers in the EXPANSION parameter, then Genstat will automatically define it to be a TSM; its component variates will be set up to have the length defined by the corresponding model in the TSM parameter.

The expansion does not change the transformation parameter of the model, nor the constant term, nor the innovation variance. If the model you supplied contains non-zero differencing orders, then the generalized model does not satisfy the stationarity constraint on the parameters; neither does the constant term have the same interpretation as it had in the supplied model.

The expansion of transfer-function models exactly parallels that of ARIMA models.

P.W.L.
G.T.W.

12 Extending Genstat

No computer program can do everything that its users may require. Genstat has been designed to make available many statistical techniques, and to handle data in many different ways; but there will always be techniques not provided, if only because new techniques are continually being invented. Consequently several methods are available to allow you to extend Genstat. With the command language described in this Manual you can write programs to do many tasks not provided as standard directives. Alternatively, you can add to Genstat at a lower level, in the general-purpose language Fortran 77. You can thus, for example, include modules from Fortran subroutine libraries dealing with specialized techniques. Or you can write your own Fortran subprograms for new methods of analysis.

Section 12.1 describes how to extend Genstat by writing programs in the Genstat command language. We also give further details about the Genstat Procedure Library.

For complicated tasks, it will almost always take less of your time to write a program in the Genstat language than in a general-purpose language like Fortran, though the Genstat program may use more computing time. However, if a task is to be repeated many times, or if it uses a lot of computer resources, it may be more economical to program at least some of the task in Fortran. Sections 12.2 to 12.4 describe three ways of adding your own Fortran subprograms into Genstat. The best method for a particular application depends on the complexity of the work done by your subprograms, how convenient you want to make them to use, and how Genstat has been implemented on your computer.

The OWN directive, described in 12.2, is available in all versions of Genstat. A skeleton Fortran subroutine is supplied with Genstat; this is called whenever you give an OWN statement. OWN has a fixed and limited syntax, and consequently you need make only simple modifications to the subroutine – so that it calls your subprograms and passes the information that they use, to and from Genstat. You also need to relink the Genstat system to include the resulting subprogram and your subprograms – an easy task on some computers, but more difficult on others. The meaning of "relink" is explained in 12.2.6.

Section 12.3 describes the PASS directive, which provides a means of accessing subprograms without having to link them into Genstat. It has an even more limited syntax than OWN, but is nevertheless adequate for most tasks. However PASS may not have been implemented on your computer.

The third method of adding subprograms is to define new directives (12.4). This method provides the full Genstat syntax for implementing your task, but you must have more knowledge of the internal workings of Genstat. You will also need to relink Genstat, and have access to the language-definition file.

The final section, 12.5, shows how Genstat can communicate with other computer programs, including the operating system itself.

12.1 Writing programs in the Genstat language

The quickest and easiest way to extend Genstat is to write a program in the Genstat command language. To use Genstat is in many ways easier than to write a special-purpose program in, say, Fortran or Pascal.

Operations like calculations (5.1) and data input or output (4.1 and 4.2) operate on complete data structures, and so you do not need to write loops over their elements. Also there are Genstat directives to provide many of the basic operations that are required in statistical analyses: for example, eigenvalue and singular value decompositions (5.7) are a necessary part of many multivariate techniques, and the calculation of sums of squares and products (5.7.3 and 9.6.1) is fundamental to regression. Many methods of analysis can be expressed as series of matrix equations, and these are readily translated into Genstat expressions so that you can use them in, for example, CALCULATE (5.1.1) or FITNONLINEAR (8.6.2).

Genstat directives are designed to be linked together. Virtually anything that can be printed from a Genstat analysis can also be stored in a data structure within Genstat, so that you can use it as input to some other statement. Thus, for example, matrices of sums of squares and products can be saved from analysis of covariance; you can then use them as input for an eigenvalue decomposition, and for other matrix operations, to provide a multivariate analysis of variance. Even if you do not want to do anything as complicated as this, you will find it useful to be able to save information from analyses. For example, you may want simply to save tables of means (9.6.1) or of predictions (8.3.4 and 8.4.4) for plotting, or for printing in the form required for a publication; or you may want to examine the residuals according to some newly suggested method.

The ability to write loops that are executed until some logical condition is fulfilled (6.2.1 and 6.2.4) allows you to program techniques that rely on the iterative solution of systems of equations, as in the E-M algorithm (Dempster, Laird, and Rubin 1977). The block-if construction (6.2.2) allows you to make any program general, so that different sets of statements are executed according to the form of a particular set of data.

You might want to make programs available to other users, especially if the techniques are new or the presentation of the results is in a novel format. One way in which you can do this is by writing an article for the *Genstat Newsletter*, listing your program and describing what it does. The article should be submitted to the Genstat Newsletter Editor, either at Rothamsted or at the Numerical Algorithms Group, Oxford (see the Preface).

12.1 Writing programs in the Genstat language

The program will be more convenient for other users if you form it into a procedure, as described in 6.3.1. And if you do write a procedure, two further possibilities open up. Your computer site may have its own procedure library that is attached to Genstat automatically whenever any user at your site runs Genstat (see 6.3). By putting your procedure into that library, you can make it available automatically to any user of your computer. The site library is likely to be controlled by the Genstat Representative at your site. Alternatively, you could submit the procedure to the official Genstat Procedure Library. This library is distributed to all Genstat sites, and is also attached automatically whenever and wherever Genstat is run. So your procedure would be available to any Genstat user.

12.1.1 The Genstat Procedure Library

The Genstat Procedure Library contains procedures submitted by users, or by the developers of Genstat. It is controlled by an Editorial Board, who check that the procedures are useful and reliable, and maintain standards for the documentation. You can obtain the Guidelines for submitting procedures by writing to the Secretary of the Genstat Procedure Library, at Rothamsted.

Using a procedure in the Genstat Library is very simple. The Library is attached automatically to your program whenever you run Genstat. When Genstat encounters a statement with a name that is not one of the standard Genstat directive names, it first checks whether you already have a procedure of that name stored in your program; if not it looks in the procedure libraries that are attached to your program, checking the Genstat Library last of all. After locating the required procedure, Genstat reads it in, if necessary, and then executes it (see 6.3). Thus you do not have to do any more than you would to use a Genstat directive. Furthermore, as the other libraries are examined before the official Library, you can use your own procedures in preference to the official ones if you so wish.

At the time of writing this Manual, Genstat 5 is not generally available and so there are not as yet many procedures in the Library. However the Library will be regularly redistributed to sites, each time in an expanded form. Thus the contents and the exact details of individual procedures will change as the Library continues to develop. Consequently, the information about the Library is available from within the Library itself. There is a procedure called LIBHELP which gives you general information about the form and contents of the Library, and tells you how to find out about the facilities and syntax of the individual procedures. The default output from LIBHELP tells you about the syntax of LIBHELP itself, and so to get started you should type just the statement:

  LIBHELP

Another procedure in the Library is called NOTICE. This has a PRINT option with settings that include 'errors' for information about any aspects of Genstat that do not work correctly, and 'news' for details of recent developments concerning Genstat (conferences, new issues of the Newsletter, and so on).

The procedures that will be included in the Library from the outset include the following techniques: correspondence analysis, generalized Procrustes analysis, canonical correlation analysis, discriminant analysis, analysis of asymmetric matrices, generation of orthogonal polynomials, estimation of variance components from a hierarchical analysis of variance, estimation of LD50s from probit analysis, combination of information from some stratified experimental designs, and the analysis of repeated-measures designs.

12.2 Adding Fortran subprograms with the OWN directive

To implement the OWN directive, you must get access to the OWN subroutine and edit it to call your subprograms. This subroutine is written in Fortran 77 and is distributed with all versions of Genstat. In its distributed form, OWN does no more than carry out a few checks on the data structures included in an OWN statement in a Genstat program. But it is constructed so that you can easily edit it to make the checks specific for your tasks, and to call your own subprograms.

12.2.1 The OWN directive

OWN
Does work specified in Fortran subprograms supplied and linked by the user.

Option

SELECT	= scalar	Sets a switch, designed to allow OWN to be used for many applications; standard set-up assumes a scalar in the range 0–9; default 0

Parameters

IN	= pointers	Supplies input structures, which must have values, needed by the auxiliary subprograms
OUT	= pointers	Supplies output structures whose values or attributes are to be defined by the auxiliary subprograms

---

The IN parameter allows you to pass values of data structures into your subprograms. OWN will check these input structures before calling your subprograms, to ensure that they are of the right type and length for your program, and that they have been assigned values. The OUT parameter copies values calculated by your subprograms into Genstat data structures. You can arrange to define the type and length of these output structures either before or after calling your subprograms.

If the setting of the IN parameter has more than one identifier, the OWN directive will call your subprograms more than once. Each time it will make available to your subprograms the values of one structure in the IN parameter, and will take information from the subprograms and put them into the corresponding structure in the OUT parameter. Therefore, to pass several structures at a time to your subprograms, you must put the structures into pointers. For example,

OWN IN=!P(A1,A2,A3),!P(B1,B2,B3); OUT=X,Y

will call your subprograms twice, passing information about A1, A2, A3, and X the first time, and about B1, B2, B3, and Y the second time. It does this because !P(A1,A2,A3), for example, is a single structure.

If you want to pass just one pointer to your subprograms, you must ensure that OWN does not treat the pointer as a set of structures each of which is to be passed. You can do this by constructing another pointer to hold just the identifier of the pointer that you want to pass; for example:

POINTER [VALUES=A,B,C] S1

OWN IN=!P(S1)

The SELECT option allows you to call any number of subprograms independently. Thus, you can set up OWN so that the statements OWN [SELECT=1] and OWN [SELECT=2] do totally unrelated tasks. The standard version of the OWN subroutine deals only with the default value, 0, of SELECT. You will need to extend the OWN subroutine considerably if you want to provide for alternative values of SELECT. But you should be able to use much of the Fortran that deals with the default setting.

12.2.2 Modifying the OWN subroutine

The standard form of the OWN subroutine carries out several checks on the structures that are given in an OWN statement, but does not actually assign values to any of them. The first check is on the number of structures in the parameters. As described in 12.2.1, an item in the parameter list is treated either as a pointer to several structures, or as a single structure; thus, the setting IN=!P(A,B,C) will provide three input structures to be passed to your subprograms, whereas IN=X, where X is not a pointer, will provide one structure. In the standard OWN subroutine, the Fortran parameter constants MININ and MAXIN are set to 0 and 10 respectively. The subroutine checks that the numbers of structures to be passed to your subprograms via the IN parameter lie in this range. The Fortran parameter constants MINOU and MAXOU are used similarly for checks on the number of output structures.

To illustrate the changes that need to be made to implement a technique with OWN, we describe what you would have to do to get the OWN subroutine to do a very simple task – one that could easily be done by CALCULATE already. This example provides the essential information for more complex and practical tasks, without our having to confuse the explanation of OWN by presenting the details of such tasks. You will find it helpful to have a copy of the OWN subroutine while reading this section.

We modify OWN so that it operates on three input structures – a variate and two scalars – and one output structure, a variate. The result of

OWN IN = !P(V,S,M); OUT = W

will be to shift, square, and scale the values of V; that is, to calculate

W = M*(V+S)**2

for each value of the variate V.

The first modification to subroutine OWN is to specify that there are precisely three input structures and one output structure:

PARAMETER (MININ=3, MAXIN=3, MINOU=1, MAXOU=1)

The second check that OWN does is on the type and mode of both input and output structures. In its standard form, OWN checks that all such structures are variates. You can change this by editing the Fortran data-initialization statement that gives initial values to the arrays JTYPIN and JTYPOU; these arrays are of length MAXIN and MAXOU respectively. The possible types are given in comments in the OWN subroutine, and are listed there as well: you can specify all settings by Fortran parameter constants that are already defined in the subroutine.

Structure	Type	Structure	Type
Scalar	TYPSCA	Expression	TYPEXP
Factor	TYPFAC	Formula	TYPFOR
Variate	TYPVAR	Pointer	TYPPOI
Matrix	TYPMAT	Table	TYPTAB
Diagonal matrix	TYPDIA	Lrv	TYPLRV
Symmetric matrix	TYPSYM	Tsm	TYPTSM

Thus, for our example, we need to edit the Fortran data-initialization statement to specify:

DATA JTYPIN/TYPVAR,2*TYPSCA/, JTYPOU/TYPVAR/

This is all that is needed to set up the checks on the structures. The other Fortran statements already in the OWN subroutine access the input and output structures and check them against the values put into JTYPIN and JTYPOU.

To use text, or an SSPM, or system structures, read the *Genstat Implementers Manual* (Harding and Simpson 1987) for the necessary details.

You can check other attributes of the input structures if you want. The number

12.2 Adding FORTRAN subprograms with the OWN directive

of values of each structure is stored in the local array NVALIN, and the origin of the values in array OVALIN (see 12.2.3 for information about storage of values). The other attributes are all made available to the OWN subroutine in arrays of the Fortran common /DATAC/ by the call to subroutine VECSET just before label 1 of the OWN subroutine. The subroutine is called for each input structure in turn in a DO loop, controlled by the integer variable K2IN (which counts the input structures). Here are the other attributes that you can check or access:

Number of levels of a factor	NLEV(1)
Reference number of levels variate	LEVELS(1)
Number of rows of a matrix or symmetric matrix	NROW(1)
Number of columns of a matrix	NCOL(1)
Margin indicator of a table (0 no, 1 yes)	MARGIN(1)
Reference number of classification pointer of a table	CLASS(1)
Type indicator of a TSM (1 arima, 2 transfer)	TSMTYP(1)

Thus, if you want to store in a variable called NRMAT the number of rows of the matrix that is the seventh input structure, you should insert this Fortran statement after the call to VECSET:

```
IF (K2IN.EQ.7) NRMAT = NROW(1)
```

Any unset attribute has the "missing value", which is actually a large negative integer whose value is in the variable IMV in common /SYSCON/. For details of other attributes, see the *Genstat Implementers Manual*.

If an attribute of one Genstat structure is the reference number of another Genstat structure, you may need to let the OWN subroutine access this latter structure as well. You can do this with subroutine GETATT, which makes the attributes of any structure available in the arrays of common /DATAC/ as above. You can have information on up to three structures available at once, in the three "banks" of information corresponding to the three elements of each array in common /DATAC/. For example, these Fortran statements would access the origin of the levels of a factor that is already in bank 1 following the call to VECSET before label 1:

```
I = LEVELS(1)
IF (I.NE.IMV) THEN
    CALL GETATT(2,I,*1000,*1000)
    OLEV = VALOR(2)
ENDIF
```

Here, the subscript 1 of LEVELS refers to the factor in bank 1, and the subscript 2 of VALOR refers to the structure in bank 2. This second structure holds the information about the origin of the levels because the variable I in the call to GETATT refers to the levels variate of the factor.

The Fortran logical array PREDEF specifies whether Genstat is to set up output structures before calling your subprograms. It is probably best to set them up in advance if you know the lengths of the output structures in advance: your subprograms will then be able to put values directly into the value blocks of the structures, instead of having to use workspace (12.2.3). In the standard form of OWN, the array PREDEF is initialized to .TRUE. so that all Genstat structures are set up before calling the subprograms: this is satisfactory for our example. But, for example, if you had three output structures, the first to be defined before and the others afterwards, you would need to edit the Fortran data-initialization statement to specify:

DATA PREDEF/.TRUE.,2*.FALSE./

The lengths of the output structures must be defined somewhere – either before or after invoking your subprograms. If PREDEF(I) is .TRUE., then you should define NVALOU(I) in the loop ending with label 2 in the OWN subroutine; otherwise you should define it in the loop ending with label 3. For example, we want the length of the output structure to be the same as that of the first input structure, and so the Fortran statement

NVALOU(1) = NVALIN(1)

must go in the loop ending with label 2.

If you wanted to set the length of a second output structure to a value, NCALC say, formed by your subprogram, you should give the Fortran statement

IF (K2OUT.EQ.2) NVALOU(K2OUT) = NCALC

in the loop ending with label 3. You must define other essential attributes of some structures at the same time as defining their length. These attributes are precisely the ones listed above for checking input structures.

If any output structures are defined after calling your subprograms, you must also set the corresponding elements of the array OVALOU to specify where you have stored the values of the structures (see 12.2.3).

Finally you must insert a call to your subprograms before the loop ending with label 3: remember that the OWN subroutine is merely a link between Genstat and your subprograms. Information about the input and output structures is in the common /OWNCOM/, and so you can include this common in your subprograms to avoid passing the information as formal parameters of your subprograms. The standard version of /OWNCOM/ contains these variables in this order:

NIN	number of input structures formed
NOUT	number of output structures formed
NVALIN()	number of values of each input structure
NVALOU()	number of values of each output structure
OVALIN()	origin of values of each input structure
OVALOU()	origin of values of each output structure defined in advance
NWORK	length of workspace (see 12.2.3)

12.2 Adding FORTRAN subprograms with the OWN directive

OWORK   origin of workspace (see 12.2.3)

How to interpret and use these is described below.

You should put extra variables in this common block to pass other attributes of structures if needed — for example, number of rows and columns of matrices. The actual call to your subprograms can then be very simple, as for example

CALL SQUARE

though you may want to add an alternate return for dealing with diagnostics (12.2.4):

CALL SQUARE(*1000)

12.2.3 Values of data structures and workspace

The values of all numerical data structures are stored in Fortran blank common, which contains several equivalenced arrays of different mode. Values of data structures have the modes:

mode	array	Genstat structure
real	RDATA	scalar, variate, matrix, diagonal matrix, symmetric matrix, table
integer	IDATA	factor, expression, pointer, LRV, TSM

In the Fortran subprograms, the block of values of a structure is referred to by its origin and length. Thus the first value of a variate with origin OV and length NV is RDATA(OV+1) and the last value is RDATA(OV+NV). You can therefore refer to values of input structures in your subprograms by using these arrays, so long as you include blank common (you can copy it from subroutine OWN). Similarly, you can assign values to output structures that have previously been defined.

If some of the output structures are not predefined, or if you need to set up extra workspace for your subprograms, you can reserve a block of values in the data arrays. The standard version of OWN sets the variable LWORK (in /OWNCOM/) to zero, so that no workspace is reserved. By modifying this to n, say, you can reserve n values in the RDATA array: the origin will be OWORK, set up by the OWN subroutine just after label 2. If you want other modes of workspace, you should change the definition of JMODWK, using the following Fortran parameter constants available in that subroutine:

real	MODER
integer	MODEI
long real	MODELR

Long real values are in the array DATA in blank common.

Missing values for all modes of data structures are recorded as large negative values. They are equal to the values of variables available in common /SYSCON/:

real	RMV
integer	IMV
long real	DMV

We can thus do the calculation required for the simple squaring transformation with this subroutine:

```
      SUBROUTINE SQUARE
      (blank common)
      (common /OWNCOM/)
      (common /SYSCON/)
      SHIFT = RDATA(OVALIN(2) + 1)
      RMULT = RDATA(OVALIN(3) + 1)
      DO 1 K = 1,NVALIN(1)
C        Check for missing values
         R = RDATA(OVALIN(1) + K)
         IF (R.EQ.RMV) THEN
            S = RMV
         ELSE
            S = RMULT * (R + SHIFT)**2
         ENDIF
         RDATA(OVALOU(1) + K) = S
1     CONTINUE
      RETURN
      END
```

12.2.4 Extra diagnostic messages, and other output

The standard OWN subroutine will produce Genstat diagnostics if the checks on the input or output structures fail, or if there is not enough workspace. These diagnostics are the standard ones with codes VA, SX, and SP, and are dealt with by a section at the end of the OWN subroutine.

You can define your own diagnostics, using the code ZZ. You are not allowed to edit the standard file of error messages that stores the one-line definitions of each diagnostic code. However, you can edit subroutine ZZDIAG which is distributed with Genstat. This prints extra messages after a ZZ diagnostic; instructions for editing the subroutine are contained as comments in it. To cause a diagnostic to be printed, you must set the variable DIAG in common /DIAGPK/ and exit from the OWN subroutine. Here is an example, where a diagnostic is triggered if a variate has any negative values:

```
      DO 11 K2 = 1,NVALIN(1)
         IF (RDATA(OVALIN(1)+K2) .LT. 0.0) GO TO 1001
11    CONTINUE
```

```
    ...
1001 DIAG = KDGZZ+1
     GO TO 1000
```

The variable KDGZZ is also in common /DIAGPK/, and stores the origin of codes for OWN diagnostics: you can define up to 30, corresponding to settings KDGZZ+1 up to KDGZZ+30 for DIAG. At label 1000 in subroutine OWN, an auxiliary routine is called to update trace information, and the subroutine is then left by an alternate return. Execution of Genstat statements following a faulted OWN statement will continue only if the run is interactive, as for any other diagnostic.

Output from your subprograms is most easily arranged by storing the information that you want in data structures, and printing these with a PRINT statement after the OWN statement. Alternatively, you can give Fortran WRITE statements; there are standard routines in Genstat for outputting numbers and strings, but they are not described here. You should use the correct Fortran unit numbers for output, and this varies between implementations of Genstat. The Fortran unit number of the current output channel is in the integer QWW in common /PERIPH/. Remember that a Fortran unit number is not the same as a Genstat channel number (Chapter 4).

12.2.5 Using OWN with the FITNONLINEAR directive

The FITNONLINEAR directive (8.6.2) has options OWN, INOWN, and OUTOWN which allow you to use Fortran rather than Genstat to program the calculation of models. If the OWN option is set, subroutine OWN will be called at each step of the iterative process to form values for the explanatory variates, the variate of fitted values, or the scalar function value – depending on the type of model specified in FITNONLINEAR. You should set the INOWN and OUTOWN options to pass data structures to the OWN subroutine, such as the parameters of the model and the variates or scalars to store the calculated values.

The FITNONLINEAR directive and the OWN directive give you very similar types of access to the OWN subroutine. For example, if you have set up the OWN subroutine to calculate values of an explanatory variate from values of two scalar parameters, you could put this statement in a program:

    OWN IN = !P(A,B); OUT = X

The same version of the OWN subroutine could be used also in fitting a model for which the values of the parameters A and B are optimized:

    FITNONLINEAR [OWN = 0; INOWN = A,B; OUTOWN = X] X

However, as pointed out in 8.6.2, you cannot arrange for your subprograms to be called several times at each step of the search process: Genstat treats the whole setting of the INOWN option as a single pointer structure, and similarly the setting of OUTOWN.

For the sake of efficiency, you should avoid all the checking within subroutine OWN after the first step of the search in FITNONLINEAR. The OWN subroutine has a formal parameter NPCALL which Genstat automatically sets, before entry, to the number of previous calls to OWN with the current statement. You can by-pass all the Fortran checking statements if all the information about input and output structures is passed by common /OWNCOM/ rather than by formal parameters, and if the attributes of the structures do not change after the first call: thus for the second and subsequent entries to OWN, your subprograms would be called immediately, without any checking. For example, if your main subprogram is called MYPROG, you should modify OWN so that the first executable Fortran statements are:

```
IF (NPCALL.GT.0) THEN
  CALL MYPROG
  RETURN
END IF
```

You must also include a call to MYPROG further down the OWN subroutine, as before, to deal with the first pass through OWN when the value of NPCALL is 0.

12.2.6 Relinking Genstat

How to relink Genstat depends on the operating system of your computer, but the next example shows what is involved. We illustrate the process for the VMS operating system for VAX computers, in which the linking commands are particularly simple.

The Genstat system as distributed includes the object code, produced by the Fortran compiler, in an object library. Say this library has filename:

[GENSTAT]GNLIB.OLB

The main module of Genstat and the block data module will be separate from this library; say they have filenames:

[GENSTAT]MAIN.OBJ and BD.OBJ

Assume that your version of subroutine OWN calls a single subprogram called MYPROG, and that the compiled versions (object modules) of these subprograms have filenames:

[MYFILES]OWN.OBJ and MYPROG.OBJ

You form the compiled version of your subprograms with the Fortran compiler on your computer. Then, working in the directory [MYFILES], you can form a new version of Genstat, called MYGENSTAT.EXE say, by giving the DCL command:

```
$ LINK/EXEC=MYGENSTAT -
OWN,MYPROG,[GENSTAT]MAIN,BD,GNLIB/LIBRARY
```

To run the new system, you must set up a "foreign" command. Assuming that

12.3 Running an external program with the PASS directive

the directory MYFILES is on device DRA01, a command MYGEN can be defined by:

$ MYGEN := = "$DRA01:[MYFILES]MYGENSTAT.EXE"

You can then use the command MYGEN in precisely the same way as the usual command GENSTAT.

12.3 Running an external program with the PASS directive

On some computers, you can arrange that one program, such as Genstat, calls for another to be executed, passing information directly between the two. You can then cause Genstat to execute your own subprograms without having to modify Genstat in any way.

To find out if the PASS directive has been implemented in your version, type

PASS

in any Genstat program. You may immediately get the message:

The PASS directive is not implemented

Alternatively, you will get a Genstat diagnostic telling you that Genstat has failed to initiate a sub-process: this means that PASS has been implemented.

If the PASS directive has not been implemented, there is no point trying to implement it yourself: you might just as well make use of the OWN directive, because you would have to relink Genstat anyway. However, you can achieve the effect of PASS by using the SUSPEND directive (12.5), if that has been implemented on your computer. First, open a file (4.3.1) and PRINT the values of the structures that you want to send to your own program. You can use a character file, or, for faster communication, an unformatted file (4.6.4). Then CLOSE the file (4.3.4), and give a SUSPEND statement (12.5) to return to the operating system. You can then run your own program, accessing data from the file, and perhaps putting values back into it or another file. Then return to Genstat, and use READ to access the results of your program. An example of this method is shown in 12.5.

To use the PASS directive when it is available, you must first get access to the GNPASS program which is distributed with Genstat. You then form an executable program consisting of GNPASS, slightly modified as detailed below, and your subprograms. The GNPASS program deals with communication with Genstat, and passes information to and from your subprograms.

12.3.1 The PASS directive

PASS
Does work specified in Fortran subprograms supplied by the user, but not linked into Genstat. This directive may not be available on some computers.

Option		
NAME	= text	Filename of external executable program; default GNPASS
Parameter		
	pointers	Structures whose values are to be passed to the external program, and returned.

You can pass the values of any data structures except texts. All the structures needed by your subprograms must be combined in a pointer structure, unless only one structure is needed and it is not a pointer (this rule is the same as for OWN, 12.2.1). The structures must have values before you include them in a PASS statement; if you want to use some of the structures to store results from your subprograms, you must initialize them to some arbitrary values, such as zero or missing. If you specify several pointers in a PASS statement, your subprograms will be invoked several times, to deal in turn with each set of structures stored by the pointers. However, the values of the structures in all the pointers are copied before any work is done by your subprograms. Thus, if you want to operate with PASS on the results of a previous operation by PASS, you must give two PASS statements with one pointer each rather than one statement with two pointers.

For example, to use PASS to carry out the transformation shown above to illustrate the OWN directive, you could give the following statements:

```
SCALAR S,M; VALUE=2,10
VARIATE [VALUES=1...10] V
& [VALUES=10(*)] W
PASS !P(V,S,M,W)
```

Numbers can be used in place of scalars, as usual in Genstat statements:

```
PASS !P(V,2,10,W)
```

To transform the values in both V, as above, and another variate X, with values 10...50 say, you could give the statements:

```
VARIATE [VALUES=41(*)] Y
PASS !P(V,2,10,W),!P(X,2,10,Y)
```

The NAME option can be set to specify the filename of the executable program formed from the GNPASS program and your subprograms. The setting must be recognizable as the correct filename by the operating system of your computer; by default, the name GNPASS will be assumed.

12.3.2 Modifying the GNPASS program

The distributed form of the GNPASS program, if available on your computer, consists of Fortran statements that receive information from Genstat, sent by a PASS statement, and send it back again unchanged. To make it do the task that you require, you need to edit the program to call your subprograms between receiving the information and sending it back, and to provide the necessary subprograms to modify the information. The position for the call to your subprograms is indicated by comments in the GNPASS program.

The values of a set of data structures sent by PASS are stored by the GNPASS program in arrays in a common called /DATA/. Values of scalars, variates, matrices, and tables are in an array called RDATA(), and values of factors are in IDATA(), coded as 1 ... (number of levels). There are also arrays DDATA() and CDATA() for use with internal system structures, but these are not described here. The /DATA/ common also contains integer arrays called MODE(), NVALUE(), and ORIGIN(). These store, for each structure in the set, the mode of the structure, coded as 2 for RDATA() and 3 for IDATA(), the number of values, and the origin of the values in the relevant array. Thus, the statements

    VARIATE [VALUES=1...10] V
    PASS V

will result in MODE(1)=2, NVALUE(1)=10, ORIGIN(1)=0, RDATA(1)=1.0, RDATA(2)=2.0, and so on. Thus, to refer to the data values in your subprograms, you must copy the /DATA/ common into the necessary subprograms and use the arrays as described. You must not change the values stored in the MODE(), NVALUE(), or ORIGIN() arrays.

For illustration, we show how to implement the calculations required for the simple squaring transformation discussed above. First, modify the GNPASS program by including the Fortran statement:

    CALL SQUARE

Then modify the subroutine SQUARE (12.2.3) to become as follows:

```
    SUBROUTINE SQUARE
    (common /DATA/)
    SHIFT = RDATA(ORIGIN(2)+1)
    RMULT = RDATA(ORIGIN(3)+1)
    DO 1 K1=1,NVALUE(1)
      R = RDATA(ORIGIN(1)+K1)
      RDATA(ORIGIN(4)+K1) = RMULT * (R + SHIFT)**2
  1 CONTINUE
    RETURN
    END
```

No checks are made here for missing values. However, we can use the values of variables RMV and IMV in common /DATA/ to make a check:

```
    IF (R .EQ. RMV) THEN
       S = RMV
    ELSE
       S = RMULT * (R + SHIFT)**2
    END IF
    RDATA(ORIGIN(4)+K1) = S
```

The standard version of the /DATA/ common imposes limits on the number of structures passed in one pointer (100) and the total number of values of any mode (10000). You can change these limits by editing the definition of the Fortran parameter constants MSTRC, MRDATA, and MIDATA at the beginning of the common specification – both in GNPASS and in the copies in your subprograms. The other Fortran statements in GNPASS will take account of whatever values are defined, and check the number of values sent by Genstat against the limits. Thus the only limit on the number of values that you can pass is the limit on the internal storage within Genstat itself.

After editing the GNPASS program and writing or amending your subprograms, you must then compile them with the Fortran compiler on your computer, and link them into an executable program. For example, with a VAX/VMS operating system, this can be done by:

```
    $ FOR GNPASS.FOR,SQUARE.FOR
    $ LINK/EXEC=SQUARE.EXE GNPASS.OBJ,SQUARE.OBJ
```

Then you can use the program from within Genstat: it is not necessary to change Genstat itself in any way. Here is an example:

EXAMPLE 12.3.2

```
  1   VARIATE [VALUES=1...10] V
  2   & [VALUES=10(*)] W
  3   PASS !P(V,2,10,W)
  4   PRINT V,W; DECIMALS=0

                  V             W
                  1            90
                  2           160
                  3           250
                  4           360
                  5           490
                  6           640
                  7           810
                  8          1000
                  9          1210
                 10          1440
```

12.4 Defining new directives

You can include any subprograms in Genstat, and design directives to control their execution. You must first get access to the Genstat *language-definition file*, which is distributed with Genstat. You next specify the new directives in your own version of the language-definition file, using the DEFINE directive (12.4.1) and the OPTION and PARAMETER directives (6.3). Then modify the subroutine EXTRAD, which is also distributed with Genstat (12.4.2), and relink Genstat to include your subprograms (12.2.6). Finally, you must arrange that your version of Genstat will reference your own language-definition file rather than the standard one. You are then ready to run Genstat to create a *binary language-definition file* and call your subprograms.

12.4.1 The DEFINE directive

DEFINE
Defines the form of a Genstat directive. The DEFINE directive can be used only in the language-definition file.

Option

CODE	= scalar	Code number to be used for the directive

Parameter

	text	Name of the directive

You must include DEFINE statements in the *language-definition file*. This is a character file attached to Genstat, and distributed with it, that contains definitions of all the standard directives in the Genstat language. You can copy this file and edit the copy to include further definitions. Then you must ensure that your relinked version of Genstat will refer to the edited version of the language-definition file: the local documentation should tell you how to do this.

Genstat will also refer to a *binary language-definition file*; you must ensure that this reference is to a new filename, otherwise the old file will be used and your version of the character language-definition file will be ignored. When you first run Genstat after relinking it to include your subprograms, Genstat will create a new binary language-definition file from your version of the character language-definition file, and then stop. This process of creating the binary file can take a long time, because about 140 directive definitions have to be processed; but once it has been produced, Genstat will use it automatically for all later jobs.

The name you assign to a new directive must not clash with an existing name; that is, the first four letters of the new name must not be the same as the first four letters of a standard directive name. Thus you cannot define a directive called CONTRAST because there is already one called CONTOUR. But if a directive name –

new or old – has fewer than four letters, then Genstat extends it with spaces. So you can define a directive called TAB, even though there are already directives called TABLE and TABULATE; likewise you could define a directive called OWNER and it would not be mistaken for the standard OWN directive. You must also be aware of possible conflicts with procedure names in your job, and in procedure libraries (1.7.1 and 12.1).

You must assign a unique code number to each new directive. Use the integers 1501 onwards to avoid all standard codes: the Genstat interpreter will deduct the origin 1500 from the code before passing it to subroutine EXTRAD.

After giving the name of the new directive with a DEFINE statement, you define its option and parameters by an OPTION and a PARAMETER statement. The rules for these are mainly described in 6.3.1. The main addition is that the OPTION directive in a language-definition file is allowed an extra parameter. This is the VALUES parameter, which comes after the MODE parameter, and specifies a list of allowed settings for options of mode W (word). Mode W is available only for the definition of new directives; it is just like mode T (text), except that matching of words is not dependent on case: the word !W(no) is the same as !W(NO). Another extra mode is I, which indicates an integer value. The setting of VALUES must be * if the mode is not W.

Here is an example defining a new directive called SQUARE which forms a new variate by squaring the values of an old one; it can also add a shift before squaring, and can change the sign of the result:

```
DEFINE [CODE=1501] 'SQUARE'
PARAMETER 'OLD','NEW'; MODE=P
OPTION 'SHIFT','CHANGESIGN'; MODE=V,W; VALUES=*,!W(NO,YES);\
   DEFAULT=0,!W(NO)
```

12.4.2 Modifying the EXTRAD subroutine

The EXTRAD subroutine is distributed with Genstat, and is already able to interpret any directive that you can define with the DEFINE directive. Thus, it accesses the individual data structures that have been set in the options and parameters of a statement, and stores their attributes in Fortran arrays.

To change EXTRAD, you must first edit the definition of the Fortran parameter constant MPARAM to correspond to the number of Genstat parameters defined in the PARAMETER statement of your directive definition. Thus, for the directive SQUARE defined above, we would change the definition of MPARAM in the EXTRAD subroutine to be:

```
PARAMETER (MPARAM=2)
```

Similarly, you must edit the Fortran data-initialization statement for the Fortran array MOODPR, which stores the modes of each parameter (see 6.3.1 for a list of

12.4 Defining new directives

possible modes). Fortran parameter constants are available for this: MOODP for mode P, MOODV for mode V, and so on. For example:

    DATA MOODPR/MOODP,MOODP/

Next, you must insert Fortran statements to check the setting of the options: this goes within the DO loop that has control variable K1OPT and finishes at label 10. At each pass of the loop, the integer variable OPTVAL holds the setting of the option. If the option has mode I and the setting is a single value, OPTVAL will have this value. For mode W, if the setting is a single value, OPTVAL will be the position of the value in the list of possible values specified in the OPTION directive. For example, suppose the first setting of CHANGESIGN is 'no' and the second is 'yes'. You can cause an integer variable MULT to be -1 if the setting is 'yes', and to be $+1$ otherwise, by these Fortran statements:

    IF (K1OPT.EQ.2) THEN
       MULT = 1
       IF (OPTVAL.EQ.2) MULT = -1
    ENDIF

If the option has mode other than I or W, or if the option setting has more than one value, the value of OPTVAL will be the reference number of a structure. You can use the GETATT subroutine to access details of the structure, as described in 12.2.2. For example:

    IF (K1OPT.EQ.2) THEN
       CALL GETATT(1,OPTVAL,*1000,*1000)
       SHIFT = RDATA(VALOR(1) + 1)
    END IF

The default setting of SHIFT here is a scalar with value 0. If you have not explicitly defined a default for an option, and if the directive is used without the option being set, then the setting will be automatically put at 0. In that circumstance, you would not want to call GETATT.

You might want to get Genstat to check the type of the structure, and report an error (12.2.4) if it is not what it should be: this would protect the directive against misuse. For example:

    IF (TYPE(1). NE. TYPSCA) GO TO 101
       ...
    101 DIAG = KDGZZ+1
        RETURN 1

Accessing the parameter settings is similar. The DO loop with control variable K1PAR, ending at label 21, cycles through each parameter in the directive. The inner loop with control variable K2STRC, ending at label 20, cycles through the settings of each parameter; the integer variable LLIST holds the number of settings, and the integer variable PLIST the position of the setting of the parameter. For mode P, IDATA(PLIST) will be the reference number of the structure that was set in the parameter; for mode V, RDATA(PLIST) is the actual real number in the setting. For other modes you will need to read the *Genstat Implementers Manual*.

For example, to access the variates in directive SQUARE, you should put:

```
CALL GETATT(1,IDATA(PLIST),*1000,*1000)
IF (TYPE(1) .NE. TYPVAR) GO TO 102
IF (VALOR(1). EQ. IMV) GO TO 103
IF (K1PAR. EQ. 1) THEN
  NIN(K2STRC) = NVAL(1)
  OIN(K2STRC) = VALOR(1)
ELSE
  IF (NVAL(1) .NE. NIN(K2STRC)) GO TO 104
  OOUT(K2STRC) = VALOR(1)
END IF
```

The final part of the definition of the directive is the main one. You must arrange for it to call whatever subprograms you want it to execute. These will operate on the values that you have supplied, and will pass information by Fortran formal parameters or by commons. An example is:

```
CALL SQUARE (NIN,OIN,OOUT,LLIST,SHIFT,MULT)
```

The subroutine SQUARE would therefore look like this; the blank common can be copied from subroutine EXTRAD:

```
      SUBROUTINE SQUARE (NIN,OIN,OOUT,LLIST,SHIFT,MULT)
      (blank common)
      (common /SYSCON/)
      INTEGER NIN(LLIST),OIN(LLIST),OOUT(LLIST),MULT
      REAL SHIFT
      DO 2 K1=1,LLIST
        DO 1 K2=1,NIN(K1)
          R = RDATA(OIN(K1)+K2)
C         Check for missing values
          IF (R.EQ.RMV) THEN
            S = RMV
          ELSE
            S = MULT * (RDATA(OIN(K1)+K2) + SHIFT)**2
          END IF
          RDATA(OOUT(K1)+K2) = S
    1   CONTINUE
    2 CONTINUE
      RETURN
      END
```

If you want to define more than one new directive, then you must make more extensive modifications to the EXTRAD subroutine. The formal parameter JDIR holds the code number of the directive, minus 1500, and so you can use this to branch within EXTRAD. The Fortran variables specifying the numbers and types of options and parameters may then have to be replaced by arrays, holding the values for each of the directives that you require.

If your subprograms are new and lengthy, you will probably need to "debug" them. You will probably find the DUMP directive (2.5.1) useful here; also you will find that the statement

SET [DIAGNOSTICS = faults,warnings,extra]

gives you useful, though voluminous, information after faults are encountered.

12.5 Communicating with other programs

Genstat is designed as a general statistical package, and so contains facilities for most of the statistical methods you need; but there are many other possible requirements. Some of these will use information that you can produce with Genstat; others will generate information that you can analyse with Genstat. Therefore you need to be able to connect Genstat to other programs.

One example involves data bases. These are collections of information on a computer, carefully structured to allow convenient access using a specially tailored command language. If you want to use Genstat to analyse data from a data base, you will need to extract the information using the data-base language, analyse it with Genstat, and then perhaps add some of the results to the data base.

The simplest method of communication between programs is via files. One program can extract data and store it in a file – either in character form or in binary form (4.6.4); this program might be written in the data-base language. A second program, written in the Genstat language, can then read that file, process the data, and perhaps form another file; and so on.

If you are content to work step-by-step, first running one program then another, your purpose will be served by the facilities described in Chapter 4 for storing and accessing information in files. However, you may prefer to run programs concurrently. Some computer systems make it easy to run several programs at the same time and to make links between them; but others do not. Therefore your computer may not let you take advantage of the provision that Genstat makes for such activities. To find out if the implementation of Genstat on your computer does let you do this you should type

SUSPEND

in an interactive run of Genstat. If you get a message saying that SUSPEND is not available, then you must either use the step-by-step method, or else arrange for SUSPEND to be implemented (which for some computers might not be possible).

The SUSPEND directive, when implemented, allows you to suspend a run of

Genstat and return to the operating system. You can then give operating-system commands, including commands to run other programs, eventually returning to Genstat at the point in your program where you left off.

SUSPEND
Suspends execution of Genstat to carry out commands in the operating system. This directive is not available on some computers.

Options

SYSTEM	= text	Commands for the operating system; default: prompt for commands (interactive mode only)
CONTINUE	= string	Whether to continue execution of Genstat without waiting for commands to complete (no, yes); default n

No parameters

If SUSPEND is implemented, and you type

    SUSPEND

then you will get a message saying that Genstat has been suspended, and a prompt "SUSPEND>" for an operating-system command. In most implementations, you will also be told what command in the operating system will return you to Genstat. For example, with the VMS system on VAX computers you do this by typing

    LOGOUT

The next example uses SUSPEND with the VAX/VMS operating system. Genstat puts some data in a file, then another program called PROC processes the data and forms another file, and finally Genstat accesses the new data.

12.5 Communicating with other programs

EXAMPLE 12.5a

```
> OPEN 'Result1'; CHANNEL=2; FILETYPE=output
> PRINT [CHANNEL=2] Data[]
> CLOSE 2; FILETYPE=output
> SUSPEND
```

You have suspended Genstat, and can now give commands to the operating system after the prompt SUSPEND>; type LOGOUT to return to Genstat.

```
SUSPEND> PROC Result1 Result2
SUSPEND> LOGOUT
GENSTATUSER logged out at 11:04 on 05-MAR-1987

> OPEN 'Result2'; CHANNEL=2; FILETYPE=input
> READ [CHANNEL=2; END=*] Data[]
```

The option SYSTEM allows you to give operating-system commands without explicitly returning to the operating system. Here is an example, doing the same as the previous example:

EXAMPLE 12.5b

```
> PRINT [CHANNEL=2] Data[]
> CLOSE 2; FILETYPE=output
> SUSPEND [SYSTEM='PROC Result1 Result2']

> OPEN 'Result2'; CHANNEL=2; FILETYPE=input
> READ [CHANNEL=2; END=*] Data[]
```

If you do not want to wait for an operating-system command to be executed before returning to Genstat, then you can set the CONTINUE option together with the SYSTEM option. You can do this only if a single operating-system command is given; that is, if the text set in the SYSTEM option contains only one string. This would not be a good idea in the above example, because the result of the processing program is required by the Genstat statement following the SUSPEND statement. But here is an example where an output file is entered in a line-printer queue during a Genstat program:

```
OPEN 'Result1'; CHANNEL=2; FILETYPE=output
PRINT [CHANNEL=2] Data[]
CLOSE 2; FILETYPE=output
SUSPEND [SYSTEM='PRINT/QUEUE=LP2 Result1'; CONTINUE=yes]
CALCULATE Newdata[1] = Data[1] * Data[2]
```

The CALCULATE statement may start executing before the file Result1 has been queued for printing; whether it does or not depends on how much work the computer is doing. If the operating-system command also produces messages to the terminal, they may get interleaved with any output from Genstat that is coming to the terminal. In that case, it might be best to avoid using the CONTINUE option.

P.W.L.
R.W.P.

Appendix 1   Reference summary

A1.1.1 Terminology

Brackets	must occur in pairs, as follows:
round brackets ( )	are used to enclose a list of numbers to be pre- or post-multiplied or to enclose the arguments of a function, they also occur in expressions;
square brackets [ ]	are used to enclose a list of option settings or to enclose the suffix list of a pointer; also, when preceded by $, they enclose lists of unit names or numbers for a qualified identifier;
curly brackets { }	are each synonymous with the corresponding square bracket.
Character	The characters used to form Genstat statements are a subset of those available on most computers. For the Genstat language they are classified as brackets, digits, letters, separators, simple operators, or special symbols.
Comment	consists of any series of characters that the computer can represent, enclosed by double quotes (''); comments are ignored and can appear anywhere in a Genstat program.
Data structure	You can declare and name structures of various types (such as scalars, variates, and matrices) to contain input (i.e. data) or output to and from Genstat.
Diagonal matrix	is a data structure that stores the diagonal elements of a square matrix whose other values are all zero.
Digits	are the numerical characters 0 to 9.
Directive	is a standard form of instruction in the Genstat language requesting a particular action or analysis. All Genstat 5 directives have the same syntax: see Statement below.
Directive name	is a system word used to request a particular action or analysis from Genstat. Directive names may be abbreviated to four characters; if characters 5–8 are given, they must match the standard form, e.g. TREATMENTS can be written as TREA, TREAT, TREATM, and so on, but not TREATS.
Expression	is an arithmetic expression consisting of lists and functions separated by operators; an expression data structure stores a Genstat expression.

Factor	is a data structure that specifies an allocation of the units into groups. It is thus a vector that, unlike the variate or the text, takes only a limited set of values, one for each group. The groups are referred to by numbers known as levels, or you can define textual labels.
Formula	is a model formula of lists and operators defining the list of model terms involved in an analysis; a formula data structure stores a Genstat formula.
Function	denotes a standard operation in an expression or formula, with the form "*function-name* (*sequence of lists and/or expressions separated by ;*)". The function-name is a system word and may be abbreviated to four characters; if characters 5–8 are given, they must match the standard form.
Identifier	is the name given to a particular data structure within a Genstat program. The first character of an identifier must be a letter; any others can be either letters or digits. Only the first eight characters are significant; subsequent characters are ignored. Directive SET allows you to specify whether or not the case of the letters (small or capital) is to be significant; e.g. whether LENGTH is the same as Length.
Item	is a number, a string, an identifier, a system word, a missing value, or an operator.
Letters	are the upper-case (capital) letters A to Z, the lower-case letters a to z, the underline symbol (_), and the percent character (%).
List	is a sequence of items separated by commas. In an identifier list, each item is an identifier or an unnamed structure, while number or string lists contain numbers or strings respectively. Lists can contain pre- or post-multipliers. Identifier and number lists can contain progressions.
LRV	is a compound data structure storing latent roots and vectors, mainly used in multivariate analysis.
Macro	is a Genstat text structure containing a section of a Genstat program. The text must have an unsuffixed identifier. It can be substituted into the program, by giving its identifier, preceded by a contiguous pair of substitution symbols (##).
Matrix	is a data structure that stores a rectangular array of numbers.
Missing value	is denoted within a Genstat program by one asterisk (*). When reading data, a series of contiguous asterisks or an

	asterisk followed by letters or digits are also treated as missing values, and other characters can also be defined to represent missing values.
Number	is a sequence of digits, optionally containing a decimal point (.). The sequence can be preceded by a sign (+ or -) and can be followed by an exponent: i.e. the letter E or D (in upper or lower case) optionally followed by spaces, then a sequence of digits optionally preceded by a sign.
Operator	is a symbol or symbols denoting an operation in an expression or formula:
simple	+ − * / . = < >
compound	** −* −/ // *+ .EQ. == .NE. /= <> .LE. <= .GE. >= .LT. .GT. .EQS. .NES. .IN. .NI. .IS. .ISNT. .AND. .OR. .EOR. .NOT. Only + − * / . −/ −* and // may occur in formulae, while . −* −/ and // cannot occur (as operators) in expressions.
Options	specify arguments that are global within a Genstat statement: i.e. they apply to all the items in the parameter list(s). Often, but not always, options have default values and so need not be specified.
Option name	is a system word that identifies a particular option setting. It can be abbreviated to the minimum number of characters required to distinguish it from the options that precede it in the prescribed order for the directive or procedure concerned; for Genstat directives, four characters are always sufficient.
Option sequence	is a list of option settings separated by semi-colons (;).
Option setting	has the form "*option name* = *list, expression or formula*"; "*option name* =" can be omitted if the settings are given in the prescribed order for the directive or procedure concerned: i.e. the name may be omitted for the first setting if this is for the first prescribed option, and for subsequent settings if the previous setting was for the option immediately before the current one in the prescribed order.
Parameters	specify parallel lists of arguments for a statement: i.e. the statement (with its option settings) operates for the first item in each list, then the second, and so on. The number of times that this happens is determined by the length of the parameter list that is first in the prescribed order for the directive or procedure concerned. Subsequent lists are recycled if they are shorter than the first list.

Parameter name	is a system word that identifies which parameter is being set. It may be abbreviated to the minimum number of characters required to distinguish it from the parameters that precede it in the prescribed order for the directive or procedure concerned; for Genstat directives, four characters are always sufficient.
Parameter sequence	is a list of parameter settings separated by semi-colons (;).
Parameter setting	has the form "*parameter name = list, expression or formula*" "*parameter name =*" can be omitted if the settings are given in the prescribed order for the directive or procedure concerned: i.e. the name may be omitted for the first setting if this is for the first prescribed parameter, and for subsequent settings if the previous setting was for the parameter immediately before the current one in the prescribed order. For directives or procedures with only a single parameter, no parameter name is defined.
Pointer	is a data structure that stores a series of identifiers, pointing to other data structures.
Post-multiplier	This occurs immediately after the second of a pair of brackets enclosing a list of identifiers, numbers, or strings, and has the effect of repeating the entire list, as a whole, the specified number of times. The post-multiplier may be a number, or the substitution symbol (#) followed by a single-valued numerical structure.
Pre-multiplier	This occurs immediately before the initial bracket of a pair enclosing a list of identifiers, numbers, or strings and has the effect of repeating each item, in turn, the specified number of times. The pre-multiplier may be a number, or the substution symbol (#) followed by a single-valued numerical structure.
Procedure	This is a structure that contains Genstat statements, and fulfils the role of the subroutine in the Genstat language. The use (or call) of a procedure looks just like a Genstat directive. All data structures within the procedure are local (i.e. they cannot be referenced, or confused, with data structures outside the procedure); input and output structures for the procedure are defined by option and parameter settings in the procedure call.
Procedure name	is a letter followed by letters and/or digits. Only the first eight characters are significant; subsequent characters are ignored. The case of the letters (small or capital) is also ignored.

Program	is a series of statements, ending with the statement STOP.
Progression	Lists of numbers ascending or descending with equal increments can be specified succinctly using the form *number, number ... number* where the first two numbers define the first two elements in the list (and thus the increment) and the list ends with the value beyond which the third number would be passed. For lists with an increment of plus or minus one, the second number can be omitted, to give the form *number ... number*.
Punctuation symbols	The Genstat 5 punctuation symbols are:
colon (:)	indicates the end of a statement;
comma (,)	separates items;
double quote (")	is used to show the beginning and end of a comment;
equals (=)	separates an option name or parameter name from its setting;
newline	is synonymous with colon, by default, but directive SET can request that it be ignored;
semi-colon (;)	separates lists;
single quote (')	is used to show the beginning and end of a string (left single quote (`) is synonymous with single quote);
space	can appear between items or can be omitted altogether if the items are already separated by another punctuation symbol, a bracket, an operator, or an ampersand.
Qualified identifier	These may occur in an identifier list within an expression to define subsets of the values of a data structure. The form is "*identifier $ qualifier*", where the qualifier is a sequence of identifier lists enclosed in square brackets. For factors, variates, and texts, the qualifier has a single list, each element of which defines a subset of the vector concerned. For matrices there are two lists running in parallel, one for each dimension. Tables cannot be qualified. The elements of the qualifier lists can be scalars, numbers, variates, quoted strings, or texts.
Scalar	is a data structure that stores a single number.
Special symbols	The special symbols in Genstat 5 are as follows:
ampersand (&)	repeats the previous statement name (unless that statement contained a syntax error) and any option settings that are not explicitly changed;
asterisk (*)	denotes a missing value (and is also used as an operator);
backslash (\)	is the continuation symbol, typed at the end of a line to indicate that the current statement continues onto the next line (this is unnecessary when directive SET has been used

	to specify that newline is to be ignored);
dollar ($)	precedes a list of unit names or numbers (enclosed in square brackets) that define subsets of a factor, variate, matrix, symmetric matrix, diagonal matrix, or text;
exclamation mark (!)	indicates an unnamed structure (vertical bar (|) is synonymous with exclamation mark);
hash (#)	is the substitution symbol and is followed by the identifier of the structure whose values are to be inserted at that point in a Genstat statement (the substitution takes place immediately before the statement is executed); a pair of contiguous substitution symbols (##) is used to introduce a macro.
SSPM	is a compound structure storing sums of squares and products, means and ancillary information for use in regression and multivariate analysis.
Statement	is an instruction in the Genstat language; it has the form *"statement-name [option-sequence] parameter-sequence terminator"*. If no option settings are given, the square brackets can be omitted. The terminator is colon (:), ampersand (&) or newline (unless directive SET has indicated that this is to be ignored).
Statement name	is the name of either a directive or a procedure.
String	is a sequence of characters forming one unit (or line) of a Genstat text structure. In most contexts, the string must be quoted: i.e. enclosed in single quotes ('). Quoted strings may contain any of the characters available on the computer. However, if single quote ('), double quote ("), or the continuation symbol (\) are required as characters within a quoted string, they must each be typed twice to distinguish this use from their action in, respectively, terminating the string, introducing a comment within the string, or indicating continuation. Newline within a quoted string is taken to terminate the current (quoted) string and begin another one, unless the newline is within a comment or preceded by an (unduplicated) continuation symbol (\), or unless directive SET has specified that newline is to be ignored. Unquoted strings can occur in unnamed texts, or in option or parameter settings where you have to specify a particular string from a prescribed set of alternatives; an unquoted string must have a letter as its first character and contain only letters or digits.
Subset selection	An identifier list within an expression may contain

	qualified identifiers, each defining a list of subsets of the values of the identifier concerned.
Suffixes	Elements of pointers can be referred to by suffixes. Each suffix takes the form of an identifier list enclosed in square brackets; the list can contain numbers, scalars, or variates to reference an element or elements by number, or texts or quoted strings to reference by label. A null list within the brackets is taken to mean all the elements of the pointer in turn. Where a pointer has other pointers as its elements, their elements can be referred to in the same way, and so the original identifier may be followed by several suffix lists each contained in its own pair of square brackets; these define a list of elements, one for each combination of an element from each suffix list, taking the combinations in an order in which the last list cycles through its elements fastest, then the next to last list, and so on.
Symmetric matrix	is a data structure that stores the lower triangle (including the diagonal) of a symmetric square matrix.
System word	is a letter followed by letters and/or digits with a special meaning within the Genstat language, e.g. directive, option, parameter, or function names. The case of the letters (small/capital) is not significant; the abbreviation rules vary according to context.
Table	is a data structure that stores a multi-dimensional array of numbers, each dimension classified by a factor. Thus a table can be used to hold a summary of data that are classified (by the factors) into groups.
Text	is a data structure that stores a series of strings, each one representing a line of textual information.
TSM	is a compound data structure storing a model for use in Box-Jenkins modelling of time series.
Unnamed structures	An identifier list may contain unnamed variates, texts, pointers, expressions, or formulae. An unnamed structure consists of an exclamation mark, followed by the type code, and then the contents contained in round brackets. The type code is E, F, P, T, or V for expression, formula, pointer, text, or variate, respectively; if no code is given, variate is assumed by default.
Variate	is a data structure that stores a series of numbers.
Vector	is a series of values, notionally arranged in a column. Genstat has three different types of vector: factors, texts, and variates.

A1.1.2 Data structures

DIAGONALMATRIX
Declares one or more diagonal matrix data structures.
Options

ROWS	= scalar, vector, or pointer	Number of rows, or labels for rows (and columns); default *
VALUES	= numbers	Values for all the diagonal matrices; default *
MODIFY	= string	Whether to modify (instead of redefining) existing structures (no, yes); default n

Parameters

IDENTIFIER	= identifiers	Identifiers of the diagonal matrices
VALUES	= identifiers	Values for each diagonal matrix
DECIMALS	= scalars	Number of decimal places for printing
EXTRA	= texts	Extra text associated with each identifier

DUMMY
Declares one or more dummy data structures.
Options

VALUE	= identifier	Value for all the dummies; default *
MODIFY	= string	Whether to modify (instead of redefining) existing structures (no, yes); default n

Parameters

IDENTIFIER	= identifiers	Identifiers of the dummies
VALUE	= identifiers	Value for each dummy
EXTRA	= texts	Extra text associated with each identifier

EXPRESSION
Declares one or more expression data structures.

Options

| VALUE | = expression | Value for all the expressions; default * |
| MODIFY | = string | Whether to modify (instead of redefining) existing structures (no, yes); default n |

Parameters

IDENTIFIER	= identifiers	Identifiers of the expressions
VALUE	= expressions	Value for each expression
EXTRA	= texts	Extra text associated with each identifier

FACTOR
Declares one or more factor data structures.
Options

NVALUES	= scalar or vector	Number of units, or vector of labels; default * takes the setting from the preceding UNITS statement, if any
LEVELS	= scalar or variate	Number of levels, or series of numbers which will be used to refer to levels in the program; default *
VALUES	= numbers	Values for all the factors, given as levels; default *
LABELS	= text	Labels for levels, for input and output; default *
MODIFY	= string	Whether to modify (instead of redefining) existing structures (no, yes); default n

Parameters

IDENTIFIER	= identifiers	Identifiers of the factors
VALUES	= identifiers	Values for each factor, specified as levels or labels
DECIMALS	= scalars	Number of decimals for printing levels
CHARACTERS	= scalars	Number of characters for printing labels
EXTRA	= texts	Extra text associated with each identifier

FORMULA
Declares one or more formula data structures.

Options

VALUE	= formula	Value for all the formulae; default *
MODIFY	= string	Whether to modify (instead of redefining) existing structures (no, yes); default n

Parameters

IDENTIFIER	= identifiers	Identifiers of the formulae
VALUE	= formulae	Value for each formula
EXTRA	= texts	Extra text associated with each identifier

LRV

Declares one or more LRV data structures.

Options

ROWS	= scalar, vector, or pointer	Number of rows, or row labels, for the matrix; default *
COLUMNS	= scalar, vector, or pointer	Number of columns, or column labels, for matrix and diagonal matrix; default *

Parameters

IDENTIFIER	= identifiers	Identifiers of the LRVs
VECTORS	= matrices	Matrix to contain the latent vectors for each LRV
ROOTS	= diagonal matrices	Diagonal matrix to contain the latent roots for each LRV
TRACE	= scalars	Trace of the matrix

MATRIX

Declares one or more matrix data structures.

Options

ROWS	= scalar, vector, or pointer	Number of rows, or labels for rows; default *
COLUMNS	= scalar, vector, or pointer	Number of columns, or labels for columns; default *

VALUES	= numbers	Values for all the matrices; default *
MODIFY	= string	Whether to modify (instead of redefining) existing structures (no, yes); default n

Parameters

IDENTIFIER	= identifiers	Identifiers of the matrices
VALUES	= identifiers	Values for each matrix
DECIMALS	= scalars	Number of decimal places for printing
EXTRA	= texts	Extra text associated with each identifier

POINTER

Declares one or more pointer data structures.

Options

NVALUES	= scalar or text	Number of values, or labels for values; default *
VALUES	= identifiers	Values for all the pointers; default *
SUFFIXES	= variate	Defines an integer number for each of the suffixes; default * indicates that the numbers 1,2,... are to be used
MODIFY	= string	Whether to modify (instead of redefining) existing structures (no, yes); default n

Parameters

IDENTIFIER	= identifiers	Identifiers of the pointers
VALUES	= pointers	Values for each pointer
EXTRA	= texts	Extra text associated with each identifier

SCALAR

Declares one or more scalar data structures.

Options

| VALUE | = scalar | Value for all the scalars; default is a missing value |
| MODIFY | = string | Whether to modify (instead of redefining) existing structures (no, yes); default n |

Parameters

IDENTIFIER	= identifiers	Identifiers of the scalars
VALUE	= scalars	Value for each scalar
DECIMALS	= scalars	Number of decimal places for printing
EXTRA	= texts	Extra text associated with each identifier

SSPM

Declares one or more SSPM data structures.

Options

TERMS	= model formula	Terms for which sums of squares and products are to be calculated; default *
FACTORIAL	= scalar	Maximum number of vectors in a term; default 3
FULL	= string	Full factor parameterization (no, yes); default n
GROUPS	= factor	Groups for within-group SSPMs; default *
DF	= scalar	Number of degrees of freedom for sums of squares; default *

Parameters

IDENTIFIER	= identifiers	Identifiers of the SSPMs
SSP	= symmetric matrices	Symmetric matrix to contain the sums of squares and products for each SSPM
MEANS	= variates	Variate to contain the means for each SSPM
NUNITS	= scalars	Number of units or sum of weights
WMEANS	= pointers	Pointers to variates of group means

SYMMETRICMATRIX

Declares one or more symmetric matrix data structures.

Options

ROWS	= scalar, vector, or pointer	Number of rows, or labels for rows and columns); default *
VALUES	= numbers	Values for all the symmetric matrices; default *
MODIFY	= string	Whether to modify (instead of redefining) existing structures (no, yes); default n

Parameters

IDENTIFIER	= identifiers	Identifiers of the symmetric matrices
VALUES	= identifiers	Values for each symmetric matrix
DECIMALS	= scalars	Number of decimal places for printing
EXTRA	= texts	Extra text associated with each identifier

TABLE

Declares one or more table data structures.

Options

CLASSIFICATION	= factors	Factors classifying the tables; default *
MARGINS	= string	Whether to add margins (no, yes); default n
VALUES	= numbers	Values for all the tables; default *
MODIFY	= string	Whether to modify (instead of redefining) existing structures (no, yes); default n

Parameters

IDENTIFIER	= identifiers	Identifiers of the tables
VALUES	= identifiers	Values for each table

DECIMALS	= scalars	Number of decimal places for printing
EXTRA	= texts	Extra text associated with each identifier
UNKNOWN	= identifiers	Identifier for scalar to hold summary of unclassified data associated with each table

TEXT

Declares one or more text data structures.

Options

NVALUES	= scalar or vector	Number of strings, or vector of labels; default * takes the setting from the preceding UNITS statement, if any
VALUES	= strings	Values for all the texts; default *
MODIFY	= string	Whether to modify (instead of redefining) existing structures (no, yes); default n

Parameters

IDENTIFIER	= identifiers	Identifiers of the texts
VALUES	= texts	Values for each text
CHARACTERS	= scalars	Numbers of characters of the lines of each text to be printed by default
EXTRA	= texts	Extra text associated with each identifier

TSM

Declares one or more TSM data structures.

Option

MODEL	= string	Type of model (arima, transfer); default a

Parameters

IDENTIFIER	= identifiers	Identifiers of the TSMs
ORDERS	= variates	Orders of the autoregressive, integrated, and moving average parts of each TSM
PARAMETERS	= variates	Parameters of each TSM
LAGS	= variates	Lags, if not default

VARIATE

Declares one or more variate data structures.

Options

NVALUES	= scalar or vector	Number of units, or vector of labels; default * takes the setting from the preceding UNITS statement, if any
VALUES	= numbers	Values for all the variates; default *
MODIFY	= string	Whether to modify (instead of redefining) existing structures (no, yes); default n

Parameters

IDENTIFIER	= identifiers	Identifiers of the variates
VALUES	= identifiers	Values for each variate
DECIMALS	= scalars	Number of decimal places for output
EXTRA	= texts	Extra text associated with each identifier

A1.1.3 Control structures

BREAK

Suspends execution of the statements in the current channel or control structure and takes subsequent statements from the channel specified.

Option

CHANNEL	= scalar	Channel number; default current input channel

No parameters

CASE

Introduces a "multiple-selection" control structure.

No options

Parameter

	expression	Expression which is evaluated to an integer, indicating which set of statements to execute

DEBUG

Puts an implicit BREAK statement after the current statement and after every NSTATEMENTS subsequent statements, until an ENDDEBUG is reached.

Options

CHANNEL	= scalar	Channel number; default current input channel
NSTATEMENTS	= scalar	Number of statements between breaks; default 1

No parameters

ELSE

Introduces the default set of statements in block-if or in multiple-selection control structures.

No options or parameters

ELSIF

Introduces a set of alternative statements in a block-if control structure.

No options

Parameter

	expression	Logical expression to indicate whether or not the set of statements is to be executed.

ENDBREAK

Returns to the original channel or control structure and continues execution.

No options or parameters

Appendix 1: Reference summary 635

ENDCASE

Indicates the end of a "multiple-selection" control structure.

No options or parameters

ENDDEBUG

Cancels a DEBUG statement.

No options or parameters

ENDFOR

Indicates the end of the contents of a loop.

No options or parameters

ENDIF

Indicates the end of a block-if control structure.

No options or parameters

ENDJOB

Ends a Genstat job.

No options or parameters

ENDPROCEDURE

Indicates the end of the contents of a Genstat procedure.

No options or parameters

EXIT

Exits from a control structure.

Options

NTIMES	= scalar	Number of control structures, N, to exit; default 1. If N exceeds the number of control structures of the specified type that are currently active, the exit is to the end of the outer one; while for N negative, the exit is to the end of the $-$N'th structure (in order of execution)
CONTROL	= string	Type of control structure to exit (job, for, if, case, procedure); default f
REPEAT	= string	Whether to go to the next set of parameters on exit from a FOR loop (no, yes); default n

Parameter

	expression	Logical expression controlling whether or not an exit takes place.

FOR

Introduces a loop; subsequent statements define the contents of the loop, which is terminated by the directive ENDFOR.

Options

NTIMES	= scalar	Number of times to execute the loop; default is to execute as many times as the length of the first parameter list or once if the first list is null
COMPILE	= string	Whether to execute each statement as it is compiled or to compile all statements before execution as a block (each, all); default a

Appendix 1: Reference summary 637

Parameters

| | dummies | are set up implicitly by the statement; each dummy appears to be a parameter |

IF

Introduces a block-if control structure.

No options

Parameter

| | expression | Logical expression, indicating whether or not to execute the first set of statements. |

JOB

Starts a Genstat job.

Options

INPRINT	= strings	Printing of input as in PRINT option of INPUT (statements, macros, procedures, unchanged); default u
OUTPRINT	= strings	Additions to output as in PRINT option of OUTPUT (dots, page, unchanged); default u
DIAGNOSTIC	= strings	Output to be printed for a Genstat diagnostic (warnings, faults, extra); default f,w
ERRORS	= scalar	Limit on number of error diagnostics that may occur before the job is abandoned; default * i.e. no limit

Parameter

| | text | Name to identify the job |

OPTION

Defines the options of a Genstat procedure.

No options

Parameters

NAME	= texts	Names of the options
MODE	= strings	Mode of each option (e, f, p, t, v, as for unnamed structures); default p
DEFAULT	= identifiers	Default values for each option

OR

Introduces a set of alternative statements in a "multiple-selection" control structure.

No options or parameters

PARAMETER

Defines the parameters of a Genstat procedure.

No options

Parameters

NAME	= texts	Names of the parameters
MODE	= strings	Mode of each parameter (e, f, p, t, v, as for unnamed structures); default p

PROCEDURE

Introduces a Genstat procedure.

No options

Parameter

text	Name of the procedure

Appendix 1: Reference summary 639

STOP

Ends a Genstat program.

No options or parameters

A1.2.1 Input and output

CLOSE

Closes files.

No options

Parameters

CHANNEL	= scalars	Numbers of the channels to which the files are attached
FILETYPE	= strings	Type of each file (input, output, unformatted, backingstore, procedurelibrary, graphics); if omitted, input is assumed

---

COPY

Forms a transcript of a job.

Option

PRINT	= strings	What to transcribe (statements, output); default s

Parameter

scalar	Channel number of output file

DISPLAY

Reprints the last diagnostic.

Option

PRINT	= string	What information to print (diagnostic); default d

No parameters

DUMP

Prints information about data structures, and internal system information.

Options

PRINT	= strings	What information to print about structures (attributes, values, identifiers); default a
CHANNEL	= scalar	Channel number of output file; default current output file
INFORMATION	= string	What information to print for each structure (brief, full); default b
TYPE	= strings	Which types of structure to include in addition to those in the parameter list (all, diagonalmatrix, dummy, expression, factor, formula, LRV, matrix, pointer, scalar, SSPM, symmetricmatrix, table, text, TSM, variate); default * i.e. none
COMMON	= strings	Which internal Fortran commons to display (all, banks, compl, diagpk, direct, fncon, inout, input, mainac, output, periph, print, root, syscon, wsp); default * i.e. none
SYSTEM	= string	Whether to display Genstat system structures (no, yes); default n

Parameter

	identifiers	Structures whose information is to be printed

Appendix 1: Reference summary 641

HELP

Prints details about the Genstat language and environment. When no parameters are given, HELP gives information on how to use it. When used interactively, at each stage HELP responds with a series of menus to allow choice of information. Responding with a star (*) causes information on all the displayed words to be printed, carriage-return (⟨RETURN⟩) causes the previous menu to be displayed, and colon (:) is used to exit from HELP.

No options

Parameter

strings	Directive names or keywords indexing the desired details.

INPUT

Specifies the input file from which to take further statements.

Options

PRINT	= strings	What output to generate from statements in the file (statements, macros, procedures, unchanged); default s
REWIND	= string	Whether to rewind the file (no, yes); default n

Parameter

scalar	Channel number of input file

OPEN

Opens files.

No options

Parameters

NAME	= texts	External names of the files

CHANNEL	= scalars	Channel number to be used to refer to each file in other statements; numbers for each type of file are independent
FILETYPE	= strings	Type of each file (input, output, unformatted, backingstore, procedurelibrary, graphics); if omitted, input is assumed
WIDTH	= scalars	Maximum width of a record in each file; if omitted, 80 is assumed for input files, the full line-printer width (usually 132) for output files
PAGE	= scalars	Number of lines per page (relevant only for output files)
ACCESS	= string	Allowed type of access for unformatted or backing-store files (readonly, writeonly, both); if unspecified, b is assumed

OUTPUT

Defines where output is to be stored or displayed.

Options

PRINT	= strings	Additions to output (dots, page, unchanged); default d,p
DIAGNOSTIC	= strings	What diagnostic printing is required (faults, warnings, extra); default f,w
WIDTH	= scalar	Limit on number of characters per record; default width of output file

Parameter

scalar	Channel number of output file

PAGE

Moves to the top of the next page of an output file.

Option

| CHANNEL | = scalars | Channel number of file; default * i.e. current output file |

No parameters

PRINT

Prints data in tabular format in an output file, unformatted file, or text.

Options

CHANNEL	= identifier	Channel number of file, or identifier of a text to store output; default current output file
SERIAL	= string	Whether structures are to be printed in serial order, i.e. all values of the first structure, then all of the second, and so on (no, yes); default n, i.e. values in parallel
IPRINT	= string	What identifier (if any) to print for the structure (identifier, associatedidentifier), for a table associatedidentifier prints the identifier of the variate from which the table was formed, IPRINT = * suppresses the identifier altogether; default i
RLPRINT	= strings	What row labels to print (labels, integers); default l
CLPRINT	= strings	What column labels to print (labels, integers); default l
RLWIDTH	= scalar	Field width for row labels; default 13
INDENTATION	= scalar	Number of spaces to leave before the first character in the line; default 0
WIDTH	= scalar	Last allowed position for characters in the line; default width of current output file
SQUASH	= string	Whether to omit blank lines in the layout of values (no, yes); default n
MISSING	= text	What to print for missing value; default '*'

ORIENTATION	= string	How to print vectors or pointers (down, across); default d, i.e. down the page
PERMUTE	= vector	Permutation of table classifiers; default *
INTERLEAVE	= scalar	Level of classification at which structures are to be parallel; default highest level, e.g. values for tables are interleaved within all the classifying factors
NDOWN	= scalar	Number of table classifiers to be printed down the page; default all but one unless there is only one, when it is printed down
PUNKNOWN	= string	When to print unknown cells of tables (present, always, zero, missing, never); default p
UNFORMATTED	= string	Whether file is unformatted (no, yes); default n
REWIND	= string	Whether to rewind unformatted file before printing (no,yes); default n

Parameters

STRUCTURE	= identifiers	Structures to be printed
FIELDWIDTH	= scalars	Field width in which to print the values of each structure (a negative value $-N$ prints numbers in E-format in width N); if omitted, a default is determined (for numbers, this is usually 12; for text, the width is one more character than the longest line)
DECIMALS	= scalars	Number of decimal places for numbers; if omitted, a default is determined which prints the mean absolute value to four significant figures
SKIP	= scalars or variates	Number of spaces to leave before each value of a structure (* means newline before structure)

JUSTIFICATION	= strings	How to position values within the field (right, left); if omitted, right is assumed
MNAME	= strings	Name to print for table margins (Margin, Total, Nobservd, Mean, Minimum, Maximum, Variance, Count, Median); if omitted, "Margin" is printed

READ

Reads data from an input file, an unformatted file, or a text.

Options

PRINT	= strings	What to print (data, errors, summary); default e,s
CHANNEL	= identifier	Channel number of file, or text structure from which to read data; default current file
SERIAL	= string	Whether structures are in serial order, i.e. all values of the first structure, then all of the second, and so on (no, yes); default n, i.e. values in parallel
SETNVALUES	= string	Whether to set number of values of structures from the number of values read (no, yes); default n
LAYOUT	= string	How values are presented (separated, fixedfield); default s
END	= text	What string terminates data (* means there is no terminator); default ':'
SEQUENTIAL	= scalar	To store the number of units read (negative if terminator is met); default *
ADD	= string	Whether to add values to existing values (no, yes); default n (Only available in serial read)
MISSING	= text	What character represents missing values; default '*'
SKIP	= scalar	Number of characters (LAYOUT=f)

		or values (LAYOUT=s) to be skipped between units (∗ means skip to next record); default 0 (Only available in parallel read)
BLANK	= string	Interpretation of blank fields with LAYOUT=f (missing, zero, error); default m
JUSTIFIED	= string	How values are to be assumed justified with LAYOUT=f (right, left, both, neither); default r
ERRORS	= scalar	How many errors to allow in the data before reporting a fault rather than a warning, a negative setting, −N, causes reading of data to stop after the N'th error; default 0
FORMAT	= variate	Allows a format to be specified for situations where the layout varies for different units, option SKIP and parameters FIELDWIDTH and SKIP are then ignored (in the variate: 0 switches to fixed format; 0.1, 0.2, 0.3, or 0.4 to free format with space, comma, colon, or semi-colon respectively as separators; ∗ skips to the beginning of the next line; in fixed format, a positive integer N indicates an item in a field width of N, −N skips N characters; in free format, N indicates N items, −N skips N items); default ∗
QUIT	= scalar	Channel number of file to return to after a fatal error; default ∗ i.e. current input file
UNFORMATTED	= string	Whether file is unformatted (no, yes); default n
REWIND	= string	Whether to rewind the file before reading (no,yes); default n

Parameters

STRUCTURE	= identifiers	Structures into which to read the data

FIELDWIDTH	= scalars	Field width from which to read values of each structure (LAYOUT=f only)
DECIMALS	= scalars	Number of decimal places for numerical data containing no decimal points
SKIP	= scalars	Number of values (LAYOUT=s) or characters (LAYOUT=f) to skip before reading a value
FREPRESENTATION	= string	How factor values are represented (labels, levels, ordinals); if omitted, levels are assumed

RETURN

Returns to a previous input stream (text vector or input channel).

Option

NTIMES	= scalar	Number of streams to ascend; default 1

No parameters

SKIP

Skips lines in input or output files.

Options

CHANNEL	= scalar	Channel number of file; default current file
FILETYPE	= string	Type of the file concerned (input, output); default i

Parameter

	identifier	How many lines to skip; for input files, a text means skip until the contents of the text have been found, these contents are then the next characters to be read

A1.2.2 Graphics

AXES

Defines the axes in each window of high-quality graphs and contours.

Option

EQUAL	= string	Whether/how to make axes equal (no, scale, lower, upper); default n

Parameters

WINDOW	= scalars	Numbers of the windows
YTITLE	= texts	Title for the y-axis in each window
XTITLE	= texts	Title for the x-axis in each window
YLOWER	= scalars	Lower bound for y-axis
YUPPER	= scalars	Upper bound for y-axis
XLOWER	= scalars	Lower bound for x-axis
XUPPER	= scalars	Upper bound for x-axis
YINTEGER	= strings	Whether y-labels integral (no, yes)
XINTEGER	= strings	Whether x-labels integral (no, yes)
YMARKS	= scalars or variates	Distance between each tick mark on y-axis (scalar) or positions of the marks (variate)
XMARKS	= scalars or variates	Distance between each tick mark on x-axis (scalar) or positions of the marks (variate)
YLABELS	= texts	Labels at each mark on y-axis
XLABELS	= texts	Labels at each mark on x-axis
YORIGIN	= scalars	Position on y-axis at which x-axis is drawn
XORIGIN	= scalars	Position on x-axis at which y-axis is drawn
STYLE	= strings	Style of axes (none, x, y, xy, box, grid)

CONTOUR

Produces contour maps of two-way arrays of numbers (on the terminal/printer).

Options

CHANNEL	= scalar	Channel number of output file; default is current output file
INTERVAL	= scalar or variate	Contour interval for scaling (scalar) or positions of the contours (variate); default * i.e. determined automatically
TITLE	= text	General title; default *
YTITLE	= text	Title for y-axis; default *
XTITLE	= text	Title for x-axis; default *
YLOWER	= scalar	Lower bound for y-axis; default 0
YUPPER	= scalar	Upper bound for y-axis; default 1
XLOWER	= scalar	Lower bound for x-axis; default 0
XUPPER	= scalar	Upper bound for x-axis; default 1
YINTEGER	= string	Whether y-labels integral (no, yes); default n
XINTEGER	= string	Whether x-labels integral (no, yes); default n
LOWERCUTOFF	= scalar	Lower cut-off for array values; default *
UPPERCUTOFF	= scalar	Upper cut-off for array values; default *

Parameters

GRID	= identifiers	Pointers (of variates representing the columns of a data matrix), matrices, or two-way tables specifying values on a regular grid
DESCRIPTION	= texts	Annotation for key

DCONTOUR

Draws contour plots on a plotter or graphics monitor.

Options

INTERVAL	= scalar or variate	Contour interval for scaling (scalar) or positions of the contours (variate); default * i.e. determined automatically

TITLE	= text	General title; default *
WINDOW	= scalar	Window number for the plots; default 1
KEYWINDOW	= scalar	Window number for the key (zero for no key); default 2
LOWERCUTOFF	= scalar	Lower cut-off for array values; default *
UPPERCUTOFF	= scalar	Upper cut-off for array values; default *
SCREEN	= string	Whether to clear the screen before plotting or to continue plotting on the old screen (clear, keep); default c

Parameters

GRID	= identifiers	Pointers (of variates representing the columns of a data matrix), matrices, or two-way tables specifying values on a regular grid
PEN	= scalars	Pen number to be used for the contours of each grid
DESCRIPTION	= texts	Annotation for key

DEVICE

Switches between (high-quality) graphics devices.

No options

Parameter

scalar	Device number

DGRAPH

Draws graphs on a plotter or graphics monitor.

Options

TITLE	= text	General title; default *
WINDOW	= scalar	Window number for the graphs; default 1

KEYWINDOW	= scalar	Window number for the key (zero for no key); default 2
SCREEN	= string	Whether to clear the screen before plotting or to continue plotting on the old screen (clear, keep); default c

Parameters

Y	= identifiers	Vertical coordinates
X	= identifiers	Horizontal coordinates
PEN	= scalars	Pen number for each graph
DESCRIPTION	= texts	Annotation for key
YLOWER	= identifiers	Lower values for vertical bars
YUPPER	= identifiers	Upper values for vertical bars
XLOWER	= identifiers	Lower values for horizontal bars
XUPPER	= identifiers	Upper values for horizontal bars

DHISTOGRAM

Draws histograms on a plotter or graphics monitor.

Options

TITLE	= text	General title; default *
WINDOW	= scalar	Window number for the histograms; default 1
KEYWINDOW	= scalar	Window number for the key (zero for no key); default 2
LIMITS	= variate	Variates of group limits for classifying variates into groups; default *
NGROUPS	= scalar	When LIMITS is not specified, this defines the number of groups into which a DATA variate is to be classified; default is the integer value nearest to the square root of the number of values in the variate
LABELS	= text	Group labels; default *
APPEND	= string	Whether or not the bars of the histograms are appended together (no, yes); default n

| SCREEN | = string | Whether to clear the screen before plotting or to continue plotting on the old screen (clear, keep); default c |
|---|---|---|劫
Parameters		
DATA	= identifiers	Data for the histograms; these can be either a factor indicating the group to which each unit belongs, a variate whose values are to be grouped, or a one-way table giving the number of units in each group
NOBSERVATIONS	= tables	One-way table to save numbers in the groups
GROUPS	= factors	Factor to save groups defined from a variate
PEN	= scalars	Pen number for each histogram
DESCRIPTION	= texts	Annotation for key

DPIE

Draws a pie chart on a plotter or graphics monitor.

Options

TITLE	= text	General title; default *
WINDOW	= scalar	Window number for the pie chart; default 1
KEYWINDOW	= scalar	Window number for the key (zero for no key); default 2
SCREEN	= string	Whether to clear the screen before plotting or to continue plotting on the old screen (clear, keep); default c

Parameters

SLICE	= scalars	Amounts in each of the slices (or categories)
PEN	= scalars	Pen number for each slice
DESCRIPTION	= texts	Description of each slice

FRAME

Defines the positions of windows within the frame of a high-quality graph. The positions are defined in normalized device coordinates ($[0,1] \times [0,1]$).

No options

Parameters

WINDOW	= scalars	Window numbers
YLOWER	= scalars	Lower y device coordinate for each window
YUPPER	= scalars	Upper y device coordinate for each window
XLOWER	= scalars	Lower x device coordinate for each window
XUPPER	= scalars	Upper x device coordinate for each window

GRAPH

Produces scatter and line graphs on the terminal or line printer.

Options

CHANNEL	= scalar	Channel number of output file; default is current output file
TITLE	= text	General title; default *
YTITLE	= text	Title for y-axis; default *
XTITLE	= text	Title for x-axis; default *
YLOWER	= scalar	Lower bound for y-axis; default *
YUPPER	= scalar	Upper bound for y-axis; default *
XLOWER	= scalar	Lower bound for x-axis; default *
XUPPER	= scalar	Upper bound for x-axis; default *
MULTIPLE	= variate	Numbers of plots per frame; default * i.e. all plots are on a single frame
JOIN	= string	Order in which to join points (ascending, given); default a
EQUAL	= string	Whether/how to make bounds equal (no, scale, lower, upper); default n
NROWS	= scalar	Number of rows in the frame; default * i.e. determined automatically

NCOLUMNS	= scalar	Number of columns in the frame; default * i.e. determined automatically
YINTEGER	= string	Whether y-labels integral (no, yes); default n
XINTEGER	= string	Whether x-labels integral (no, yes); default n

Parameters

Y	= identifiers	Y coordinates
X	= identifiers	X coordinates
METHOD	= strings	Type of each graph (point, line, curve, text); if unspecified, p is assumed
SYMBOLS	= factors or texts	For factor SYMBOLS, the labels (if defined), or else the levels, define plotting symbols for each unit, whereas a text defines textual information to be placed within the frame for METHOD = text or the symbol to be used for each plot for other METHOD settings; if unspecified, * is used for points, with integers 1-9 to indicate coincident points, ' and . are used for lines and curves
DESCRIPTION	= texts	Annotation for key

HISTOGRAM

Produces histograms of data on the terminal or line printer.

Options

CHANNEL	= scalar	Channel number of output file; default is the current output file
TITLE	= text	General title; default *
LIMITS	= variate	Variate of group limits for classifying variates into groups; default *
NGROUPS	= scalar	When LIMITS is not specified, this defines the numbers of groups into

		which a data variate is to be classified; default is the integer value nearest to the square root of the number of values in the variate
LABELS	= text	Group labels
SCALE	= scalar	Number of units represented by each character; default 1

Parameters

DATA	= identifiers	Data for the histograms; these can be either a factor indicating the group to which each unit belongs, a variate whose values are to be grouped, or a one-way table giving the number of units in each group
NOBSERVATIONS	= tables	One-way table to save numbers in the groups
GROUPS	= factors	Factor to save groups defined from a variate
SYMBOLS	= texts	Characters to be used to represent the bars of each histogram
DESCRIPTION	= texts	Annotation for key

PEN

Defines the properties of "pens" for high-quality graphics.

No options

Parameters

NUMBER	= scalars	Numbers associated with the pens
COLOUR	= scalars	Number of the colour used with each pen
LINESTYLE	= scalars	Style for line used by each pen when joining points (zero = no line)
METHOD	= strings	Method for determining line (point, line, monotonic, closed, open)
SYMBOLS	= identifiers	Symbols for points – scalar for special symbols, texts or factors for character symbols

JOIN	= strings	Order in which points are to be joined by each pen (ascending, given)
BRUSH	= scalars	Number of the type of area filling used with each pen when drawing pie charts or histograms

A1.2.3 Backing store

CATALOGUE

Displays the contents of a backing-store file.

Options

PRINT	= strings	What to print (subfiles, structures); default s,st
CHANNEL	= scalar	Channel number of the backing-store file; default 0, i.e. the workfile
LIST	= string	How to interpret the list of subfiles (inclusive, exclusive, all); default i
SAVESUBFILE	= text	To save the subfile identifiers; default *

Parameters

SUBFILE	= identifiers	Identifiers of subfiles in the file to be catalogued
SAVESTRUCTURE	= texts	To save the identifiers of the structures in each subfile

MERGE

Copies subfiles from backing-store files into a single file.

Options

PRINT	= string	What to print (catalogue); default *
OUTCHANNEL	= scalar	Channel number of the backing-store file where the subfiles are to be stored; default 0, i.e. the work file

| METHOD | = string | How to append subfiles to the OUT file (add, overwrite, replace); default a, i.e. clashes in subfile identifiers cause a fault (note: replace overwrites the complete file) |
| PASSWORD | = text | Password to be checked against that stored with the file; default * |

Parameters

SUBFILE	= identifiers	Identifiers of the subfiles
INCHANNEL	= scalars	Channel number of the backing-store file containing each subfile
NEWSUBFILE	= identifiers	Identifier to be used for each subfile in the new file

---

RECORD

Dumps a job so that it can later be restarted by a RESUME statement.

Option

| CHANNEL | = scalar | Channel number of the unformatted file where information is to be dumped; default 1 |

No parameters

RESUME

Restarts a recorded job.

Option

| CHANNEL | = scalar | Channel number of the unformatted file where the information was dumped; default 1 |

No parameters

RETRIEVE

Retrieves structures from a subfile.

Options

CHANNEL	= scalar	Channel number of the backing-store file containing the subfile; default 0, i.e. the workfile
SUBFILE	= identifier	Identifier of the subfile; default SUBFILE
LIST	= string	How to interpret the list of structures (inclusive, exclusive, all); default i
MERGE	= string	Whether to merge structures with those already in the job (no, yes); default n, i.e. a structure whose identifier is already in the job overwrites the existing one, unless it has a different type

Parameters

IDENTIFIER	= identifiers	Identifiers to be used for the structures after they have been retrieved
STOREDIDENTIFIER	= identifiers	Identifier under which each structure was stored

STORE

To store structures in a subfile of a backing-store file.

Options

PRINT	= string	What to print (catalogue); default *
CHANNEL	= scalar	Channel number of the backing-store file where the subfile is to be stored; default 0, i.e. the workfile
SUBFILE	= identifier	Identifier of the subfile; default SUBFILE
LIST	= string	How to interpret the list of structures (inclusive, exclusive, all); default i
METHOD	= string	How to append the subfile to the file

		(add, overwrite, replace); default a, i.e. clashes in subfile identifiers cause a fault (note: replace overwrites the complete file)
PASSWORD	= text	Password to be stored with the file; default *
PROCEDURE	= string	Whether subfile contains procedures only (no, yes); default n
Parameters		
IDENTIFIER	= identifiers	Identifiers of the structures to be stored
STOREDIDENTIFIER	= identifiers	Identifier to be used for each structure when it is stored

A1.2.4 Calculation and manipulation

ASSIGN

Sets elements of pointers and dummies.

No options

Parameters

STRUCTURE	= identifiers	Values for the dummies or pointer elements
POINTER	= dummies or pointers	Structure that is to point to each of those in the STRUCTURE list
ELEMENT	= scalars or texts	Unit or unit label indicating which pointer element is to be set; if omitted, the first element is assumed

CALCULATE

Calculates numerical values for data structures.

Options

PRINT	= string	Printed output required (summary);

			default * i.e. no printing
ZDZ		= string	Value to be given to zero divided by zero (missing, zero); default m
TOLERANCE		= scalar	If the scalar is non missing, this defines the smallest non-zero number; otherwise it accesses the default value, which is defined automatically for the computer concerned
Parameter			
		expression	Expression defining the calculations to be performed

COMBINE

Combines or omits "slices" of a multi-way data structure (table, matrix or variate).

Options

OLDSTRUCTURE	= identifier	Structure whose values are to be combined; no default i.e. this option must be set
NEWSTRUCTURE	= identifier	Structure to contain the combined values; no default i.e. this option must be set

Parameters

OLDDIMENSION	= factors or scalars	Dimension number or factor indicating a dimension of the OLDSTRUCTURE
NEWDIMENSION	= factors or scalars	Dimension number or factor indicating the corresponding dimension of the NEWSTRUCTURE; this can be omitted if the dimensions are in numerical order
OLDPOSITIONS	= pointers, texts, or variates	These define positions in each OLDDIMENSION: pointers are appropriate for matrices whose rows

		or columns are indexed by a pointer; texts are for matrices indexed by a text, variates with a textual labels vector, or tables whose OLDDIMENSION factor has labels; and variates either refer to levels of table factors or numerical labels of matrices or variates, if these are present, otherwise they give the (ordinal) number of the position. If omitted, the positions are assumed to be in (ordinal) numerical order. Margins of tables are indicated by missing values
NEWPOSITIONS	= pointers, texts or variates	These define positions in each NEWDIMENSION, specified similarly to OLDPOSITIONS; these indicate where the values from the corresponding OLDDIMENSION positions are to be entered (or added to any already entered there)
WEIGHTS	= variates	Define weights by which the values from each OLDDIMENSION coordinate are to be multiplied before they are entered in the NEWDIMENSION

CONCATENATE

Concatenates and truncates lines (units) of text structures.

Option

NEWTEXT	= text	Text to hold the concatenated/truncated lines; default is the first OLDTEXT vector

Parameters

OLDTEXT	= texts	Texts to be concatenated
WIDTH	= scalars	Number of characters to take from

	or variates	the lines of each text; if * or omitted, all the (unskipped) characters are taken
SKIP	= scalars or variates	Number of characters to skip at the left-hand side of the lines of each text; if * or omitted, none is skipped

DELETE

Deletes the attributes and values of structures.

Options

REDEFINE	= string	Whether or not to delete the attributes of the structures so that the type etc can be redefined (no, yes); default n
LIST	= string	How to interpret the list of structures (inclusive, exclusive, all); default i

Parameter

	identifiers	Structures whose values (and attributes, if requested) are to be deleted

EDIT

Edits text vectors.

Options

CHANNEL	= scalar or text	Text structure containing editor commands, or a scalar giving the number of a channel from which they are to be read; default is the current input channel
END	= text	Character(s) to indicate the end of the commands read from an input channel; default is the character colon (:)
WIDTH	= scalar	Limit on the line width of the text;

		default *
SAVE	= text	Text to save the editor commands for future use; default *
Parameters		
OLDTEXT	= texts	Texts to be edited
NEWTEXT	= texts	Text to store each edited text; if any of these is omitted, the corresponding OLDTEXT is used

EQUATE

Transfers data between structures of different sizes or types (but the same modes i.e. numerical or text) or where transfer is not from single structure to single structure.

Options

OLDFORMAT	= variate	Format for values of OLDSTRUCTURES; within the variate, a positive value N means take N values, $-N$ means skip N values and a missing value means skip to the next structure; default * i.e. take all the values in turn
NEWFORMAT	= variate	Format for values of NEWSTRUCTURES; within the variate, a positive value N means fill the next N positions, $-N$ means skip N positions and a missing value means skip to the next structure; default * i.e. fill all the positions in turn

Parameters

OLDSTRUCTURES	= identifiers	Structures whose values are to be transferred; if values of several structures are to be transferred to one item in the NEWSTRUCTURES list, they must be placed in a pointer
NEWSTRUCTURES	= identifiers	Structures to take each set of

	transferred values; if several structures are to receive values from one item in the OLDSTRUCTURES list, they must be placed in a pointer

FLRV

Forms the values of LRV structures.

Options

PRINT	= strings	Printed output required (roots, vectors); default * i.e. no printing
NROOTS	= scalar	Number of roots or vectors to print; default * i.e. print them all
SMALLEST	= string	Whether to print the smallest roots instead of the largest (no, yes); default n

Parameters

INMATRIX	= symmetric matrices	Matrices whose latent roots and vectors are to be calculated
LRV	= LRVs	LRV to store the latent roots and vectors from each INMATRIX
WMATRIX	= symmetric matrices	(Generalized) within-group sums of squares and products matrix used in forming the "two-matrix decomposition"; if any of these is omitted, it is taken to be the identity matrix, giving the usual spectral decomposition

FSSPM

Forms the values of SSPM structures.

Options

PRINT	= strings	Printed output required (correlations, wmeans, SSPM); default * i.e. no printing

WEIGHTS	= variate	Weightings for the units; default * i.e. all units with weight one
SEQUENTIAL	= scalar	Used for sequential formation of SSPMs; a positive value indicates that formation is not yet complete (see READ directive); default * i.e. not sequential
SAVE	= identifier	Regression work structure (see TERMS directive); default *

Parameter

	SSPMs	Structures to be formed

GENERATE

Generates factor values for designed experiments: with no options set, factor values are generated in standard order; the options allow pseudo-factors to be generated describing confounding in partially balanced experimental designs.

Options

TREATMENTS	= formula	Model term for which pseudo-factors are to be generated; default *
REPLICATES	= formula	Factors defining replicates of the design; default *
BLOCKS	= formula	Block term; default *

Parameter

	factors	Factors whose values are to be generated

GET

Accesses details of the "environment" of a Genstat job.

Options

ENVIRONMENT	= pointer	Pointer given unit labels inprint, outprint, diagnostic, errors, pause, newline, case and run, used to save

		the current settings of those options of SET; default *
SPECIAL	= pointer	Pointer given unit labels units, blockstructure, treatmentstructure, covariate, asave, rsave and tsave, used to save the current settings of those options of SET; default *
LAST	= text	To save the last input statement; default *
FAULT	= scalar	To save the last fault code; default *
EPS	= scalar	To obtain the value of the smallest x (on this computer) such that $1+x > 1$; default *

No parameters

GETATTRIBUTE

Accesses attributes of structures.

Option

ATTRIBUTE	= strings	Which attributes to access (nvalues, nlevels, nrows, ncolumns, type, levels, labels, nmv); default * i.e. none

Parameters

STRUCTURE	= identifiers	Structures whose attributes are to be accessed
SAVE	= pointers	Pointer to store copies of the attributes of each structure, these are labelled by the ATTRIBUTE strings

INTERPOLATE

Interpolates values at intermediate points.

Options

COS(x)	cosine of x, for x in radians.
CUM(x)	synonym of CUMULATE.
CUMULATE(x)	forms the cumulative sum of the values of x; i.e. x_1, x_1+x_2, $x_1+x_2+x_3$, and so on.
D(x)	synonym of DETERMINANT.
DET(x)	synonym of DETERMINANT.
DETERMINANT(x)	the determinant of a square or symmetric matrix.
DIFFERENCE(x;s)	forms the differences of x, i.e. $x_i - x_{i-s}$. If s is omitted, first differences are formed, as for s=1.
ELEMENTS(x; e1; e2)	forms a sub-structure of x. If x is a vector or a diagonal matrix, then only e1 should be specified; this then indicates the selected elements of x. If x is a rectangular matrix, then both e1 and e2 should be given, to specify respectively the selected rows and columns of x. For a symmetric matrix x, if the same rows and columns are to be selected (giving a symmetric matrix) then only e1 should be specified; otherwise both e1 and e2 should be given (and the result is a matrix).
EXP(x)	exponential: e^x
EXPAND(x; s)	forms a variate of length s, containing zeroes and ones; if s is omitted and the length cannot be determined from the context, the length of the current units structure, if any, is taken. The values in x specify the numbers of the units that are to contain the value one.
FED(p; s1; s2)	the F-distribution equivalent deviate for probability p ($0 < p < 1$) and (s1, s2) degrees of freedom
FPROBABILITY(x; s1; s2)	the F-ratio probability of $t < x$ with (s1, s2) degrees of freedom
FRATIO(x; s1; s2)	synonym of FPROBABILITY.
I(x)	synonym of INVERSE.
INV(x)	synonym of INVERSE.
INVERSE(x)	the inverse of a non-singular square or symmetric matrix x.
INT(x)	synonym of INTEGER.
INTEGER(x)	integer part of x: [x].
LLB(x; n; p)	synonym of LLBINOMIAL.
LLBINOMIAL(x; n; p)	log-likelihood function for the Binomial distribution; n is the sample size and p the mean proportion (or the probability).
LLG(x; m; d)	synonym of LLGAMMA.
LLGAMMA(x; m; d)	log-likelihood function for the Gamma distribution; m is the mean and d the index.

LLN(x; m; v)	synonym of LLNORMAL.
LLNORMAL(x; m; v)	log-likelihood function for the Normal distribution; m is the mean and v the variance.
LLP(x; m)	synonym of LLPOISSON.
LLPOISSON(x; m)	log-likelihood function for the Poisson distribution; m is the mean.
LOG(x)	natural logarithm of x, for $x > 0$.
LOG10(x)	logarithm to base 10 of x, for $x > 0$.
LTPRODUCT(x; y)	left transposed product of x and y: a more efficient way of calculating TRANSPOSE(x) *+ y.
MAX(x)	synonym of MAXIMUM.
MAXIMUM(x)	finds the maximum of the values in x.
MEAN(x)	forms the mean of the values of x.
MED(x)	synonym of MEDIAN.
MEDIAN(x)	finds the median of the values in x.
MIN(x)	synonym of MINIMUM.
MINIMUM(x)	finds the minimum of the values in x.
MVREPLACE(x; y)	replaces missing values in x with the values in the corresponding units of y.
NCOLUMNS(x)	gives the number of columns of x.
NED(p)	gives the Normal equivalent deviate: that is the value x that leaves a proportion p $(0 < p < 1)$ to the left of it under the standard Normal curve.
NEWLEVELS(f; x)	forms a variate from the factor f; the variate x defines a value for each level and should be the same length as the number of levels of the factor.
NLEVELS(f)	gives the number of levels of factor f.
NMV(x)	counts the number of missing values in x.
NOBSERVATIONS(x)	counts the number of observations (that is non-missing values) in x.
NORMAL(x)	the Normal probability integral: gives the probability that a random variable with a standard Normal N(0,1) distribution is less than x.
NROWS(x)	gives the number of rows of x.
NVALUES(x)	gives the number of values, including missing values, of x (that is the length of x).
PRODUCT(x; y)	forms the matrix product of x and y (that is x *+ y).
QPRODUCT(x; y)	forms the quadratic product of x and y (that is x *+ y *+ TRANSPOSE(x)), where x is a rectangular matrix or variate and y is a symmetric or diagonal matrix or a scalar.
RESTRICTION(x)	forms a variate with the value one in the units to which x

	is currently restricted.
REVERSE(x)	reverses the values of x.
ROUND(x)	rounds the values of x to the nearest integer.
RTPRODUCT(x; y)	forms the right transposed product of x and y (that is x *+ TRANSPOSE(y)).
SHIFT(x; s)	shifts the values of x by s places (to the right or left according to the sign of s). This is not a circular shift, so some positions lose their values and are given missing values.
SIN(x)	sine of x, for x in radians.
SOLUTION(x; y)	finds the solution b of the set of simultaneous linear equations x *+ b = y.
SORT(x; y)	sorts the elements of x into the order that would put the values of y into ascending order; if y is omitted, the values of x are sorted.
SQRT(x)	gives the square root of x (x ≥ 0).
SUBMAT(x)	forms sub-triangles or sub-rectangles of a rectangular or symmetric matrix. The rows and columns to be included are determined by matching the pointers indexing the resultant matrix with the pointers indexing x. (SUBMAT does not allow for indexing by variates or texts.)
SUM(x)	forms the sum of the values in x (synonym TOTAL).
TOTAL(x)	forms the total of the values in x (synonym SUM).
T(x)	synonym of TRANSPOSE.
TMAXIMA(t)	forms margins of maxima for table t.
TMEDIANS(t)	forms margins of medians for table t.
TMEANS(t)	forms margins of means for table t.
TMINIMA(t)	forms margins of minima for table t.
TNOBSERVATIONS(t)	forms margins counting the numbers of observations (non-missing values) in table t.
TNMV(t)	forms margins counting the numbers of missing values in table t.
TNVALUES(t)	forms margins counting the numbers of values (missing or non-missing) in table t.
TSUMS(t)	synonym of TTOTALS.
TTOTALS(t)	forms margins of totals for table t.
TVARIANCES(t)	forms margins of between-cell variances for table t.
TRACE(x)	calculates the trace of the square, diagonal, or symmetric matrix x (that is the sum of all its diagonal elements).
TRANSPOSE(x)	forms the transpose of a rectangular matrix x.
UNSET(d)	returns a scalar logical value according to whether or not the dummy d is set.

URAND(seed; s)	provides s uniform pseudo-random numbers in the range (0, 1). If s is not supplied and URAND cannot determine the length of the result from the context of the expression, the length of the current units structure (if any) is taken. Scalar seed provides the seed for the random numbers and must be non-zero when URAND is first used in a job; subsequently a zero value continues the sequence of random numbers.
VAR(x)	synonym of VARIANCE.
VARIANCE(x)	gives the variance of the values in x.
VMAXIMA(p)	finds the maximum of the values in each unit over the variates in pointer p.
VMEANS(p)	gives the mean of the non-missing values in each unit over the variates in pointer p.
VMEDIANS(p)	finds the median of the values in each unit over the variates in pointer p.
VMINIMA(p)	finds the minimum of the values in each unit over the variates in pointer p.
VNMV(p)	counts the number of missing values in each unit over the variates in pointer p.
VNOBSERVATIONS(p)	counts the number of observations (non-missing values) in each unit over the variates in pointer p.
VNVALUES(p)	gives the number of values in each unit over the variates in pointer p (that is the number of values of p).
VSUMS(p)	gives the sum of the non-missing values in each unit over the variates in pointer p (synonym VTOTALS).
VTOTALS(p)	gives the total of the non-missing values in each unit over the variates in pointer p (synonym VSUMS).
VVARIANCES(p)	gives the variance of the non-missing values in each unit over the variates in pointer p.

A1.2.6 Commands for the EDIT directive

Commands to the editor may be given in upper or lower case. Any number of commands may be given on one line, each command being separated by at least one space character. The string delimiter (/) may be any character other than space or a digit.

A	inserts the next line of text, within the current line, immediately after the marker.
B	breaks the current line at the marker; text to the left of the marker becomes a line in its own right, immediately before the current line, which then contains just the text to the right of the marker.

C	cancels any changes made to the current line, restoring it to its original form.
D	deletes the current line, placing the marker at the start of the next line.
D+n	deletes n lines, starting at the current line.
D+	is a synonym for D or D+1.
D+*	deletes from the current line to the end of the text.
D*	is a synonym for D+*.
D−	deletes the current line, placing the marker at the start of the previous line.
D−n	deletes the current and n−1 previous lines.
D−*	deletes the current line and all previous lines.
D+/str/	deletes the current line and all subsequent lines until the next line containing the character string str.
D/str/	is a synonym for D+/str/.
D−/str/	deletes the current line and all previous lines until one containing the character string str.
F/id/	insert the contents of the text with identifier id, immediately before the current line.
G+/str$_1$/str$_2$/	substitutes string str$_2$ for all occurrences of string str$_1$ found after the marker on the current and subsequent lines, and moves the marker to the end of the text.
G/str$_1$/str$_2$/	is a synonym for G+/str$_1$/str$_2$/.
G−/str$_1$/str$_2$/	substitutes string str$_2$ for all occurrences of string str$_1$ found before the marker on the current and previous lines, and moves the marker to the start of the text.
I/str/	inserts string str on a new line immediately before the current line.
L	moves the marker to the start of the next line.
L+n	moves the marker to the start of the nth line after the current line.
L+	is a synonym for L or L+1.
L−n	moves the marker to the start of the nth line before the current line.
L−	is a synonym for L−1.
L+*	moves the marker to the start of a notional line after the last line of the text.
L*	is a synonym for L+*.
L−*	moves the marker to the start of a notional line before the first line of the text.
L+/str/	moves the marker to the position immediately before the next occurrence of the character string str.
L/str/	is a synonym for L+/str/.
L−/str/	as L+/str/ but finds the last previous occurrence of str.

P	moves the marker one character to the right along the current line.
P+n	moves the marker n characters to the right along the current line.
P+	is a synonym for P or P+1.
P+*	moves the marker to the position immediately after the last non-blank character of the current line.
P*	is a synonym for P+*.
P−n	moves the marker n characters to the left along the current line.
P−	is a synonym for P−1.
P−*	moves the marker to be immediately before the first non-blank character of the current line.
Pn	moves the marker to the column n of the current line.
Q	abandons editing without saving the results.
R+/str_1/str_2/	substitutes character string str_2 for the next occurrence of character string str_1 after the marker on the current or subsequent lines, and moves the marker to the position immediately after str_2.
R/str_1/str_2/	is a synonym for R+/str_1/str_2/.
R−/str_1/str_2/	substitutes string str_2 for the nearest occurrence of string str_1 before the marker on the current or previous lines; the marker moves to be immediately before string str_2.
S/str_1/str_2/	substitutes character string str_2 for the next occurrence of character string str_1 after the marker within the current line, and moves the marker to the position immediately after str_2.
V+c	sets verification on, causing the current line to be displayed if editing interactively, and changes the marker character, which indicates the position from which further editing will take place, to c; by default the marker character is >.
Vc	is a synonym for V+c.
V+	sets verification on and leaves the marker character unchanged.
V	is a synonym for V+.
V−	turns verification off.
(commands)n	repeats the commands within the brackets n times; the brackets can be nested up to 10 deep, however the complete sequence must all be on a single line.
(commands)*	repeats the commands within the brackets until the end of the text.

A1.3.1 Regression analysis

ADD
Adds extra terms to a linear, generalized linear, or nonlinear model.

ADD
Adds extra terms to a linear, generalized linear, or nonlinear model.

Options

PRINT	= strings	What to print (model, summary, accumulated, estimates, correlations, fittedvalues, monitoring); default m,s,e
NONLINEAR	= string	How to treat nonlinear parameters between groups (common, separate, unchanged); default u
CONSTANT	= string	How to treat constant (estimate, omit, unchanged); default u
FACTORIAL	= scalar	Limit for expansion of model terms; default 3
POOL	= string	Whether to pool ss in accumulated summary between all terms fitted in a linear model (no, yes); default n
DENOMINATOR	= string	Whether to base ratios in accumulated summary on rms from model with smallest residual ss or smallest residual ms (ss, ms); default s
NOMESSAGE	= strings	Which warning messages to suppress (dispersion, leverage, residual, aliasing, marginality); default *

Parameter

	formula	List of explanatory variates and factors, or model formula

DROP
Drops terms from a linear, generalized linear, or nonlinear model.

Options

PRINT	= strings	What to print (model, summary, accumulated, estimates, correlations, fittedvalues, monitoring); default m,s,e
NONLINEAR	= string	How to treat nonlinear parameters between groups (common, separate, unchanged); default u
CONSTANT	= string	How to treat constant (estimate,

		omit, unchanged); default u
FACTORIAL	= scalar	Limit for expansion of model terms; default 3
POOL	= string	Whether to pool ss in accumulated summary between all terms fitted in a linear model (no, yes); default n
DENOMINATOR	= string	Whether to base ratios in accumulated summary on rms from model with smallest residual ss or smallest residual ms (ss, ms); default s
NOMESSAGE	= strings	Which warning messages to suppress (dispersion, leverage, residual, aliasing, marginality); default *

Parameter

	formula	List of explanatory variates and factors, or model formula

FIT
Fits a linear or generalized linear regression model.

Options

PRINT	= strings	What to print (model, summary, accumulated, estimates, correlations, fittedvalues, monitoring); default m,s,e
CONSTANT	= string	How to treat constant (estimate, omit); default e
FACTORIAL	= scalar	Limit for expansion of model terms; default 3
POOL	= string	Whether to pool ss in accumulated summary between all terms fitted in a linear model (no, yes); default n
DENOMINATOR	= string	Whether to base ratios in accumulated summary on rms from model with smallest residual ss or smallest residual ms (ss, ms); default s
NOMESSAGE	= strings	Which warning messages to suppress

Appendix 1: Reference summary 683

		(dispersion, leverage, residual, aliasing, marginality); default *
Parameter		
	formula	List of explanatory variates and factors, or model formula

FITCURVE
Fits a standard nonlinear regression model.

Options

PRINT	= strings	What to print (model, summary, accumulated, estimates, correlations, fittedvalues, monitoring); default m,s,e
CURVE	= string	Type of curve (exponential, dexponential, cexponential, lexponential, logistic, glogistic, gompertz, ldl, qdl, qdq); default e
SENSE	= string	Sense of curve (right, left); default r
ORIGIN	= scalar	Constrained origin; default *
NONLINEAR	= string	How to treat nonlinear parameters between groups (common, separate); default c
CONSTANT	= string	How to treat constant (estimate, omit); default e
FACTORIAL	= scalar	Limit for expansion of model terms; default 3
POOL	= string	Whether to pool ss in accumulated summary between all terms fitted in a linear model (no, yes); default n
DENOMINATOR	= string	Whether to base ratios in accumulated summary on rms from model with smallest residual ss or smallest residual ms (ss, ms); default s

NOMESSAGE	= strings	Which warning messages to suppress (dispersion, leverage, residual, aliasing, marginality); default *
Parameter		
	formula	Explanatory variate, list of variate and factor, or variate*factor

---

FITNONLINEAR
Fits a nonlinear regression model or optimizes a scalar function.

Options

PRINT	= strings	What to print (model, summary, accumulated, estimates, correlations, fittedvalues, monitoring, grid); default m,s,e (default g if NGRIDLINES is set)
CALCULATION	= expression or pointer	Calculation of fitted values or of explanatory variates involving nonlinear parameters; default * (only valid if OWN set)
OWN	= scalar	Option setting for OWN directive if this is to be used rather than CALCULATE; default *
CONSTANT	= string	How to treat constant (estimate, omit); default e
FACTORIAL	= scalar	Limit for expansion of model terms; default 3
POOL	= string	Whether to pool ss in accumulated summary between all terms fitted in a linear model (no, yes); default n
DENOMINATOR	= string	Whether to base ratios in accumulated summary on rms from model with smallest residual ss or smallest residual ms (ss, ms); default s
NOMESSAGE	= strings	Which warning messages to suppress (dispersion, leverage, residual, aliasing, marginality); default *
NGRIDLINES	= scalar	Number of values of each parameter

		for a grid of function evaluations; default *
SELINEAR	= string	Whether to calculate s.e.s for linear parameters (no, yes); default n
INOWN	= identifiers	Setting to be used for the IN parameter of OWN if used in place of CALCULATE; default *
OUTOWN	= identifiers	Setting to be used for the OUT parameter of OWN if used in place of CALCULATE; default *
Parameters		
	formula	List of explanatory variates and/or one factor to be used in linear regression, within nonlinear optimization

MODEL

Defines the response variate(s) and the type of model to be fitted for linear, generalized linear, and nonlinear regression models.

Options

DISTRIBUTION	= string	Distribution of the response variable (normal, poisson, binomial, gamma, inversenormal, multinomial); default n
LINK	= string	Link function (canonical, identity, logarithm, logit, reciprocal, power, squareroot, probit, complementaryloglog); default c i.e. i for DIST=n, l for DIST=p, logit for DIST=b, r for DIST=g, p for DIST=i, * for DIST=m
EXPONENT	= scalar	Exponent for power link; default -2
DISPERSION	= scalar	Value of dispersion parameter in calculation of s.e.s etc; default * for DIST=n,g,i, 1 for DIST=p,b,m
WEIGHTS	= variate	Variate of weights for weighted regression; default *
OFFSET	= variate	Offset variate to be included in

		model; default *
GROUPS	= factor	Factor defining the groups for within-groups linear regression; default *
RMETHOD	= string	Type of residuals to form, if any, after each model is fitted (deviance, Pearson); default d
FUNCTION	= scalar	Scalar whose value is to be minimized by calculation; default *
SAVE	= identifier	To name regression save structure; default *

Parameters

Y	= variates	Response variates; only the first is used in generalized linear and nonlinear models
NBINOMIAL	= variate	Variate of binomial totals
RESIDUALS	= variates	To save residuals for each response variate after fitting a model
FITTEDVALUES	= variates	To save fitted values, and provide fitted values if no terms are given in FITNONLINEAR

PREDICT

Forms predictions from a linear or generalized linear model.

Options

PRINT	= string	What to print (description,predictions,se); default d,p
CHANNEL	= scalar	Channel number for output; default * i.e. current output channel
COMBINATIONS	= string	Which combinations of factors in the current model to include (all, present); default a
ADJUSTMENT	= string	Type of adjustment (marginal, equal); default m
WEIGHTS	= table	Weights classified by some or all standardizing factors; default *
METHOD	= string	Method of forming margin (mean,

		total); default m
ALIASING	= string	How to deal with aliased parameters (fault, ignore); default f
PREDICTIONS	= table or pointer	To save tables of predictions for each y variate; default *
SE	= table or pointer	To save tables of standard errors of predictions for each y variate; default *
VCOVARIANCE	= symmetric matrix or pointer	To save variance-covariance matrices of predictions for each y variate; default *
SAVE	= identifier	Specifies save structure of model to display; default * i.e. that of the latest model fitted

Parameters

CLASSIFY	= vectors	Variates and/or factors to classify table of predictions
LEVELS	= variates or scalars	To specify values of variates, levels of factors

---

RCYCLE

Controls iterative fitting of generalized linear and nonlinear models and specifies parameters, bounds etc for nonlinear models.

Options

MAXCYCLE	= scalar	Maximum number of iterations; default * gives 10 for generalized linear, 20 for nonlinear models
TOLERANCE	= scalar	Convergence criterion; default 0.0004
FITTEDVALUES	= variate	Initial fitted values for generalized linear model; default *
METHOD	= string	Algorithm for fitting nonlinear model (GaussNewton, NewtonRaphson); default G, but N for scalar minimization

Parameters

PARAMETER	= scalars	Nonlinear parameters in the model

LOWER	= scalars	Lower bound for each parameter
UPPER	= scalars	Upper bound for each parameter
STEPLENGTH	= scalars	Initial step length for each parameter
INITIAL	= scalars	Initial value for each parameter

RDISPLAY
Displays the fit of a linear, generalized linear, or nonlinear model.

Options

PRINT	= strings	What to print (model, summary, accumulated, estimates, correlations, fittedvalues); default m,s,e
CHANNEL	= scalar	Channel number for output; default * i.e. current output channel
DENOMINATOR	= string	Whether to base ratios in accumulated summary on rms from model with smallest residual ss or smallest residual ms (ss, ms); default s
NOMESSAGE	= strings	Which warning messages to suppress (dispersion, leverage, residual); default *
SAVE	= identifier	Specifies save structure of model to display; default * i.e. that of the latest model fitted

No parameters

RKEEP
Stores results from a linear, generalized linear, or nonlinear model.

Options

EXPAND	= string	Whether to put estimates in the order defined by the maximal model (no, yes); default n
SAVE	= identifier	Specifies save structure of model; default * i.e. that of the latest model fitted

Parameters

Y	= variates	Y variates for which results are to be saved
RESIDUALS	= variates	Standardized residual for each y variate
FITTEDVALUES	= variates	Fitted values for each y variate
LEVERAGES	= variate	Leverages of the units for each y variate
ESTIMATES	= variates	Estimates of parameters for each y variate
SE	= variates	Standard errors of the estimates
INVERSE	= symmetric matrix	Inverse matrix from a linear or generalized linear model, second derivative matrix from a nonlinear model
VCOVARIANCE	= symmetric matrix	Variance-covariance matrix of parameters
DEVIANCE	= scalars	Residual ss or deviance
DF	= scalar	Residual degrees of freedom
TERMS	= pointer or formula	Fitted terms
ITERATIVEWEIGHTS	= variate	Iterative weights from a generalized linear model
LINEARPREDICTOR	= variate	Linear predictor from a generalized linear model
EXIT	= scalar	Exit status from nonlinear model
GRADIENTS	= pointer	Derivatives of fitted values with respect to parameters in a nonlinear model
GRID	= variate	Grid of function or deviance values from a nonlinear model

STEP
Selects a term to include in or exclude from a linear or generalized linear model according to the ratio of residual mean squares.

Options

PRINT	= strings	What to print (model, summary, accumulated, estimates, correlations, fittedvalues, monitoring, changes); default m,s,e,ch
FACTORIAL	= scalar	Limit for expansion of model terms; default 3
POOL	= string	Whether to pool ss in accumulated summary between all terms fitted in a linear model (no, yes); default n
DENOMINATOR	= string	Whether to base ratios in accumulated summary on rms from model with smallest residual ss or smallest residual ms (ss, ms); default s
NOMESSAGE	= strings	Which warning messages to suppress (dispersion, leverage, residual, aliasing, marginality); default *
INRATIO	= scalar	Criterion for inclusion of terms; default 1.0
OUTRATIO	= scalar	Criterion for exclusion of terms; default 1.0

Parameter

	formula	List of explanatory variates and factors, or model formula

SWITCH
Adds terms to, or drops them from, a linear, generalized linear, or nonlinear model.

Options

PRINT	= strings	What to print (model, summary, accumulated, estimates, correlations, fittedvalues, monitoring); default m,s,e
NONLINEAR	= string	How to treat nonlinear parameters between groups (common, separate, unchanged); default u

CONSTANT	= string	How to treat constant (estimate, omit, unchanged); default u
FACTORIAL	= scalar	Limit for expansion of model terms; default 3
POOL	= string	Whether to pool ss in accumulated summary between all terms fitted in a linear model (no, yes); default n
DENOMINATOR	= string	Whether to base ratios in accumulated summary on rms from model with smallest residual ss or smallest residual ms (ss, ms); default s
NOMESSAGE	= strings	Which warning messages to suppress (dispersion, leverage, residual, aliasing, marginality); default *

Parameter

	formula	List of explanatory variates and factors, or model formula

---

TERMS
Specifies a maximal model, containing all terms to be used in subsequent linear, generalized linear, and nonlinear models.

Options

PRINT	= strings	What to print (correlations, SSPM, wmeans); default *
FACTORIAL	= scalar	Limit for expansion of model terms; default 3
FULL	= string	Whether to assign all possible parameters to factors and interactions (no, yes); default n
SSPM	= sspm	Gives sums of squares and products on which to base calculations
TOLERANCE	= scalar	Criterion for testing for linear dependence; default * gives 10*EPS or 10000*EPS depending on the computer's precision, where EPS is smallest real value r with 1+r

		greater than 1 on the computer)
Parameter		
	formula	List of explanatory variates and factors, or model formula

---

TRY
Displays results of single-term changes to a linear or generalized linear model.

Options

PRINT	= strings	What to print (model, summary, accumulated, estimates, correlations, fittedvalues, monitoring); default m,s,e
FACTORIAL	= scalar	Limit for expansion of model terms; default 3
POOL	= string	Whether to pool ss in accumulated summary between all terms fitted in a linear model (no, yes); default n
DENOMINATOR	= string	Whether to base ratios in accumulated summary on rms from model with smallest residual ss or smallest residual ms (ss, ms); default s
NOMESSAGE	= strings	Which warning messages to suppress (dispersion, leverage, residual, aliasing, marginality); default *

Parameter

	formula	List of explanatory variates and factors, or model formula

A1.3.2 Analysis of balanced experiments

ADISPLAY
Displays further output from analyses produced by ANOVA.

Options

PRINT	= strings	Output from the analyses of the y-variates, adjusted for any covariates (aovtable, information, covariates, effects, residuals, contrasts, means, %cv, missingvalues); default * i.e. no printing
UPRINT	= strings	Output from the unadjusted analyses of the y-variates (aovtable, information, effects, residuals, contrasts, means, %cv, missingvalues); default * i.e. no printing
CPRINT	= strings	Output from the analyses of the covariates, if any (aovtable, information, effects, residuals, contrasts, means, %cv, missingvalues); default * i.e. no printing
CHANNEL	= scalar	Channel number for output; default * i.e. current output channel
PFACTORIAL	= scalar	Limit on number of factors in printed tables of means or effects; default 9
PCONTRASTS	= scalar	Limit on order of printed contrasts; default 9
PDEVIATIONS	= scalar	Limit on number of factors in a treatment term whose deviations from the fitted contrasts are to be printed; default 9
FPROBABILITY	= string	Printing of probabilities for variance ratios in the aov table (no, yes); default n
SE	= string	Standard errors to be printed with tables of means, SE = * requests s.e.'s to be omitted (differences, means); default d
TWOLEVEL	= string	Representation of effects in 2**N experiments (responses, Yates, effects); default r

Parameters

RESIDUALS	= variates	Variate to save residuals from each analysis
FITTEDVALUES	= variates	Variate to save fitted values
SAVE	= identifiers	Save structure (from ANOVA) storing details of each analysis from which information is to be displayed; if omitted, output is from the most recent ANOVA

AKEEP
Copies information from an ANOVA analysis into Genstat data structures.

Options

FACTORIAL	= scalar	Limit on number of factors in a model term; default 3
STRATUM	= formula	Model term of the lowest stratum to be searched for effects; default * implies the lowest stratum
SUPPRESSHIGHER	= string	Whether to suppress the searching of higher strata if a term is not found in STRATUM (no, yes); default n
SAVE	= identifier	SAVE structure (from ANOVA) storing details of the analysis; default * gives that from the most recent ANOVA

Parameters

TERMS	= formula	Model terms for which information is required
MEANS	= tables	Table to store means for each term (treatment terms only)
EFFECTS	= tables	Table to store effects (treatment terms only)
PARTIALEFFECTS	= tables	Table to store partial effects (treatment terms only)
REPLICATIONS	= tables	Table to store replications

RESIDUALS	= tables	Table to store residuals (block terms only)
DF	= scalars	Number of degrees of freedom for each term
SS	= scalars	Sum of squares for each term
EFFICIENCY	= scalars	Efficiency factor for each term
VARIANCE	= scalars	Unit variance for the effects of each term
CREGRESSION	= variates	Estimated regression coefficients for the covariates in the specified stratum
CSSP	= symmetric matrices	Covariate sums of squares and products in the specified stratum

ANOVA

Analyses y-variates by analysis of variance according to the model defined by earlier BLOCKSTRUCTURE, COVARIATE, and TREATMENTSTRUCTURE statements.

Options

PRINT	= strings	Output from the analyses of the y-variates, adjusted for any covariates (aovtable, information, covariates, effects, residuals, contrasts, means, %cv, missingvalues); default a,i,c,m,mi
UPRINT	= strings	Output from the unadjusted analyses of the y-variates (aovtable, information, effects, residuals, contrasts, means, %cv, missingvalues); default * i.e. no printing
CPRINT	= strings	Output from the analyses of the covariates, if any (aovtable, information, effects, residuals, contrasts, means, %cv, missingvalues); default * i.e. no printing
FACTORIAL	= scalar	Limit on number of factors in a treatment term; default 3
CONTRASTS	= scalar	Limit on the order of a contrast of a treatment term; default 4
DEVIATIONS	= scalar	Limit on the number of factors in a

		treatment term for the deviations from its fitted contrasts to be retained in the model; default 9
PFACTORIAL	= scalar	Limit on number of factors in printed tables of means or effects; default 9
PCONTRASTS	= scalar	Limit on order of printed contrasts; default 9
PDEVIATIONS	= scalar	Limit on number of factors in a treatment term whose deviations from the fitted contrasts are to be printed; default 9
FPROBABILITY	= string	Printing of probabilities for variance ratios (no, yes); default n
SE	= string	Standard errors to be printed with tables of means, SE = * requests s.e.'s to be omitted (differences, means); default d
TWOLEVEL	= string	Representation of effects in 2**N experiments (responses, Yates, effects); default r
DESIGN	= pointer	Stores details of the design for use in subsequent analyses; default *
WEIGHT	= variate	Weights for each unit; default * i.e. all units with weight one
ORTHOGONAL	= string	Whether or not design to be assumed orthogonal (no, yes, compulsory); default n
SEED	= scalar	Seed for random numbers to generate dummy variate for determining the design; default 12345
TOLERANCES	= variate	Tolerances for zero in various contexts; default * i.e. appropriate zero values assumed for the computer concerned
MAXCYCLE	= scalar	maximum number of iterations for estimating missing values; default 20

Parameters

Y	= variates	Variates to be analysed
RESIDUALS	= variates	Variate to save residuals for each y

Appendix 1: Reference summary

		variate
FITTEDVALUES	= variates	Variate to save fitted values
SAVE	= identifiers	Save details of each analysis for use in subsequent ADISPLAY or AKEEP statements

BLOCKSTRUCTURE
Defines the blocking structure of the design and hence the strata and the error terms.

No options

Parameter

	formula	Block model (defines the strata or error terms for subsequent ANOVA statements)

COVARIATE
Specifies covariates for use in subsequent ANOVA statements.

No options

Parameter

	variates	Covariates

TREATMENTSTRUCTURE
Specifies the treatment terms to be fitted by subsequent ANOVA statements.

No options

Parameter

	formula	Treatment formula, specifies the treatment model terms to be fitted by subsequent ANOVAs

A1.3.3 Functions for use in treatment formulae

Function name	Description
POL(f; s; v)	indicates that the effects of factor f are to be partitioned into polynomial components (linear, quadratic etc) up to order s, where s is a scalar containing an integer between 1 and 4; variate v defines a numerical value for each level

Appendix 1: Reference summary

	of the factor, if omitted, the factor levels themselves are used.
POLND(f; s; v)	has the same effect as POL, except that no Dev components are fitted for factor f in interactions.
REG(f; s; m)	indicates that the effects of factor f are to be partitioned into the contrasts specified by the first s rows of the matrix m, where s is a scalar containing an integer between 1 and 7.
REGND(f; s; v)	has the same effect as REG, except that no Dev components are fitted for factor f in interactions.

A1.3.4 Multivariate and cluster analysis

ADDPOINTS
Adds points for new objects to a principal coordinates analysis.

Option

PRINT	= strings	Printed output required (coordinates, residuals); default * i.e. no output

Parameters

NEWDISTANCES	= matrices	Squared distances of the new objects from the original points
LRV	= LRVs	Latent roots and vectors from the PCO analysis
CENTROID	= diagonal matrices	Centroid distances from the PCO analysis matrices
COORDINATES	= matrices	Saves the coordinates of the additional points in the space of the original points
RESIDUALS	= matrices	Saves the residuals of the new objects from that space

---

CLUSTER
Forms a non-hierarchical classification.

Options

PRINT	= strings	Printed output required (criterion,

DATA	= matrix or pointer		optimum, units, typical, initial); default * i.e. no output Data from which the classification is formed, supplied as a units-by-variates matrix or as a pointer containing the variates of the data matrix
CRITERION	= string		Criterion for clustering (sums, predictive, within, Mahalanobis); default s
INTERCHANGE	= string		Permitted moves between groups (transfer, swop); default t (implies s also)
START	= factor		Initial classification; default * i.e. splits the units, in order, into NGROUPS classes of nearly equal size

Parameters

NGROUPS	= scalars	Numbers of classes into which the units are to be classified: note, the values of the scalars must be in descending order
GROUPS	= factors	Saves the classification formed for each number of classes

CVA
Performs canonical variates analysis.

Options

PRINT	= strings	Printed output required (roots, loadings, means, residuals, distances, tests); default * i.e. no output
NROOTS	= scalar	Number of latent roots for printed output; default * requests them all to be printed
SMALLEST	= string	Whether to print the smallest roots instead of the largest (no, yes); default n

Parameters

WSSPM	= SSPMs	Within-group sums of squares and products, means etc (input for the analyses)
LRV	= LRVs	Loadings, roots, and trace from each analysis
SCORES	= matrices	Canonical variate means
RESIDUALS	= matrices	Distances of the means from the dimensions fitted in each analysis
DISTANCES	= symmetric matrices	Inter-group-mean Mahalanobis distances

FACROTATE

Rotates factors from a principal components or canonical variates analysis according to either the varimax or quartimax criterion.

Options

PRINT	= strings	Printed output required (communalities, rotation); default * i.e. no output
METHOD	= string	Criterion (varimax, quartimax); default v

Parameters

OLDLOADINGS	= matrices	Original loadings
NEWLOADINGS	= matrices	Rotated loadings for each set of OLDLOADINGS

FSIMILARITY

Forms a similarity matrix or a between-group-elements similarity matrix or prints a similarity matrix.

Options

PRINT	= string	Printed output required (similarity); default * i.e. no output
STYLE	= string	Print percentage similarities in full or just the 10% digit (full, abbreviated); default f
METHOD	= string	Form similarity matrix or

SIMILARITY	= matrix	rectangular between-group-element similarity matrix (similarity, betweengroupsimilarity); default s Input or output matrix of similarities; default *
GROUPS	= factor	Grouping of units into two groups for between-group-element similarity matrix; default *
PERMUTATION	= variate	Permutation of units (possibly from HCLUSTER) for order in which units of the similarity matrix are printed; default *
UNITS	= text or variate	Unit names to label the rows of the similarity matrix; default *

Parameters

DATA	= variates	The data values
TEST	= scalars	Test type, defining how each variate is treated in the calculation of the similarity between each unit
RANGE	= scalars	Range of possible values of each variate; if omitted, the observed range is taken

HCLUSTER
Performs hierarchical cluster analysis.

Options

PRINT	= strings	Printed output required (dendrogram, amalgamations); default * i.e. no output
METHOD	= string	Criterion for forming clusters (singlelink, nearestneighbour, completelink, furthestneighbour, averagelink, mediansort, groupaverage); default s
CTHRESHOLD	= scalar	Clustering threshold at which to print formation of clusters; default * i.e. determined automatically

Parameters

SIMILARITY	= symmetric matrices	Input similarity matrix for each cluster analysis
GTHRESHOLD	= scalars	Grouping threshold where groups are formed from the dendrogram
GROUPS	= factors	Stores the groups formed
PERMUTATION	= variates	Permutation order of the units on the dendrogram
AMALGAMATIONS	= matrices	To store linked list of amalgamations

HDISPLAY

Displays results ancillary to hierarchical cluster analyses: matrix of mean similarities between and within groups, a set of nearest neighbours for each unit, a minimum spanning tree, and the most typical elements from each group.

Options

PRINT	= strings	Printed output required (neighbours, tree, typicalelements, gsimilarities); default t

Parameters

SIMILARITY	= symmetric matrices	Input similarity matrix for each cluster analysis
NNEIGHBOURS	= scalars	Number of nearest neighbours to be printed
NEIGHBOURS	= matrices	Matrix to store nearest neighbours of each unit
GROUPS	= factors	Indicates the groupings of the units (for calculating typical elements and mean similarities between groups)
TREE	= matrices	To store the minimum spanning tree (as a series of links and corresponding lengths)
GSIMILARITY	= symmetric matrices	To store similarities between groups

HLIST

Lists the data matrix in abbreviated form.

Options

GROUPS	= factor	Defines groupings of the units; used to split the printed table at appropriate places and to label the groups; default *
UNITS	= text or variate	Names for the rows (i.e. units) of the table; default *

Parameters

DATA	= variates	The data values
TEST	= scalars	Test type, defining how each variate is treated for calculation of the similarity between each unit
RANGE	= scalars	Range of possible values of each variate; if omitted, the observed range is taken

HSUMMARIZE

Forms and prints a group by levels table for each test together with appropriate summary statistics for each group.

Option

GROUPS	= factor	Factor defining the groups; no default i.e. this option must be specified

Parameters

DATA	= variates	The data values
TEST	= scalars	Test type, defining how each variate is treated for calculation of the similarity between each unit
RANGE	= scalars	Range of possible values of each variate; if omitted, the observed range is taken

PCO

Performs principal coordinates analysis, also principal components and canonical variates analysis (but with different weighting from that used in CVA) as special cases.

Options

PRINT	= strings	Printed output required (roots, scores, loadings, residuals, centroid, distances); default * i.e. no output
NROOTS	= scalar	Number of latent roots for printed output; default * requests them all to be printed
SMALLEST	= string	Whether to print the smallest roots instead of the largest (no, yes); default n

Parameters

DATA	= identifiers	These can be specified either as a symmetric matrix of similarities or transformed distances or, for the canonical variate analysis, as an SSPM containing within-group sums of squares and products etc. or, for principal components analysis, as a pointer containing the variates of the data matrix
LRV	= LRVs	Latent vectors (i.e. coordinates or scores), roots and trace from each analysis
CENTROID	= diagonal matrices	Squared distances of the units from their centroid
RESIDUALS	= matrices	Distances of the units from the fitted space
LOADINGS	= matrices	Principal component loadings, or canonical variate loadings
DISTANCES	= symmetric matrices	Computed inter-unit distances calculated from the variates of a data matrix, or inter-group Mahalanobis distances calculated from a within-group SSPM

PCP

Performs principal components analysis.

Options

PRINT	= strings	Printed output required (loadings, roots, residuals, scores, tests); default * i.e. no output
NROOTS	= scalar	Number of latent roots for printed output; default * requests them all to be printed
SMALLEST	= string	Whether to print the smallest roots instead of the largest (no, yes); default n
METHOD	= string	Whether to use sums of squares or correlations (ssp, correlation); default s

Parameters

DATA	= pointers or SSPMs	Pointer of variates forming the data matrix or SSPM giving their sums of squares and products (or correlations) etc
LRV	= LRVs	To store the principal component loadings, roots and trace from each analysis
SSPM	= SSPMs	To store the computed sum-of-squares-and-products or correlation matrix
SCORES	= matrices	To store the principal component scores
RESIDUALS	= matrices	To store residuals from the dimensions fitted in the analysis (i.e. number of columns of the SCORES matrix, or as defined by the NROOTS option)

REDUCE

Forms a reduced similarity matrix (referring to the GROUPS instead of the original units).

Option

| PRINT | = string | Printed output required (similarities); default * i.e. no output |
| METHOD | = string | Method used to form the reduced similarity matrix (first, last, mean, minimum, maximum, zigzag); default f |

Parameters

SIMILARITY	= symmetric matrices	Input similarity matrix
REDUCEDSIMILARITY	= symmetric matrices	Output (reduced) similarity matrix
GROUPS	= factors	Factor defining the groups
PERMUTATION	= variates	Permutation order of units (for METHOD = f, l, or z)

RELATE

Relates the observed values on a set of variates to the results of a principal coordinates analysis.

Options

| COORDINATES | = matrices | Points in reduced space; no default i.e. this option must be specified |
| NROOTS | = scalar | Number of latent roots for printed output; default * requests them all to be printed |

Parameters

DATA	= variates	The data values
TEST	= scalars	Test type, defining how each variate is treated for calculation of the similarity
RANGE	= scalars	Range of possible values of each variate; if omitted, the observed range is taken

Appendix 1: Reference summary 707

ROTATE
Does a Procrustes rotation of one configuration of points to fit another.

Options

PRINT	= strings	Printed output required (rotations, coordinates, residuals, sums); default * i.e. no output
SCALING	= string	Whether or not isotropic scaling is allowed (no, yes); default n
STANDARDIZE	= strings	Whether to centre the configurations (at the origin), and/or to normalize them (to unit sum of squares) prior to rotation (centre, normalize); default c,n
SUPPRESS	= string	Whether to suppress reflection (no, yes); default n

Parameters

XINPUT	= matrices	Inputs the fixed configuration
YINPUT	= matrices	Inputs the configuration to be fitted
XOUTPUT	= matrices	To store the (standardized) fixed configuration
YOUTPUT	= matrices	To store the fitted configuration
ROTATION	= matrices	To store the rotation matrix
RESIDUALS	= matrices	To store distances between the (standardized) fixed and fitted configurations
RSS	= scalars	To store the residual sum of squares

A1.3.5 Time series

CORRELATE
Forms correlations between variates, autocorrelations of variates, and lagged cross-correlations between variates.

Options

PRINT	= strings	What to print (correlations, autocorrelations, partialcorrelations, crosscorrelations); default *

GRAPH	= strings	What to display with graphs (autocorrelations, partialcorrelations, crosscorrelations); default *
MAXLAG	= scalar	Maximum lag for results; default * i.e. value inferred from variates to save results

Parameters

SERIES	= variates	Variates from which to form correlations
LAGGEDSERIES	= variates	Series to be lagged to form cross-correlations with first series
AUTOCORRELATIONS	= variates	To save autocorrelations, or to provide them to form partial autocorrelations if SERIES = *
PARTIALCORRELATIONS	= variates	To save partial autocorrelations
CROSSCORRELATIONS	= variates	To save cross-correlations
TEST	= scalars	To save test statistics
VARIANCES	= variates	To save prediction error variances
COEFFICIENTS	= variates	To save prediction coefficients

ESTIMATE
Estimates parameters in Box-Jenkins models for time series.

Options

PRINT	= strings	What to print (model, summary, estimates, correlations, monitoring); default m,s,e
LIKELIHOOD	= string	Method of likelihood calculation (exact, leastsquares, marginal); default e
CONSTANT	= string	How to treat constant (estimate, fix); default e
RECYCLE	= string	Whether to continue from previous estimation (no, yes); default n
WEIGHTS	= variate	Weights; default *
MVREPLACE	= string	Whether to replace missing values by their estimates (no, yes); default n
FIX	= variate	Defines constraints on parameters

Appendix 1: Reference summary 709

		(ordered as in each model, tf models first): zeros fix parameters, parameters with equal numbers are constrained to be equal; default *
MAXCYCLE	= scalar	Maximum number of iterations; default 15
TOLERANCE	= scalar	Criterion for convergence; default 0.0004
SAVE	= identifier	To name save structure, or supply save structure with transfer-functions; default * i.e. transfer-functions taken from latest model

Parameters

SERIES	= variate	Time series to be modelled (output series)
TSM	= TSM	Model for output series
BOXCOXMETHOD	= string	How to treat transformation parameter in output series (fix, estimate); if omitted, transformation parameter is fixed
RESIDUALS	= variate	To save residual series

FILTER
Filters time series by time-series models.

Option

PRINT	= strings	What to print (series); default *

Parameters

OLDSERIES	= variates	Time series to be filtered
NEWSERIES	= variates	To save filtered series
FILTER	= TSMs	Models to filter with respect to
ARIMA	= TSMs	ARIMA models for time series

FORECAST
Forecasts future values of a time series.

Options

PRINT	= strings	What to print (forecasts, sfe, limits); default f,l
CHANNEL	= scalar	Channel number for output; default * i.e. current output channel
ORIGIN	= scalar	Number of known values in FORECAST variate; default 0
MAXLEAD	= scalar	Maximum lead time beyond the origin, i.e. number of forecasts; default * i.e. value taken from length of FORECAST variate
PROBABILITY	= scalar	Probability level for confidence limits; default 0.9
UPDATE	= string	Whether existing forecasts are to be updated (no, yes); default n
FORECAST	= variate	To save forecasts of output series; default *
SETRANSFORM	= variate	To save standard errors of forecasts (on transformed scale, if defined); default *
LOWER	= variate	To save lower confidence limits; default *
UPPER	= variate	To save upper confidence limits; default *
SFE	= variate	To save standardized forecast errors; default *
COMPONENTS	= pointer	Variates to save components of forecast
SAVE	= identifier	Save structure to supply fitted model; default * i.e. that from last model fitted

Parameters

FUTURE	= variates	Future values of input series
METHOD	= strings	How to treat future values of input series (observations, forecasts); if omitted, they are treated as observations

Appendix 1: Reference summary

FOURIER
Calculates cosine or Fourier transforms of real or complex series.

Option

PRINT	= strings	What to print (transforms); default *

Parameters

SERIES	= variates	Real part of each input series
ISERIES	= variates	Imaginary part of each input series
TRANSFORM	= variates	To save real part of each output series
ITRANSFORM	= variates	To save imaginary part of each output series
PERIODOGRAM	= variates	To save periodogram of each transform

FTSM
Forms preliminary estimates of parameters in time-series models.

Option

PRINT	= strings	What to print (models); default *

Parameters

TSM	= TSMs	Models whose parameters are to be estimated
CORRELATIONS	= variates	Auto- or cross-correlations on which to base estimates for each model
BOXCOXTRANSFORM	= scalars	Box-Cox transformation parameter
CONSTANT	= scalars	Constant term
VARIANCE	= scalars	Variance of ARIMA model, or ratio of input variance to output variance for transfer model

TDISPLAY
Allows further display after an analysis by ESTIMATE.

Options

PRINT	= strings	What to print (model, summary, estimates, correlations); default m,s,e
CHANNEL	= scalar	Channel number for output; default * i.e. current output channel
SAVE	= identifier	Save structure to supply fitted model; default * i.e. that from last model fitted

No parameters

TKEEP
Saves results after an analysis by ESTIMATE.

Option

SAVE	= identifier	Save structure to supply fitted model; default * i.e. that from last model fitted

Parameters

OUTPUTSERIES	= variate	Output series to which model was fitted
RESIDUALS	= variate	Residual series
ESTIMATES	= variate	Estimates of parameters
SE	= variate	Standard errors of estimates
INVERSE	= symmetric matrix	Inverse matrix
VCOVARIANCE	= symmetric matrix	Variance-covariance matrix of parameters
DEVIANCE	= scalar	Residual deviance
DF	= scalar	Residual degrees of freedom
MVESTIMATES	= variate	Estimates of missing values in series
SEMV	= variate	S.e.s of estimates of missing values
COMPONENTS	= pointer	Variates to save components of output series

TRANSFERFUNCTION
Specifies input series and transfer function models for subsequent estimation of a model for an output series.

Option

SAVE	= identifier	To name time-series save structure; default *

Parameters

SERIES	= variates	Input time series
TRANSFERFUNCTION	= TSMs	Transfer-function models; if omitted, model with 1 moving-average parameter, lag 0
BOXCOXMETHOD	= strings	How to treat transformation parameters (fix, estimate); if omitted, they are treated as fixed
PRIORMETHOD	= strings	How to treat prior values (fix, estimate); if omitted, they are treated as fixed
ARIMA	= TSMs	ARIMA models for input series

TSUMMARIZE
Displays characteristics of time series models.

Options

PRINT	= strings	What to print (autocorrelations, expansion, impulse, piweights, psiweights); default *
GRAPH	= strings	What to display with graphs (autocorrelations, impulse, piweights, psiweights); default *
MAXLAG	= scalar	Maximum lag for results; default 0

Parameters

TSM	= TSMs	Models to be displayed
AUTOCORRELATIONS	= variates	To save theoretical autocorrelations
IMPULSERESPONSE	= variates	To save impulse-response function
STEPFUNCTION	= variates	To save step function from impulse
PIWEIGHTS	= variates	To save pi-weights

PSIWEIGHTS	= variates	To save psi-weights
EXPANSION	= TSMs	To save expanded models
VARIANCE	= scalars	To save variance of each TSM

Appendix 2 Diagnostics

Below we list the error and warning diagnostics that can be given by Genstat. Usually there will also be a few lines of explanation to help you to correct the fault (see 2.1.2). The diagnostics are categorized by a two-letter code, according to the area of Genstat involved; within that category they are identified by a two-digit number. You can access the code for the last fault that occurred in your program using option FAULT of the GET directive. This returns an integer value, calculated as the two-digit number for the diagnostic plus a value identifying the category. The category values, for example zero for AN, are given below in brackets.

AN Analysis of Variance (0)

1. Design unbalanced – cannot be analysed by ANOVA
2. Not an equally replicated orthogonal design
3. Too many mutually non-orthogonal model terms for ANOVA
4. Too many different factors in formulae
5. Invalid function or operator in BLOCK, TREATMENT, or AKEEP model formula
6. Pseudo factors allowed only in TREATMENT formulae
7. Contrast functions allowed only in TREATMENT formulae
8. Incorrect number of arguments in function
9. Incorrect dimensions in 3rd argument of function
10. Illogical compound contrast
11. Unbalanced contrasts
12. Missing values not allowed in factors in designed experiments
13. Missing or negative weight
14. Insufficient units are non-missing and have positive weights
15. DESIGN control structure is out of date
16. Partial aliasing
17. Partial confounding
18. Negative residual degrees of freedom
19. Numerical failure in ANOVA
20. AKEEP term (from formula/STRAT option) was not in BLOCK/TREAT formulae
21. STRATUM option of AKEEP must specify exactly one term
22. Term in STRATUM option of AKEEP was not in the BLOCK model formula
23. Covariate coefficients are saved only for BLOCK terms
24. Covariate coefficients cannot be saved (none in the original ANOVA)
25. Residuals can be saved only for BLOCK terms

26 Means can be saved only for TREATMENT terms
27 Means cannot be saved unless STRAT option is set to the bottom stratum
28 Effects can be saved only for TREATMENT terms
29 Partial effects can be saved only for TREATMENT terms
30 No save structure available for ADISP or AKEEP
31 Matrix inversion failure in analysis of covariance
32 BLOCK formula for RANDOMIZE must contain only operators * or /
33 For RANDOMIZE each block-factor combination must have replication one
34 EXCLUDEd factor is not part of the BLOCK formula
35 Operators not allowed in arguments of contrast functions

BS Structure storage and retrieval (40)

1 Repeated identifier or attempt to reorder suffices within pointer
2 Sub-file not present
3 Illegal renaming, first parameter has a value set missing
4 File or sub-file identifier omitted
5 Attempt to give a pointer a new suffix
6 Sub-file wrong type
7 File created by an earlier version of Genstat
8 Sub-file or structure not present
9 Attempt to write to a protected file (INCORRECT PASSWORD)
10 I/O error or mark 4 file
11 Clash of structure types
12 Attempt to give a procedure an illegal name
13 Attempt to give a procedure a directive name
14 Subfile name already in target file (could be system reserved name)
15 Output channel is also an input channel
16 Attempt to add 2 subfiles with the same name to target file
17 Userfile not found
18 Wrong kind of file attached
19 Length of units vector has been changed

CA Calculations (60)

1 List of r.h.s. operands longer than list to hold results
2 Too many or too few arguments in a function
3 Use of function which cannot accept missing values
4 Wrong factor in classifying set
5 Sub-matrix labels not subset of matrix labels
6 Invalid exponentiation
7 Invalid value for argument of function

8 Matrix not positive semi-definite
9 Values for interpolation not monotonically increasing
10 Address from restrict vector out of range
11 Tables have incompatible margins
12 Too many operands in arithmetic expression
13 Attempt to use operation only valid on a square matrix
14 Order of matrix non-positive
15 RESTRICT has generated a null list
16 Matrix singular
17 Invalid operation
18 Attempt to divide by zero
19 Structures of incompatible types occuring together
20 URAND not previously called with non-zero seed
21 Too many operands in GENERATE
22 Scalar values in GENERATE parameter list must be positive
23 GENERATED pseudo factors are not balanced
24 Number of units of GENERATED factors does not give an exact number of reps
25 OLDSTRUCTURE or NEWSTRUCTURE not specified for COMBINE
26 Attempt to declare a factor by default
27 ELEMENTED matrix dimension too small or too large
28 ELEMENTING address out of range for structure <1 or $>$ nval
29 Attempt to COMBINE a dimension more than once
30 OLDDIMENSION factor is not in the classification set of the OLDSTRUCTURE
31 NEWDIMENSIONS must be factors unless NEWSTRUCTURE table already declared
32 OLD/NEWDIMENSION number out of range
33 Text given for OLD/NEWPOSITIONS but the dimension is not labelled
34 OLD/NEWPOSITION not found
35 OLD/NEWPOSITION out of range
36 Pointer given for OLD/NEWPOSITIONS but dimension not indexed by a pointer
37 Both values missing in MVREPLACE
38 Number of values of result from URAND/EXPAND is undefined
39 Too many dimensions for qualified identifier
40 For METHOD=interval, NEWVALUES variate must have values
41 For METHOD=value, NEWINTERVALS variate must have values
42 Too few values for cubic interpolation
43 Repeated factor in classifying set of table
44 Columns of structure pair are unequal
45 Rows of structure pair are unequal
46 Number of columns of matrix1 do not match number of rows of matrix2
47 Address list cannot be formed
48 Function given structure of wrong type
49 Elemented/qualified identifiers not allowed in function

50 Error in EDIT
51 Elemented structure must have its number of values declared
52 Embedded assignment of factors not allowed

CL Cluster Analysis (120)

 1 Data structure of incompatible dimensions
 2 GROUP parameter missing or not set for REDUCE directive
 3 Range check fails in forming similarity matrix
 4 Too few units for sorting
 5 Negative or missing values in similarity matrix
 6 Not all units are on the minimum spanning tree
 7 Invalid values in permutation vector < 1 or > nval
 8 Page width on output channel too narrow to print similarity matrix
 9 GTHRESHOLD parameter ignored as GROUPS not set
10 GROUPS parameter ignored as GTHRESHOLD not set
11 For METHOD = gsimilarities GROUPS parameter must be set
12 NROOTS option of RELATE is greater than 10. Maximum of 10 allowed
13 Number of columns of neighbours matrix is not a multiple of 2
14 Page width too narrow for abbreviated form of data matrix
15 Neighbours matrix cannot be defined

FI File handling (140)

1 Attempt to read past end of file
2 (Fortran) failure while reading file
3 System cannot find input file on current channel
4 System cannot close file
5 Record has over flowed input buffer (limit is 160 characters)
6 Error when accessing unformatted file (writing to readonly file?)
7 Unable to get reply from terminal
8 Unable to initiate a sub-process for SUSPEND or PASS

HG High quality Graphics (150)

1 Unset window
2 Invalid or missing window number
3 Invalid window size (eg lower $\geq$ upper)
4 Invalid key window number
5 Invalid key window size (eg lower $\geq$ upper)
6 Unset pen number
7 Illegal pen number

8 Illegal colour number
9 Illegal line style
10 Illegal method
11 Illegal symbol number
12 Illegal join setting
13 Illegal integer setting
14 Illegal style setting
15 Nothing to contour
16 There must be at least three rows and columns
17 Data contains missing values
18 Not a two way table
19 Number of levels < 1
20 Key window full
21 Illegal brush number
22 Slice must be a single variate
23 Truncation of text in key window
24
25 Internal graphics error
26 Negative value for pie chart
27 Less than 2 or greater than 180 elements
28 Internal I/O error
29 Error in NAG Graphical Supplement
30 Window outside of device limits
31 Missing parameter
32 Invalid device number
33 Missing device number
34 No graphics devices
35 File name longer than 128 chars

IO Input and output (190)

1 Data element terminates before end of fixed-format
2 Integer part of a number too large
3 Real number too large
4 Invalid character in element in fixed-format read
5 Blank field in data
6 Illegal leading blanks found in data
7 End of data, missing and separator characters must be distinct
8 Too many errors in data
9 Number of values not a multiple of number of structures
10 Channel number out of range
11 Errors in data values

12 Unexpected characters at end of data (end string not present?)
13 Attempt to skip over end string in input data
14 No format information supplied for fixed-format read
15 Incompatible structures in parallel read
16 Required data separator not found in free-format read
17 Unexpected unit structure value
18 Unexpected factor value
19 Null filename
20 Channel already open. Channel number or filetype incorrect?
21 Channel not open, request to close ignored
22 Fortran error when opening or closing a file
23 Current I/O channel cannot be closed
24 Closing terminal I/O is deemed irreverent and can only lead to excommunication!
25 Channel for input or output has not been opened, or has been terminated
26 Data cannot be read sequentially from current channel or text
27 End character must be present if length is to be set from data
28 Only variates, factors and texts can be read with SETN=Y
29 The size of a pointer must be declared before its values can be read
30 To add to existing values, data must be in serial order
31 A format must indicate at least one field to be read
32 Values punched as labels, but no LABELS vector found
33 Illegal trailing blanks found in data
34 Unquoted integer read as value of text

LP Lineprinter graphics (230)

1 All values missing
2 Invalid setting of scale option is ignored
3 Limits variate contains some missing values
4 Table has margins
5 Table has more than one classifying factor
6 Table has no classifying factor
7 X-axis title or Y-axis title has more than one line, – first line used
8 Only two classifying factors are allowed for tables being contoured
9 Page width too narrow for contour
10 Invalid setting of multiple option
11 Number of rows is less than four
12 Number of columns is less than four
13 Missing values not allowed in structure
14 Repetition in limits variate

MV Multivariate Analysis (250)

1 Input structure missing
2 Input structure of wrong type
3 Input structure has no values
4 Input structure has some missing values
5 Input structure has inconsistent number of values
6 Input factor has inconsistent number of levels
7 Output structure of wrong type
8 Output structure has inconsistent size
9 Output structure has inconsistent number of rows
10 Output structure has inconsistent number of columns
11 Output factor has inconsistent number of levels
12
13 More latent roots requested than can be printed
14 Latent root computation needed more than 30 iterations
15
16 Numbers of groups requested not in descending order
17 Values of variates must be binary for maximal predictive cluster
18 Determinant is zero

OP Optimization (270)

1 Optimization process has not converged
2 Unsuccessful optimization: a parameter has gone out of bounds
3 Unsuccessful optimization: the function appears to be constant
4 Unsuccessful optimization: no progress can be made
5 Standard errors are not available, because information matrix is singular
6 No parameters have been specified (use directive RCYCLE)
7 No model calculations have been specified (use option CALCULATION)
8 There are too many parameters for general function optimization
9 No scalar was specified for value of general function (use MODEL [FUNC])
10 Fitted values or general function value not changed by model calculations
11 There are not enough non-missing data to fit the model
12 Invalid model formula for FITNONLINEAR (only variates and 1 factor allowed)
13 Invalid model formula for FITCURVE (only 1 variate and 1 factor allowed)
14 Invalid error distribution
15 Invalid attempt to reverse sense of curve
16 Invalid attempt to constrain origin: constraint is ignored
17 Invalid attempt to omit constant: constant is retained
18 The constant must be omitted with Poisson likelihood
19 The asymptote has been reversed

20 Fitted curve is close to, or has reached limiting form
21 The fitted curve has a vertical asymptote within the data range
22 A parameter has been included twice

PC Program Control (310)

1 Expression missing in CASE, IF, or EXIT statement
2 Invalid setting of NTIMES option
3 Attempt to reset active dummy
4 Invalid control structure: unmatched ENDFOR statement
5 Invalid control structure: unmatched ELSIF statement
6 Invalid control structure: ELSIF follows ELSE statement
7 Invalid control structure: unmatched ELSE statement
8 Invalid control structure: multiple ELSE statements
9 Invalid control structure: unmatched ENDIF statement
10 Invalid control structure: unmatched OR statement
11 Invalid control structure: OR follows ELSE statement
12 Invalid control structure: unmatched ENDCASE statement
13 Attempt to define procedure within a control structure
14 OPTIONS statement out of sequence
15 PARAMETERS statement out of sequence
16 ENDPROCEDURE statement out of sequence
17 Invalid control structure: open FOR loop
18 Invalid control structure: open IF block
19 Invalid control structure: open CASE block
20 Invalid control structure: PROCEDURE definition incomplete
21 Invalid control structure: PROCEDURE execution incomplete
22 Invalid control structure: missing ENDBREAK statement
23 Repeated option or parameter names
24 Procedure fault
25 Attempt to redefine directive name
26 Invalid procedure name
27 FOR loops not allowed within a BREAK sequence
28 ENDBREAK statement out of sequence – ignored
29 ENDDEBUG statement out of sequence – ignored

RE Regression (340)

1 The number of degrees of freedom has not been specified
2 Model formula contains an operator not valid in regression models
3 Model formula contains too many variates and factors
4 Interactions between variates are not valid in regression model formulae

5 The SAVE option of FSSPM has changed during sequential accumulation
6 There are no units with values present for all vectors and positive weight
7 Numbers of binomial trials have not been supplied
8 The distribution is invalid for the type of model chosen
9 Model cannot be analysed within groups because it needs iterative fitting
10 Regression models cannot be fitted until a MODEL statement is given
11 No response variate has been specified in the MODEL statement
12 The SSPM does not include information on a current response variate
13 The parameter list was ignored because an SSPM was supplied
14 An option of TERMS was ignored because an SSPM was supplied
15 An option of MODEL was overridden because an SSPM was supplied
16 There is not enough space to form all sums of squares and products
17 A response variate cannot be used as an explanatory variate as well
18 Regression with an offset variate cannot be based on an SSPM
19 Iteratively fitted model cannot be based on a sequentially formed SSPM
20 The PRINT option was ignored because a saved SSP was supplied
21 Single precision SSPM values may not be sufficient for accurate regression
22 Statements to modify regression model must be preceded by a TERMS statement
23 A term is not in the maximal model defined by the TERMS statement
24 The constant cannot be dropped from a within-groups model
25 A value of the response variate is incompatible with the model
26 A fitted value predicted by the model is out of valid range for the model
27 Iterative fitting process has diverged, probably because of extreme data
28 Iterative weights have become 0, probably because a fitted value is 0
29 The iterative fitting process has not converged
30 STEP and TRY statements cannot be used with nonlinear models and curves
31 There are no degrees of freedom to calculate residual mean squares
32 Results cannot be displayed or kept because no model has been fitted
33 PREDICT statements can be used only for linear or generalized linear models
34 REPARAMETERIZE statements can be used only for nonlinear models or curves
35 Requested results from the current model are not available
36 Predictions cannot be formed
37 Factors to classify predictions do not match those classifying weight table
38 A structure in the Y parameter is not a response variate

SP Data space (390)

1 No room to expand directory: delete numerical structures
2 No room for structure attributes: delete numerical structures
3 Character data space full: delete text structures
4 Numerical data space full: delete numerical structures

5 No room to compile list: delete structures
 6 Compiler tables full – simplify list or expression
 7 Input Character Buffer full: use fewer input streams
 8 Input Ascii Buffer full: use fewer input streams
 9 Inadequate page width for printing

SX Syntax (410)

 1 Dummy already declared
 2 "&" invalid – no previous directive has been established
 3 Illegal repetition of classifying factors
 4 Unknown directive name
 5 Unknown option or parameter name
 6 Invalid option value
 7 Error in option syntax
 8 Valid single value not found when expected
 9 Invalid element in list
 10 Invalid character in element (possibly unprintable, such as TAB)
 11 Invalid function name
 12 Incompatible adjacent elements (e.g. comma missing)
 13 Syntactic error (e.g. unmatched brackets)
 14 Invalid type or value of identifier in number list
 15 Invalid progression (e.g. 15,17...13)
 16 Too many operators in an unbracketed expression
 17 Invalid operator
 18 Value to be substituted in integer is too big
 19 Attempt to substitute structure of incompatible type
 20 Too many options or parameters
 21 Zero list length
 22 Precisely one element expected in parameter list
 23 Invalid item in parameter list
 24 Missing value not permitted in parameter list
 25 Parameter has invalid values
 26 No option values supplied
 27 Both option and parameter missing from UNIT statement
 28 Too many parameters in list
 29 Option values supplied when not required
 30 Default not in values list
 31 Default structure has no values
 32 Default structure of wrong type
 33 Attempt to suffix non-pointer
 34 Invalid suffix structure

35 Pointer has no text labels
36 Option has too many settings
37 Option has too few settings
38 Qualifier value outside valid range
39 ## not followed by a simple identifier
40 Too many items in a subsequent parameter list
41 List of expressions not allowed
42 Option setting of invalid type or value
43 Wrong number of qualifiers
44 Qualifier not numerical
45 Parent of substructure not adequately defined
46 Parent of substructure has no values
47 Parameter list missing
48 Format all skips
49 First parameter list not longest: extra elements will be ignored
50 Unexpected characters after end of command
51 A qualifier cannot contain a null list

SY Probable Program Errors (470)

1 Program error
2 NBANK insufficient
3 Invalid parameter
4 Illegal attempt to substitute pointer values
5 Wrong number of options or parameters
6 Illegal bank number
7 Structure number out of range
8 Illegal type number
9 Packet fault
10 Execution stack pointer error
11 No space for TIDYUP
12 Cannot form index
13 Type substructure found
14 Directory stack pointer error
15 Incompatible data blocks
16 Incorrect option mode
17 Incorrect option type
18 Option setting out of range
19 Fatal Fortran error, rest of job ignored
20 Invalid directive number
21 Invalid option list
22 Invalid parameter list

23 Cannot find directive definition
24 Cannot open bootstrap file
25 Cannot close bootstrap file
26 Substructure expected but type incorrect
27 Bootstrap error
28 Character address error
29 Failure to set up subprocess by SUSPEND

TB Tabulate (510)

1 No classifying set in TABULATE directive
2 No tables or variates in TABULATE directive
3 No named tables in sequential TABULATE
4 Too many factors in classifying set
5 Incompatible classifying sets
6 Mixture of tables with and without margins

TS Time Series (520)

1 The number of orders in a time-series model is wrong
2 An order in a time-series model is invalid
3 The number of parameters in a time-series model is wrong
4 A parameter in a time-series model is invalid
5 A group of parameters in a time-series model is invalid
6 The number of lags in a time-series model is wrong
7 A lag in a time-series model is invalid
8 A time-series model is of the wrong type (arima or transfer-function)
9 The autocorrelation function is not positive definite
10 A time-series model is non-stationary
11 The maximum lag has not been defined
12 A value of the imaginary part of a sequence is invalid: set to zero
13 Invalid restriction for time series: units must be contiguous
14 Solution of equations has failed in calculation of filtered series
15 Preliminary estimation has not been completed
16 Marginal likelihood is not defined: LIKELIHOOD option reset to EXACT
17 Transformation cannot be applied due to negative values
18 The setting of the FIX option has the wrong number of values
19 There are no residual degrees of freedom
20 The iterative estimation process has failed to progress towards a solution
21 The iterative estimation process has not converged
22 TDISPLAY, TKEEP, and FORECAST must be preceded by an ESTIMATE statement
23 If set, the OUTPUTSERIES parameter must supply the output series identifier

24 Future values of all input series must be supplied in FORECAST
25 Origin of updated forecast is less than origin of previous forecast

VA Structures and their attributes (570)

1 Incompatible restrictions on set of structures
2 Attributes not set (e.g. length of structure)
3 Invalid value(s)
4 Values not set
5 Too many values
6 Too few values (including null subset from RESTRICT)
7 Invalid repetition of values in suffix list.
8 Attempt to change the type of an identifier
9 Dummy unset
10 Null identifier list
11 Invalid or incompatible type(s)
12 Invalid or incompatible mode(s)
13 Invalid or incompatible numbers of values
14 Invalid or incompatible number of factor levels
15 Invalid factor value
16 Attempt to assign values to a matrix or table of unknown size
17 Attempt to assign values to a factor without specifying number of levels
18 All values missing
19 Inconsistent structure(s)
20 Invalid or incompatible classifying set(s)
21 Attempt to define LRV with more columns than rows
22 Repetition of values in factor levels vector
23 Repetition of values in factor labels vector

References

PREFACE

Digby, P.G.N., Galwey, N.W., and Lane, P.W. (1988). Genstat 5: a second course. Oxford University Press.

Lane, P.W., Galwey, N.W., and Alvey, N.G. (1988). Genstat 5: an introduction. Oxford University Press.

CHAPTER 5

Bowdler, H., Martin, R.S., Reinsch, C., and Wilkinson, J.H. (1968). The QR and QL algorithms for symmetric matrices. Numerische Mathematik 11, 293–306.

Digby, P.G.N. and Kempton, R.A. (1987). Multivariate analysis of ecological communities. Chapman and Hall, London.

Eckart, C. and Young, G. (1936). The approximation of one matrix by another of lower rank. Psychometrika 1, 211–8.

Golub, G.H. and Reinsch, C. (1971). Singular value decomposition and least squares solutions. Numerische Mathematik 14, 403–20.

Herraman, C. (1968). Algorithm AS 12: Sums of squares and products matrix. Applied Statistics 17, 289–92

Martin, R.S., Reinsch, C., and Wilkinson, J.H. (1968). Householders tridiagonalisation of a symmetric matrix. Numerische Mathematik 11, 181–95.

Press, W.H., Flannery, B.P., Teukolsky, S.A., and Vetterling, W.T. (1986). Numerical recipes: the art of scientific computing. Cambridge University Press.

Rao, C.R. (1973). Linear statistical inference and its applications. Wiley, New York.

Wichmann, B.A. and Hill, I.D. (1982). An efficient and portable pseudo-random number generator. Applied Statistics 31, 188–90.

CHAPTER 7

Butland, J. (1980). A method of interpolating reasonably-shaped curves through any data. Proceedings of Computer Graphics 80, 409–22.

McConalogue, D.J. (1970). A quasi-intrinsic scheme for passing a smooth curve through a discrete set of points. Computer Journal 13, 392–6.

CHAPTER 8

Bouvier, A., Gelis, F., Huet, S., Messean, A., and Neveu, P. (1985). CS-NL. Laboratoire de Biometrie, INRA-CNRZ, Jouy-en-Josas.

Cook, R.D. and Weisberg, S. (1982). Residuals and influence in regression. Chapman and Hall, New York.

Draper, N.R. and Smith, H. (1981). Applied regression analysis (2nd Edition). Wiley, New York.

Finney, D.J. (1978). Statistical method in biological assay (3rd Edition). Griffin, London.

Grewal, R.S. (1952). A method of testing analgesics in mice. British Journal of Pharmacology and Chemotherapy 7, 433–7.

Koch, G.G. and Tolley, H.D. (1975). A generalized modified chi-squared analysis of categorical data from a complex dilution experiment. Biometrics 31, 59–92.

McCullagh, P. and Nelder, J.A. (1983). Generalized linear models. Chapman and Hall, New York.

Ratkowsky, D.A. (1983). Nonlinear regression analysis. Dekker, New York.

Ross, G.J.S. (1987). MLP manual. Numerical Algorithms Group, Oxford.

Snedecor, G.W. and Cochran, W.G. (1980). Statistical methods (7th Edition). Iowa State University Press.

Sprent, P. (1969). Models in regression and related topics. Methuen, London.

Stuart, A. (1953). The estimation and comparison of strengths of association in contingency tables. Biometrika 40, 105–10.

CHAPTER 9

Armitage, P. (1974). Statistical methods in medical research. Blackwell, Oxford.

Bartlett, M.S. (1937). Some examples of statistical methods of research in agriculture and applied biology (with discussion). Journal of the Royal Statistical Society Supplement 4, 137–83.

Cochran, W.G. and Cox, G.M. (1957). Experimental designs (2nd Edition). Wiley, New York.

Cox, D.R. (1958). Planning of experiments. Wiley, New York.

Healy, M.J.R. and Westmacott, M.H. (1956). Missing values in experiments analysed on automatic computers. Applied Statistics 5, 203–6.

John, J.A. and Quenouille, M.H. (1977). Experiments: design and analysis. Griffin, London.

John, P.W.M. (1971). Statistical design and analysis of experiments. Macmillan, New York.

James, A.T. and Wilkinson, G.N. (1971). Factorisation of the residual operator and canonical decomposition of non-orthogonal factors in analysis of variance. Biometrika 58, 279–94.

Kempthorne, O. (1952). The design and analysis of experiments. Wiley, New York.

Nelder, J.A. (1965a). The analysis of randomized experiments with orthogonal block structure. I Block structure and the null analysis of variance. Proceedings of the Royal Society A 283, 147–62.

Nelder, J.A. (1965b). The analysis of randomized experiments with orthogonal block structure. II Treatment structure and the general analysis of variance. Proceedings of the Royal Society A 283, 163–78.

Payne, R.W. and Wilkinson, G.N. (1977). A general algorithm for analysis of variance. Applied Statistics 26, 251–60.

Preece, D.A. (1971). Iterative procedures for missing values in experiments. Technometrics 13, 743–53.

Rogers, C.E. (1973). Algorithm AS 65: Interpreting structure formulae. Applied Statistics 22, 414–24.

Snedecor, G.W. and Cochran, W.G. (1980). Statistical methods (7th Edition). Iowa State University Press.

Wilkinson, G.N. (1970). A general recursive algorithm for analysis of variance. Biometrika 57, 19–46.

Wilkinson, G.N. and Rogers, C.E. (1973). Symbolic description of factorial models for analysis of variance. Applied Statistics 22, 392–9.

Wilkinson, G.N. (1957). The analysis of covariance with incomplete data. Biometrics 13, 363–72.

Yates, F. (1936). Incomplete randomized blocks. Annals of Eugenics 7, 121–40.

Yates, F. (1937). The design and analysis of factorial experiments. Technical Communication No. 35 of the Commonwealth Bureau of Soils. Commonwealth Agricultural Bureaux, Farnham Royal.

CHAPTER 10

Bartlett, M.S. (1938). Further aspects of the theory of multiple regression. Proceedings of the Cambridge Philosophical Society 34, 33–40.

Bladon, S. (1986). The new observer's book of automobiles. Frederick Warne, Harmondsworth.

Cooley, W.W. and Lohnes, P.R. (1971). Multivariate data analysis. Wiley, New York.

Digby, P.G.N. and Gower, J.C. (1981). Ordination between- and within-groups applied to soil classification. In Down-to-earth statistics: solutions looking for geological problems (ed. D.F. Merriam), Syracuse University Geology Contribution 8. Syracuse University.

Digby, P.G.N. and Kempton, R.A. (1987). Multivariate analysis of ecological communities. Chapman and Hall, London.

Doran, J.E. and Hodson, F.R. (1975). Mathematics and computers in archaeology. Edinburgh University Press.

Eckart, C. and Young, G. (1936). The approximation of one matrix by another of lower rank. Psychometrika 1, 211–8.

Friedman, H.P. and Rubin, J. (1967). On some invariant criteria for grouping data. Journal of the American Statistical Association 62, 1159–86.

Gordon, A.D. (1981). Classification: methods for the exploratory analysis of multivariate data. Chapman and Hall, London.

Gower, J.C. (1966). Some distance properties of latent root and vector methods used in multivariate analysis. Biometrika 53, 325–38.

Gower, J.C. (1968). Adding a point to vector diagrams in multivariate analysis. Biometrika 55, 582–5.

Gower, J.C. (1971). A general coefficient of similarity and some of its properties. Biometrics 27, 857–71.

Gower, J.C. (1974). Maximal predictive classification. Biometrics 30, 643–54.

Gower, J.C. (1975). Algorithm AS 82: The determinant of an orthogonal matrix. Applied Statistics 24, 150–3.

Gower, J.C. (1985a). Multivariate analysis: ordination, multidimensional scaling and allied topics. In Handbook of applicable mathematics (ed. W. Ledermann), Statistics Vol. VIB (ed. E. Lloyd). Wiley, Chichester.

Gower, J.C. (1985b). Measures of similarity, dissimilarity, and distance. In Encyclopaedia of statistical sciences, Volume V (eds. S. Kotz, N.L. Johnson, and C.B. Read), pp 397–405. Wiley, New York.

Gower, J.C. and Legendre, P. (1986). Metric and Euclidean properties of dissimilarity coefficients. Journal of Classification 3, 5–48.

Gower, J.C. and Ross, G.J.S. (1969). Minimum spanning trees and single linkage cluster analysis. Applied Statistics 18, 54–64.

Lawley, D.N. and Maxwell, A.E. (1971). Further analysis as a statistical method (2nd Edition). Butterworths, London.

Mardia, K.V., Kent, J.T., and Bibby, J.N. (1979). Multivariate analysis. Academic Press, London.

Nathanson, J.A. (1971). An application of multivariate analysis in astronomy. Applied Statistics 20, 239–49.

Rayner, J.H. (1966). Classification of soils by numerical methods. Journal of Soil Science 17, 79–92.

CHAPTER 11

Anderson, O.D. (1976). Time series analysis and forecasting. Butterworths, London.

Bloomfield, P. (1976). Fourier analysis of time series: an introduction. Wiley, New York.

Box, G.E.P. and Jenkins, G.M. (1970). Time series analysis, forecasting and control. Holden-Day, San Francisco.

de Boor, C. (1980). FFT as nested multiplication, with a twist. SIAM Journal of Scientific and Statistical Computing 1, 173–8.

Jenkins, G.M. and Watts, D.G. (1968). Spectral analysis and its applications. Holden-Day, San Francisco.

Nelson, C.R. (1973). Applied time series analysis for managerial forecasting. Holden-Day, San Francisco.

Shi-fang, L., Shi-guang, L., Shu-hua, Y., Shao-zhong, Y., and Yuan-xi, L. (1977). Analysis of periodicity in the irregular rotation of the Earth. Chinese Astronomy 1, 221–7.

Tunnicliffe Wilson, G. (1987). On the use of marginal likelihood in time-series model estimation. Journal of The Royal Statistical Society B (submitted).

CHAPTER 12

Dempster, A.P., Laird, N.M., and Rubin, D.B. (1977). Maximum likelihood from incomplete data via the EM algorithm (with discussion). Journal of the Royal Statistical Society B 39, 1–38.

Harding, S.A. and Simpson, H.R. (1987). Genstat implementers manual. Rothamsted Experimental Station, Harpenden.

Index

abandoning analysis 241
abbreviation
   directive name 22
   function name 18, 165
   identifier 12
   option and parameter
      names 23
   option and parameter
      settings 26
   statements 23
ABS() 165
absorbtion in regression 308
accumulated analysis of
   deviance 360
accumulated analysis of variance
   table 318, 325
ADD 321, 324, 371
ADDPOINTS 490
ADISPLAY 401
adjusted analysis of variance
   table 422
adjusted means 406
adjusted R^2, *see* percentage
   variance accounted for
AKEEP 433
aliased parameters—prediction
   and 345
aliasing
   in analysis of variance 448
   effect on TRY 326
   of explanatory variables in
      regression 319
   of parameters in
      regression 334
alphabetic sorting 185
ampersand 9, 26
analysis of covariance 418–24
analysis of designed
   experiments 389–448
   see also individual entries
analysis of deviance 356
analysis of parallelism 371
analysis of variance 398–401
   algorithm 446
   display of further
      output 401–10
   hierarchical 598
   missing values in 424

save structure 42, 47, 401
saving information
   from 433–6
   weighted 400
   see also analysis of designed
      experiments
analysis of variance table 393,
   422
   accumulated 318, 325
   in regression 310, 318, 324
.AND. 13, 143
ANG() 177
angular transformation 177
ANGULAR() 177
ANOVA 397
ARCCOS() 165
ARCSIN() 165
argument 24
   function 18, 164
   statement 22
ARIMA model 548
   autocorrelation function 535,
      591
   multiple-seasonal 554
   non-seasonal 552
   printing 551
   saving theoretical 591, 592
   seasonal 553
   weights 592
ARIMA modelling 548–69
   displaying 556, 560–1
   forecasting 564–9
   initial estimates 587–9
   model fitting 554–60
   model specification
      548–54
   saving 555, 556, 560, 562–3
   selecting models 558
   see also autocorrelated errors;
      time series modelling;
      transfer function modelling
arithmetic operations, *see*
   numerical calculations
arithmetic operators 12, 141
arithmetic progression 14, 16
array, *see* table
ASCII characters 8
ASSIGN 207

assigning values to structures—in
   declaration 60
   see also data input; READ
assignment operator 13, 140,
   142, 153, 155, 175
asterisk 9, 32
   see also missing value
asymmetric matrices—analysis
   of 598
asymptotic regression, *see*
   exponential curve -ordinary
attributes 56, 167, 600
   definition 60
   deleting 53
   inspecting 55, 56
   modifying 61
   saving 56
autocatalytic curve, *see* logistic
   curve—ordinary
autocorrelated errors 569–77
   forecasting 576–7
   initial estimates 589–90
   model fitting 571, 574
   model specification 571, 573–4
   see also ARIMA modelling;
      time series modelling;
      transfer function modelling
autocorrelation 535
   see also partial autocorrelation
autocorrelation function 535
   saving 591
autoregressive model, *see*
   ARIMA modelling
AXES 275

back-forecasts—in ARIMA
   modelling 556
backing-store 121–37
   file protection 127
   retrieving 127
   storing 124
backing-store file
   adding subfiles 122
   displaying the contents of 130
   merging 133
   opening and closing 115, 117
   overwriting 127
   workfile 124

backslash 9, 11, 22
backward elimination 329
backward shift operator 552
balanced design 440, 447
banks 601
batch mode 6, 35, 37, 41
binary file 87, 615
binary language-definition
    file 611
blank lines in PRINT 105, 108
block data module 606
block if 236, 237
block of statements—repetition
    of 235
block structure 390, 395,
    410–18
block terms 412
BLOCKSTRUCTURE 410
    null 394
    save structure 42, 47
bounds for iterative fitting, see
    iterative fitting
Box and Jenkins method 533
Box–Cox power
    transformation 508, 552
    effect on deviance 558
    parameter estimation 555, 573
brackets 8, 14, 17
BREAK 250

CALCULATE 140
calculations 139–230
calibration 313
canonical axes, see canonical
    dimensions
canonical correlation
    analysis 598
canonical dimensions 464, 466
canonical link 353
canonical variates analysis 449,
    460–6, 469, 481, 487
    adding points 481
    Mahalanobis distance and 521
    printing results 450–2, 464
    rotation 466–70
    saving results 450–2, 464
carriage-return—as
    punctuation 9, 11, 22
carriage-return key 5, 6
CASE 239

case of letters
    control of significance of 41
    in new directive names 611
    in option/parameter name of a
        procedure 246
    significance of 11, 12, 13
case selection, see multiple
    selection structure
CATALOGUE 130
catalogues—types of 130
cataloguing 130–3
CED( ) 177
centroid 480
channel number 89, 115, 605
character file 87, 607, 615
characters 8
chi-square distribution 177
chi-square equivalent deviate
    177
CHISQ( ) 177
Choleski decomposition 170
CHOLESKI( ) 170
CIRCULATE( ) 165
city block similarity
    coefficient 472
clasification, see non-hierachical
    classification
classifying factors 223, 229
    combining 194
    omitting levels of 194
    weighted combination of 196
classifying set of a table 76
    see also table
CLOSE 117
CLUSTER 511
cluster analysis 449
    see also hierarchical cluster
        analysis
coefficient of similarity, see
    similarity coefficient
coefficient of variation 410
colinearity of explanatory
    variables in regression 319
colon 5, 9, 22, 88
column labels—printing 107
combination of information 406,
    444, 598
COMBINE 193
comma 5, 9, 10, 20
comment 5, 11, 22
commons—Fortran, see Fortran
    commons
communalities of variables 468
communication with external
    programs 138, 615

compaction
    of identifier lists 16
    of number lists 14
    of programs 27
    of qualified identifier lists 19
completely randomized
    design 390, 417
compound operators 8
compound structure
    definition 79
    formation and
        manipulation 208–20
computer time 558, 595
CONCATENATE 198
confounding—partial 437, 448
conjugate sequence 547
constant parameter—omission of
    in regression 312
constraints
    in analysis of variance 407
    on parameters in curve
        fitting 368
    in regression
        parameterization 332, 334
contingency table 348, 351
continuation character, see
    backslash
CONTOUR 265
contour maps, see high-quality
    contours
    see also line-printer contours
contrasts 427–33
control observations—prediction
    and 345
control of PRINT layout, see data
    output
control structure 234–43
    abandoning, see exit
    block if 236–8
    for loop 234–6
    job, see job
    multiple selection 238–40
    procedure, see procedure
conventions for examples 29
convergence
    criterion for in iterative
        fitting 359, 377
    failure in iterative fitting 377
convolutions 546
COPY 42
    copying a job 42
    switching off 43
CORRELATE 534
correlation 534
correlation coefficient 534

Index

correlation matrix
  between variables 320, 535
  see also CORRMAT()
  of parameter estimates 312, 359, 369
  saving 535
correlogram 535, 537
correspondence analysis 598
CORRMAT() 170
COS() 165
cosine transformation of time series 543
cov. ef., *see* covariance efficiency factor
covariable, *see* explanatory variable
covariance efficiency factor 421
COVARIATE 421
  save structure 42, 47
covariate analysis, *see* analysis of covariance
covariates 390, 420
critical exponential curve 367
cross-correlation 539–41
crossing operator
  in formulae 21, 396
  in randomization 189
  in regression 332
Crouts method 170
cubic interpolation, *see* interpolation
CUM() 165
CUMULATE() 165
cumulative data input 99
cumulative formation of SSPMs 220
curly brackets 8
currency symbol (£), *see* hash
currency symbol ($), *see* dollar
current regression model 317
curve fitting, *see* nonlinear curves
CV%, *see* coefficient of variation
CVA 460
Czekanowski similarity coefficient 478

D() 170
data
  free and fixed format 93
  serial and parallel 88
data bases 615
data errors 92, 101

data input 2, 7, 87–102, 116, 206
  on declaration of structure 87, 88
  errors in 92, 101
  formatted 94
  from another channel 7, 88, 89, 116
  missing values 93
  non-numeric data 97
  reordering values on input 100
  rewinding files on 93
  scaling values 100
  sequential 99
  skipping values in 94, 95
  systematic order 206
  see also READ, reading
data manipulation 139–230
data output 102–13, 116
  default annotation 104
  default format 105
  justification 105
  labelling of structures 107
  missing values 109
  mixing named and unnamed structures 112
  orientation of tables 110
  orientation of vectors 109
  page margins and 108
  rewinding files 111
  serial and parallel 107
  skipping spaces and lines in 105, 112, 116
  supression of blank lines in 108
  to a text structure 107
  to another channel 107
  using PRINT 102
  see also PRINT 102
data separator 88, 93
data structure 59–85
  declaration, *see* declaration
  deleting 53
  dimensionality 157, 174
  inspecting 55–6
  internal storage of values 603
  modes of values 59, 603, 609
  operating on elements of 174
  as a qualifiers 159
  redefining 53, 61, 128
  reference number 613
  saving attributes of 56
  storage and retieval 121–30
  transferring values between 179
  trees of 71

type 58, 59
  see also individual structure entries
data structure attributes, *see* attributes
data structure identifier 11
data structure storage 121–30
data summary 92, 145
data terminator 88, 93, 94
data transfer—skipping values in 182
DCL command 606
DCONTOUR 296
DEBUG 252
decimal places
  in data input 100
  in data output 105, 113
  defining in structure declaration 60
  textual output and 105
decimal point 10
declaration 59–60
  defining decimal places in 60
  defining extra text with 60
  effect on PRINT 113
  explicit 3, 59
  implicit 3, 141, 148, 154, 155
  multiple 59
  of vector and pointer data structures 64
DEFINE 611
degrees of freedom 393
  in regression 310, 318
  saving from analysis of variance 435
  saving from ARIMA modelling 563
DELETE 53
deletion operators—in formulae 21
dendrogram 450, 496
dependent variable, *see* response variable
derivatives of fitted values—saving 371
design
  analysable by ANOVA 390
  balanced 447
  completely randomized 390, 391, 417
  factorial plus added control 397, 418, 432
  latin square 190, 416
  lattice square 417

design (cont'd)
  non-orthogonal 436
  partially-balanced lattice 443
  quasi-latin square 417
  randomized block 189, 411, 417, 418
  with several error terms 410–18
  simple lattice 443
  with a single error term 391–410
  split-plot 412, 417
  see also individual entries
design matrix—in regression 304, 316
design randomization 189
designed experiments—analysis of 389–448
DET( ) 170
DETERMINANT( ) 170
deviance 355–60
  in ARIMA modelling 557, 563
  saving from regression 316
  scaled 356
  total in glms 359
deviance ratios 356
deviance residuals, see residuals
DEVICE 283
device, see graphical device
DF, see degrees of freedom
DGRAPH 290
DHISTOGRAM 287
diagnostics 35–8, 40, 232
  user supplied 604
diagonal matrix 73–4
  generalized inverse 170
DIAGONALMATRIX 73
DIFFERENCE( ) 165
differences 165
differencing operator 552
digit 8
dilution assays 351, 354
dimensionality in multivariate analysis 210
dimensionality of a data structure 157, 174
directive
  codes 612, 615
  definition 1
  defining new 610–15
  name 4, 22, 611
discriminant analysis 598
dispersion parameter 351, 352, 356

DISPLAY 38
dissimilarity 470
  from similarity 478
distance—as a measure of association 480
divergence of parameter estimates
  curve fitting 370
  glms 357
  nonlinear models 377
dollar 9, 19
dot operator—in formulae 21, 397
dots—control of printing of 40, 121, 232
double equals, see .EQ.
double exponential curve 367
double precision accuracy 321
double quote 5, 9, 11
DPIE 301
DROP 321, 324, 371
DSSP structure 221
DUMMY 62
dummy 62, 207
  in a procedure 245
  in expressions 142
  in for loop 235
  in lists 17
dummy analysis 391, 400, 448
dummy functions 174
DUMP 54
dumping
  data structures 54
  a job 135
dyadic operators 19

echoing input and output 42–3
ecological similarity coefficient 473
EDIT 200
  commands 201–6
effective standard error 407, 422, 429
effects 392, 407, 422, 435
  partial 435, 446
efficiency factor 435, 437–40, 447
eigenvalue, see latent root
eigenvector, see latent vector
element-by-element action of operators 141, 150
element-by-element multiplication of matrices 152

elements of data structures 174
ELEMENTS( ) 157, 171, 174
ellipsis (...) 14, 16
ELSE 237, 239
ELSIF 236
EM algorithm 596
embedded assignment 161
empirical distribution functions—fitting 384
end of data marker 88, 92, 93, 94
ENDBREAK 251
ENDCASE 239
ENDDEBUG 253
ENDFOR 235
ENDIF 237
ENDJOB 233
ENDPROCEDURE 246
environment 31–58
  effect of JOB on 231
  high-quality graphics 275–86
  initial default settings 40
  inspecting settings 33, 46, 605
  procedures and 248
  setting 38–42, 232
.EOR. 13, 143
EPS 48, 321
.EQ. (= =) 13, 142
.EQS. 13, 142, 155
equals sign 9
EQUATE 179
error 35–8
  codes 48, App 2
  control of permitted number 41, 232
  in the Genstat program 597
  when reading data 92, 101
  see also faults and warnings
error distribution—in glms 351
error model—for a designed experiment 412
ESE, see effective standard error
ESTIMATE 550, 554, 564, 574, 582
euclidean similarity coefficient 473
examples—conventions 29
exclamation mark 9
EXIT 241
exit from control structure 241–3
exit status
  curve fitting 370
  nonlinear models 377

Index

exiting from Genstat 233
EXP( ) 165
EXPAND( ) 52, 175
experimental design, *see* design
explanatory variables 303, 338
explicit declaration of data structures, *see* declaration
exponent codes 10
exponential curves 366–7
expression 16, 18–20, 174
  calculations and 140–6
  factors in 143
  matrices in 149–53
  as option or parameter setting 25
  order of evaluation 155
  rules for building 18
  scalars in 146–9
  tables in 153–5
  vectors in 146–9
expression data structure 63
EXPRESSION 63
extension of Genstat 595
external files—accessing 113
external programs
  communicating with 615–18
  running from within Genstat 607–10
extra text—defining in structure declaration 60
extreme observations 311, 357

f-distribution 177
f-distribution equivalent deviate 177
FACROTATE 467
factor
  calculations with 148
  in curve fitting 365
  definition from a text or variate 186
  in expressions 143
  generation of values in systematic order 205
  maximum number in formulae 399, 433
  in nonlinear model fitting 381
  operations on 205–7
  in regression 331
  setting default length 42, 43
  standard order 206
  transferring values between 183

FACTOR 67
factor data structure 67–9
factor rotation 466–70
factorial experiments 389
factorial plus added control 397, 418, 432
fault 35–8
  codes 48, App 2
  supression of output from 40, 232
  in table operations 154
FED( ) 177
fieldwidth 93, 105
file
  backing-store 123
  backing-store workfile 124
  binary 87, 615
  binary language-definition 611
  character 87, 607, 615
  procedure library 249
  read-only 115
  terminal as 87, 113
  unformatted 137, 607
  userfile 123
  write-only 115
file handling
  accessing external 113–18, 615–18
  closing 117
  opening 114
  rewinding 93, 111
file protection—backing-store and 127
filename 608
filter 583, 585
FILTER 584
first-order balance 440, 447
FIT 309
  effect of TERMS on 324
FITCURVE 364
FITNONLINEAR 379
  with user supplied models 605
fitted values
  in regression 312
  saving derivatives of 371
  saving from ANOVA 401, 403
  saving from regression 307, 316
fixed format data 93–5, 96
FLRV 213
FOR 234
for loops 234–6
  exiting from 243
FORECAST 563, 576, 582

foreign command 606
format—of statements 9
format variate
  in data input 94, 96
  in data transfer 182
formatted data 93–5
FORMULA 64
formula 16, 18, 20–1
formula data structure 64
formula operators 13
formulae
  in designed experiments 395
  in regression 332
fortran 77 595
fortran commons
  Blank 603
  /DATA/ 609
  /DATAC/ 601
  /DIAGPK/ 604
  /OWNCOM/ 602, 603, 606
  /SYSCON/ 601, 603
  inspecting 56
fortran compiler 606, 610
fortran subprograms 598
fortran subroutine
  EXTRAD 612–15
  GETATT 601, 613
  VECSET 601
  ZZDIAG 604–5
fortran type codes 600
fortran unit numbers 605
fortran WRITE statements 605
forward selection 329
FOURIER 542
Fourier transform
  of a complex series 542, 546
  orthogonal matrix transformation 546
  saving 542
Fourier transformation 541–8
  definition 545, 546, 548
  general form 546
  index 534, 541
  inverse transformation 547
  of conjugate sequence 547
  of time series 545
  order 545, 546, 548
  order of and speed 542
  restricted units and 542
FPROBABILITY( ) 177
FRAME 284
FRATIO( ) 177
free format data 88, 93, 95
FSIMILARITY 471

FSSPM 218
FTSM 587
function
  argument 18, 164
  name abbreviation 18, 165
  syntax 18
  see also individual entries
function minimization 375–6, 385–8
functions in expressions 164–78
  dummy 174
  matrix 170–2
  scalar 167–8
  statistical 176–8
  table 173–4
  transformations 165–6
  variate 168–70

.GE. (> =) 13, 142
general nonlinear models 385–8
generalized inverse 170, 211, 217
  Moore–Penrose 211, 217
generalized linear models 347–61
  fitting 358–61
  theory 347–58
generalized logistic curve 367
generalized Procrustes analysis 598
GENERATE 205
generation of factor levels 205–7
generation of pseudo-factors 444–5
Genstat
  block data module 606
  as a command language 596
  errors 597
  exiting from 233
  extension of 595–618
  language-definition file 611
  main module 606
  object code 606
  object library 606
  recent developments 597
  relinking 606
  running 5
Genstat Newsletter 596
Genstat procedure library, *see* procedure library
Genstat program, *see* program
GET 45
GETATTRIBUTE 56
GHOST 274

GINO 274
GKS 274
GLM, *see* generalized linear model
GNPASS program 607
  modifying 609
Gompertz curve 367
grand mean—definition 392
graph, *see* high-quality graphs
graph, *see* line-printer graphs
GRAPH 261
graphical device 255, 274, 283
graphical display 255–302
graphical pen 275, 278, 292
graphical window 274, 284
graphics
  high-quality 255, 273–302
    see also high-quality graphics
  line-printer 255–73
    see also line-printer graphics
  metafile 274
grids
  in function evaluation 386
  in likelihood evaluation 382
grouped data 67, 83
  and linear regression 330
  forming groups 186–7
growth curves, *see* logistic curves
.GT. (>) 13, 142

hash 9, 17, 71, 80
  double 27
HCLUSTER 496
HDISPLAY 502
HELP 31–5
  environmental information 605
  on libraries, *see* LIBHELP procedure
  high-quality graphics 285
heterogeneity factor 352
hierarchical analysis of variance 598
hierarchical cluster analysis 495–509
  cutting the dendogram 497
  interpreting results 499–509
  methods of 496, 498
  minimum spanning tree 498, 506
  printing data matrix 499
  printing results 497, 503, 507
  saving results 497, 503, 505
  typical group member 504

high-quality contours 284–5, 296–300
  environment 273–84
  see also high-quality graphics
high-quality graphics 255, 273–302
  axes 275–8
  brush patterns 280–2
  colours 278
  contour maps 296–300
  device, *see* graphical device
  environment 273–84, 85
  graphs 290–6
  histograms 287–90
  keys 285, 286
  legend 288
  pen, *see* graphical pen
  pie-charts 301–2
  screen clearing 285
  window, *see* graphical window
high-quality graphs 284–5, 290–6
  environment 273–84
  linestyle codes 278
  symbol codes 279
  see also high-quality graphics
high-quality histograms 284–5, 287–90
  environment 273–84
  see also high-quality graphics
high-quality pie-charts, *see* pie-charts
HISTOGRAM 256
histogram, *see* line-printer histograms
  see also high-quality histograms
HLIST 499
Householder transformations 209, 214
HSUMMARIZE 507
hyperbola, *see* quadratic-divided-by-linear curve
  rectangular, *see* linear-divided-by-linear curve

I() 170
IDATA array 603
identifier 10, 11
  inspecting 55
  qualified, *see* qualified identifier

suffixed, *see* suffixed identifier
identifier list 15, 16, 25
identifier operators 13
identity matrix 73
IF 236
implementers manual 614
implicit declaration
   of pointers 12, 70
   of other data structures, *see*
      declaration
impulse-response function 589, 592
.IN. 13, 143, 155
independent variable, *see*
   explanatory variable
index vector—in sorting 184
indexing files, *see* cataloguing
influential observations—in regression 311
information summary 437
innovation variance 550
innovations 548
input 87–102
   from text structures 118
   see also data input
INPUT 118
input channel 89, 113, 115–20
instability in regression 319
INT() 165
INTEGER() 165
interaction 389, 392, 395
interaction of factors in regression 331
interactive mode 5, 32, 36, 41
intercept in regression, *see* constant parameter
interleaving tables when printing 110
intermediate structure 155
internal storage—limit on 610
INTERPOLATE 161
interpolation 162–4
INV() 170
inverse—generalized, *see* generalized inverse
inverse cubic interpolation, *see* interpolation
inverse exponential curve, *see* logistic curve
inverse linear interpolation, *see* interpolation
INVERSE() 170, 171
invoking Genstat 5
.IS. 13, 142, 155

.ISNT. 13, 142, 155
item—definition 9
iteration—controlling in model fitting 358
iterative fitting 361–77
iterative model 359
iterative weights—saving from regression 361

Jaccard similarity coefficient 478
JOB 231
job
   definition 231
   naming 232
   restarting 136
   storing 135–6
job control 231–54
joining lines of text 198
justification of text 105, 107

lag zero 539
lags—of a time series model 548
latent roots 80, 81, 213–18
   and dimensionality 451
   formation of 213–18
   as percentage of the trace 214, 451
latent vectors 80, 81, 213–18
   formation of 213–18
latin square design 416
   randomization of 190
lattice—partially-balanced 443
lattice square design 417
law-like relationship 303
.LE. (< =) 13, 142
least squares approximation of rank r 210
least squares estimation 391
legend 288
letter 8
leverage 308, 311, 312, 369, 371
   in glms 359
LIBHELP procedure 597
library, *see* procedure library
likelihood
   evaluation in nonlinear models 377, 382
   evaluation in time series modelling 556–60, 574, 582
   exact in ARIMA modelling 557

   least-squares in ARIMA modelling 557
   marginal in ARIMA modelling 574
line plus exponential curve 367
line-printer contours 255–6, 265–72
line-printer graphics 255–73
   contour maps 265–72
   graphs 261–5
   histograms 256–61
   see also individual entries
line-printer graphs 255–6, 261–5
line-printer histograms 255–6, 256–61
linear contrast 427
linear interploation, *see* interpolation
linear model
   generalized *see* generalized linear model
   ordinary 350
linear predictor 352, 361
linear regression
   generalized 347–61
   grouped data 330–47
   multiple 316–30
   simple 304–16
linear-divided-by-linear-curve 368
link function 352, 353
list 13
   compaction 14, 16, 19
   in expression 144
   see also number list, string list and identifier list
LLB() 177
LLBINOMIAL() 177
LLG() 177
LLGAMMA() 177
LLN() 177
LLNORMAL() 177
LLP() 177
LLPOISSON() 177
local documentation 5
log linear model 348, 351, 354, 356
LOG() 166
log-likelihood functions 176–7
LOG10() 166
logical expression 142
   in block if 237
   and exiting 243
   restricting vectors with 51

logical inclusion and non-inclusion 143
logical operator 13, 143
logistic curves 367
logit transformation 305
loops, see for loops
LRV 80
   data structure 80
   formation of 213
.LT. (<) 13, 142
LTPRODUCT() 170

macro 27–8
Mahalanobis distance 509, 521
   squared 460, 481
main effect 389, 392, 395
   in regression 331
MARGIN 226
margin of a table 78, 106, 153, 154, 173, 223, 226–8
marginal summaries—in tables 78
marginal weights—in prediction 342
marginality—in regression 334
matrix
   calculations with 149–53
   combining rows/columns 197
   decompositions 208–21
   diagonal, see diagonal matrix
   element-by-element multiplication of 152
   generalized inverse, see generalized inverse
   latent roots, see latent roots
   latent vectors, see latent vectors
   rank of 218
   symmetric, see symmetric matrix
MATRIX 72
matrix data structures 71–5
matrix functions 170–2
MAX() 168
maximal model 320, 356
maximum
   length of a statement 22
   number of factors in formulae 399, 433
   number of non-orthogonal terms 447

number of parameters in a general function 376
number of structures in a pointer 610
number of values of each mode 610
maximum likelihood 376
MAXIMUM() 168
mean square 310, 318, 393
MEAN() 168
means
   from analysis of covariance 406
   from analysis of variance 405, 431, 435
   standard errors for 406, 422, 440
   measure of association 449, 479, 480
   forming 470, 477
   see also similarity matrix
MED() 168
median alphabetic values 187
median values of grouped data 187
MEDIAN() 168
MERGE 133
merging backing-store files 133–5
messages 35
   in regression 313, 334
   user supplied 604
   when curve fitting 369
metric scaling, see principal coordinates analysis
Michaelis-Menten law 368
MIN() 168
MINIMUM() 168
Minkowski similarity coefficient 479
minus sign—in output of comments 7
missing identifier—in qualified identifier 160
missing value indicator 10, 93, 12, 97, 100, 109
missing values 168, 609
   in ARIMA modelling 551, 558
   in calculations 141, 142, 144, 145
   in designed experiments 401, 424–6
   in functions 169
   in interpolation 163

   in lists 14, 15
   in regression 305, 307, 320
   in SSPMs 219
   in tables 79
   in tabulation 224
   in time series 536
   internal storage 601, 603
Mitscherlich curve, see exponential curve
MLP 375
mode
   of data structure values 59, 603, 609
   of option/parameter in a procedure 246
   I 612, 613
   W 612, 613
mode setting 41
MODEL 306, 351, 368
model
   description in regression 310
   specification in GLMs 351
model formula
   order of terms in 332
   see also formula
model term 20, 395
modified Gauss–Newton method of optimization 376
modified Newton method 361, 376
module
   block data 606
   main 606
monadic operators 18
monitoring iterative fitting 369
Moore-Penrose inverse 211, 217
moving average model, see ARIMA modelling
MS, see mean square
multi-input transfer function models, see transfer function
multidimensional scaling, see ordination
multiple comparisons 311
multiple diagnostics 37
multiple regression 316–30
multiple seasonal ARIMA model, see ARIMA model
multiple selection control structure 238–41
multiplier 17
multivariate analysis 449–531
   consistency when saving results 451

graphs and 466
printing results from 450-2
saving results from 450-2
see also individual entries
multivariate analysis of
  variance 596
multivariate data—types of 449
MVREPLACE( ) 166

NCOLUMNS( ) 168
.NE. (/= or < >) 13, 142
NED( ) 177
.NES. 13, 142, 155
nested loops, *see* for loops
nesting of control structures 236
nesting operator
  in formulae 21, 396
  in regression 189, 332
NEWLEVELS( ) 148, 166
newline, *see* carriage-return
.NI. 13, 143, 155
NLEVELS( ) 168 N
NMV( ) 168
NOBSERVATIONS( ) 168
non-constant variance 305, 308
non-hierachical
  classification 509-21
  binary data 510, 516
  criteria 509, 512
  initial classification 510, 512, 520
  number of groups 512
  printing results 511
  saving classifications 512
non-orthogonal designs 436
non-orthogonal
  treatments 445-6, 447
non-seasonal ARIMA model, *see*
  ARIMA model
nonlinear curve fitting 361-75
nonlinear curves
  error distribution 368
  imposing constraints on 368
  method of fitting 361
  standards 366
  through the origin 368
  weighted fitting 369
nonlinear model 375-88
  likelihood evaluation 377, 382
  maximum number of
    parameters 376
  method of fitting 380
  specifying 376
  see also iterative fitting

nonlinear regression 375-88
  with no linear
    parameters 382-8
normal distribution 178
normal equivalent deviate 177
NORMAL( ) 178
normalized device
  coordinates 274, 284
.NOT. 13, 143
NOTICE procedure 597
NROWS( ) 168
null model—in regression 310
null settings—of options and
  parameters 24
null suffix list 70
number 10
number list 14, 16, 26
numerical calculations 139
NVALUES option—in data
  structure declaration 64
NVALUES( ) 168

object code 606
object library 606
offset—in regression 308
offset variable 354, 369
one-matrix problem 213
OPEN 89, 114
operating system 607, 608, 616
  giving commands from within
    Genstat 617
  VMS 606, 610, 616
operator 8, 10, 12-13, 18-19,
    155, 396
  precedence 19, 20, 141
optimization—methods of 376
optimizing a function 375-88
OPTION 245, 612
option name 4, 23, 25
option of a procedure 244-6
option setting 4, 22, 24, 25, 26
options
  in statements 4
  syntax 22
  use of 25
OR 239
.OR. 13, 143
order
  constraints in ARIMA
    models 552, 553
  moving average 548
  of cosine transformation 543
  of differencing 548

of fourier transformation 542,
  545, 546, 548
of seasonal autoregression 553
of seasonal differencing 553
of seasonal moving
  average 553
of tranfer function
  models 580, 581
order of option and parameter
  names 23
ordered classifying set of a table,
  see table
ordination 479-95
orthogonal design—analysis
  of 400, 448
orthogonal polynomials 427, 598
orthogonal Procrustes rotation,
  see Procrustes rotation
other programs—communicating
  with 138, 615-18
ouput channel 107, 113
outliers—in regression 311
OUTPUT 120
output 87, 102-13
  control of diagnostic printing
    in 121
  control of dots and pages 40,
    121, 232
  pausing in 41
  to a file 113-21
  see also data output
output buffer—default
  length 107
output channel—number
  allowed 115
over-parameterization in
  regression 334, 335
overdispersion 352
overwriting—backing store
  and 126, 129, 135
OWN 598
  modifying the subroutine 599
  in nonlinear fitting 381, 382

PAGE 117
page—control in output 40, 121,
  232
page margins—when
  printing 108
page size—in output files 115
paging 117
parallel arguments—of
  statements 24

parallel curve fitting 365
parallel data 88
parallel printing 107
parallel regressions 330
PARAMETER 245
parameter name 23–5
parameter of a procedure 244–5
parameter setting 2, 22, 24, 25, 26
parameterization of regression 332–7
parameters in statements 4, 22, 24
parameters
  in a time series model 548
  in regression 303, 304, 316, 325
partial aliasing—in analysis of variance 448
partial autocorrelation 537–9
partial confounding 437, 448
partial effects 435, 446
partially-balanced lattice design 443
Parzen window 543
PASS 607, 607
pause—return to operating system 616
pausing 250
  in output 41
PCO 481
PCP 452
Pearson residuals, see residuals
PEN 278
penalty function—in iterative fitting 377
percentage data 305
percentage variance accounted for 310, 359
periodogram 545
permutations 188
pi-weights 592
pictures, see high-quality graphics
pie-charts 284–5, 301–2
  environment 273–84
  see also high-quality graphics
pivot sweep—in analysis of variance 447
pling, see exclamation mark
plots—in designed experiments 413
plots, see graphics
point plot 262

POINTER 69
pointer 16, 64
  manipulation 207
  and OWN 599
  reading values of 98
  retrieval from backing-store 129
  setting values of 207
pointer data structure 69–71
pointer value, see suffixed identifier
POL() 427
POLND() 431
polynomial contrasts 427
pooling sums of squares in regression 318
pooling terms 360, 369
pooling variability—in curve fitting 365
post-multiplier in a list 17
pound, see hash
powers—of numbers 10
pre-multiplier in a list 17
precedence—of operators 141
PREDICT 337, 361
predicted means 392
prediction
  in generalized linear models 361–3
  in regression 337–47
  in time-series 563–9, 576–7, 582–3
primary arguments—of statements 24, 144
principal axes, see principal dimensions
principal components
  analysis 452–60, 466–70, 521, 485
  printing results 450–2, 455
  Pythagorean distance 521
  rotation 466–70
  saving results 450–2, 456
  efficiently 485
principal coordinate
  analysis 479–81, 481–90, 490–5
  adding points 481, 490–3
  interpreting results 493–5
  printing results 450–2, 481, 483
  saving results 450–2, 488, 491
  special cases 480
principal dimensions 466

PRINT 102
printing 102–13
  control of 40, 232
  data, see data output and PRINT
  last diagnostic 38
  pausing in 41
  skipping values and 116–7
  tables 106, 109–11
  to another channel 120–1
printing results of analyses 26
  see also individual entries
probit analysis 351, 352, 598
PROCEDURE 244
procedure 27, 242, 243–9
  backing store and 124, 127, 128
  using 243–4, 246–7
  writing 244–9
procedure library 243, 249–50, 597–8
  official Genstat 243, 249, 597
  site 243, 249, 597
Procrustes rotation 521–31
  alternative rotations 524
  centring 523, 527
  criteria 522
  printing results 526
  reflection 525, 527
  rotation 523
  saving results 527
  scaling 523, 524, 527
  translation 523
PRODUCT() 152, 170, 171
program 1, 231
  compaction using procedures and macros 27–9
  debugging 250–4
  separation into jobs 231–3
progression 14, 16
prompt 6, 32, 33
proportional data 351
proportional replication 431
pseudo-factors 432, 441–5, 448
pseudo-inverse, see Moore–Penrose inverse
pseudo-terms—in analysis of variance model 441
psi-weights 592
punctuation 5, 8–9

Q-techniques 449, 470
QL algorithm 214

Index

QPRODUCT( ) 171
QR algorithm 209
quadratic-divided-by-linear curve 368
quadratic-divided-by-quadratic curve 368
qualified identifier 19–20, 157–61, 174
qualitative variables in regression 330
quantitative variables in regression 330
quartiles 187
quasi-latin square design 417
quotation marks 9
quote—double, *see* double quote
quote—single, *see* single quote
quoted string 11
   see also string
quotes—when reading textual data 97

R^2 statistic, *see* percentage variance accounted for
R-techniques 449, 452
random numbers 178
random permutation 188–93
randomization—importance of for designed experiments 389
RANDOMIZE 188
randomized block design 411, 417, 418
   randomization of 189
ratios of polynomial curves 368
RCYCLE 358, 376
RDATA array 603
RDISPLAY 313
   effect of TERMS on 324
re-analysis sweeps in analysis of variance 447
READ 89
read-only files 115
reading 87–102
   unformatted files 137
   from another channel 118–20
   textual data 97–9
   data, *see also* data input
RECORD 135
record length 115
recovering information from last analysis 47

rectangular hyperbola, *see* linear-divided-by-linear curve
recursive procedures 248
redeclaring a pointer 70
REDUCE 476
REG( ) 429
REGND( ) 431
regression 303–88
   autocorrelated errors, *see* autocorrelated errors
   curve fitting 361–88
   display of further output 313–14
   general nonlinear 375–88
   generalized linear 347–61
   grouped data 330–47
   linear 304–16
   multiple linear 316–30
   output to a different channel 314
   parallel 330
   parameterization 332–7
   save structure 42, 47, 309, 314
   stepwise, *see* stepwise regression
   see also individual entries
RELATE 493
relational expression 142
relational operators 13, 141
relinking Genstat 606
reordering data values on input 100
repeated measures—analysis of 598
repetition
   of a statement 26
   of block of statements 235
   of items in a list 17
resetting the environment 38, 232
residual degrees of freedom
   saving from ANOVA 436
   saving from regression 316
residual sum of squares
   saving from ANOVA 436
   saving from regression 316, 318
residuals
   in the linear model 391
   in regression 308, 312, 360, 369
   Pearson 308, 360
   saving from analysis of variance 401, 403, 436

saving from ARIMA modelling 562
saving from multivariate analysis 451
saving from regression 307, 316
standardized deviance 308, 360
response 408
response variable 303
restricted units—principal coordinate analysis and 485
RESTRICT 49
   effect on PRINT 113
restricted units 169, 170, 176, 188, 199, 200, 219
   analysis of variance and 390
   ARIMA modelling and 551
   canonical variates analysis and 461
   calculations and 147, 148
   Fourier transformation and 542
   principal components and 459
   printing and 113
   similarity matrix and 472
   regression and 305, 307, 320
   see also individual entries
RESTRICTION( ) 52, 176
restrictions on vectors 49–52, 148, 157
RESUME 136
retrieval
   of data from unformatted files 137–8
   of data structures 122, 127–30
   of a job 136
   structure-renaming and 128
RETRIEVE 127
RETURN 119
⟨RETURN⟩, *see* carriage-return key
REVERSE( ) 166
rewinding
   input files 93
   output files 111
   unformatted files 138
RKEEP 314, 370, 377
   effect of TERMS on 324
ROTATE 525
round brackets 8
ROUND( ) 166
rounding errors 146
RTPRODUCT( ) 171
running Genstat 5

sample autocorrelation function 535
sample cross correlation 540
sample partial autocorrelation function 537
save structure
  accessing 42, 47
  for analysis of variance 434
  for regression 309, 314
  for time series 560, 574
saving
  attributes 56
  catalogue of contents of a subfile 130
  catalogue of subfiles of a backing-store file 131
  environmental settings 46
  fault codes 48
  fitted values from ANOVA 401, 403
  fitted values from regression 307
  model details from ANOVA 400
  residuals from ANOVA 401, 403
  residuals from regression 307
  results from multivariate analysis 450, 451
  space 48, 308, 321
  a statement 48
  variate of restricted units 52
SCALAR 61
scalar data structure 16, 61–2
scalar functions 167–9
scalars—calculations with 146–9, 151
scaled deviance 356
scaling data values on input 100
scatter plot 262
scientific notation 10
scream, see exclamation mark
SE, see standard error
searching, see iterative fitting
searching for nonlinear parameter values 366
seasonal ARIMA model, see ARIMA model
seasonal period 553
seasonal transfer function model, see transfer function model
secondary arguments—of statements 24
SED, see standard error
seed
  in analysis of variance 448

for random numbers 178
  in randomization 188
self correlation, see autocorrelation
semicolon 5, 9, 164
sequential data input 99
sequential formation of SSPMs 220
sequential tabulation 225
serial data 88
serial printing 107
seriation, see ordination
SET 38, 615
settings—of options and parameters 22
sharp, see hash
SHIFT( ) 166
shriek, see exclamation mark
sigmoid curve, see logistic curve
sign of a number 10
similarity coefficient 470–3
  city block 472
  Czekanowski 478
  ecological 473
  Euclidean 473
  from Euclidean distance 478
  Jaccard 472, 478
  Minkowski 479
  non-standard 477
  simple matching 472, 478
similarity matrix 450, 470–9
  between groups similarities 473, 475
  from distances 480
  reduced 475–9
  type codes 472–3
simple lattice design 443
simple matching similarity coefficient 472, 478
simple operators 8
SIN( ) 166
single quote 9, 11
single-precision accuracy—and regression 321
singular value decomposition 209–12, 218
singular values of a matrix 209
singular vectors of a matrix 209
site procedure library, see procedure library
skeleton analysis of variance table 401
SKIP 116

skipping lines 116
skipping to next page 117
slash operator—in formulae 21, 396
smallest number 48
smoothed means 431
smoothing time series, see time series modelling
SOLUTION( ) 171
SORT 183
SORT( ) 166
sorting 166, 183–8
  text 185
space character 9
spaces—in numbers 10
special symbols 9
spectral analysis 541
spectral decomposition 213, 218
spectrum - from autocorrelations 543
split plot design 412, 417
SQRT( ) 166
square brackets 8, 16, 19, 21
square matrix 71, 170
SS, see sum of squares
Ss. Div., see sum of squares divisor
SSPM 82
  displaying with TERMS 320
  formation of 218, 220
  in multivariate analysis 449
  within groups 462
SSPM data structure 82–3
stagewise regression 329
standard curves—in regression 366
standard errors
  in analysis of variance 406
  in nonlinear fitting 382
  in regression 307
standard errors of parameter estimates 311, 359, 369
standard errors of predictions 341, 347
standard order 206
standardized residuals, see residuals
star, see asterisk
statement 1, 4, 21–3
  repetition 26
  saving 48
  name 22
statistical functions 176–9
STEP 327

stepwise regression 327–30
STOP 2, 233
STORE 124
storing, see backing-store
stratified designs 410
  combination of
    information 598
stratum 412
string 10, 11
string list 11, 15, 26
structure storage 121
structure-renaming
  on retrieval 128
  on storage 125
sub-plots 413
sub-pointer 70
subfile 123–4
  examining contents 130–4
  moving between backing-store
    files 134–5
  retrieval 127–30
  storage 124–7
SUBMAT( ) 171
subprograms supplied by the
  user 598
subsets of values, see qualified
  identifier
subsets of vectors, see restrictions
  on vectors
substitution 17, 27
substitution symbol, see hash
suffix list 16, 70
suffix number 208
suffixed identifier 12, 70
  backing store and 123, 124, 129
  implicit pointer declaration
    and 70
  see also pointer
sum of squares
  in analysis of variance 392,
    393, 435
  in regression 310, 318
sum of squares and products
  matrix
    formation of 218
    within groups 83
  see also SSPM
sum of squares divisor 429
SUM( ) 168
summarizing effects of variables
  in regression 338
summary analysis of variance—in
  regression 310, 318, 324
summary of data

  in calculations 145
  in READ 92
summary tables 222
summary values of data
  structures 167
suppression of printing—using
  the PRINT option 26
supression of messages in
  regression 313
SUSPEND 616
SVD 209
sweep operations—in analysis of
  variance 447
SWITCH 321, 371
  effect of TERMS on 324
symmetric matrix 74–5
SYMMETRICMATRIX 75
syntax 1–30
  of EDIT commands 201–6
  of options and parameters 22
  of statements 4, 21
system faults 37
system information 31, 56
system word 10, 12
systematic data input 206

T( ) 171
tab character 9
TABLE 76
table
  adding margins 227
  cell 76, 79, 223
  classifying set 76, 153, 154,
    156
  expressions with 153–5
  formation and
    manipulation 222–30
  margins 78, 153, 154, 173,
    227–30, 223
  printing 106–11
  reclasification 193–7
  sequential formation 225
  see also classifying factors
table data structure 76–9
table functions 173–4
tables of means—with ANOVA,
  see means
TABULATE 222
TDISPLAY 560
temporary files in backing-
  store 124
terminal—as a file 87, 113
termination of a statement 5, 22,
  27

terminator
  control of ⟨RETURN⟩ as 41
  of data 88, 93
terminology 1–30
TERMS 221, 319
TEXT 66
text
  as a qualifier 159
  calculations with 149
  in PASS 608
  macros in 27
  manipulation of 197–207
  reading 97–9
  unnamed 16
text data structure 66–7
  reading from 99, 118
  writing to 107, 108
text sorting 185–6
text substitution 17
tidying space 53
time series 533–94
  cosine transformation 543–4
  Fourier transformation
    545–8
  index 533
  missing values in 536
  pre-whitening 584
  restrictions on units 537
  save structure 42, 47, 560, 574
time series model 548
  generalized form 592–4
  summarizing 590–2
time series modelling
  constraining parameters 575
  deviance 575
  displaying further output 575
  displaying models 587
  filtering 583–7
  forecasting future values 565,
    576
  incorporating further
    observations 566–7, 576
  initial parameter
    estimates 587–90
  likelihood evaluation 557, 560,
    574, 582
  see also ARIMA modelling
  see also autocorrelated errors
  see also transfer function
    modelling
TKEEP 561, 582
TMAXIMA( ) 173
TMEANS( ) 173
TMEDIANS( ) 173

TMINIMA() 173
TNMV() 173
TNOBSERVATIONS() 173
TNVALUES() 173
tolerance
   in ANOVA 426, 448
   in calculations 146
   in fitting ARIMA models 560
   in iterative fitting 359
   in regression 321
TOTAL() 168
trace 214, 451
TRACE() 171
transcribing a job, *see* copying a job
transfer function model 577–80
transfer function
   modelling 580–3
   displaying 560–1
   forecasting 582–3
   initial parameter estimates 589–90
   model fitting 577–9, 582
   model specification 580–1
   saving 582
   see also ARIMA modelling
   see also autocorrelated errors
   see also time series modelling
TRANSFERFUNCTION 573, 581
transferring input and output control 118
transformations 165–7, 177, 179
transients 581
TRANSPOSE() 171
treatment—in a designed experiment 389, 445–6
treatment contrasts 426–33
treatment effect 392
treatment formula 395–7
treatment sum of squares 392
TREATMENTSTRUCTURE 394
   save structure 42, 47
trees of data structures 71
true value 142
TRY 326
TSM 84, 550
   in ARIMA modelling 551
   in transfer function modelling 580
TSM data structure 84
TSUMMARIZE 590
TSUMS() 173
TTOTALS() 173

TVARIANCES() 173
two-matrix problem 214
type codes
   for data structures 58
   for similarity matrices 472
   for unnamed structures 16
   Fortran code for data structures 600

unbalanced design 443
unequal variance, *see* non-constant variance
unformatted file 115, 117, 135–38, 607
unit labels—in regression 312
unit structure—see block structure
unit values as labels 44
unit variances—saving from analysis of variance 435
UNITS 43–5
units structure 42, 43–5
unknown cell—of a table 79, 223
unnamed parameters 4
unnamed scalar—ellipsis (...) with 16
unnamed structure 16, 54, 125, 128
unquoted string 11
   see also string
UNSET() 174, 248
unstable regression 319
unsuffixed identifier 11
URAND() 176, 178
userfile 123

values
   assigning on declaration 60
   copying between data structures 179
   in intepolation 162
   recycling 181
VAR() 168
variance—estimation in regression 307
variance components—estimation 598
variance function—in glms 351
variance ratio
   in analysis of variance 393
   in regression 318, 325, 355

VARIANCE() 168
VARIATE 65
variate
   calculations with 140–7, 147–9, 151
   cumulative data totals on input 100
   setting default length 42, 43
   data structure 16, 65
   functions 168–70
VAX computers 606
vector 64
   see also factor
   see also text
   see also variate
vertical bar 9
VMAXIMA() 169
VMEANS() 169
VMEDIANS() 169
VMINIMA() 169
VMS operating system, *see* operating system
VNMV() 169
VNOBSERVATIONS() 169
VNVALUES() 169
VR, *see* variance ratio
VSUMS() 169
VTOTALS() 169
VVARIANCES() 169

warnings 35–8
   codes 48, App 2
   supression of output from 40, 232
weighted analysis of variance 400
weighted prediction in regression 343
weighted regression 307
   nonlinear 369
weighted replication 406
whole-plots 413
window—Parzen 543
workfile
   backing-store 124
   unformatted 137
workspace 6, 602, 603
   used in a job 6
write-only files 115
writing to unformatted files 137

x-variable, *see* explanatory variable

y-variable, *see* response variable

zero divided by zero 145
zigzag method 476

A.E.A.